Proceedings of the 7th Symposium

Frequency Standards and Metrology

Proceedings of the 7th Symposium Frequency Standards and Metrology

Asilomar Conference Grounds, Pacific Grove, CA, USA
5 – 11 October 2008

editor
Lute Maleki
OEwaves Inc, USA

World Scientific

NEW JERSEY • LONDON • SINGAPORE • BEIJING • SHANGHAI • HONG KONG • TAIPEI • CHENNAI

Published by

World Scientific Publishing Co. Pte. Ltd.
5 Toh Tuck Link, Singapore 596224
USA office: 27 Warren Street, Suite 401-402, Hackensack, NJ 07601
UK office: 57 Shelton Street, Covent Garden, London WC2H 9HE

British Library Cataloguing-in-Publication Data
A catalogue record for this book is available from the British Library.

FREQUENCY STANDARDS AND METROLOGY
Proceedings of the 7th Symposium

ISBN-13 978-981-283-821-6
ISBN-10 981-283-821-X

Printed in Singapore by World Scientific Printers

GOVERNING BOARD

Jacques Vanier, University of Montreal, Canada
Claude Audoin, Paris Observatoire – SYRTE, France
Andrea de Marchi, Italy
James C. Bergquist, NIST, USA
Patrick Gill, NPL, United Kingdom

SYMPOSIUM CHAIR

Lute Maleki, OEwaves

INTERNATIONAL STEERING COMMITTEE

Claude Audoin, France
Andreas Bauch, Germany
V. S. Bagnato, Brazil
James Berquist, USA
Patrick Gill, UK
Massimo Inguscio, Italy
H. Katori, Japan
Steve Lamoreaux, USA
Alan Madej, Canada
Andrea de Marchi, Italy
Shin-ichi Ohshima, Japan
V.G. Pal'chikov, Russia
Fritz Riehle, Germany
Christopher Salomon, France
Stephan Schiller, Germany
Michael Tobar, Australia
Jacques Vanier, Canada
Yiqiu Wang, China
Bruce Warrington, Australia
Jun Ye, USA

LOCAL ORGANIZING COMMITTEE

James Camparo, Aerospace Corp., USA
Bernardo Jaduszliwer, Aerospace Corp., USA
John Prestage, JPL, USA
Robert Tjoelker, JPL, USA
Nan Yu, JPL, USA

SPONSORING/SUPPORTING ORGANIZATIONS

Jet Propulsion Laboratory
Pasadena, California, USA

The Aerospace Corporation
El Segundo, California, USA

National Institute of Standards and Technology (NIST)
Boulder, Colorado, USA

CONFERENCE SECRETARY

Allyson Beatrice

REGISTRATION COORDINATOR

Barbara McGivney

PREFACE

The fields of frequency standards and frequency metrology have a special inter-connection to physics. Metrology and standards allow exacting measurements that explore the farthest reach of scientific models and theories. The most well known example is the atomic clock, which plays an important role in test of relativity, and gravitation. On the other hand, the same standards and metrological techniques depend on advances in various fields of physics to realize higher performance and capabilities. This duplex relationship has made frequency standards and metrology extremely rich. The International Symposium for Standards and Metrology has been organized for the past 37 years to provide a forum where researchers in diverse fields connected with frequency standards and frequency metrology gather to discuss the latest advances in the fields, and explore future directions.

I first learned about the International Symposium on Frequency Standards and Metrology (ISFSM) in 1980, shortly after I had switched research to work in the field of atomic frequency standards. This highly anticipated event was planned for 1981, and my supervisor, who had attended a previous symposium, was extremely excited. Despite my best efforts, I was too junior to secure a travel approval to Aussoi, but I wasn't too disappointed until after my supervisor returned and related all the stories about talks, people, and the cutting edge research that was discussed and disclosed in the symposium. Seven years later I finally had the opportunity to attend the 4th ISFSM in Ancona. I remember vividly, as I was then a relative newcomer to the field, how eager I was about the prospect of learning the latest information and meeting the experts and luminaries of the field. After attending the conference, I left Ancona with a great sense of energy, and was charged by all the latest information that the talks and many hours of one on one interaction with many experts attending the symposium provided. The meeting was key in motivating thoughts on how to expand our research at JPL, and how to connect with other fields related to frequency standards and metrology. After Ancona, I attended the other symposia in Woods Hole and St Andrews, both of which extended and strengthened the already impressive experience of Ancona.

It was against this backdrop that when asked by Patrick Gill, the chair of the 6th Symposium, to organize the 7th Symposium, I was both excited about the opportunity, and keenly aware of the challenges to maintain the tradition of ISFSM in providing a venue and program that is faithful to its tradition and reputation.

The meeting that was held during October 6-9, 2008 in Asilomar Conference Center, Pacific Grove, California stayed faithful to the tradition of ISFSM in that 125 participants gathered to interact and discuss the latest developments in the field. Among the subjects that have developed new advances since the 6th symposium, optical lattice clocks, quantum controlled ion clock, and advanced trapped ion clocks stand out. Significant advances were also made in "traditional" clocks, such as cold atom fountains and microwave clocks. Clocks being developed for space applications have advanced significantly, as have miniature atomic clocks based on coherent population trapping (CTP). New and promising directions in optical combs and microwave oscillators based on miniature optical whispering gallery mode resonators were identified. The latest advances in gravity gradiometer, a close relative of the atomic clock, were discussed.

All in all, the symposium was a testament to the health of a field that is nearly as old as physics itself, and has its modern roots in the Harrison clock.

In organizing the conference, I began with assembling a team of extremely competent and dedicated people, and by seeking support from JPL, NIST and Aerospace Corp. The support that was generously extended by the three institutions allowed an early establishment of a web site, and selection of a venue that would be both serene and conducive to informal interaction amongst the participants. Asilomar is well known for its location on the Monterey peninsula, providing a pleasant environment and good food.

The program was organized with the generous help from a distinguished international steering committee, followed by many hours of help from the Local Organizing Committee. The arrangements to reserve Asilomar, and all the registration matters were competently performed by Synergistic Management, Inc. But the ingredient that pulled all these pieces together, while handling the trauma of working with a demanding conference chair was the conference secretary, Allyson Beatrice of OEwaves, Inc. Allyson began helping with the initial announcement for the conference and continued her support with registration, organization of the program, and collection of papers for these proceedings. The conference could simply not have been successful without her expert support and committed care for smoothing out all thorny issues that were encountered.

Lute Maleki
Conference Chairman
Pasadena, CA

Historical information about the *Symposium on Frequency Standards and Metrology*

The series of meetings was initiated by Jacques Vanier and Helmut Hellwig in consultation with several other scientists in particular R.F.C. Vessot and C. Audoin.

The **first meeting** was held in Canada at the Pavillon de la Forêt Montmorency, Université Laval, Quebec, Canada. The dates of the meeting were August 30 - September 1, 1971. The organizer and chairman was Prof. Jacques Vanier, Laval University. The number of participants was 57. At the time, the meeting was modestly called **Frequency Standards and Metrology Seminar**. Upon request from European participants the name was changed from Seminar to Symposium in later meetings. Proceedings of the meeting (soft covers) were edited by H. Hellwig and J. Vanier, published by Les Presses de l'Université Laval.

The second meeting, called **2nd Frequency Standards and Metrology Symposium** was held at the ski resort, Copper Mountain, Colorado, USA. The dates of the meeting were July 5 to 8, 1976. The organizer and chairman was H. Hellwig, National Bureau of Standards. The number of participants was 94. The Proceedings of the meeting (soft covers) were assembled and published by NBS.

The **Third Symposium on Frequency Standards and Metrology** was held at the Centre Paul Langevin, Aussois, France. The dates of the meeting were October 12-15 1981. The meeting was organized and chaired by Dr Claude Audoin. The Proceedings were published as a hard cover volume of the Journal de Physique Colloque C-8, supplement to No12 Dec. 1981. The number of participants was 120.

The fourth meeting was called: **Frequency Standards and Metrology, Fourth Symposium**. The meeting was held at the University of Ancona, Ancona, Italy. The organizer and chairman was Prof. Andrea De Marchi. The dates of the meeting were September 5-9, 1988. The proceedings were edited by A. De Marchi, and published (hard cover) by Springer Verlag. The number of participants was 165.

The **Fifth Symposium on Frequency Standards and Metrology** was held at Woods Hole, MA, USA. The organizer and chairman was Dr. James C. Bergquist, NIST, Boulder. The dates of the meeting were October 15-19, 1995. The proceedings were edited by J. C. Bergquist and published (hard cover) by World Scientific. The number of participants was 158.

The **6th Symposium on Frequency Standards and Metrology** was held at the University of Saint Andrews, Scotland. The organizer and Chairman was Dr Patrick Gill of National Physical Laboratories, Middlesex, UK. The dates of the meeting were September 9-14, 1991. The proceeding were edited by P. Gill and published (hard cover) by World Scientific. The number of participants was 178.

The symposium has been extremely successful in providing a forum where ideas related to the development of new atomic frequency standards were presented, as well as in encouraging exchanges within the community. The work of several scientists participating in past symposia has been recognized recently with the granting of a Nobel prize.

A charter was established in 1987, providing guidance to organizers and chairmen of future symposia. The symposium is managed by a governing board whose members are the organizers of all prior symposia. The role of the governing board is essentially limited to the selection of the symposium organizer for the next meeting and providing the selected organizer with general guidance. The present members of the governing board are:

Jacques Vanier
Claude Audoin
Andrea De Marchi
Jim Bergquist
Patrick Gill
Lute Maleki, *chairman*

The last organizer will automatically become the Chairman of the Governing Board. The charter was given in the Proceedings of the Ancona meeting, pages 483-484.

Jacques Vanier
Université de Montréal,
Canada

EXIT

Contents

Part II: Frequency & Metrology

Part IV: Optical Clocks I: Lattice Clocks

Part V: Optical Clocks II: Ion Clocks

Part VI: Optical Frequency Combs

Part VII: Atomic Microwave Standards

Part XI: Time Scales

Part XII: Interferometers

Part XIII: New Directions

Part I

Fundamental Physics

VARIATION OF FUNDAMENTAL CONSTANTS FROM THE BIG BANG TO ATOMIC CLOCKS: THEORY AND OBSERVATIONS

V. V. Flambaum and J. C. Berengut

School of Physics, University of New South Wales,
Sydney, NSW 2052, Australia

Theories unifying gravity with other interactions suggest the possibility of temporal and spatial variation of the fundamental "constants" in an expanding Universe. In this review we discuss the effects of variation of the fine-structure constant and fundamental masses on measurements covering the lifespan of the Universe from a few minutes after Big Bang to the present time. Measurements give controversial results, including some hints for variation in Big Bang nucleosynthesis and quasar absorption spectra data. Furthermore there are very promising methods to search for the variation of fundamental constants by comparison of different atomic clocks. Huge enhancements of the relative variation effects happen in transitions between accidentally degenerate nuclear, atomic, and molecular energy levels.

1. Introduction

It is widely believed that the Standard Model of elementary particles is a low-energy manifestation of a more complete theory that unifies gravity with the other interactions (electromagnetic, weak nuclear, and strong nuclear forces). Many well-motivated extensions to the Standard Model include variation of fundamental constants as a possibility, or even as a necessity in an expanding Universe (see e.g. review 1). Additionally, the fundamental constants may be slightly different near massive bodies (see e.g. review 2). There are some hints for the variation of different fundamental constants in quasar absorption spectra[3–8] and Big Bang nucleosynthesis[9,10] data. However, a majority of publications report limits on the variations (see e.g. recent reviews 11,12).

We can only detect variation of dimensionless fundamental constants. Here we will discuss variation of the fine structure constant α, the proton g-factor g_p, and the dimensionless ratios $X_e = m_e/\Lambda_{QCD}$ and $X_q = m_q/\Lambda_{QCD}$ where m_e and m_q are the electron and quark masses, and Λ_{QCD}

is the quantum chromodynamics (QCD) scale, defined as the position of the Landau pole in the logarithm of the running strong coupling constant, $\alpha_s(r) \sim 1/\ln(\Lambda_{QCD} r/\hbar c)$. The proton mass m_p is proportional to Λ_{QCD}, therefore the relative variation of $\mu = m_e/m_p$ is equal to the relative variation of $X_e = m_e/\Lambda_{QCD}$ (if we neglect a small contribution of quark masses ($m_q \sim 5$ MeV) to the proton mass, $m_p = 938$ MeV). In the Standard Model electron and quark masses are proportional to the vacuum expectation value of the Higgs field.

The hypothetical unification of all interactions implies that variations of different fundamental constants may be related.[13–17] For example, grand unification theories predict

$$\frac{\delta X_q}{X_q} \sim 35 \frac{\delta \alpha}{\alpha} . \tag{1}$$

The coefficient here is model-dependent, but large values are generic for models in which variations come from high energy scales (for a simple explanation see Ref. 10). If these ideas are correct, the variation in $X_{e,q} = m_{e,q}/\Lambda_{QCD}$ or $\mu = m_e/M_p$ may be easier to detect than the variation in α.

In Sec. 2 we discuss variation of fundamental constants in Big Bang nucleosynthesis. The factor of three disagreement between the calculations and measurements of the BBN abundance of ^{7}Li may, in principle, be explained by the variation of m_q/Λ_{QCD} at the level of $\sim 10^{-2}$; there have also been several publications studying variation of α and gravitational constant in BBN which we do not consider here, see review 1. The claim of detection of variation of fundamental constants based on the Oklo data in Ref. 18 is not confirmed by recent studies[19–21] which give a stringent limit on the possible variation of the resonance in ^{150}Sm during the last two billion years (Sec. 3). We discuss the variation of fundamental constants in atomic (Sec. 4) and molecular (Sec. 5) spectra, which can be used both in laboratory-based clocks as well as over cosmological timescales by comparison with astrophysical data. Additionally, we discuss possible enhancements in these systems as well as in nuclear clocks (Sec. 6) and ultracold systems near Feshbach resonances (Sec. 7).

2. Big Bang nucleosynthesis

The result of our work[9] suggested that a reduced deuteron binding energy of $\Delta Q/Q = -0.019 \pm 0.005$ would yield a better fit to observational data (the WMAP value of barion-to-photon ratio η and measured primordial

^{2}H, ^{4}He, and ^{7}Li abundances) for Big Bang Nucleosynthesis. Using our calculations[22] we obtained an estimate of the strange quark mass variation.

Recently Dent, Stern, and Wetterich[23] calculated the sensitivity of BBN abundances for ^{2}H, ^{4}He and ^{7}Li to the variation of binding energies of ^{2,3}H, ^{3,4}He, ^{6,7}Li and ^{7}Be in a linear approximation. We calculated the dependence of these binding energies on the light quark mass variation and estimated the sensitivity of BBN yields to variation of the quark mass.[10,24–26] Then we used the observational data to obtain the following equations for ^{2}H, ^{4}He and ^{7}Li:[10]

$$1 + 7.7x = 1.07 \pm 0.15 \ , \qquad (2)$$

$$1 - 0.95x = 1.005 \pm 0.036 \ , \qquad (3)$$

$$1 - 50x = 0.33 \pm 0.11 \ , \qquad (4)$$

where $x = \delta X_q/X_q$. These equations yield 3 consistent values of x: 0.009 ± 0.019, -0.005 ± 0.038 and 0.013 ± 0.002. The statistically weighted average of $\delta X_q/X_q = 0.013 \pm 0.002$ is dominated by the ^{7}Li data. Allowing for the theoretical uncertanties, we should understand this BBN result as $\delta X_q/X_q = K \cdot (0.013 \pm 0.002)$ where $K \sim 1$ and the expected accuracy in K is about a factor of 2. Note that here we neglected effects of the strange quark mass variation. A rough estimate of these effects on BBN due to the deuteron binding energy variation was made in Refs. 9,22.

We have performed a preliminary, more accurate calculation that takes into account the effect of the ^{8}Be binding energy variation (which is not included in Ref. 23), the variation of the virtual ^{1}S$_0(np)$ level, and nonlinear corrections in x which are important for ^{7}Li (with η given by WMAP, the discrepancy between ^{7}Li abundances predicted by theory and inferred from observations is now a factor of $2.4 - 4.3$ at the $4 - 5\sigma$ level[27]). Taking the observed $^7\mathrm{Li/H} = 1 - 2 \times 10^{-10}$, we obtain a variation of $\delta X_q/X_q = 0.015 \pm 0.003$.

3. Oklo natural nuclear reactor

The results from the Oklo natural nuclear reactor are based on the measurement of the position of a very low energy resonance ($E_r = 0.1$ eV) in neutron capture by ^{149}Sm nucleus. The shift of this resonance induced by variation of α was estimated a long time ago in Refs. 28,29. Recently we performed a rough estimate of the effect of the variation of $X_q = m_q/\Lambda_{QCD}$.[9,22,30] The final result is

$$\delta E_r \approx 10^6 eV(\frac{\delta\alpha}{\alpha} - 10\frac{\delta X_q}{X_q} + 100\frac{\delta X_s}{X_s}) \qquad (5)$$

with $m_q = (m_u + m_d)/2$ and m_s the strange quark mass. Refs. 19–21 found that $|\delta E_r| < 0.1$ eV. This gives us a limit

$$|0.01\frac{\delta\alpha}{\alpha} - 0.1\frac{\delta X_q}{X_q} + \frac{\delta X_s}{X_s}| < 10^{-9} \tag{6}$$

The contribution of α variation to this equation is very small and should be neglected since the accuracy of the calculation of the main term is low. Thus, the Oklo data can not give any limit on the variation of α. Assuming linear time dependence during last 2 billion years we obtain an estimate $|\dot{X}_s/X_s| < 10^{-18}$ yr^{-1}.

4. Atomic spectra

4.1. *Comparison of quasar absorption spectra with laboratory spectra*

Savedoff[31] proposed the use of a fine structure multiplet to perform measurements of α variation by comparison of cosmic and laboratory optical spectra. There were numerous works successfully implementing this "alkali-doublet" method (see review 1).

Later we developed a different approach: the many-multiplet method.[32,33] The relative value of any relativistic corrections to atomic transition frequencies is proportional to α^2. These corrections can exceed the fine structure interval between the excited levels by an order of magnitude (for example, an *s*-wave electron does not have the spin-orbit splitting but it has the maximal relativistic correction to energy). The relativistic corrections vary very strongly from atom to atom and can have opposite signs in different transitions (for example, in *s*-*p* versus *d*-*p* transitions). Thus, any variation of α could be revealed by comparing different transitions in different atoms in cosmic and laboratory spectra. The number of spectral lines of different elements involved is 1-2 orders of magnitude larger. This method improves the sensitivity to any variation of α by more than an order of magnitude compared to the alkali-doublet method.

Relativistic many-body calculations are used to reveal the dependence of atomic frequencies on α for a range of atomic species observed in quasar absorption spectra[32–38] (a 2004 summary may be found in Ref. 39). It is convenient to present results for the transition frequencies as functions of α^2 in the form

$$\omega = \omega_0 + qx, \tag{7}$$

where $x = (\frac{\alpha}{\alpha_0})^2 - 1 \approx \frac{2\delta\alpha}{\alpha}$ and ω_0 is a laboratory frequency of a particular transition. We stress that the second term contributes only if α deviates from the laboratory value α_0. We performed accurate many-body calculations of the coefficients q for all transtions of astrophysical interest (strong E1 transtions from the ground state) in Mg I, Mg II, Fe I, Fe II, Cr II, Ni II, Al II, Al III, Si II, Zn II, Mn II, and many other atoms and ions which are seen in quasar absorption spectra, but have not yet been used in the quasar measurements because of the absence of accurate UV transition laboratory wavelengths. For a "shopping list" of needed measurements, see Ref. 39. It is very important that this set of transitions contains three large classes: positive shifters (large positive coefficients $q > 1000$ cm^{-1}), negative shifters (large negative coefficients $q < -1000$ cm^{-1}) and anchor lines with small values of q. This gives us an excellent control of systematic errors since systematic effects do not "know" about sign and magnitude of q. Comparison of cosmic frequencies ω and laboratory frequencies ω_0 allows us to measure $\frac{\delta\alpha}{\alpha}$.

Analysis of three independent samples of data contaning 143 absorption systems spread over redshift range $0.2 < z < 4.2$ gives[5] $\frac{\delta\alpha}{\alpha} = (-0.543 \pm 0.116) \times 10^{-5}$. If one assumes the linear time-dependence of α, the fit of the data gives $d\ln\alpha/dt = (6.40 \pm 1.35) \times 10^{-16}$ yr^{-1} (over time interval about 12 billion years). A very extensive search for possible systematic errors has shown that known systematic effects cannot explain the result.[40]

Our method and calculations[32–37] were used by two other groups[41–43] to analyse data obtained on the VLT. However, they have not detected any variation of α. Recently, the results of Ref. 41 were questioned in a reanalysis of the same spectral data.[6,7] The re-analysis revealed flawed parameter estimation methods; a more accurate fit gives $\frac{\delta\alpha}{\alpha} = (-0.64 \pm 0.36) \times 10^{-5}$ (instead of the $\frac{\delta\alpha}{\alpha} = (-0.06 \pm 0.06) \times 10^{-5}$ reported in Ref. 41). However, even this revised result may require further revision.

One systematic effect that is still not completely excluded is that the effect of α variation may be imitated by a large change in relative isotope abundance during last 10 billion years. Spurious observation of variation in α due to a change in the relative isotope abundance of any one element has been ruled out. Nevertheless, an improbable "conspiracy" of several elements could mimic the observed effect. We have performed very complicated calculations of these isotopic shifts.[44–49] However, as shown in Ref. 49, calculations in atoms and ions with an open d-shell (like Fe II, Ni II, Cr II, Mn II, Ti II) are difficult, and our accuracy may be very low. Therefore measurements for at least a few lines are needed in order to benchmark calculations. Additionally, these measurements are needed to study the evolution of isotope abundances in the Universe, and to test models of nuclear reactions in stars and supernovae.

A comparison of the hyperfine transition in atomic hydrogen with optical transitions in ions was done in Refs. 50,51. This method allows one to study time-variation of the parameter $F = \alpha^2 \mu g_p$. Analysis of 9 quasar spectra with redshifts $0.23 \le z \le 2.35$ gave

$$\delta F/F = (6.3 \pm 9.9) \times 10^{-6}, \tag{8}$$

$$\dot{F}/F = (-6 \pm 12) \times 10^{-16}\ \mathrm{yr}^{-1}. \tag{9}$$

4.2. *Optical atomic clocks*

Optical clocks also include transitions which have positive, negative or small constributions of the relativistic corrections to frequencies. We used the same methods of relativistic many-body calculations used in the quasar absorption studies to calculate the dependence on α of different clocks.[33,34,52–55] A 2004 summary of the results for the coefficients q is presented in Ref. 56. The q coefficients for optical clock transitions may be substantially larger than in cosmic transitions since the clock transitions are often in heavy atoms (Hg II, Yb II, Yb III, etc.) while cosmic spectra contain mostly light atoms lines ($Z < 33$). The relativistic effects are proportional to $(Z\alpha)^2$.

4.3. *Enhancement of the effect of α-variation in atoms*

An enhancement of the relative effect of α-variation can be obtained in transitions between the almost degenerate levels in Dy atom.[34,55] These levels move in opposite directions if α varies. The relative variation may be presented as $\delta\omega/\omega = K\delta\alpha/\alpha$ where the coefficient K exceeds 10^8 ($q = 30,000\ \mathrm{cm}^{-1}$, $\omega \sim 10^{-4}\ \mathrm{cm}^{-1}$). Specific values of $K = 2q/\omega$ are different for different hyperfine components and isotopes which have different ω. An experiment is currently underway to place limits on α variation using this transition.[57,58] The current limit is $\dot{\alpha}/\alpha = (-2.7 \pm 2.6) \times 10^{-15}\ \mathrm{yr}^{-1}$. Unfortunately, one of the levels has quite a large linewidth and this limits the accuracy. Several other enhanced effects of α variation in atoms have been calculated.[59,60]

4.4. *Atomic microwave clocks*

Hyperfine microwave transitions may be used to search for α-variation.[61] Karshenboim[62] has pointed out that measurements of ratios of hyperfine structure intervals in different atoms are also sensitive to variations in nuclear magnetic moments. However, the magnetic moments are not the fundamental parameters and cannot be directly compared with any theory of the variations. Atomic and nuclear calculations are needed for the interpretation of the measurements. We have performed both atomic calculations of the α-dependence[33,34,52–56] and nuclear calculations of the X_q-dependence[63] (see also Ref. 26) for all microwave transitions of current experimental interest including hyperfine transitions in ^{133}Cs, ^{87}Rb, ^{171}Yb$^+$, ^{199}Hg$^+$, ^{111}Cd, ^{129}Xe, ^{139}La, ^{1}H, ^{2}H and ^{3}He. The results for the dependence of the transition frequencies on variation of α, $X_e = m_e/\Lambda_{QCD}$ and $X_q = m_q/\Lambda_{QCD}$ are presented in Ref. 63. Also, one can find there experimental limits on these variations which follow from the recent measurements. The accuracy is approaching 10^{-15} per year. This may be compared to the sensitivity $\sim 10^{-5} - 10^{-6}$ per 10^{10} years obtained using the quasar absorption spectra.

According to Ref. 63 the frequency ratio Y of the 282 nm ^{199}Hg$^+$ optical clock transition to the ground state hyperfine transition in ^{133}Cs has the following dependence on the fundamental constants:

$$\dot{Y}/Y = -6\dot{\alpha}/\alpha - \dot{\mu}/\mu - 0.01\dot{X}_q/X_q \tag{10}$$

This ratio has been measured[64] as $\dot{Y}/Y = (0.37 \pm 0.39) \times 10^{-15}$ yr^{-1}. Assuming a linear time dependence we obtained from the quasar result[65] (see Sec. 5.2) $\dot{\mu}/\mu = \dot{X}_e/X_e = (1 \pm 3) \times 10^{-16}$ yr^{-1}. Combining this result and the atomic clock result for Y gives the best present limit on the variation of α:

$$\dot{\alpha}/\alpha = (-0.8 \pm 0.8) \times 10^{-16}\ \mathrm{yr}^{-1}\ . \tag{11}$$

Here we neglected the small ($\sim 1\%$) contribution of X_q.

5. Molecular spectra

Recently we wrote a review[66] about search for the variation of the fundamental constants in quasar and laboratory molecular spectra. One recent noteworthy result is a limit on cosmological variation of μ in quasar spectra using molecular hydrogen transitions in the Ly-α forest.[67] By carefully controlling systematics, the authors obtained $\Delta\mu/\mu = (2.6 \pm 3.0) \times 10^{-6}$ at redshifts $z \approx 2.6 - 3.0$. Below we present several examples related to our works.

5.1. *Comparison of hydrogen hyperfine and molecular rotational quasar spectra*

The frequency of the hydrogenic hyperfine line is proportional to $\alpha^2\mu g_p$; molecular rotational frequencies are proportional to μ. Comparison places limits on the variation of the parameter $F = \alpha^2 g_p$.[68] Recently a similar analysis was repeated by Murphy *et al.*[69] using more accurate data for the same object at $z = 0.247$ and for a more distant object at $z = 0.6847$, and the following limits for the relative variation of F were obtained:

$$\delta(\ln F) = (-2.0 \pm 4.4) \times 10^{-6} \tag{12}$$

$$\delta(\ln F) = (-1.6 \pm 5.4) \times 10^{-6} \tag{13}$$

The object at $z = 0.6847$ is associated with the gravitational lens toward quasar B0218+357 and corresponds to lookback time ~ 6.5 Gyr.

5.2. *Enhancement of variation of μ in inversion spectrum of ammonia and limit from quasar spectra*

A few years ago van Veldhoven *et al.* suggested using a decelerated molecular beam of ND_3 to search for the variation of μ in laboratory experiments.[70] The ammonia molecule has a pyramidal shape and the inversion frequency depends on the exponentially small tunneling of three hydrogens (or deuteriums) through the potential barrier. Because of that, it is very sensitive to any changes of the parameters of the system, particularly to the reduced mass for this vibrational mode. This fact was used in[65] to place the best limit on the cosmological variation of μ.

The inversion vibrational mode of ammonia is described by a double well potential with the first two vibrational levels lying below the barrier. Because of the tunneling, these two levels are split in inversion doublets. The lower doublet corresponds to wavelength $\lambda \approx 1.25$ cm and is used in ammonia masers. Molecular rotation leads to the centrifugal distortion of the potential curve, therefore the inversion splitting depends on the rotational angular momentum J and its projection on the molecular symmetry axis K:

$$\omega_{\text{inv}}(J,K) = \omega_{\text{inv}}^0 - c_1\left[J(J+1) - K^2\right] + c_2 K^2 + \cdots, \tag{14}$$

where we omitted terms with higher powers of J and K. Numerically, $\omega_{\text{inv}}^0 \approx$ 23.787 GHz, $c_1 \approx 151.3$ MHz, and $c_2 \approx 59.7$ MHz.

In addition to the rotational structure (14) the inversion spectrum has a much smaller hyperfine structure. For the main nitrogen isotope ^{14}N, the hyperfine structure is dominated by the electric quadrupole interaction ($\sim$ 1 MHz). Because of the dipole selection rule $\Delta K = 0$ the levels with $J = K$ are metastable. In astrophysics the lines with $J = K$ are also narrower and stronger than others, but the hyperfine structure for spectra with high redshifts is still not resolved. We obtained the following results for NH_3[65] (in atomic units):

$$\delta(\ln \omega^0_{\rm inv}) \approx 4.46\ \delta(\ln \mu) \tag{15}$$
$$\delta(\ln c_{1,2}) = 5.1\ \delta(\ln \mu)\,. \tag{16}$$

For ND_3 the inversion frequency is 15 times smaller and this leads to a higher relative sensitivity of the inversion frequency to μ:

$$\delta(\ln \omega^0_{\rm inv}) \approx 5.7\ \delta(\ln \mu) \tag{17}$$
$$\delta(\ln c_{1,2}) = 6.2\ \delta(\ln \mu)\,. \tag{18}$$

We see that the inversion frequency $\omega^0_{\rm inv}$ and the rotational intervals $\omega_{\rm inv}(J_1, K_1) - \omega_{\rm inv}(J_2, K_2)$ have different dependencies on the constant μ. In principle, this allows one to study time-variation of μ by comparing different intervals in the inversion spectrum of ammonia. For example, if we compare the rotational interval to the inversion frequency, then Eqs. (15) and (16) give:

$$\delta \ln \left([\omega_{\rm inv}(J_1, K_1) - \omega_{\rm inv}(J_2, K_2)]/\omega^0_{\rm inv}\right) = 0.6\ \delta(\ln \mu)\,. \tag{19}$$

The relative effects are substantially larger if we compare the inversion transitions with the transitions between the quadrupole and magnetic hyperfine components. However, in practice this method will not work because of the smallness of the hyperfine structure compared to typical line widths in astrophysics.

We compared the inversion spectrum of NH_3 with rotational spectra of other molecules, where

$$\frac{\delta\omega_{\rm rot}}{\omega_{\rm rot}} = \frac{\delta\mu}{\mu}\,. \tag{20}$$

High precision data on the redshifts of NH_3 inversion lines exist for the previously mentioned object B0218+357 at $z \approx 0.6847$.[71] Comparing them with the redshifts of rotational lines of CO, HCO^+, and HCN molecules from Ref. 72 one can get the following limit:

$$\frac{\delta\mu}{\mu} = \frac{\delta X_e}{X_e} = (-0.6 \pm 1.9) \times 10^{-6}. \tag{21}$$

Assuming a linear time dependence over the 6.5 Gyr (corresponding to $z \approx 0.68$), we obtain the most stringent present limit for the variation of μ and X_e:[65]

$$\dot{\mu}/\mu = \dot{X}_e/X_e = (1 \pm 3) \times 10^{-16}\ \mathrm{yr}^{-1}\,. \tag{22}$$

This result is combined with atomic clock results (Sec. 4.4) to give the best current limits on variation of α (Equation 11).

5.3. *Proposals for enhanced effects in diatomic molecules*

In transitions between very close narrow levels of different natures in diatomic molecules the relative effects of the variation may be enhanced by several orders of magnitude. Such levels may occur due to cancelation between the hyperfine and rotational structures,[73] or between the fine and vibrational structures of the electronic ground state.[74] The intervals between the levels are conveniently located in microwave frequency range and the level widths are very small, typically $\sim 10^{-2}$ Hz.

5.3.1. *Molecules with cancelation between hyperfine structure and rotational intervals*

Consider diatomic molecules with unpaired electron and ground state $^2\Sigma$, for example, LaS, LaO, LuS, LuO, YbF.[75] The hyperfine interval Δ_{hfs} is proportional to $\alpha^2 Z F_{\mathrm{rel}}(\alpha Z)\mu g_{\mathrm{nuc}}$, where F_{rel} is additional relativistic (Casimir) factor. The rotational interval $\Delta_{\mathrm{rot}} \sim \mu$ is roughly independent of α. If we find a molecule with $\Delta_{\mathrm{hfs}} \approx \Delta_{\mathrm{rot}}$ the splitting ω between hyperfine and rotational levels will depend on the following combination

$$\omega \sim \left[\alpha^2 F_{\mathrm{rel}}(\alpha Z)\, g_{\mathrm{nuc}} - \mathrm{const}\right]\,. \tag{23}$$

Relative variation is then given by

$$\frac{\delta\omega}{\omega} \approx \frac{\Delta_{\mathrm{hfs}}}{\omega}\left[(2+K)\,\frac{\delta\alpha}{\alpha} + \frac{\delta g_{\mathrm{nuc}}}{g_{\mathrm{nuc}}}\right], \tag{24}$$

where the factor K comes from variation of $F_{\mathrm{rel}}(\alpha Z)$, and for $Z \sim 50$, $K \approx 1$. Using data from Ref. 75 one can find[73] that $\omega = (0.002 \pm 0.010)\ \mathrm{cm}^{-1}$ for $^{139}\mathrm{La}^{32}\mathrm{S}$. Note that for $\omega = 0.002\ \mathrm{cm}^{-1}$ the relative frequency shift is:

$$\frac{\delta\omega}{\omega} \approx 600\,\frac{\delta\alpha}{\alpha}\,. \tag{25}$$

5.3.2. *Molecules with cancelation between fine structure and vibrational intervals*

The fine structure interval ω_f rapidly grows with nuclear charge Z

$$\omega_f \sim Z^2\alpha^2\,, \tag{26}$$

while the vibration energy quantum decreases with the atomic mass:

$$\omega_{\text{vib}} \sim M_r^{-1/2}\mu^{1/2}\,, \tag{27}$$

where the reduced mass for the molecular vibration is $M_r m_p$. Therefore, we obtain an equation $Z = Z(M_r, v)$ for the lines on the Z–M_r plane where we can expect approximate cancelation between the fine structure and vibrational intervals:

$$\omega = \omega_f - v\,\omega_{\text{vib}} \approx 0\,, \quad v = 1, 2, ... \tag{28}$$

Using Eqs. (26–28) it is easy to extract the dependence of the transition frequency on the fundamental constants:

$$\frac{\delta\omega}{\omega} = \frac{1}{\omega}\left(2\omega_f\frac{\delta\alpha}{\alpha} + \frac{v}{2}\omega_{\text{vib}}\frac{\delta\mu}{\mu}\right) \approx K\left(2\frac{\delta\alpha}{\alpha} + \frac{1}{2}\frac{\delta\mu}{\mu}\right), \tag{29}$$

where the enhancement factor $K = \frac{\omega_f}{\omega}$ determines the relative frequency shift for the given change of fundamental constants. Large values of K suggest potentially favorable cases for making experiment, because it is usually preferable to have larger relative shifts. However, there is no strict rule that larger K is always better. In some cases, such as very close levels, this factor may become irrelevant. Thus, it is also important to consider the absolute values of the shifts and compare them to the linewidths of the corresponding transitions.

Because there are a finite number of molecules we cannot have $\omega = 0$ exactly. However, a large number of molecules have $\omega/\omega_f \ll 1$ and $|K| \gg 1$. Moreover, an additional "fine tuning" may be achieved by selection of isotopes and rotational, Ω-doublet, and hyperfine components. Therefore, we have two large manifolds, the first one built on the electron fine-structure excited state and the second one built on the vibrational excited state. If these manifolds overlap one may select two or more transitions with different signs of ω. In this case the expected sign of the $|\omega|$-variation must be different (since the variation $\delta\omega$ has the same sign) and one can eliminate some systematic effects. Such control of systematic effects was used for transitions between close levels in two dysprosium isotopes.[57,58] The sign of the energy difference between two levels belonging to different electron configurations is different in ^{163}Dy and ^{162}Dy.

Among the interesting molecules where the ground state is split in two fine-structure levels and (28) is approximately fulfilled, there are Cl_2^+ (enhancement $K = 1600$), SiBr ($K = 360$), CuS ($K = 24$) and IrC ($K = 160$). The list of molecules is not complete because of the lack of data in Ref. 75. The molecules Cl_2^+ and SiBr are particularly interesting. For both of them the frequency ω defined by (28) is of the order of 1 cm^{-1} and comparable to the rotational constant B. Reducing ω further by the proper choice of isotopes, rotational quantum number J and hyperfine components, we may expect $K \sim 10^3 - 10^5$. New dedicated measurements are needed to determined exact values of the transition frequencies and find the best transitions. However, it is easy to find necessary accuracy of the frequency shift measurements. According to (29) the expected frequency shift is

$$\delta\omega = 2\omega_f \left(\frac{\delta\alpha}{\alpha} + \frac{1}{4}\frac{\delta\mu}{\mu} \right) \tag{30}$$

Assuming $\delta\alpha/\alpha \sim 10^{-15}$ and $\omega_f \sim 500\ \text{cm}^{-1}$, we obtain $\delta\omega \sim 10^{-12}\ \text{cm}^{-1} \sim 3 \times 10^{-2}$ Hz, which is larger than the natural width $\sim 10^{-2}$ Hz. (In order to obtain similar sensitivity comparing hyperfine transition frequencies for Cs and Rb one has to measure the shift at the $\sim 10^{-5}$ Hz level.)

5.3.3. *Molecular ion HfF^+*

The ion HfF^+ and other similar ions are considered by Cornell's group in JILA for experiments to search for the electric dipole moment (EDM) of the electron. Recent calculations[76] suggest that the ground state of this ion is $^1\Sigma^+$ and the first excited state $^3\Delta_1$ lies only 1633 cm^{-1} higher. Calculated vibrational frequencies for these two states are 790 and 746 cm^{-1} respectively. For these parameters the vibrational level $v = 3$ of the ground state is only 10 cm^{-1} from the $v = 1$ level of the state $^3\Delta_1$. Thus, instead of (28) we now have:

$$\omega = \omega_{\text{el}} + \frac{3}{2}\omega_{\text{vib}}^{(1)} - \frac{7}{2}\omega_{\text{vib}}^{(0)} \approx 0\,, \tag{31}$$

where superscripts 0 and 1 correspond to the ground and excited electronic states. The electronic transition ω_{el} is not a fine structure transition and (26) is not applicable. Instead, we can write:

$$\omega_{\text{el}} = \omega_{\text{el},0} + qx\,, \quad x = \alpha^2/\alpha_0^2 - 1\,. \tag{32}$$

Our estimate is[74]

$$\frac{\delta\omega}{\omega} \approx \left(\frac{2q}{\omega}\frac{\delta\alpha}{\alpha} + \frac{\omega_{\rm el}}{2\omega}\frac{\delta\mu}{\mu}\right) \approx \left(2000\frac{\delta\alpha}{\alpha} + 80\frac{\delta\mu}{\mu}\right), \tag{33}$$

$$\delta\omega \approx 20000 \text{ cm}^{-1}(\delta\alpha/\alpha + 0.04\delta\mu/\mu)\,. \tag{34}$$

Assuming $\delta\alpha/\alpha \sim 10^{-15}$ we obtain $\delta\omega \sim 0.6$ Hz. The natural width is about 2 Hz.

We also present the result for transition between close levels in Cs_2 molecule suggested in.[77,78] Our estimate is:[66]

$$\delta\omega \approx (-240\frac{\delta\alpha}{\alpha} - 1600\frac{\delta\mu}{\mu})cm^{-1}\,, \tag{35}$$

6. Enhanced effect of variation of α and strong interaction in UV transition of ^{229}Th nucleus (nuclear clock)

A very narrow level (7.6 ± 0.5) eV above the ground state exists in ^{229}Th nucleus.[79] The position of this level was determined from the energy differences of many high-energy γ-transitions to the ground and excited states. The subtraction produces the large uncertainty in the position of the 7.6 eV excited state. The width of this level is estimated to be about 10^{-4} Hz.[80] This would explain why it is so hard to find the direct radiation in this very weak transition. However, the search for the direct radiation continues.[81]

The ^{229}Th transition is very narrow and can be investigated with laser spectroscopy. This makes ^{229}Th a possible reference for an optical clock of very high accuracy, and opens a new possibility for a laboratory search for the variation of the fundamental constants.[82] As is shown in Ref. 83 there is an additional very important advantage: the relative effects of variation of α and m_q/Λ_{QCD} are enhanced by 5 orders of magnitude. This estimate has been confirmed recently by more accurate calculations.[84,85]

A rough estimate for the relative variation of the ^{229}Th transition frequency is

$$\frac{\delta\omega}{\omega} \approx 10^5(0.1\frac{\delta\alpha}{\alpha} + \frac{\delta X_q}{X_q}) \tag{36}$$

Therefore, the experiment would have the potential of improving the sensitivity to temporal variation of the fundamental constants by many orders of magnitude. Indeed, we obtain the following energy shift in the 7.6 eV ^{229}Th transition:

$$\delta\omega \approx \frac{\delta X_q}{X_q} MeV \tag{37}$$

This corresponds to the frequency shift $\delta\nu \approx 3 \cdot 10^{20} \delta X_q / X_q$ Hz. The width of this transition is 10^{-4} Hz so one may hope to get the sensitivity to the variation of X_q about 10^{-24} per year. This is 10^{10} times better than the current atomic clock limit on the variation of X_q.

Note that there are other narrow low-energy levels in nuclei, e.g. 76 eV level in ^{235}U with lifetime 26.6 minutes (see, e.g. Ref. 82). One may expect a similar enhancement there. Unfortunately, this level cannot be reached with usual lasers. In principle, it may be investigated using a free-electron laser or synchrotron radiation. However, the accuracy of the frequency measurements is much lower in this case.

7. Enhancement of variation of fundamental constants in ultracold atom and molecule systems near Feshbach resonances

Scattering length A, which can be measured in Bose-Einstein condensate and Feshbach molecule experiments, is extremely sensitive to the variation of the electron-to-proton mass ratio $\mu = m_e/m_p$ or $X_e = m_e/\Lambda_{QCD}$:[86]

$$\frac{\delta A}{A} = K \frac{\delta \mu}{\mu} = K \frac{\delta X_e}{X_e}, \tag{38}$$

where K is the enhancement factor. For example, for Cs-Cs collisions we obtained $K \sim 400$. With the Feshbach resonance, however, one is given the flexibility to adjust position of the resonance using external fields. Near a narrow magnetic or an optical Feshbach resonance the enhancement factor K may be increased by many orders of magnitude.

Acknowledgments

The authors are grateful to E. Shuryak and M. Kozlov for valuable contributions to this review. This work is supported by the Australian Research Council.

References

1. J.-P. Uzan, *Rev. Mod. Phys.* **75**, 403 (2003).
2. V. V. Flambaum and E. Shuryak arXiv:physics/0701220.
3. J. K. Webb *et al.*, *Phys. Rev. Lett.* **82**, 884 (1999).
4. J. K. Webb *et al.*, *Phys. Rev. Lett.* **87**, 091301 (2001).
5. M. T. Murphy, J. K. Webb and V. V. Flambaum, *Mon. Not. R. Astron. Soc.* **345**, 609 (2003).

6. M. T. Murphy, J. K. Webb and V. V. Flambaum, *Phys. Rev. Lett.* **99**, 239001 (2007).
7. M. T. Murphy, J. K. Webb and V. V. Flambaum, *Mon. Not. R. Astron. Soc.* **384**, 1053 (2008).
8. E. Reynold *et al.*, *Phys. Rev. Lett.* **96**, 151101 (2006).
9. V. F. Dmitriev, V. V. Flambaum and J. K. Webb, *Phys. Rev. D* **69**, 063506 (2004).
10. V. V. Flambaum and R. B. Wiringa, *Phys. Rev. C* **76**, 054002 (2007).
11. V. V. Flambaum, *Int. J. Mod. Phys. A* **22**, 4937 (2007).
12. S. N. Lea, *Rep. Prog. Phys.* **70**, 1473 (2007).
13. W. J. Marciano, *Phys. Rev. Lett.* **52**, 489 (1984).
14. X. Calmet and H. Fritzsch, *Eur. Phys. J. C* **24**, 639 (2002).
15. P. Langacker, G. Segre and M. J. Strassler, *Phys. Lett. B* **528**, 121 (2002).
16. C. Wetterich, *JCAP* **0310**, 02 (2003).
17. T. Dent and M. Fairbairn, *Nucl. Phys. B* **653**, 256 (2003).
18. S. K. Lamoreaux and J. R. Torgerson, *Phys. Rev. D* **69**, 121701 (2004).
19. C. R. Gould, E. I. Sharapov and S. K. Lamoreaux, *Phys. Rev. C* **74**, 024607 (2006).
20. Yu. V. Petrov *et al.*, *Phys. Rev. C* **74**, 064610 (2006).
21. Y. Fujii *et al.*, *Nucl. Phys. B* **573**, 377 (2000).
22. V. V. Flambaum and E. V. Shuryak, *Phys. Rev. D* **67**, 083507 (2003).
23. T. Dent, S. Stern and C. Wetterich, *Phys. Rev. D* **76**, 063513 (2007).
24. V. V. Flambaum, A. Höll, P. Jaikumar, C. D. Roberts and S. V. Wright, *Few-Body Syst.* **38**, 31 (2006).
25. A. Höll, P. Maris, C. D. Roberts and S. V. Wright arXiv:nucl-th/0512048.
26. V. V. Flambaum, D. B. Leinweber, A. W. Thomas and R. D. Young, *Phys. Rev. D* **69**, 115006 (2004).
27. R. H. Cyburt, B. D. Fields and K. A. Olive arXiv:0808.2818.
28. A. I. Shlyakhter, *Nature* **264**, 340 (1976).
29. T. Damour and F. J. Dyson, *Nucl. Phys. B* **480**, 37 (1996).
30. V. V. Flambaum and E. V. Shuryak, *Phys. Rev. D* **65**, 103503 (2002).
31. M. P. Savedoff, *Nature* **178**, 689 (1956).
32. V. A. Dzuba, V. V. Flambaum and J. K. Webb, *Phys. Rev. Lett.* **82**, 888 (1999).
33. V. A. Dzuba, V. V. Flambaum and J. K. Webb, *Phys. Rev. A* **59**, 230 (1999).
34. V. A. Dzuba, V. V. Flambaum and M. V. Marchenko, *Phys. Rev. A* **68**, 022506 (2003).
35. V. A. Dzuba and V. V. Flambaum, *Phys. Rev. A* **71**, 052509 (2005).
36. V. A. Dzuba, V. V. Flambaum, M. G. Kozlov and M. Marchenko, *Phys. Rev. A* **66**, 022501 (2002).
37. J. C. Berengut, V. A. Dzuba, V. V. Flambaum and M. V. Marchenko, *Phys. Rev. A* **70**, 064101 (2004).
38. A. Dzuba and V. V. Flambaum, *Phys. Rev. A* **77**, 012514 (2008).
39. J. C. Berengut, V. A. Dzuba, V. V. Flambaum, M. G. Kozlov, M. V. Marchenko, M. T. Murphy and J. K. Webb arXiv:physics/0408017.
40. M. T. Murphy, J. K. Webb, V. V. Flambaum, C. W. Churchill and J. X.

Prochaska, *Mon. Not. R. Astron. Soc.* **327**, 1223 (2001).
41. R. Srianand, H. Chand, P. Petitjean and B. Aracil, *Phys. Rev. Lett.* **92**, 121302 (2004).
42. S. A. Levshakov *et al.*, *Astron. Astrophys.* **434**, 827 (2005).
43. S. A. Levshakov *et al.*, *Astron. Astrophys.* **449**, 879 (2006).
44. M. G. Kozlov, V. A. Korol, J. C. Berengut, V. A. Dzuba and V. V. Flambaum, *Phys. Rev. A* **70**, 062108 (2004).
45. J. C. Berengut, V. A.Dzuba, V. V. Flambaum and M. G. Kozlov, *Phys. Rev. A* **69**, 044102 (2004).
46. J. C. Berengut, V. A. Dzuba and V. V. Flambaum, *Phys. Rev. A* **68**, 022502 (2003).
47. J. C. Berengut, V. V. Flambaum and M. G. Kozlov, *Phys. Rev. A* **72**, 044501 (2005).
48. J. C. Berengut, V. V. Flambaum and M. G. Kozlov, *Phys. Rev. A* **73**, 012504 (2006).
49. J. C. Berengut, V. V. Flambaum and M. G. Kozlov, *J. Phys. B* **41**, 235702 (2008).
50. P. Tzanavaris, J. K. Webb, M. T. Murphy, V. V. Flambaum and S. J. Curran, *Phys. Rev. Lett.* **95**, 041301 (2005).
51. P. Tzanavaris, J. K. Webb, M. T. Murphy, V. V. Flambaum and S. J. Curran, *Mon. Not. R. Astron. Soc.* **374**, 634 (2007).
52. V. A. Dzuba and V. V. Flambaum, *Phys. Rev. A* **61**, 034502 (2000).
53. E. J. Angstmann, V. V. Flambaum and S. G. Karshenboim, *Phys. Rev. A* **70**, 044104 (2004).
54. E. J. Angstmann, V. A. Dzuba and V. V. Flambaum, *Phys. Rev. A* **70**, 014102 (2004).
55. V. A. Dzuba and V. V. Flambaum, *Phys. Rev. A* **77**, 012515 (2008).
56. E. J. Angstmann, V. A. Dzuba and V. V. Flambaum arXiv:physics/0407141.
57. A. T. Nguyen, D. Budker, S. K. Lamoreaux and J. R. Torgerson, *Phys. Rev. A.* **69**, 022105 (2004).
58. A. Cingöz *et al.*, *Phys. Rev. Lett.* **98**, 040801 (2007).
59. V. A. Dzuba and V. V. Flambaum, *Phys. Rev. A* **72**, 052514 (2005).
60. E. J. Angstmann, V. A. Dzuba, V. V. Flambaum, S. G. Karshenboim and A. Yu. Nevsky, *J. Phys. B* **39**, 1937 (2006), physics/0511180.
61. J. D. Prestage, R. L. Tjoelker and L. Maleki, *Phys. Rev. Lett.* **74**, 3511 (1995).
62. S. G. Karshenboim, *Can. J. Phys.* **78**, 639 (2000).
63. V. V. Flambaum and A. F. Tedesco, *Phys. Rev. C* **73**, 055501 (2006).
64. T. M. Fortier *et al.*, *Phys. Rev. Lett.* **98**, 070801 (2007).
65. V. V. Flambaum and M. G. Kozlov, *Phys. Rev. Lett.* **98**, 240801 (2007).
66. V. V. Flambaum and M. G. Kozlov arXiv:0711.4536.
67. J. A. King, J. K. Webb, M. T. Murphy and R. F. Carswell, *accepted to Phys. Rev. Lett.* arXiv:0807.4366.
68. M. J. Drinkwater, J. K. Webb, J. D. Barrow and V. V. Flambaum, *Mon. Not. R. Astron. Soc.* **295**, 457 (1998).
69. M. T. Murphy *et al.*, *Mon. Not. R. Astron. Soc.* **327**, 1244 (2001).
70. J. van Veldhoven *et al.*, *Eur. Phys. J. D* **31**, 337 (2004).

71. C. Henkel *et al.*, *Astron. Astrophys.* **440**, 893 (2005).
72. F. Combes and T. Wiklind, *Astrophys. J.* **486**, L79 (1997).
73. V. V. Flambaum, *Phys. Rev. A* **73**, 034101 (2006).
74. V. V. Flambaum and M. G. Kozlov, *Phys. Rev. Lett.* **99**, 150801 (2007).
75. K. P. Huber and G. Herzberg, *Constants of Diatomic Molecules* (Van Nostrand, New York, 1979).
76. A. N. Petrov, N. S. Mosyagin, T. A. Isaev and A. V. Titov, *Phys. Rev. A* **76**, 030501(R) (2007).
77. D. DeMille, invited talk at 35th Meeting of the Division of Atomic, Molecular and Optical Physics, Tucson, Arizona (2004).
78. D. DeMille *et al.*, *Phys. Rev. Lett.* **100**, 043202 (2008).
79. B. R. Beck *et al.*, *Phys. Rev. Lett.* **98**, 142501 (2007).
80. E. V. Tkalya, A. N. Zherikhin and V. I. Zhudov, *Phys. Rev. C* **61**, 064308 (2000).
81. Z.-T. Lu, E. Peik and D. Habs, private communications.
82. E. Peik and Chr. Tamm, *Europhys. Lett.* **61**, 181 (2003).
83. V. V. Flambaum, *Phys. Rev. Lett.* **97**, 092502 (2006).
84. X.-T. He and Z.-Z. Ren, *J. Phys. G* **34**, 1611 (2006).
85. V. V. Flambaum and R. B. Wiringa arXiv:0807.4943.
86. C. Chin and V. V. Flambaum, *Phys. Rev. Lett.* **96**, 230801 (2006).

ALPHA-DOT OR NOT: COMPARISON OF TWO SINGLE ATOM OPTICAL CLOCKS

T. ROSENBAND*, D.B. HUME, C.-W. CHOU, J.C.J. KOELEMEIJ†, A. BRUSCH‡, S. BICKMAN, W. H. OSKAY¶, T.M. FORTIER, J.E. STALNAKER§, S.A. DIDDAMS, N.R. NEWBURY, W.C. SWANN, W.M. ITANO, D.J. WINELAND, AND J.C. BERGQUIST

National Institute of Standards and Technology,
325 Broadway, Boulder, CO 80305
** E-mail: trosen@boulder.nist.gov*

Repeated measurements of the frequency ratio of $^{199}Hg^+$ and $^{27}Al^+$ single-atom optical clocks over the course of a year yield a constraint on the possible present-era temporal variation of the fine-structure constant α. The time variation of the measured ratio corresponds to a time variation in the fine structure constant of $\dot{\alpha}/\alpha = (-1.6 \pm 2.3) \times 10^{-17}$/year, consistent with no change. The frequency ratio of these clocks was measured with a fractional uncertainty of 5.2×10^{-17}. Stability simulations for optical clocks whose probe period is limited by 1/f-noise in the laser local oscillator provide an estimate of the optimal probe period, as well as a modified expression for the theoretical clock stability.

1. Introduction

Recent analyses and modeling of astrophysical and geophysical data have produced conflicting reports of temporal and spatial variations of the fundamental "constants". The conflicting evidence of change, and its important ramifications if true, has triggered a number of experimental efforts toward finding unambiguous evidence of such a variation.[1,2] Much of the experimental work so far has been focused on a search for a variation of the fine structure constant α. The astrophysical and geophysical studies mentioned above attain high sensitivity to small temporal changes in α by observ-

†Present address: Dept. of Physics and Astronomy, Vrije Universiteit Amsterdam, The Netherlands
‡Present address: Niels Bohr Institute, University of Copenhagen, Denmark
¶Present address: Stanford Research Systems, Sunnyvale, CA
§Present address: Dept. of Physics and Astronomy, Oberlin College, Oberlin, OH

ing physical phenomena that occurred long ago and comparing these with the corresponding phenomena today. The prominent examples include the geochemical remains of the Oklo natural nuclear reactor[3] and the astronomical observations of spectroscopic absorption lines from distant interstellar clouds with a quasar used as the light source.[4] However, the analysis of data from Oklo, as well as that from distant quasars, is complicated by difficult quantitative interpretations that have produced discrepancies (see refs. 5 and 6 for example).

Since the energies of optical atomic transitions can be parameterized simply in terms of the fine structure constant α, two or more well stabilized and accurate optical clocks offer an alternate approach to the measurement of $d\alpha/dt$ – repetitive measurements of their frequency ratio(s) over time. At NIST, the Hg^+/Al^+ clock ratio measurements have now reached a precision sufficiently high that a year-long search for a linear time variation of α, or other fundamental constants, can be performed with a sensitivity higher than that offered by the geological and cosmological observations for the present era. High accuracy frequency comparison experiments between atomic clocks are also useful to search for violations of Local Position Invariance (LPI), which is a fundamental tenet of Einstein's Theory of General Relativity. A check for possible violations of LPI can be made by looking for correlations between the frequency ratio of two atomic clocks and the change of the gravitational potential due, for example, to the revolution of the Earth around the Sun.[7,8]

We refer the reader elsewhere for a more comprehensive description of the Hg^+ and Al^+ single-ion, optical clocks (see, for example, refs. 9 and 10). Likewise, the significant contributions to the systematic frequency uncertainty of each optical clock have been described previously.[11,12] Here we briefly summarize the dominant frequency uncertainties of each of the two clocks (Sec. 2 and 3) and then present a unitless measurement of their frequency ratio that spans a year. From the results of these highly accurate frequency ratio measurements, an estimate of the temporal stability of the constant α can be provided.

2. Summary of Hg^+ Clock Systematics

The residual motion of the laser-cooled Hg^+ ion gives rise to a fractional frequency uncertainty associated with the relativistic time dilation shift. The uncertainty due to the secular motion is constrained to be below 4×10^{-18} when the ion is cooled to near the Doppler cooling limit; the uncertainty due to residual micro-motion is also constrained to be below 4×10^{-18} when the trap is well compensated to minimize RF micro-motion from stray static electric fields. The micro-motion of the trap is continuously monitored,[11] and, if need be, corrected by use of small bias voltages applied to either endcap and/or to two bias electrodes.

The ($^2S_{1/2}$ F=0) → ($^2D_{5/2}$ F=2 m_F=0) clock transition in $^{199}Hg^+$ is first-order magnetic field insensitive at zero field and the second-order sensitivity is approximately $(-1.89 \times 10^{10}$ Hz $T^{-2})B^2$, where B is the magnetic field. The quadratic field shift of the clock transition due to the small (8 μT) applied quantization field is monitored during the measurement run by occasionally interleaving a frequency measurement of the first-order field sensitive ($^2S_{1/2}$ F=0) → ($^2D_{5/2}$ F=2 m_F=2) component with the regular clock cycles. Slow variations of the magnetic field presently limit the uncertainty of the DC quadratic Zeeman shift to 5×10^{-18}. The maximum AC Zeeman shift caused by any asymmetry of the RF currents flowing in the trap electrodes is conservatively estimated to be less than 10^{-17} by assuming that there is no more than a 25 % imbalance of the RF currents.[12]

The electronic charge density of the $^2D_{5/2}$ state has an electric quadrupole moment that can produce an energy shift in the presence of a static electric field gradient. The magnitude and sign of the shift depend on the relative orientation of the field gradient and the applied field, which offers a route to its cancellation.[13] Typically, the fractional frequency shift is less than 10^{-16}, but because we neither know, nor control, the ambient field gradient, the uncertainty is equally as large. However, the quadrupole shift, as well as its uncertainty, can be effectively eliminated by averaging the clock frequency for three mutually perpendicular field quantization axes.[13] In our realization of this scheme, the orientation of the three axes and the probe light polarization were chosen to give the same scattering rate for the clock transition over the three directions of the applied magnetic field.[11] Changing the field direction at a regular interval (typically, every 300 s) and steering the frequency of the clock laser to resonance with the ion at each field setting are automated. The average clock frequency for the three field directions has zero quadrupole shift with a residual uncertainty of 10^{-17}. The remaining uncertainty is due to the slight nonorthogonality of the three applied fields.

The cryogenic surfaces in the Hg^+ system, adsorb all gases with the possible exception of He, so we have mounted a small cell of compressed charcoal at the bottom of the dewar that cryo-pumps the He. However, because of its limited pumping speed for He and the possibility of He leaks through, for example, the indium seals, the background He pressure could rise to a level that could cause a significant (10^{-18}) pressure shift. An upper limit to the He partial pressure was set at 7×10^{-10} Pa by using a quadrupole mass analyzer, which is conservatively estimated to give a maximum fractional frequency shift[12] of 4×10^{-18}.

Another potential shift arises from the pulsed interrogation and detection sequence. If the clock atoms were to move synchronously with each clock cycle, the clock frequency would suffer a first-order Doppler shift with a non-zero average. This synchronized movement could have different sources, for example, the creation of photoelectric charges on ion trap electrodes due to the scattering of UV light, or, simply, motion of the trapped atom(s) with the opening or closing of mechanical shutters. The systematic shift from this effect can be substantial: a synchronized velocity of only 10 nm/s would cause a fractional frequency shift of 3×10^{-17}. The shift can be averaged away by alternately probing the optical clock resonance with collinear but counter-propagating beams. We detected no statistically significant frequency difference between the two probe directions, but an imbalance in the statistical weight of the two probe directions leads to a fractional uncertainty of 7×10^{-18} in the frequency ratio reported here.[12]

A frequency shift can be caused by the thermal loading of any acousto-optic modulator (AOM) in the clock beam path that is turned on and off, such as the stepping AOM used here to probe either side of the clock resonance. The shift appears to be a thermal effect[14] that scales linearly with RF power, but diminishes as the pulse duty-cycle approaches unity. For Hg^+ the typical level of applied power during the probe "on" phase gives a maximum fractional frequency shift of 6×10^{-18}.

3. Quantum-logic Al^+ clock

Single-ion clocks based on the $^1S_0 \leftrightarrow {}^3P_0$ transition of singly ionized group IIIA atoms were proposed by Dehmelt.[15] Of the suggested species (B^+, Al^+, Ga^+, In^+, Tl^+), a single-ion clock based on In^+ was investigated in several experiments.[16,17] The other group IIIA ions were considered impractical, due to the fact that their laser-cooling and state-detection atomic-transitions have wavelengths in the deep UV. For the indium ion, this problem was avoided by laser-cooling and detecting fluorescence from the $^1S_0 \leftrightarrow {}^3P_1$ transition. Yet, this approach is a compromise, because the scattering rate of the weakly allowed $^1S_0 \leftrightarrow {}^3P_1$ transition ($\gamma = 2\pi \times 360$ kHz) limits the efficiency of laser-cooling and fluorescence detection.

Quantum-logic spectroscopy[18,19] avoids this problem by shifting the burden of laser-cooling and fluorescence state detection to a second (logic) ion in the same trap that has an allowed electronic transition ($\gamma \approx 10^8$ Hz). Strong Coulomb coupling between the clock and cooling ions allows for sympathetic laser cooling of the clock ion. Likewise, the coupled motion of the two ions allows efficient transfer of the clock ion's internal quantum state to the logic ion, where it can be efficiently measured.[20] Thus, a strongly allowed transition in the clock species is not required, which greatly expands the list of potential clock ions. The Coulomb interaction of the clock and logic ion suggests the use of a linear Paul trap to confine the ions, so that an RF nodal line exists with a simultaneous micro-motion null for both ions. In addition, the logic ion will generally produce a significant electric-field gradient on the clock ion. This favors the use of a clock transition with negligible electric-quadrupole shift.

In this article we report the results of a quantum-logic $^{27}Al^+$ clock that relies on $^9Be^+$ for sympathetic cooling and state detection. A more recent experiment combines $^{27}Al^+$ with $^{25}Mg^+$, where a quality factor for the atomic resonance of $Q = 3.5\times10^{14}$ was achieved.[21] For the Al^+ / Be^+ clock, the systematic frequency shifts have been evaluated[12] with a fractional frequency uncertainty of 2.3×10^{-17}. Here the dominant uncertainty is due to radial micro-motion, which occurs when slowly varying electric fields in the ion trap force the ion away from the RF nodal line. These electric fields vary for two main reasons. First, electrons are sprayed through the ion trap when the ions are loaded by electron-impact ionization (typically they must be reloaded every few hours). These electrons slowly discharge during the course of the clock measurements ($\tau \approx 15$ minutes). Second, stray light from the laser beams (267 nm) that drive the Al^+ $^1S_0 \leftrightarrow {}^3P_1$ and $^1S_0 \leftrightarrow {}^3P_0$ transitions causes some parts of the ion trap to emit photoelectrons, leading to slowly varying electric fields in the ion trap. We apply real-time computer feedback to detect and eliminate the resulting micro-motion. Spot checks show that these feedback loops generally keep the magnitude of the micro-motion frequency shift below 2×10^{-17}.

The small differential polarizability between the two clock states leads to a room-temperature black-body radiation shift of $(-8\pm3)\times10^{-18}$. As the ion-trap used here operates at an elevated temperature, the black-body shift for these measurements was $(-12\pm5)\times10^{-18}$. The differential polarizability also couples to the electric field of the clock laser beam, whose probe period of 100 ms corresponds to an electric field amplitude in the laser beam of of about $E_0 = 0.2$ V/cm. This leads to a Stark shift of much less than 1 mHz through allowed transitions and coupling to the $^3P_1$ level. Because the frequency produced by the $^{27}Al^+$ clock is the average of the ($^1S_0$ F$= 5/2$

$m_F=\pm5/2) \leftrightarrow (^3P_0$ F= 5/2 $m_F=\pm5/2)$ transition frequencies near 1 Gauss magnetic field, the average can only be shifted by an imbalance in the σ_+ and σ_- polarization components in the probe beam. Such imbalances are minimized by use of π-polarized probe light that travels perpendicular to the quantization field. The effective field amplitude of this unbalanced Stark shifting field is $E_s = E_\sigma \sin\epsilon$, where $E_\sigma \leq 0.03E_0$ is the unbalanced circular polarization component of the beam, whose amplitude can be determined with a crossed polarizer, and $\epsilon \leq 0.1$ is the maximum misalignment angle of the beam. These bounds yield a maximum Stark shift of about 10^{-5} Hz. Stark shifts due to the probe beam were also tested experimentally. An increase in the probe laser power by a factor of 10000 (1 ms probe period) caused no noticable clock shifts at the 1 Hz level, thereby reducing the uncertainty of this shift to the level of 10^{-19} in units of fractional frequency.

4. The Frequency Ratio of Al^+ to Hg^+

Figure 4(A) shows the fractional ratio instability (Allan deviation) of the two optical standards for a typical measurement run. For integration periods longer than 100 s, when the frequencies of the clocks are fully steered by the atoms, the fractional instability is $3.9 \times 10^{-15}(\tau/\mathrm{s})^{-1/2}$. The fractional frequency instability has been improved by nearly a factor of two with respect to measurements that were made in early 2006, primarily due to decreasing the dead time and improving the duty cycle of the Hg^+ optical clock as well as incorporating a better laser reference cavity, a better cooling process and longer probe periods for the Al^+ optical clock.

Figure 4(B) shows the frequency ratio measurements of Al^+ to Hg^+ from December 2006 to November 2007. The full accuracy evaluation for the optical clocks that include corrections for the first order Doppler shifts synchronized with the probe periods has been carried out only for the last four points. The weighted average of the frequency ratio from these last four measurements is[12] $f_{Al}/f_{Hg} = 1.052871833148990438(55)$. The systematic fractional frequency uncertainties of Hg^+ (1.9×10^{-17}) and Al^+ (2.3×10^{-17}) contribute by nearly the same amount to the overall fractional uncertainty of their ratio (5.2×10^{-17}), which is dominated by a statistical uncertainty of (4.3×10^{-17}).

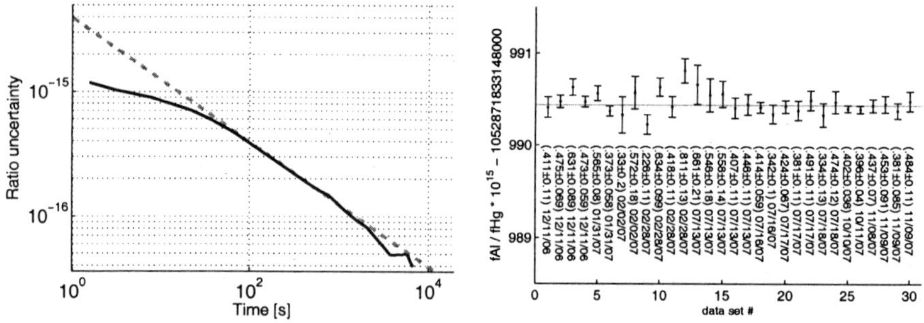

(A) (B)

Fig. 1. (A) Allan deviation[22] of a frequency ratio measurement (11 000 s total). The dashed line represents a $1/\sqrt{\tau}$ slope, beginning at 3.9×10^{-15} for 1 s. (B) History of frequency ratio measurements of the $^{199}Hg^{+}$ and $^{27}Al^{+}$ frequency standards. Error bars are statistical. Only the last four points are used in the ratio reported here,[12] as the systematic shifts for the earlier data were not as well controlled. The fractional digits of the ratio, scaled by 10^{15}, as well as the uncertainty, and the measurement date (month/day/year) are shown on the plot.

5. Test of the temporal stability of α

The fine structure constant α is the natural scaling factor for the energies involved in atomic spectroscopy. Transition frequencies from electronic (gross), fine (FS) and hyperfine (HFS) structure can be written as a function of a nonrelativistic part, which depends only on the structure involved in the transition, and a relativistic part $F_j(\alpha)$, which depends on the specific transition j that is considered. Thus the transition frequencies can be written in the following manner:[23–25]

$$f_j(el) = R_y F_j(\alpha) \tag{1}$$

$$f_j(FS) = \alpha^2 R_y F_j(\alpha) \tag{2}$$

$$f_j(HFS) = \alpha^2 (\mu/\mu_B) R_y F_j(\alpha), \tag{3}$$

where R_y is the Rydberg constant, μ/μ_B is the ratio between the magnetic moment of the nucleus μ and the Bohr magneton μ_B.

Sensitivity to any change of R_y (temporal or otherwise) is lost in a ratio measurement, since R_y is common to all transition frequencies. The measurement of the frequency ratio of any two electronic transitions in an atom or pair of atoms returns the simplest, most direct measure of any temporal change of α. However, the measurement of the ratio of electronic transitions is not necessarily very sensitive if the precision of the measurements is not high, and/or if the corresponding $F_j(\alpha)$ functions are not strongly disparate. $F_j(\alpha)$ contains the dependence of the specific transition involved, can be either positive or negative, and tends to be larger for transitions involving heavier atoms (where relativistic corrections play a more important role). Values of $F_j(\alpha)$ have been calculated for several atomic and molecular transitions of cosmological and laboratory interest, including[4,26] Hg^+ and Al^+, as well as for several HFS transitions in the microwave spectral region.[4]

Multiple measurements of the frequency ratio $r = f_{Al}/f_{Hg}$ of the optical-clock transitions of Al^+ and Hg^+ spanning about 1 year (December 2006 to November 2007) give the following result:[12]

$$(\frac{d}{dt}r)/r = (-5.3 \pm 7.9) \times 10^{-17}\text{yr}^{-1}. \tag{4}$$

From Eq. 4, the temporal variation of α can be expressed as

$$N(\frac{d}{dt}\alpha)/\alpha = (\frac{d}{dt}r)/r, \tag{5}$$

where $N = N_{Al+} - N_{Hg+} = 3.2$ and $N_{Al+} \approx 0.008$. Hence, $(\frac{d}{dt}\alpha)/\alpha$ is constrained to $(-1.6 \pm 2.3) \times 10^{-17}\text{yr}^{-1}$.

This, the most stringent limit to any present-era linear change in α, is achieved primarily because of the large negative relativistic contribution for Hg^+ and the accuracy of the measurement data. The intermediate points were excluded, since the evaluation of their systematic uncertainties was incomplete. The tighter constraint made on any temporal variation of α can be used together with the 10-year record of absolute frequency measurements of the Hg^+ optical clock to tighten the coupled constraint on the possible temporal variation of the Cs magnetic moment, μ_{Cs}, to[12] $(\frac{d}{dt}\mu_{Cs})/\mu_{Cs} = (-1.9 \pm 4.0) \times 10^{-16}\text{yr}^{-1}$.

6. Stability of optical atomic clocks

There have been several studies of optical clock stability,[27–29] but the stability limits of the newest generation of optical atomic clocks have not yet been fully explored. For these clocks the coherence times of the atoms that compose the clocks are likely to be significantly longer than the coherence times of the lasers that drive them. This is the current situation for Al^+, where the natural linewidth of the $^1S_0 \leftrightarrow {}^3P_0$ transition is 8 mHz, while the probe-laser linewidth is of order 100 mHz.[30] Neutral-atom-based lattice clocks[31] are in a similar situation,[32] as they are also based on doubly-forbidden $^1S_0 \leftrightarrow {}^3P_0$ transitions with natural linewidths of order 10 mHz. If an isotope without nuclear spin is chosen, such as ^{174}Yb, the natural linewidth can be made arbitrarily small.[33,34]

While atomic resonances that are extremely narrow can be chosen for optical clocks, a fundamental noise mechanism in the probe lasers that drive these resonances has been identified, and it cannot be reduced so easily. Numata *et al.*[35] have found that thermomechanical noise in the length of the Fabry-Perot laser stabilization cavities leads to inevitable $1/f$ noise in the frequency of the laser local oscillator. What then is the stability of optical atomic clocks that are limited by this type of laser decoherence, and not by atomic decoherence? To address this question, some results of discrete-time numerical simulations are presented.

In this computer simulation the Ramsey method is used to probe a collection of N atoms, and dead-time is neglected, unless otherwise noted. Let t_R be the Ramsey evolution period. The discrete times of the simulations are $t_i = it_R$, where i is a non-negative integer that denotes the number of preceding measurement cycles. First, a noise signal n_i with $1/f$ spectral-density is generated[36] by appropriately filtering a white-noise signal in the Fourier-domain. Subsequently, the measurement-feedback cycles of the atomic clock are simulated by the iterative application of equations 6-8 on a computer.

Let c_i be the constant frequency correction applied to the clock's laser local oscillator (LO) during the time interval t_i, and let the initial correction be $c_0 = 0$. Then the clock's mean output frequency f_i during this time interval is the sum of the atomic resonance frequency ν_0, the noise frequency n_i, and correction frequency c_i:

$$f_i = \nu_0 + c_i + n_i \tag{6}$$

This corresponds to a clock phase error of

$$\phi_i = 2\pi t_R(c_i + n_i), \tag{7}$$

which is detected through its effect on the atomic response to a Ramsey interrogation that begins at time t_i and ends at t_{i+1}. The response function is the probability $p_i = R(\phi_i)$ (see Eq. 9) that each atom is in the excited state after the Ramsey interrogation is completed. For N atoms, the total number of excitations is a binomial random variable[37] X_i that represents the number of successes in N tries of probability p_i. Finally, the clock's frequency correction for the next time step is calculated from the inverse atomic response function $R^{-1}(p_i)$, with a feedback gain of G:

$$c_{i+1} = c_i + \frac{G}{2\pi t_R} R^{-1}(X_i/N) \tag{8}$$

This iterative sequence is repeated many times, until the clock's long-term stability becomes apparent, i.e. the coefficient σ_0 of the stability asymptote $\sigma_y(\tau) = \sigma_0(\tau/\mathrm{s})^{-1/2}$ can be determined.

For the atomic response function $R(\phi)$ we take a sine wave, corresponding to the Ramsey signal from two infinitesimally short $\pi/2$-pulses at frequency f_i separated in time by t_R, where the second pulse is phase-shifted by 90° with respect to the first:

$$R(\phi) = \frac{1}{2}(1 + \sin\phi) \tag{9}$$

$$R^{-1}(p) = \arcsin(2p - 1) \tag{10}$$

Here $R^{-1}(R(\phi)) = \phi$ for the interval $-\frac{\pi}{2} < \phi \leq \frac{\pi}{2}$. It should be noted that when the Ramsey period is maximized, the LO noise n_i may sometimes cause ϕ_i to fall outside this interval of unique invertibility. In this case, the frequency correction applied in Eq. 8 contains an "inversion error" in addition to the quantum projection noise[37] inherent in the random variable X_i. For this reason, atomic clocks with large atom number N and low projection noise will generally choose a somewhat shorter probe period t_R than single-ion clocks, where $N = 1$ and the projection noise is high when all other parameters are equal.

We have numerically simulated the performance of atomic clocks with different atom number N (see Fig. 2), and also shown the expected clock stability according to the well known expression[37]

$$\sigma_y(\tau) = \frac{1}{2\pi\nu\sqrt{N\tau t_R}}. \tag{11}$$

For these simulations, a laser noise floor of $\sigma_f(\tau) = 1$ Hz is assumed, which corresponds to laser frequency noise with a one-sided power-spectral-density of[22] $S_f(f) = \frac{\sigma_f^2}{2f\log 2}$. The fractional frequency uncertainty $\sigma_y(\tau)$ and the frequency uncertainty $\sigma_f(\tau)$ are related by $\sigma_y(\tau) = \nu^{-1}\sigma_f(\tau)$, where ν is the clock's oscillation frequency. For the optimum Ramsey probe period we find $t_R = 0.07/\sigma_f$ and $t_R = 0.04/\sigma_f$ for $N = 1$ and $N = 10^4$ respectively. When this decoherence-limited Ramsey period is inserted into Eq. 11, we find

$$\sigma_y(\tau) = a(N)\left(\frac{\sigma_{y0}}{N\nu\tau}\right)^{1/2}, \tag{12}$$

where $a(N)$ is a numerical factor that depends very weakly on the atom number ($a(1) \approx 0.6$, $a(10^4) \approx 0.8$), and $\sigma_{y0} = \nu^{-1}\sigma_f$ is the fractional-frequency stability floor of the probe laser. Significantly, the clock stability in Eq. 12 improves as $\nu^{-1/2}$, while the improvement in Eq. 11 is ν^{-1}.

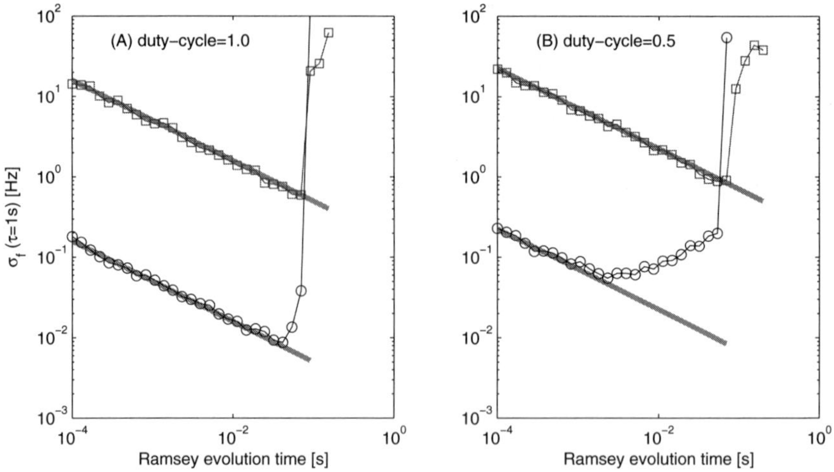

Fig. 2. Dependence of clock stability on Ramsey probe period, given a local oscillator with an Allan deviation noise floor of 1 Hz. Simulation results are shown for $N = 10^4$ atoms (○), and $N = 1$ atoms (□). Thick lines show the theoretical stability given by Eq. 11, which assumes a perfect LO. The vertical axis shows the asymptotic clock stability, extrapolated to $\tau = 1$ s. Optimal probe periods are $t_R(N = 1) \approx 0.07$ s, and $t_R(N = 10^4) \approx 0.04$ s. The optimal feedback gain was found to be $G = -0.2$ for one atom, and $G = -1$ for 10^4 atoms. At long Ramsey periods, the simulated clock diverges because of LO noise. In (B) the results of the same simulation with a reduced duty-cycle of 0.5 are shown. As expected,[27] the loss of stability is greater for $N = 10^4$ than for $N = 1$.

This shows that entangled states[38] where the effective Ramsey evolution frequency is $\nu' = N\nu$ and the effective atom number is $N' = 1$ will not significantly improve the stability of the optical atomic clocks considered here, because $N'\nu' = N\nu$. The same conclusion was reached in a separate analysis.[39]

We have performed analogous simulations for the Rabi interrogation method, and find a slight increase in the optimal probe period. However, the Ramsey method still outperforms the Rabi method, due to its sharper frequency discriminant.

7. Conclusions

We have measured the frequency ratio of two single-ion optical clocks with an accuracy of 5.2×10^{-17}. The historical record of these measurements over the course of a year shows no significant temporal variation, and indicates that the present rate of change of the fine-structure constant is no more than a few parts in 10^{17} fractionally. The measurement uncertainty is dominated by statistical uncertainty due to clock instability, and we have explored numerically how $1/f$ frequency noise in the clock's probe laser limits the achievable stability for the Al^+ clock.

Acknowledgements

This work was supported by ONR, IARPA, and NIST. J.C.J.K. acknowledges support from the Netherlands Organisation for Scientific Research (NWO). We thank M. A. Lombardi, A. Ludlow, and D. R. Smith for their careful reading of this manuscript. This work is a contribution of NIST, and is not subject to U.S. copyright.

References

1. V. V. Flambaum, *Int. J. Mod. Phys. A* **22**, 4937 (2007).
2. S. N. Lea, *Reports on Progress in Physics* **70**, 1473 (2007).
3. T. Damour and F. Dyson, *Nuclear Physics B* **480**, 37 (1996).
4. V. A. Dzuba, V. V. Flambaum and J. K. Webb, *Phys. Rev. A* **59**, 230 (1999).
5. M. T. Murphy, J. K. Webb and V. V. Flambaum, *Mon. Not. Roy. Astron. Soc.* **345**, p. 609 (2003).
6. S. A. Levshakov, M. Centurión, P. Molaro, S. D'Odorico, D. Reimers, R. Quast and M. Pollmann, *Astronomy and Astrophysics* **449**, 879 (2006).
7. T. M. Fortier, N. Ashby, J. C. Bergquist, M. J. Delaney, S. A. Diddams, T. P. Heavner, L. Hollberg, W. M. Itano, S. R. Jefferts, K. Kim, F. Levi, L. Lorini, W. H. Oskay, T. E. Parker, J. Shirley and J. E. Stalnaker, *Physical Review Letters* **98**, p. 070801 (2007).
8. S. Blatt, A. D. Ludlow, G. K. Campbell, J. W. Thomsen, T. Zelevinsky, M. M. Boyd, J. Ye, X. Baillard, M. Fouché, R. L. Targat, A. Brusch, P. Lemonde, M. Takamoto, F.-L. Hong, H. Katori and V. V. Flambaum, *Physical Review Letters* **100**, p. 140801 (2008).
9. W. H. Oskay, W. M. Itano and J. C. Bergquist, *Phys. Rev. Lett.* **94**, p. 163001 (2005).
10. T. Rosenband, P. O. Schmidt, D. B. Hume, W. M. Itano, T. M. Fortier, J. E. Stalnaker, K. Kim, S. A. Diddams, J. C. J. Koelemeij, J. C. Bergquist and D. J. Wineland, *Phys. Rev. Lett.* **98**, p. 220801 (2007).
11. W. H. Oskay, S. A. Diddams, E. A. Donley, T. M. Fortier, T. P. Heavner, L. Hollberg, W. M. Itano, S. R. Jefferts, M. J. Delaney, K. Kim, F. Levi, T. E. Parker and J. C. Bergquist, *Phys. Rev. Lett.* **97**, p. 020801 (2006).
12. T. Rosenband, D. B. Hume, P. O. Schmidt, C. W. Chou, A. Brusch, L. Lorini, W. H. Oskay, R. E. Drullinger, T. M. Fortier, J. E. Stalnaker, S. A. Diddams, N. R. N. W. C. Swann, W. M. Itano, D. J. Wineland and J. C. Bergquist, *Science* **319**, p. 1808 And online supplementary materials (http://www.sciencemag.org/cgi/content/full/1154622/DC1) (2008).
13. W. M. Itano, *J. Res. NIST* **105**, p. 829 (2000).

14. C. Degenhardt, T. Nazarova, C. Lisdat, H. Stoehr, U. Sterr and F. Riehle, *IEEE Trans. Inst. Meas.* **54**, p. 771 (2005).
15. H. G. Dehmelt, *IEEE Trans. Inst. Meas.* **31**, p. 83 (1982).
16. J. A. Sherman, W. Trimble, S. Metz, W. Nagourney and N. Fortson, Progress on indium and barium single ion optical frequency standards (2005), arXiv.org:physics/0504013.
17. T. Liu, Y. Wang, V. Elman, A. Stejskal, Y. Zhao, J. Zhang, Z. Lu, L. Wang, R. Dumke, T. Becker and H. Walther, *Frequency Control Symposium, 2007 Joint with the 21st European Frequency and Time Forum. IEEE International* , 407 (2007).
18. D. J. Wineland, J. C. Bergquist, J. J. Bollinger, R. E. Drullinger and W. M. Itano, in *Proceedings of the 6th Symposium on Frequency Standards and Metrology*, ed. P. Gill (World Scientific, 2002).
19. P. O. Schmidt, T. Rosenband, C. Langer, W. M. Itano, J. C. Bergquist and D. J. Wineland, *Science* **309**, p. 749 (2005).
20. D. B. Hume, T. Rosenband and D. J. Wineland, *Phys. Rev. Lett.* **99**, p. 120502 (2007).
21. C.-W. Chou *et al.*, Manuscript in preparation.
22. D. B. Sullivan, D. Allan, D. A. Howe and F. L. Walls, *Characterization of Clocks and Oscillators*, Tech. Rep. 1337, NIST (1990).
23. S. G. Karshenboim, *Can. J. Phys.* **78**, 639 (2000).
24. T. M. Fortier, N. Ashby, J. C. Bergquist, M. J. Delaney, S. A. Diddams, T. P. Heavner, L. Hollberg, W. M. Itano, S. R. Jefferts, K. Kim, F. Levi, L. Lorini, W. H. Oskay, T. E. Parker, J. Shirley and J. E. Stalnaker, *Physical Review Letters* **98**, p. 070801 (2007).
25. E. Peik *et al.*, Laboratory limits on temporal variations of fundamental constants: An update, in *Proceedings of the 11th Marcel Grossmann Meeting*, (Berlin, 2006).
26. E. J. Angstmann, V. A. Dzuba and V. V. Flambaum, *Phy. Rev. A* **70**, p. 014102 (2004).
27. A. Quessada, R. P. Kovacich, I. Courtillot, A. Clairon, G. Santarelli and P. Lemonde, *Journal of Optics B: Quantum and Semiclassical Optics* **5**, S150 (2003).
28. E. Peik, T. Schneider and C. Tamm, *Journal of Physics B: Atomic, Molecular and Optical Physics* **39**, 145 (2006).
29. E. Riis and A. G. Sinclair, *Journal of Physics B: Atomic, Molecular and Optical Physics* **37**, 4719 (2004).
30. B. C. Young, F. C. Cruz, W. M. Itano and J. C. Bergquist, *Phys. Rev. Lett.* **82**, p. 3799 (1999).
31. H. Katori, M. Takamoto, V. G. Pal'chikov and V. D. Ovsiannikov, *Phys. Rev. Lett.* **91**, p. 173005 (2003).
32. M. M. Boyd, T. Zelevinsky, A. D. Ludlow, S. M. Foreman, S. Blatt, T. Ido and J. Ye, *Science* **314**, p. 1430 (2006).
33. A. V. Taichenachev, V. I. Yudin, C. W. Oates, C. W. Hoyt, Z. W. Barber and L. Hollberg, *Physical Review Letters* **96**, p. 083001 (2006).
34. Z. W. Barber, C. W. Hoyt, C. W. Oates, L. Hollberg, A. V. Taichenachev

and V. I. Yudin, *Physical Review Letters* **96**, p. 083002 (2006).
35. K. Numata, A. Kemery and J. Camp, *Phys. Rev. Lett.* **93**, p. 250602 (2004).
36. J. L. Lennon, *Ecography* **23**, 101 (2000).
37. W. M. Itano, J. C. Bergquist, J. J. Bollinger, J. M. Gilligan, D. J. Heinzen, F. L. Moore, M. G. Raizen and D. J. Wineland, *Phys. Rev. A* **47**, p. 3554 (1993).
38. J. J. . Bollinger, W. M. Itano, D. J. Wineland and D. J. Heinzen, *Phys. Rev. A* **54**, R4649 (1996).
39. D. J. Wineland, C. Monroe, W. M. Itano, D. Leibfried, B. E. King and D. M. Meekhof, *J. Res. NIST* **103**, p. 259 (1998).

VARIATION OF THE FINE-STRUCTURE CONSTANT AND LASER COOLING OF ATOMIC DYSPROSIUM

N. A. Leefer, A. Cingöz, D. Budker*,

Department of Physics, University of California at Berkeley,
Berkeley, CA 94720-7300, USA
** E-mail: budker@berkeley.edu*

S. J. Ferrell, V. V. Yashchuk

Lawrence Berkeley National Laboratory,
Berkeley, California 94720, USA

A. Lapierre

TRIUMF National Laboratory,
Vancouver, British Columbia, V6T 2A3, Canada

A.-T Nguyen

Department of Otolaryngology, University of Pittsburgh,
Pittsburgh, Pennsylvania 15213, USA

S. K. Lamoreaux

Department of Physics, Yale University,
New Haven, Connecticut 06520-8120, USA

J. R. Torgerson

Physics Division, Los Alamos National Laboratory,
P-23, MS-H803, Los Alamos, New Mexico 87545, USA

Radio-frequency electric-dipole transitions between nearly degenerate, opposite parity levels of atomic dysprosium (Dy) were monitored over an eight-month period to search for a variation in the fine-structure constant, α. The data provide a rate of fractional temporal variation of α of $(-2.4 \pm 2.3) \times 10^{-15}$ yr^{-1} or a value of $(-7.8 \pm 5.9) \times 10^{-6}$ for k_α, the variation coefficient for α in a changing gravitational potential. All results indicate the absence of significant variation at the present level of sensitivity. We also present initial results on laser cooling of an atomic beam of dysprosium.

1. Introduction

A component of Einstein's equivalence principle (EEP) is local position invariance (LPI), which states that the laws of physics, including the values of fundamental constants, should be independent of space and time. Modern theories attempting to unify gravitation with the other fundamental interactions allow, or even predict, violations of EEP,[1] which has sparked searches for violation of LPI, and hence EEP, through searches for temporal and spatial variation of fundamental constants.

Various studies have reported results of searches for a temporal variation of the fine-structure constant ($\alpha = e^2/\hbar c$) over cosmological time scales of 10^{10} years,[2–5] geological time scales of 10^9 years,[6–8] and present day laboratory searches over the course of years.[9–15] In contrast to studies involving analyses of processes that have occurred billions of years ago, the results from laboratory searches are easier to interpret since the experiments are repeatable, and systematic uncertaintities can be studied by changing experimental conditions. The best limit on a temporal variation that is independent of assumptions regarding other constants (published after our main result in Ref. 14) was obtained by monitoring the ratio of Al^+ and Hg^+ optical transition frequencies.[15]

Another type of search for an LPI violation is a "null" gravitational red-shift experiment where two clocks with different composition are compared side by side in a changing gravitational potential.[16] Laboratory clock comparisons can be used for this type of test due to the eccentricity of Earth's orbit around the Sun, which leads to a small oscillatory component of the gravitational potential with a period of a year. In earlier work,[13,17–19] clocks of different types were compared, and the ratios of the clock rates were analyzed for a possible correlation with the gravitational potential, which led to bounds on parameters that characterize structure-dependent modifications to the clock frequencies. At the time of publication, we were able to present the best limit on k_α, the linear-variation coefficient for α in a changing gravitational potential.[20] A better result has since been published in Ref. 21.

Advances in direct comparison of single-ion optical or neutral optical lattice clocks through frequency-comb metrology promise significant improvements in sensitivity by up to three orders of magnitude, as well as simplification in interpretation of the results since such comparisons directly probe α-variation, independent of other fundamental constants (see, for example Ref. 15). In this article, we present results obtained with an alternative method of competitive sensitivity, which is also independent of

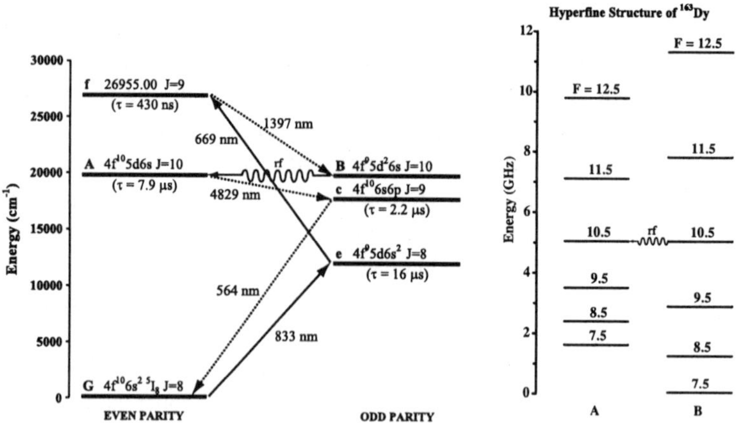

Fig. 1. Relevant levels and transitions in atomic dysprosium. Population and detection scheme: Level B is populated in a three-step process. The atoms are excited to level f in the first two steps using 833- and 699-nm laser light (solid arrows). The third step is a spontaneous decay (labeled by a short-dashed arrow) from level f to level B with ~ 30 % branching ratio. The population is transferred to level A by the rf electric field (wavy line). Atoms in level A decay to level c and then to the ground state. The fluorescence from the second decay (564 nm) is used for detection. Inset on the right: Hyperfine structure of levels A and B for ^{163}Dy. Zero energy is arbitrarily chosen to coincide with the lowest hyperfine component.

other fundamental constants. This method utilizes rf electric-dipole transitions between nearly degenerate electronic levels in atomic Dy.

2. Effect of α-variation in dysprosium

Tests for variation of α in atomic systems rely upon the fact that the sensitivity of energy levels to α is different for different energy levels. The total energy of a level can be written as

$$E = h\nu = E_0 + q(\alpha^2/\alpha_0^2 - 1), \tag{1}$$

where E_0 is the present-day energy, α_0 is the present-day value of the fine-structure constant, and q is a coefficient which determines the sensitivity to variation of α. A calculation,[22] utilizing relativistic Hartree-Fock and configuration interaction methods, found values of q for two nearly degenerate opposite-parity levels in atomic dysprosium (Fig. 1, levels A and B) that are large and of opposite sign. For the even-parity level (level A), $q_A/hc \simeq 8 \times 10^3$ cm^{-1}, while for the odd-parity level (level B), $q_B/hc \simeq -25 \times 10^3$ cm^{-1}.

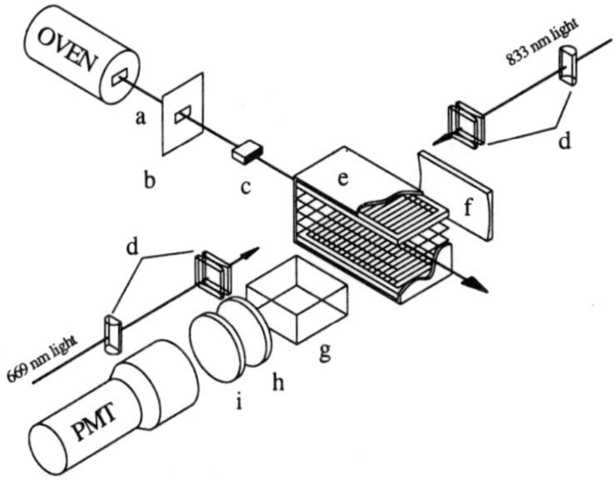

Fig. 2. Experimental setup (not shown to scale): a) Atomic beam produced by effusive oven source at $\sim$ 1500 K; b) atomic-beam collimator; c) oven light collimator; d) cylindrical lenses to diverge laser beams; e) interaction region of atoms with the electric field enclosed in a magnetic shield (not shown); f) cylindrical mirror to collect fluorescent light; g) lucite light pipe; h) interference filter; and i) short-pass filter.

3. Experimental Technique

A unique aspect of the Dy system is that the directly observable quantity is the energy difference of the levels, $\nu_B - \nu_A$, due to the fact that an electric-dipole transition can be induced between them. The time variation of the transition frequency between levels A and B is related to temporal variation of α by

$$\dot{\Delta\nu} = \dot{\nu}_B - \dot{\nu}_A = 2\frac{q_B - q_A}{h}\frac{\dot{\alpha}}{\alpha}, \tag{2}$$

where $|2(q_B - q_A)/h| \simeq 2.0 \times 10^{15}$ Hz.[22] For instance, $|\dot{\alpha}/\alpha| = 10^{-15}$/yr implies $|\dot{\Delta\nu}| \simeq 2$ Hz/yr.

A preliminary analysis of statistical and systematic uncertainties shows that the measurement of the transition frequency and the control of the systematics is feasible at a mHz level, which corresponds to an ultimate sensitivity of $|\dot{\alpha}/\alpha| \sim 10^{-18}$/yr for two measurements separated by a year's time.[23] A mHz resolution on a transition frequency of 1 GHz requires a relatively modest fractional stability of 10^{-12} for the reference frequency standard. This also means that the results are insensitive to variation of the Cs reference frequency due to changes in the values of fundamental constants, which have experimental upper limits$\sim 10^{-15}$ yr^{-1} (see, for example, Ref. 10).

The transitions utilized in the experiment are the 235-MHz (J = 10

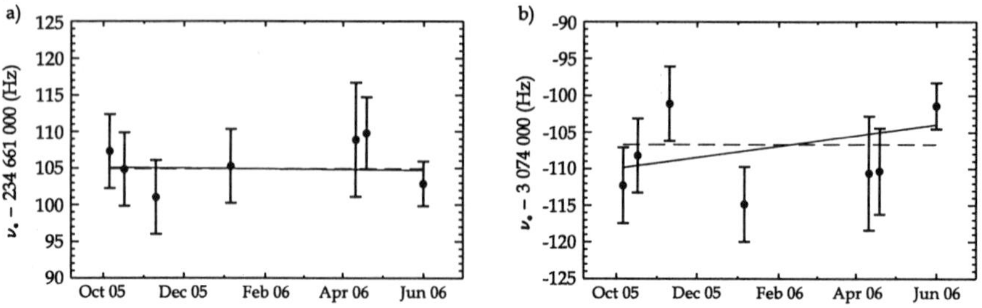

Fig. 3. Measured transition frequencies for (a) the 235-MHz and (b) 3.1-MHz transitions over an eight-month period. The data have been corrected for collisional shifts. The solid lines are the least-squares linear fits to the data. The dashed lines are the least-squares fits to a constant function. The apparent correlation between the transition frequencies for the two isotopes are due to the stray magnetic field effect.

$\to$ J = 10) transition of the isotope ^{162}Dy and the 3.1-MHz (F = 10.5 $\to$ F = 10.5) transition of ^{163}Dy. The long-lived level B (lifetime[24] $\tau >$ 200 μs) is populated via three transitions. Lasers at 833 nm and 669 nm are used for the first two transitions to a high-lying upper state, which then spontaneously decays to level B as shown in Fig. 1. The transition between levels A and B is induced by an applied rf electric field. The atoms decay from level A in two steps, and the fluorescence at 564 nm is monitored to detect the rf transition.

The rf-generation and detection system is discussed in Ref. 25. The rf field is frequency modulated by the 10 kHz reference output of a lock-in amplifier with a modulation index of 1. In order to reduce lineshape asymmetries due to drifts, the ratio of the first- and second-harmonic outputs of the lock-in amplifier is used to measure the transition frequency, which is extracted from the ratio by a two-step process described in detail in Ref. 25.

4. Results and Analysis

4.1. *Temporal Variation*

We have measured the two rf transition frequencies in the course of seven runs over eight months. Figure 3 shows the results of these measurements corrected for collisional shifts.[25] The present uncertainty is dominated by systematic effects primarily due to polarization imperfections coupled to the residual magnetic field, collisional shifts, and rf electric-field inhomogeneities. A detailed analysis of these systematic effects and a discussion of feasible levels of stability are given in Ref. 23. Error bars are due both to sta-

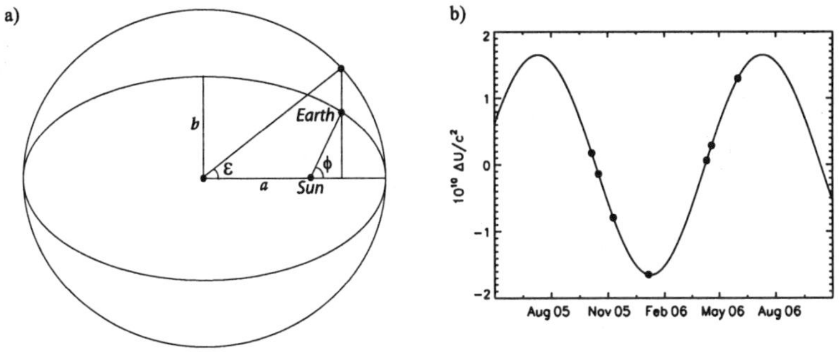

Fig. 4. a) The true anomaly, ϕ, is the angle subtended from perihelion, the point of closest approach. The eccentric anomaly, ε, is the angle between perihelion and the position of the Earth in its orbit projected onto the auxiliary circle of the ellipse (the eccentricity of the ellipse is exaggerated for clarity). b) The change in gravitational potential of the Sun at the Earth due to the ellipticity of the orbit. The points indicate the dates on which data were taken.

tistical and systematic uncertainties which have been added in quadrature. A least-squares linear fit to the data points gives slopes of -0.6 ± 6.5 Hz/yr and 9.0 ± 6.7 Hz/yr for the 235-MHz and 3.1-MHz transitions, respectively. Using an analysis presented in Ref. 14, we arrive at the final result of $\dot{\alpha}/\alpha = (-2.4 \pm 2.3) \times 10^{-15}\ \mathrm{yr}^{-1}$, consistent ($1\sigma$) with no variation at the present level of sensitivity. This result is slightly improved over what is presented in Ref. 14 due to recently improved calculations of the sensitivity coefficients, q_A and q_B.[22]

4.2. *Gravitational-Potential Dependence*

The data used in the previous analysis can also be analyzed for the dependence of α on variations of a gravitational potential. The data were taken over the course of several months, and because the Earth's orbit has a slight eccentricity, each measurement took place at a different gravitational potential.

The fractional change in α as a function of gravitational potential, $U(t)$, can be parameterized as[26]

$$\frac{\delta\alpha}{\alpha} = k_\alpha \frac{\Delta U(t)}{c^2}\,, \tag{3}$$

where k_α is the variation coefficient for α in a changing gravitational potential, c is the speed of light, and $\Delta U(t)$ is the oscillatory component of the sun's gravitational potential at the Earth (see Fig.4 and Ref. 20). The measured frequencies for each isotope are fitted to the gravitational potential

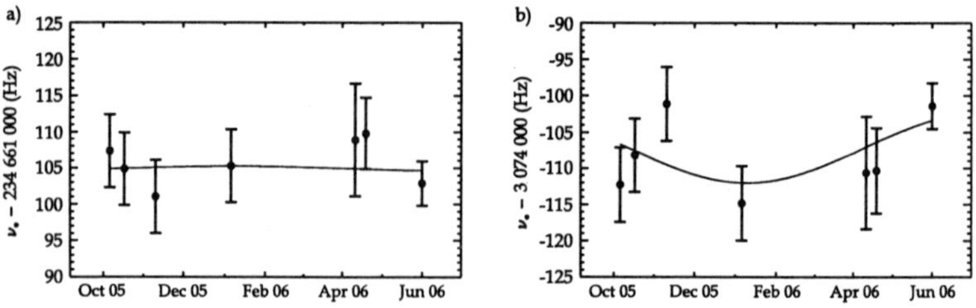

Fig. 5. Same data as in Fig. 3. Solid lines are least-squares fit to a function given by Eq.řefeq:fit where ΔU is the oscillatory component of the gravitational potential of the Sun at Earth with a period of a year.

using a two-parameter least-squares fit given by

$$\delta(\Delta\nu) - \nu^* = x_0 \frac{\Delta U(t)}{c^2} + x_1 \ , \tag{4}$$

where ν^* is an arbitrary reference frequency, and x_0 and x_1 are the fit parameters. The parameter x_1 accounts for the offset due to the reference frequency while the parameter x_0 determines the correlation between the change in gravitational potential and the transition frequency.

The parameters x_0 and x_1 were obtained by a least-squares fit to the data shown in Fig. 5. These values can be used to calculate the constraint on the parameter k_α from Eq. (3). Substituting Eq. (2) into Eq. (3), we get a relation similar to Eq. (4),

$$\delta(\Delta\nu) = \left(2\,\frac{q_B - q_A}{h}\right) k_\alpha \frac{\Delta U(t)}{c^2} \ . \tag{5}$$

The fit parameter yields $k_\alpha = (-1.0 \pm 8.2) \times 10^{-6}$ for the 235-MHz transition, $k_\alpha = (-14.6 \pm 8.4) \times 10^{-6}$ for the 3.1-MHz transition. The final result of $k_\alpha = (-7.8 \pm 5.9) \times 10^{-6}$ is obtained from the analysis presented in Ref. 20. In addition, our value of k_α, combined with other experimental constraints, is used to extract the first limits on k_e and k_q. These coefficients characterize the variation of m_e/m_p and m_q/m_p in a changing gravitational potential, where m_e, m_p, and m_q are electron, proton, and quark masses. The results are $k_e = (4.3 \pm 3.5) \times 10^{-5}$ and $k_q = (5.9 \pm 4.7) \times 10^{-5}$. As was noted in Sec. 4.1, these results are improved over those presented in Ref. 20 due to a more accurate calculation of sensitivity coefficients.[22]

5. Laser Cooling of Dysprosium

Recently, we have started to explore the possibility of laser cooling dysprosium, with the goal of generating a well collimated beam for our experiment.

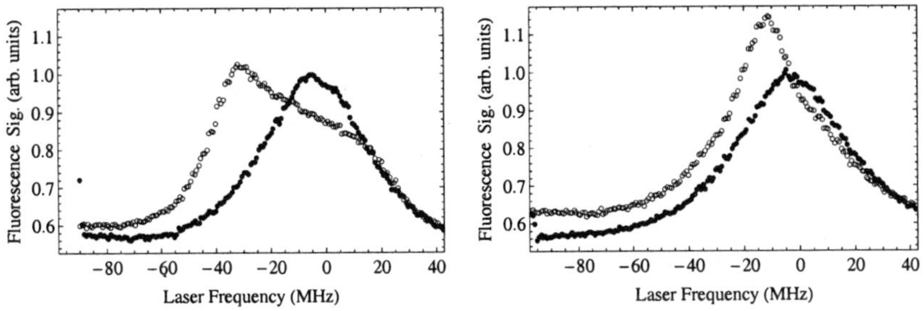

Fig. 6. Scans of the ^{164}Dy peak in the 658-nm transition with (hollow points) and without (solid points) the 421-nm pushing light incident on the atomic beam. In the scan on the left the 421-nm light was tuned to the center of the ^{164}Dy resonance, and in the scan on the right the 421-nm light was detuned from resonance by half the transverse doppler width. In both scans an alteration of the lineshape, and hence velocity distribution, is clearly apparent.

This is expected to increase beam brightness, reduce sensitivity to spatial magnetic- and electric-field inhomogeneities, and allow us to characterize shifts associated with Dy-Dy collisions.

While dysprosium has been magnetically trapped before,[27] there is no previously known effort to perform laser cooling. After a review of literature concerning excited state lifetimes,[28,29] it was decided that the best option to attempt laser cooling was to use the 421.291-nm, $4f^{10}6s^2\,(J = 8) \rightarrow 4f^{10}6s6p\,(J = 9)$ transition. With an upper state lifetime of 4.8 ns, this transition is the strongest recorded cycling transition in the visible spectrum of Dy I.

To generate the 421-nm light, 300 mW of 842-nm light from a Ti:Sapphire ring laser (Coherent 899) is frequency-doubled by a $1\times2\times10$ mm^3 periodically poled potassium titanyl phosphate (PPKTP) crystal (Raicol Crystals Ltd.) located within a resonant bow-tie cavity. The maximum output with this resonator-enhanced doubling scheme is approximately 90 mW.

In our initial attempt we choose to "push" the atomic beam, as cooling required more changes to the experimental set-up. To detect changes in the atomic beam, we use a parallel probe beam 1-cm downstream from the pushing beam. Both the pushing and probing laser beams intersect the atomic beam at ninety degrees. The 658.12-nm light for the probe laser is generated by a ring dye laser with DCM dye and resonant with the $4f^{10}6s^2(J = 8) \rightarrow 4f^95d6s^2(J = 7)$ transition. The linewidth of this transition[28] ($\approx$100 kHz) is much less than the Doppler width (40 MHz), therefore the lineshape of the probe fluorescence is dominated by the transverse velocity distribution of the atomic beam. A more detailed description of the atomic beam source is found in Ref. 30.

The results of the experiment are shown in Fig. 6. For the first scan the 421-nm light is resonant with the center of the transverse velocity distribution, resulting in a shift and broadening of the probe lineshape. In the second scan the 421-nm is red-detuned from the center of the velocity distribution, resulting in an increase in amplitude and narrowing of the lineshape. In both scans the power in the pushing beam is approximately 60 mW in a 1.1-cm diameter, collimated beam. Work is currently underway to modify the experiment for the purposes of more efficient transverse cooling.

Laser cooling of dysprosium is not trivial because of many available decay channels of the upper state of the cooling transition. However, our estimates (confirmed by more detailed atomic calculations[31]) show that the overall branching ratio into all excited states is $< 10^{-3}$. In future work, we will determine the branching ratios and devise repumping schemes as needed. In principle magneto-optical trapping, or even Bose-Einstein condensation may be possible. The latter would be of interest to the degenerate atomic gas community due to the large ground-state angular momentum and magnetic moment of Dy.

6. Conclusion

We have presented improved results of a measurement of time variation ofα and a correlation with a changing gravitational potential using atomic dysprosium. In our case the interpretation does not require comparisons with different measurements to eliminate dependence on other constants, and provides an alternative to measurements that utilize state-of-the-art atomic optical frequency clocks. In addition, the data have been used to extract a limit on the gravitational variation of α. We have also presented results on the first effort to laser cool dysprosium.

A significant improvement in our results is expected from a newly operation apparatus, which provides better control over polarization imperfections coupled to the residual magnetic field, collisional shifts, and rf electric-field inhomogeneities, as well as other systematic effects expected to be important to achieve better than 1-Hz sensitivity. Ultimately, mHz-level sensitivity may be achievable with this method.[23]

7. Acknowledgments

We would like to thank V. V. Flambaum for intellectual inspiration and collaboration on the analysis of the gravitational potential dependence of α and V. A. Dzuba for calculation of matrix elements in Dy. We are grateful to D. F. Jackson Kimball, and D. English for valuable discussions. This work was supported in part by the University of California - Los Alamos National Laboratory CLC program, NIST Precision Measurement Grant, Los Alamos National Laboratory LDRD, and by grant RFP1-06-15 from the Foundational Questions Institute (fqxi.org).

References

1. T. Damour *et al.*, *Phys. Rev. Lett.* **89**, p. 081601 (2002).
2. J. K. Webb *et al.*, *Phys. Rev. Lett.* **87**, p. 091301 (2001).
3. M. T. Murphy *et al.*, *Mon. Not. R. Astron. Soc.* **345**, p. 609 (2003).
4. R. Quast *et al.*, *Astron. Astrophys.* **415**, p. L7 (2004).
5. R. Srianand *et al.*, *Phys. Rev. Lett.* **92**, p. 121302 (2004).
6. T. Damour *et al.*, *Nucl. Phys.* **B480**, p. 37 (1996).
7. Y. Fujii *et al.*, *Nucl. Phys.* **B573**, p. 377 (2000).
8. C. R. Gould *et al.*, *Phys. Rev. C* **74**, p. 024607 (2006).
9. S. Bize *et al.*, *Phys. Rev. Lett.* **90**, p. 150802 (2003).
10. H. Marion *et al.*, *Phys. Rev. Lett.* **90**, p. 150801 (2003).
11. M. Fischer *et al.*, *Phys. Rev. Lett.* **92**, p. 230802 (2004).
12. E. Peik *et al.*, *Phys. Rev. Lett.* **93**, p. 170801 (2004).
13. T. M. Fortier *et al.*, *Phys. Rev. Lett.* **98**, p. 070801 (2007).
14. A. Cingöz *et al.*, *Phys. Rev. Lett.* **98**, p. 040801 (2007).
15. T. Rosenband *et al.*, *Science* **319**, p. 1808 (2008).
16. C. M. Will, *Theory and Exp. in Grav. Phys.* (Camb. Univ. Pr., 1981).
17. A. Godone *et al.*, *Phys. Rev. D* **51**, p. 319 (1994).
18. A. Bauch and S. Weyers, *Phys. Rev. D* **65**, p. 081101(R) (2002).
19. N. Ashby *et al.*, *Phys. Rev. Lett.* **98**, p. 070802 (2007).
20. S. J. Ferrell *et al.*, *Phys. Rev. A* **76**, p. 062104 (2007).
21. S. Blatt *et al.*, *Phys. Rev. Lett.* **100**, p. 140801 (2008).
22. V. A. Dzuba *et al.*, *Phys. Rev. A* **77**, p. 012515 (2008).
23. A. T. Nguyen *et al.*, *Phys. Rev. A* **69**, p. 022105 (2004).
24. D. Budker *et al.*, *Phys. Rev. A* **50**, 132 (1994).
25. A. Cingöz *et al.*, *Phys. Rev. A* **72**, p. 063409 (2005).
26. V. V. Flambaum, *Int. J. Mod. Phys. A* **22**, p. 4937 (2007).
27. C. I. Hancox *et al.*, *Nature* **431**, p. 281 (2004).
28. J. J. Curry *et al.*, *J. Opt. Soc. Am. B* **14**, p. 2788 (1997).
29. M. E. Wickliffe, *J. Quant. Spect. & Rad. Trans.* **66**, p. 363 (2000).
30. A. T. Nguyen *et al.*, *Phys. Rev. A* **56**, p. 3453 (1997).
31. V. Dzuba. private communication, (2008).

MEASUREMENT OF SHORT RANGE FORCES USING COLD ATOMS

F. PEREIRA DOS SANTOS*, P. WOLF, A. LANDRAGIN, M.-C. ANGONIN, P. LEMONDE, S. BIZE, and A. CLAIRON

LNE-SYRTE, CNRS UMR 8630, UPMC, Observatoire de Paris, 61 avenue de l'Observatoire, 75014 Paris, France
** E-mail: franck.pereira@obspm.fr*
http://syrte.obspm.fr/

A. LAMBRECHT, B. LAMINE, and S. REYNAUD

Laboratoire Kastler-Brossel, CNRS UMR 8552, ENS, UPMC, 4 Place Jussieu, 75252 Cedex 05, France

We describe the new project FORCA-G, which aims at studying the short range interactions between a surface and atoms trapped in its vicinity. Using cold atoms confined in the wells of an optical standing wave, the atom-surface potential will be measured with high sensitivity using atom interferometry techniques. The experiment will allow a test of gravity at short distances, which will put stringent bounds on a possible deviation from the known laws of physics. FORCA-G will also allow a measurement of the Casimir Polder interaction (QED vacuum fluctuations) with unprecedented accuracy.

Keywords: Casimir Force - Gravitation - Tests of fundamental Physics

1. Introduction

One of the frontiers of modern physics is the study of forces at very small length scales ($<$ 1 mm). Such length scales are the domain where quantum electrodynamics, QED, interactions (Casimir type forces) become important and where several recent theoretical ideas point to the possibility of new interactions related to gravity. All of physics is at present well described by the standard model that unifies three of the four fundamental interactions in nature. However, the unification of all four interactions (including gravitation) has so far proven elusive. Several recent astronomical observations indicate that our current understanding of gravitation might be incomplete or erroneous because to match those observations with present theory one has to invoke some form of additional matter and energy (dark matter and dark energy) that constitutes about 97% of the total mass of the universe. Consequently, it appears essential to experimentally test gravitation at all accessible length scales. In particular, unification theories suggest some deviations from the known laws of gravity at very short range.[1] Quite generally, such deviations from Newtonian gravity can be parameterized by a Yukawa type potential of the form $U = U_N(1 + \alpha e^{-\frac{r}{\lambda}})$ where U_N is the Newtonian potential and α and λ are the amplitude and range of the modification. Exclusion regions in the (α, λ) plane are provided by experiments scaling from laboratory size to the size of the solar system. Possible deviations are relatively well constrained over a large range of distances from a mm to the size of the solar system. However, two "open windows" remain at very small ($<$ 1 mm) and very large ($> 10^{15}$ m) range, the former being the focus of this experiment.

However, at distances below $\sim$ 30 microns the gravitational force between two neutral objects is dominated by the Casimir force. For distances of this order or below it is then Casimir force measurements which give the strongest constraints for the existence of hypothetical new forces in the $(\alpha - \lambda)$ plane. After its prediction[2] in 1948, the Casimir force between two macroscopic objects has been observed in a number of "historic" experiments which confirmed its existence and main properties. More recent measurements with largely improved accuracy have allowed for comparison between measured values of the force and theoretical predictions at the few % level, some of them have also allowed to improve constraints in the $(\alpha - \lambda)$ plane. Shortly after the prediction of the Casimir force between two parallel plates, Casimir and Polder[3] predicted the analogous attractive force between an atom and a macroscopic plane surface. More generally, the Casimir-Polder potential ($1/r^4$ dependence on distance) is the retarded part of the total QED interaction between the atom and the surface. The non-retarded part, which is dominant at short distances, is known as the "Van der Waals" potential and has a $1/r^3$ dependence. Also, at non-zero temperatures (and larger distances), the overall QED potential becomes dominated by a temperature dependent term again with $1/r^3$ dependence. This leads to an interesting phenomenological behaviour with two distance dependent crossover points from $1/r^3$ dependence to $1/r^4$ and back. At 300 K the two crossover points are situated at a few tenths of a micron and a few microns respectively (depending on the atom and surface characteristics). These features give a lot of richness to the atom-surface interaction (rather than surface-surface), that can be explored at experiments at the micron and sub-micron scale like the one proposed here.

On the experimental side, measurements at distances ranging from 10^{-8} m to 10^{-3} m have been the domain of microelectromechanical systems and of torsion balance experiments. Two major difficulties of such mechanical experiments are the exact knowledge of the geometry of the setup (distance, surface roughness, etc...) and the precise measurement of the very small forces involved. An alternative that might provide a way around those difficulties is the use of cold atoms. The experiments that have been carried out so far all confirm the theoretical predictions from QED (Van der Waals and Casimir-Polder effect) at distances ranging from a few tens of nanometers to several microns, however, none of them have yet reached the uncertainties achieved by the best mechanical measurements. Typically, experiments measuring the atom-wall QED interaction have an overall relative uncertainty at or above 10 %.[4] In this project we pursue an original scheme, first described in 5, that has the advantage of providing accurate control of the distance and accurate direct measurement of the potential (rather than the force) between an atom and a macroscopic surface.

2. Principle of the experiment

Atoms are trapped in a vertical standing wave, created by a laser far detuned from resonance. The internal atomic structure is approximated by a two-level system, with two long lived states $|g\rangle$ and $|e\rangle$ with energy difference $\hbar\omega_{eg}$. The external hamiltonian $\hat{H}_{ext}$ (kinetic energy, trapping and gravitational potential) is identical for both $|g\rangle$ and $|e\rangle$. For sufficiently large depth U_0 of the trapping potential, Landau-Zener tunnelling can be neglected. The Eigenstates of $\hat{H}_{ext}$ are then the so called Wannier-Stark (WS) states $|W\rangle$. The discrete quantum number m is the "well index" characterizing the well containing the main peak of the wave function $\langle x|W_m\rangle$. The energy separation between adjacent states is simply the change in gravitational potential between adjacent wells: $\hbar\Delta_g = m_a g\lambda_l/2$, where g is the gravity acceleration. This leads to a Wannier-Stark ladder of Eigenstates. Transitions between $|g\rangle$ and $|e\rangle$ are induced by a probe laser, which couples $|W_m, g\rangle$ to $|W_{m'}, e\rangle$ in either the same well or in neighboring wells, with coupling strengths of the same order for realistic U_0 (see figures 1 and 2). A sequence of probe laser pulses on resonance or detuned by Δ_g then provides a powerful method of spatially separating and re-combining the atoms on the WS ladder.

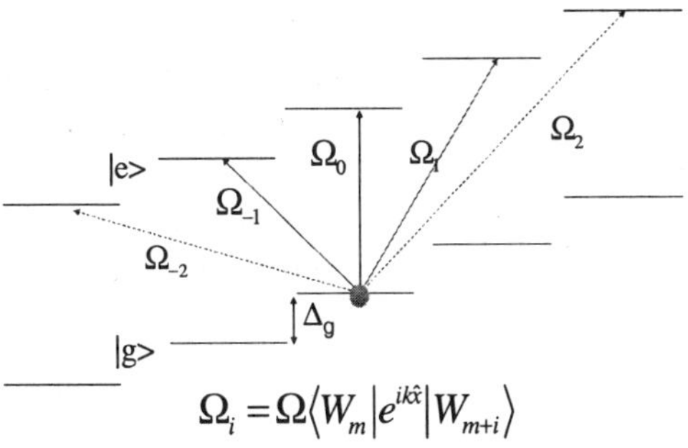

Fig. 1. Wannier-Stark ladder of states and couplings between states by the probe laser

Typically, about 10^3 to 10^4 atoms per lattice site can be trapped and cooled to a few μK, close to the reflecting mirror of the vertical trap laser. Once selected in a single vertical well (see below for details on how that is achieved), an interferometer is then created around that well using a sequence of probe laser pulses. Starting with atoms in state $|W_m, g\rangle$, a first $\pi/2$ pulse on resonance creates a superposition of $|W_m, g\rangle$ and $|W_m, e\rangle$. Next, a π pulse detuned by $+\Delta_g$ transfers atoms from $|W_m, g\rangle \rightarrow |W_{m+1}, e\rangle$ and $|W_m, e\rangle \rightarrow |W_{m-1}, g\rangle$ leaving a superposition of spatially separated states in wells $m+1$ and $m-1$. After a time T a "symmetrization" π pulse on resonance switches internal states. A time T later a π pulse detuned by $-\Delta_g$ transfers atoms back ($|W_{m+1}, g\rangle \rightarrow |W_m, e\rangle$ and $|W_{m-1}, e\rangle \rightarrow |W_m, g\rangle$) with a final $\pi/2$ pulse on resonance recombining the atoms in the initial well m, where the internal state is detected. The different energies of the states along the two paths and the initial phases of the probe laser pulses lead to an overall phase difference

$$\Delta\phi = \frac{1}{\hbar}\left(m_a g \lambda_l + U_{m+1} - U_{m-1}\right) 2T + \left(\omega_{eg}^{(m+1)} - \omega_{eg}^{(m-1)}\right) T \qquad (1)$$
$$-\phi_s^{(1)} + 2(\phi_s^{(2)} - \phi_s^{(3)} + \phi_s^{(4)}) - \phi_s^{(5)}$$

where $\omega_{eg}^{(m)}$ is the separation between internal states in well m, $\phi_s^{(i)}$ is the initial phase of the i th pulse of the probe laser and U_m an additional perturbation (QED, new interaction, stray e-m fields, etc...). The signal of interest allows the measurement of a potential U that varies over the size of the interferometer. Assuming state of the art measurement noise of atom interferometers, $\Delta\phi$ can be determined with a precision of $\approx 10^{-4}$ rad after 10^3 to 10^4 s integration. For interaction times $2T \approx 0.1$ s this corresponds to a measurement noise on $(m_a g \lambda_l + U_{m+1} - U_{m-1})/(2\pi\hbar)$ of about 1.6 10^{-4} Hz.

The experiment will be carried out with Rb atoms. Atomic states will be manipulated using two-photon Raman transitions. The momentum of the probe laser is then $k_{eff} = k_1 - k_2$, which corresponds to an effective wavelength of 390 nm. Efficient coupling between neighbouring wells is then expected to occur for $k_L \simeq k_{eff}$, which corresponds to a blue detuned trap for the lattice. In that case, $\Delta_g/2\pi$ is on the order of 500 Hz.

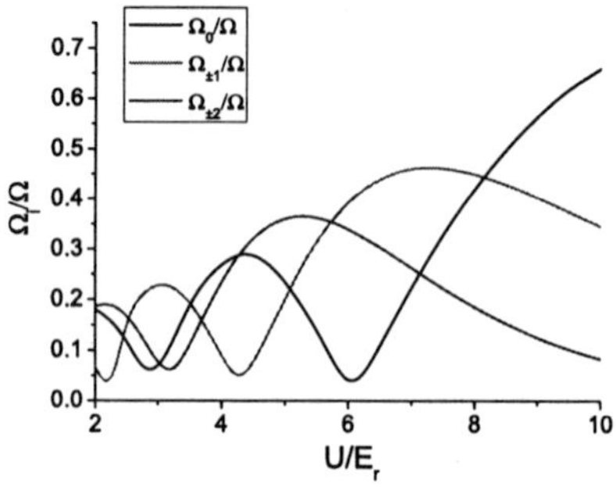

Fig. 2. Relative coupling strengths of the transitions in the same well and into the four first neighbouring wells ±1 and ±2 as a function of the lattice depth U_0.

An important feature of this experiment lies in the influence of the depth and wavelength of the lattice on the coupling strength for transitions between wells. Good couplings between adjacent wells can be realized with lattice depths on the order of a few recoil energies, and the ratio between same well - adjacent wells transitions can be adjusted at will by tuning the lattice depth and wavelength. We find a remarkable feature at 532 nm, for which high power single mode lasers exist. Figure 2 displays, for this specific lattice laser wavelength, the coupling strengths for transitions in the same well, between states in adjacent wells as well as between states separated by 2 wells.

The couplings are normalized by the coupling in the absence of the lattice. At $U \simeq 3$ Er (with Er the recoil Energy at the lattice wavelength), the coupling strength for a transition between adjacent wells is larger than for transition in the same well. If using copropagating Raman lasers, instead of counterpropagating lasers, the effective wavevector k_{eff} becomes negligible and only transitions in the same well are allowed (only $\Omega_0 \neq 0$). This suggests a strategy in addressing given transitions with very large efficiency by setting the lattice depth where one coupling strength is dominant (e.g. at

$U \simeq 3$ Er), and using depending on the transition required either counter or copropagating transitions.

With a lattice beam at 532 nm laser, an additional red detuned (progressive) beam for transverse confinement, which will be realized with an infrared laser, operating at 1030 nm. The choice of this wavelength is motivated by the possibility to additionally generate green light, at a wavelength of 515 nm, close to the lattice wavelength, by frequency doubling, which will be useful for well selection. Overlapping this second green vertical beam with the lattice beam results in a super-lattice, whose wells depth is modulated, with a characteristic period of 8 μm. This induces a position dependent resonance frequency for the transition between adjacent wells, which lifts the degeneracy between $m \to m+1$ transitions. Well selection is then achieved by applying a single Raman pulse which selectively transfers atoms from one of the wells to the adjacent well, and then clearing remaining atoms with a pusher beam.

3. Systematic effects

Many perturbations are expected to affect the measurement. We briefly detail some of the most important effects:

- Light shifts affect the measurement in two ways. First, a modification of the transition frequency ω_{eg} leads to imperfect cancelation of the second term in equation 2 if the lattice laser intensity varies in space and/or time. Second, a spatial variation of the intensity modifies the energy of the ground state $|g\rangle$, which leads to a contribution on the term $U_{m+1} - U_{m-1}$. Lattice depths of a few recoils correspond to several tens of kHz of light shift, which is 8 orders of magnitude larger than the target frequency shift accuracy. As the transition between hyperfine states is affected only by the differential shift, a reduction by 5 orders of magnitude of the first effect is expected with respect to the case where the two states would be coupled by an optical transition. Thus, a reasonable temporal and spatial stability of the laser intensity will be sufficient. As for the second effect, spatial intensity variations need to be controlled at the 10^{-8} level, which, whilst challenging, is still possible over the relatively short distances involved. For example, the results of Ref. 6 are consistent with $< 10^{-8}$ control of spatial intensity variations. Although controlling the intensity close to the surface is certainly more difficult, Ref. 6 provides a good indication that light shifts due to spatial intensity fluctuations should be controllable at the required level.

- The uncertainty of any measurement of atom-surface interactions depends crucially on the precise determination of the distance between the atom and the surface. Here, the precise knowledge of the position of the wells of the lattice with respect to the surface allows determining the atom-surface separation, which an accuracy limited by wave-fronts distortions. Over the small extension of the trapped atoms ($\approx 100\mu$m) it should be possible to control wave front flatness to about $10^{-3}\,\lambda_l$. However, interference between the trap laser and stray reflections due to surface roughness may play a non-negligible role. Other effects could come from surface roughness (≈ 0.5 nm). Therefore, we expect to control the atom-surface separation to ≈ 1 nm.

- For Rb the main limitation identified in Ref. 5 is related to collisions between atoms and resulting differential energy shifts due to a density difference in the two wells of the superposition. We will solve that problem by using a powerful trapping laser that allows a large waist (1 mm) and correspondingly low densities whilst keeping a reasonable number of atoms. With 10^3 atoms per well the density will be of order 10^{10} at/cm^3 and the corresponding absolute shift is about 7×10^{-2} Hz. The differential density can be controlled at the 10^{-3} level by selecting atoms using adiabatic passage schemes, as developed for microwave clocks.[7]

- Finally, one of the most significant error sources in the measurement reported in Ref. 4 are stray electric and magnetic fields originating from contaminations of the surface. In our case this is likely to be less of a problem: the characterization of magnetic effects is likely to be more precise than in Ref. 4 because of the absence of a magnetic trap, as the different m_F states can then be used to measure the magnetic fields "in situ". Stray electric fields remain a challenge, which we will address by applying controlled external electric fields (similar to Ref. 4,8). In particular Ref. 8 has shown that very accurate characterization of electric fields is possible and that heating of the surface can significantly reduce such fields.

4. QED potential

A rough calculation of the QED potential for Rb atoms in the optical lattice is shown in Table 1, with the expected relative uncertainty of its measurement, taking into account the uncertainties on the determination of potential (10^{-4} Hz) and distance (10^{-9} m).

At short distances the uncertainty is dominated by the uncertainty of the distance determination, at large distances by the uncertainty of the potential measurement. The optimum is situated around 6 μm where the relative uncertainty is less than 10^{-3}, which is two orders of magnitude better than the best present atomic (ie. Casimir-Polder) measurements.

Table 1. QED potential for Rb in different wells of the optical lattice at 532 nm and at $T = 300$ K, and the expected relative uncertainty in its determination.

Well No.	1	2	5	22	50
r / nm	266	532	1330	5852	13300
U_{QED} / Hz	5×10^4	3131	80	0.34	0.03
$\delta U_{QED}/U_{QED}$	0.01	8×10^{-3}	3×10^{-3}	6×10^{-4}	3×10^{-3}

It is well into the region where theoretical calculations and predictions of the effect of quantum fluctuations need to take into account a number of parameters (surface properties, atomic polarizabilities, finite temperature). For example the optimum distance (~ 6 μm) is close to the crossover between the vacuum (r^{-4}) and thermally (r^{-3}) dominated regimes, situated around 4 μm at $T = 300$ K.

5. Test of gravitation

Considering the search for new interactions related to gravity, experiments set limits in the α, λ plane. Figure 3 (from Refs. 9,11) shows the present experimental exclusion regions in the 10^{-6} m to 10^{-3} m range, together with some theoretical predictions for new interactions (shaded colored regions) and the estimated limits attainable using our proposed experimental setup assuming a 10^{-4} Hz resolution, (green and purple line) with a sapphire surface (density ~ 3980 kg/m^3).

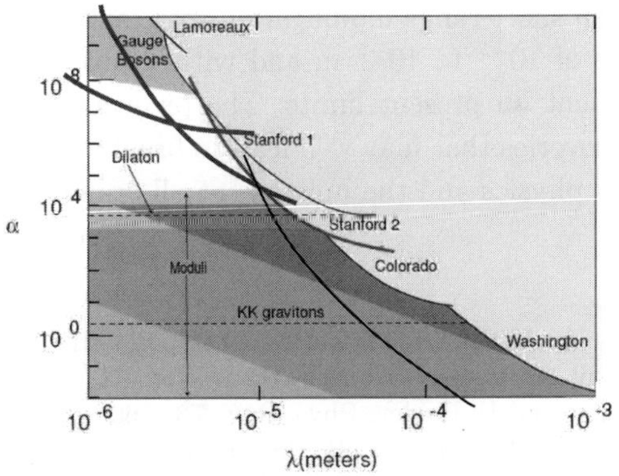

Fig. 3. Present limits on additional interactions related to gravity. The figure has been taken from Ref. 9. A black line was added to take into account more recent results obtained by the Washington group.[11] The green and purple lines show the estimated limits from FORCA-G in the first and second stage respectively (see text for details).

The most serious issue when trying to measure additional gravitational interactions is the perturbation from U_{QED}, especially at short distances. Table 1 shows that for measurements in the 3rd well one needs to correct and/or cancel the effect of U_{QED} at the 10^{-6} level for the 10^{-4} Hz uncertainty that we aim at, and even when relatively far from the surface (50th well) a correction at the % level is still required. We will address that issue in a two stage experiment, starting with an experiment at relatively large distance ($> 10\mu$m) where U_{QED} can be modeled and corrected to the 10^{-4} Hz level, and exploring shorter separations in a second stage where U_{QED} is canceled in a differential measurement between the two isotopes (^{85}Rb and ^{87}Rb), as U_{QED} is dependent on the atomic polarizability, which is the same for the two isotopes down to isotopic shifts ($\leq 10^{-6}$). The two colored lines in figure 3 show the estimated limits in the two stages. They correspond to improvements on present limits by more than three orders of magnitude in the 1 μm to 10 μm region. At shorter ranges, present limits are less well established (see e.g. the review[10]) but we expect similar improvements from our experiment down to about $\lambda \sim 100$ nm.

6. Conclusion

The proposed experiment will allow the measurement of U_{QED} (Casimir-Polder force) below the % level, well into the interesting region where refined theoretical predictions are required, and about two orders of magnitude below the best present measurements. Furthermore, it will allow setting new stringent limits on short range modifications of gravity, in the theoretically interesting range of 10^{-7} to 10^{-5} m and with up to three orders of magnitude improvement on present limits. The project therefore has a large potential for discoveries that may well lead to breakthroughs in our current understanding of physics and the universe at all scales.

References

1. S. Dimopoulos, A. A. Geraci, Phys. Rev. D **68**, 124021,(2003)
2. H. B. G. Casimir, Proc. Kon. Nederl. Akad. Wet. **51**, 793 (1948)
3. H. B. G. Casimir and P. Polder, Phys. Rev. **73**, 360, (1948)
4. J.M. Obrecht et al. Phys. Rev. Lett. **98**, 063201, (2007)
5. P. Wolf et al., Phys. Rev. A **75**, 063608 (2007)
6. P. Cladé et al., Phys. Rev. Lett. **96**, 033001, (2006)
7. F. Pereira Dos Santos et al. Phys. Rev. Lett. **89**, 233004 (2002)
8. J.M. Obrecht et al., Phys. Rev. A **75**, 062903 (2007)
9. S.J. Smullin, et al., Phys.Rev. D **72**, 122001 (2005)
10. E. G. Adelberger et al., Annu. Rev. Nucl. Part. Sci. **53**, 77 (2003)
11. D.J. Kapner et al., Phys. Rev. Lett. **98**, 021101 (2007)

ATOM INTERFEROMETRY EXPERIMENTS IN FUNDAMENTAL PHYSICS

S. W. Chiow and S. Herrmann

Physics Department, Stanford University, Stanford, CA 94305, U.S.A

S. Chu and H. Müller*

Department of Physics, University of California, Berkeley, CA 94720, U.S.A.

* *E-mail: hm@berkeley.edu*

Keywords: Atom Interferometry; Experimental Gravitation; Atom Optics; Bragg diffraction; Bloch oscillations.

1. Introduction

Light-pulse atom interferometers[1] have been used for experiments of outstanding precision, like gravimeters,[2] gravity gradiometers,[3] gyroscopes,[4] measurements of Newton's gravitational constant G,[5,6] the fine-structure constant α,[7,8] or tests of gravitational theories.[9,10] They apply the momentum $\hbar k$ of photons to direct an atom on two (or more) paths which interfere when recombined. The sensitivity of atom interferometers increases with the phase shift between the arms. This depends linearly on the momentum splitting between the interferometer arms in gravimeters or gyroscopes — or even quadratically, like in measurements of α or certain gradiometers. However, many interferometers to date are limited to a splitting of $2\hbar k$ by the use of two-photon Raman transitions. Larger splitting of up to $6\hbar k$ have been achieved with multiple two-photon pulses or Bragg diffraction in atomic beam setups[11–13] and up to $12\hbar k$ using Bloch oscillations.[14]

We have demonstrated several ways in which to increase the momentum transfer, and thereby the sensitivity of atom interferometers, and to reduce the influence of systematic effects and vibrations. The following few pages will outline these.

2. Atom Interferometry with 24-Photon-Momentum-Transfer Bragg Beam Splitters

We have demonstrate the use of up to 24-photon Bragg diffraction[15] as a beam splitter in light-pulse atom interferometers, the largest splitting in momentum space so far. Relative to the 2-photon processes used in the most sensitive present interferometers, these large momentum transfer (LMT) beam splitters increase the phase shift 12-fold for Mach-Zehnder (MZ-) and 144-fold for Ramsey-Bordé (RB-) geometries. We achieve a high visibility of the interference fringes (up to 52% for MZ or 36% for RB) and long pulse separation times and superior control of systematic effects that are typical of atomic fountain setups. As the atom's internal state is not changed, important systematic effects can cancel. Figure 1 shows a gallery of interference fringes obtained in MZ and RB geometry at momentum transfers between 12-24$\hbar k$. More details will be found in.[16]

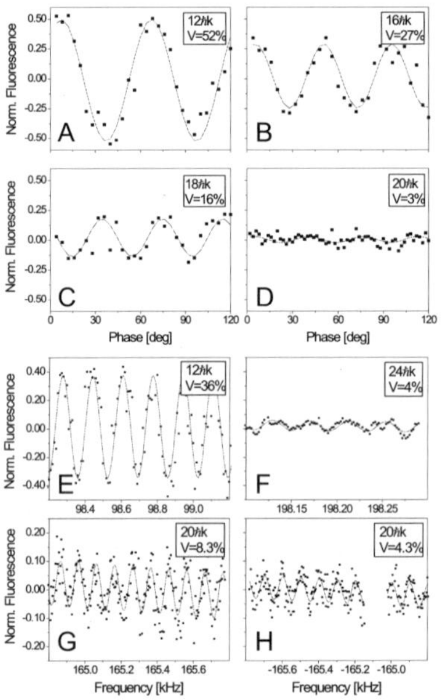

Fig. 1. A-D show MZ fringes with between 12 and 20$\hbar k$ momentum transfer; E and F are RB fringes with 12 and 24$\hbar k$. G and H show a conjugate 20$\hbar k$ RB-pair. Throughout, $T = 1\,\text{ms}$, $T' = 2\,\text{ms}$. Each data point is from a single launch (that takes 2 s), except for F, where 5-point adjacent averaging was used. The lines represent a sinewave fit.

3. Noise-Immune, Recoil-Sensitive, Large-Area Atom Interferometers

Furthermore, we have created a pair of simultaneous conjugate RB atom interferometers, see Fig. 2, left. Their sensitivity towards the photon recoil is similar, but the one towards inertial forces is reversed. That allows us to cancel the influence of gravity and, with simultaneous operation, noise.

Cancellation of vibrations between similar interferometers at separate locations has been demonstrated before.[3] In some important applications, however, the interferometers must be dissimilar so that a large differential signal can be picked up. Here, we present a method to cancel vibrational noise between dissimilar interferometers, with LMT beam splitters, see Fig. 2.

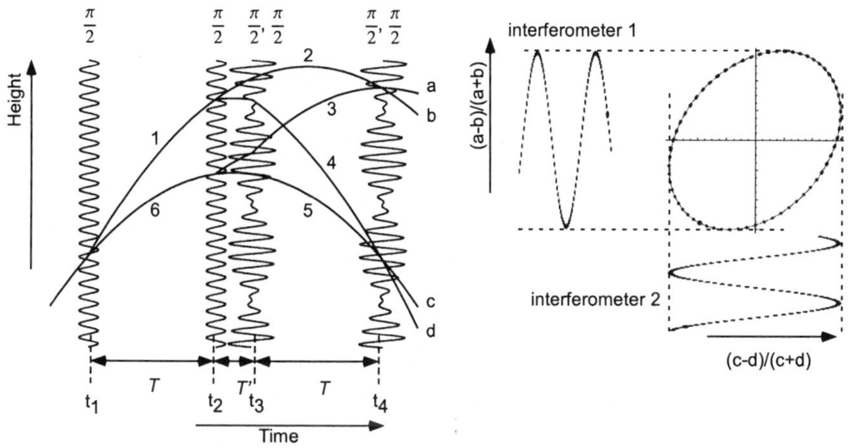

Fig. 2. Correlating the fringes of two interferometers creates an ellipse whose shape (eccentricity and major axis) allows to determine the relative phase.

The cancellation of vibrations is based on the simultaneous application of the beam splitters for the conjugate interferometers. Our experimental setup is optimized to provide the laser radiation needed with an extremely tight phase relationship; any vibrationally-induced phase shifts are thus common mode and can be taken out in an ellipse-fitting analysis of the correlation. At short pulse separation times of 1 ms, a contrast of around 25-31% is achieved at momentum transfers between $(8-20)\hbar k$, see Fig. 3 for examples. This should be compared to the theoretical contrast of 50%. Also, it is evident that the strong dependence of the contrast upon the momentum transfer, that was observed in previous LMT interferometers,[16] is absent.

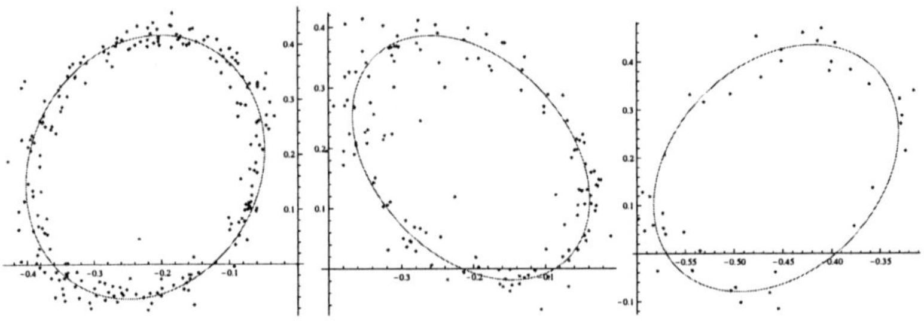

Fig. 3. A: $12\hbar k$, 1ms, 25% contrast. B: $14\hbar k$, 1ms, 25% contrast; D: $20\hbar k$, 1ms, 27%, C:

For $20\hbar k$ interferometers, about 10% contrast can be obtained at $T = 50\,\text{ms}$. Without SCIs, this is only possible at $T = 1\,\text{ms}$,[16] so the use of SCIs allows us to improve the pulse separation time T to 50 ms from 1 ms, without loss of contrast. This corresponds to a 2,500-fold increase in the enclosed area. At 70 ms, a contrast of 4.1% is still observable). This paves the path towards enhanced sensitivity in many cutting-edge applications. Examples include improved measurements of the photon recoil and the fine structure constant[7,8,17] and tests of the equivalence principle.[10]

To further confirm the applicability of our method, we have taken 15,000 pairs of data for a $10\hbar k$ interferometer with a pulse separation time of 100 ms over a 12-h period, see Fig. 4. Using Bayesian estimation,[18] we extract the differential phase to a resolution of 6.8 ppb. This is also the resolution to which the interferometers can determine $\hbar/M$; correspondingly, they are sensitive to the fine structure constant α via $\alpha^2 = (2R_\infty/c)(M/m_e)(h/M)$ to a resolution of 3.4 ppb.

4. Very large area atom interferometers by differential optical acceleration

The Bragg diffraction beam splitters used for LMT so far require extremely large laser power for increased momentum transfer. Even using our injection-locked 6.2 W Ti:sapph laser, which, we believe, is the strongest laser at a wavelength of 852 nm, we are limited to $20\hbar k$ for a reasonable contrast of the interference fringes. To increase the diffraction order, a further increase of the laser power would be required, which seems hard to

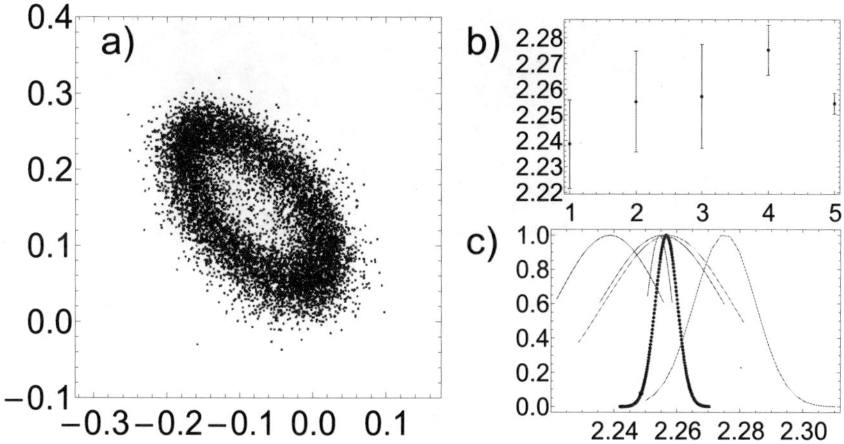

Fig. 4. a) 9958 Data pairs out of 15,000 that were taken during a 12-h session. b) Center and 1σ widths of the probability distribution c) of the differential phase. The dotted line in c) is the combined probability distribution for all data sets, $\Phi = 2.2567 \pm 0.0036$ rad.

achieve.

Adiabatic transfer[7] or Bloch oscillations of matter waves in an accelerated optical lattice[8] can been used to transfer a thousand $\hbar k$, but, so far, this only affects the common momentum of the arms, not the splitting. Here, we use it as a basis for increase the momentum *splitting* without being limited by the laser power.

To do so, we have first demonstrated the differential acceleration of atomic samples by Bloch oscillations in two superimposed optical lattices. Decoupling of the samples is due to their initial momentum separation, provided by a 4th order Bragg diffraction. A Bloch oscillation - Bragg diffraction - Bloch oscillation sequence forms a "BBB" beam splitter. Using this, we can achieved substantially increased momentum splittings. Fig. 5 shows the output of an $88\hbar k$ splitter.

Four BBB splitters make a RB atom interferometer, see Fig. 6. Two of them, running simultaneously to reject noise and systematic effects, show 15% contrast at 24-photon-momentum splitting each, see Fig. 7. This splitting is limited by a ramp generator used in this experiment, and we expect interferometers with substantially larger splitting in the near future.

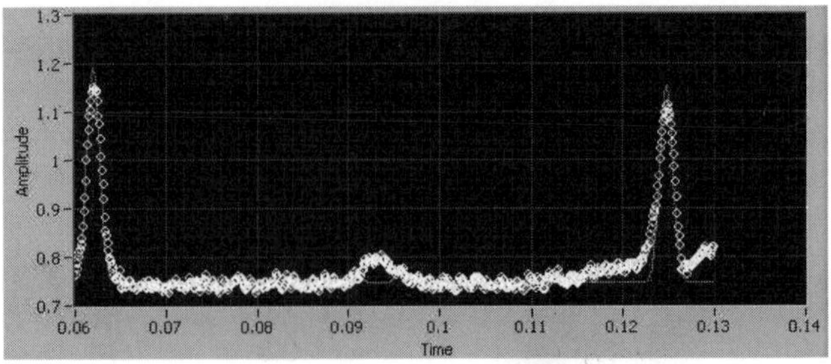

Fig. 5. Time-of-flight sheet of the output of an $88\hbar k$ beam splitter consisting of a $8\hbar k$ Bragg diffraction embedded between pairs of $40\hbar k$ Bloch oscillations. The population in the middle is the residual of the input atomic sample that was lost in the process.

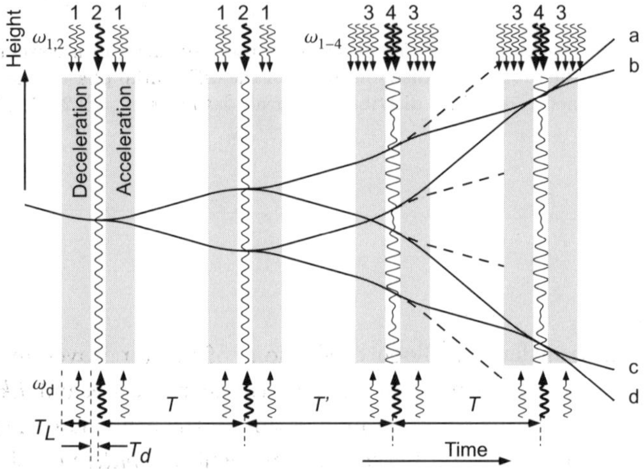

Fig. 6. Space-Time diagram of simultaneous conjugate Ramsey-Bordé BBB-Interferometers. 1: Dual optical lattice; 2: Single Bragg beam splitter; 3: Quadruple optical lattice; 4: Dual Bragg beam splitter; a-d: outputs. The dashed lines indicate trajectories that do not interfere.

5. Towards fundamental physics measurements by atom interferometry

Taken together, the advances we just reported allow for a tremendous increase in the sensitivity of atom interferometers. We will soon apply them for a measurement of the fine structure constant α at the part per billion level of accuracy. By comparison to the value derived from the electron's anomalous gyromagnetic moment $g-2$,[19] this will correspond to testing the theory of quantum electrodynamics at the highest precision ever. The

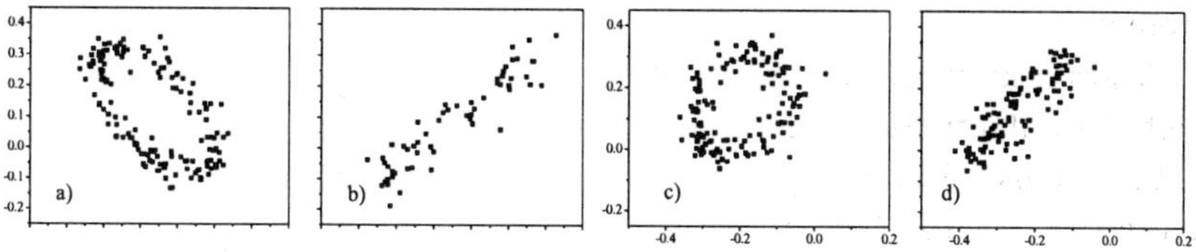

Fig. 7. Ellipses from simultaneous conjugate interferometers with Bragg-Bloch beam splitters. The x and y axes show the normalized flourescence of the upper and lower interferometer. (a) $\Delta p = 12\hbar k, C = 16.5\%$. (b) $\Delta p = 18\hbar k, C = 20.3\%$; (c) $\Delta p = 20\hbar k, C = 16.9\%$, (d) $\Delta p = 24\hbar k, C = 15.1\%$.

influence of muons and Hadrons to $g-2$ will be revealed for the first time. Moreover, this measurement would provide a limit on low-energy dark matter candidates or supersymmetric particles, and serve as a probe for the internal structure of the electron. Indeed, a measurement to 0.1 ppb would correspond to a search for physics beyond the standard model on the TeV energy scale.[20]

5.1. *Detection of gravitational waves*

Gravitational wave detection with light-pulse atom interferometry has the potential to overcome important limitations of LIGO.[21] For example, the influence of vibrations can be common-mode rejected between simultaneous atom interferometers. This will allow them to sense for low-frequency gravitational waves, for which more sources are expected to exist.

Two MZ atom interferometers with a pulse separation time T are located vertically above one another, separated by a distance L. To facilitate common-mode rejection of vibrational noise, they are addressed by the same laser beams. If ω and h are the angular frequency and amplitude of the gravitational wave, respectively, the induced phase difference between the interferometers has an amplitude of $\phi = 2nkhL\sin^2(\omega T/2)$ (where $\omega L \ll 1$ was assumed and $c = 1$). n is the number of photon momenta transferred to the atom in the beam splitter.

The required sensitivity, however, mandates increasing the area enclosed within the interferometer to 100s or 1000s of what is possible with conventional beam splitters for matter waves, i.e., two-photon Raman transitions. As described in this paper, we have already experimentally demonstrated the required critical technology, large-momentum transfer beam splitters and common-mode rejection between simultaneous large-area atom interferometers.

A gravitational wave detector on Earth with $L = 1\,\text{km}$, is inherently limited by fluctuations in the backgorund gravity gradient to $h \approx 10^{-19}$ at 1 Hz. This still allows for the detection of interesting sources that are inaccessible to LIGO.[21] Reaching this sensitivity with $n = 2$ and $k = 2\pi/852\,\text{nm}$ requires detecting a fringe shift by $\phi = 0.003\,\mu\text{rad}$, very much lower than achievable at present: With 10^{10} atoms, the shot noise limit is at 10 murad; Squezing[22] has yet to demonstrate improvements by more than a factor of 1.5 or so. Thus, only $h \approx 10^{-16}$ in one second, at best, seems realistic.

With our present 20-photon beam splitters, however, already $h \approx 10^{-17}$ in one second is possible. The BBB splitter might ultimately be scalable to 200-2000$\hbar k$, and thus enable $10^{-19} - 10^{-18}$. To be sure, reaching that requires extremely flat optical wavefronts, but LIGO-developed optics already goes a long way to achieving this. Because of the relatively moderate power requirement of the BBB splitter and our very strong injection locked laser, even the thick beams required for km-sized Rayleigh ranges should present no problem. Thus, we believe we can present a very strong case that gravitational wave detection with atom interferometry is realistic.

References

1. A.D. Cronin, J. Schmiedmayer, and D.E. Pritchard, arXiv:0712.3703v1
2. A. Peters, K.Y. Chung, and S. Chu, Nature (London) **400,** 849 (1999).
3. M.J. Snadden *et al.*, Phys. Rev. Lett. **81,** 971 (1998).
4. T.L. Gustavson, A. Landragin, and M.A. Kasevich, Class. Quantum Gravity **17,** 2385 (2000).
5. J.B. Fixler *et al,* Science **315,** 74 (2007).
6. G. Lamporesi *et al,* Phys. Rev. Lett. **100,** 050801 (2008).
7. A. Wicht *et al.*, Physica Scripta **T102,** 82 (2002).
8. P. Cladé *et al.,* Phys. Rev. Lett. **96,** 033001 (2006); Phys. Rev. A **74,** 052109 (2006).
9. H. Müller *et al.*, Phys. Rev. Lett. **100,** 031101 (2008).
10. S. Dimopoulos *et al.*, Phys. Rev. Lett. **98,** 111102 (2007).
11. J.M. McGuirk, M.J. Snadden, and M.A. Kasevich, Phys. Rev. Lett. **85,** 4498 (2000).
12. S. Gupta *et al,* Phys. Rev. Lett. **89,** 140401 (2002).
13. D.M. Giltner, R.W. McGowan, and S.A. Lee, Phys. Rev. Lett. **75,** 2638 (1995); Phys. Rev. A **52,** 3966 (1995).
14. J. Hecker Denschlag, *et al,* J. Phys. B **35,** 3095 (2002).
15. H. Müller, S.-w. Chiow, and S. Chu, Phys. Rev. A **77,** 023609 (2008).
16. H. Müller *et al,* Phys. Rev. Lett. **100,** 180405 (2008).
17. H. Müller *et al.,* Appl. Phys. B **84,** 633 (2006).

18. J. K. Stockton, X. Wu, and M. A. Kasevich, Phys. Rev. A **76,** 033613 (2007).
19. G. Gabrielse *et al.,* Phys. Rev. Lett. **97,** 030802 (2006); *ibid.* **99,** 039902 (2007).
20. Hertzog *et al.*, arxiv:07054617 (2007).
21. S. Dimopoulos, P. W. Graham, J. M. Hogan, M. A. Kasevich, and S. Rajendran, arXiv:0712.1250 (2008); arXiv:0806.2125 (2008).
22. D. Leibfried, M. D. Barrett, T. Schaetz, J. Britton, J. Chiaverini, W. M. Itano, J. D. Jost, C. Langer, D. J. Wineland, Science **304** 1476 (2008).

SPACE SCIENCE APPLICATIONS OF FREQUENCY STANDARDS AND METROLOGY

M. TINTO

Jet Propulsion Laboratory, MS 238-737,
Pasadena, California 91109, USA
** E-mail: Massimo.Tinto@jpl.nasa.gov*
www.jpl.nasa.gov

We are witnessing the beginning of a new era for space-qualified atomic clocks. These new instruments, characterized by low-mass, low-power consumption, and frequency stabilities comparable to those of ground-based clocks, will allow interplanetary spacecraft radio science experiments at unprecedented Doppler sensitivities.

By adding a digital receiver to the onboard instrumentation, it is possible to take advantage of the frequency stability of a space-qualified atomic clock by linearly combining digitally the multi-link Doppler measurements performed on the ground and onboard. This allows one to optimally suppress the frequency fluctuations of the Earth atmosphere, ionosphere and mechanical vibrations of the ground antenna (the dominant Doppler tracking noises) in the resulting Doppler data set.

After providing the general expression of the optimal combination of the ground and onboard Doppler data, we apply it, as an example, to interplanetary Doppler tracking searches for gravitational radiation. The resulting estimated sensitivity to milliHertz gravitational waves is about one order of magnitude better than that achievable by traditional two-way coherent Doppler experiments.

The Doppler tracking technique discussed in this article can be performed (at minimal additional cost) with future planned interplanetary missions, and extended to other experiments such as spacecraft Doppler tests of relativistic gravity, gravity field measurements, and radio occultations of planetary atmospheres and rings.

Keywords: Doppler tracking; Space-qualified atomic clocks; Radio science experiments

1. Introduction

Measurements of the relative velocity between the Earth and an interplanetary spacecraft, by means of coherent microwave tracking, have allowed us to study solar system bodies[1] , testing relativistic gravity[2,3] , searching for low-frequency gravitational radiation[4–6] , and other science objectives[1] . In the frequency band ($10^{-6}-10^{-2}$) Hz, the most sensitive deep-space Doppler observations to date have been limited by antenna mechanical noise (unmodeled motion of the phase center of the ground antenna) and residual post-calibration tropospheric scintillation (i.e. Doppler fluctuations caused by refractive index fluctuations in the Earth's atmosphere)[5,7] . The most sensitive observations hit the limit identified by these noise sources with an Allan standard deviation of about 3×10^{-15} for integration times of a few thousand seconds. Improved sensitivity would benefit the science disciplines listed above, but antenna mechanical noise, in particular, has seemed irreducible at reasonable cost since it would require a large, moving, steel structure which is much more rigid than that of the current ground tracking stations.

Several ideas have been proposed to reduce the antenna mechanical noise or the tropospheric noise, or both[8] . Those ideas do not involve modifications to the spacecraft, but rather use simultaneous tracking on the ground with appropriate linear combinations of those data to synthesize an observable with the ground/tropospheric noises of the *better* of the two receiving stations. If, however, additional microwave instrumentation on the spacecraft is considered, such as a space-borne, high-stability frequency standard[9,10] and a space-qualified digital receiver[11] , then an alternative method for noise suppression is possible. This involves properly combining the two one-way (spacecraft to Earth and simultaneous Earth to spacecraft) Doppler data taken onboard and on the ground in such a way to maximize the signal-to-noise ratio of the observed physical observable. Typical Doppler tracks of deep space probes have their sensitivity limited, for time scales of interest to many science objectives, by plasma or tropospheric scintillation noises[7,12] . In some observations these noises can be substantially reduced, however, when multi-link tracking at different frequencies (to calibrate the plasma noise) and independent tropospheric monitoring are available[3,13–15] . These high-precision tracks are then limited by unmodeled motion of the ground antenna and the residual error after tropospheric correction[7,13] . The resulting measured limit expressed as an Allan standard deviation is about 3×10^{-15} for time scales of about 1000 seconds[6] .

Multiple Doppler observations with different noises, or different transfer functions to the same noises, are clearly useful in identifying noise sources and minimizing (and in some cases canceling) their effects on the final observable. In particular, the use of multiple radio links (some of which driven by a high-quality space-borne frequency standard) was pioneered by R. Vessot in the Gravity Probe A sub-orbital experiment[16] . Because of mass and power constraints, however, no very high quality frequency standards have yet been flown on deep space probes.

Recent advances in clock technology indicate that a new era of space-qualified, highly stable frequency standards has started[9] , which will result into significantly improved Doppler radio science experiments. A summary of this paper is given below.

In Section 2 we present a brief overview of the theory underlying Doppler tracking experiments relying only on the two one-way Doppler data measured onboard and on the ground. Although this experimental configuration has been discussed in previous publications[16,17] , it has been shown only relatively recently[5,13] how to fully take advantage of it for improving the precision of Doppler tracking radio science experiments.

The main advantage of spacecraft Doppler tracking experiments relying on the two one-way Doppler data, over those based on two-way coherent measurements, is in their ability of exactly canceling the frequency fluctuations due to the Earth atmosphere and ionosphere, and the mechanical vibrations of the ground antenna (presently the main noise sources of Doppler tracking experiments)[5] . Depending on the specific signal one attempts to observe with this technique, it is actually possible to optimally combine the two one-way Doppler measurements and maximize the signal-to-noise ratio (SNR) of the experiment. After deriving the expression of the optimal SNR, we apply it, as an example application, to searches for gravitational waves and show that the sensitivity one can reach with this technique is about one order of magnitude better than that obtainable with two-way coherent Doppler tracking experiments.

2. The One-Way Doppler Tracking Observables

In Doppler tracking experiments a radio signal of nominal frequency ν_0 is transmitted to a spacecraft and coherently transponded back to Earth where the frequency of the received signal is compared to that of a signal referenced to a highly stable clock (typically a Hydrogen-Maser). Relative frequency changes $\Delta\nu/\nu_0$ as functions of time are then measured, as they contain the effects of the physical observable of interest.

In the case of Doppler tests of relativistic gravity, an alternative way of performing such experiments was suggested in[5,17,18] where a highly-stable clock and a digital receiver were added to the spacecraft radio instrumentation (Figure 1) and two one-way Doppler data were simultaneously recorded at the ground station and on board the spacecraft.

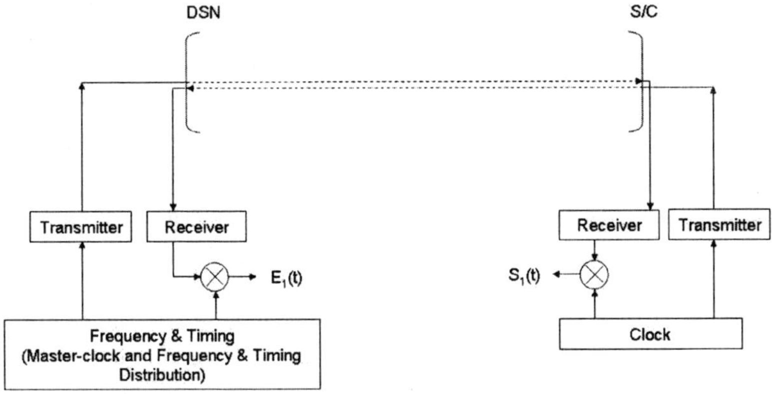

Fig. 1. Block diagram of the radio hardware at the ground antenna of the NASA Deep Space Network (DSN) and on board the spacecraft (S/C), which allows the acquisition and recording of the two Doppler data $E_1(t)$, $S_1(t)$.

If we assume, for the moment, the ground and onboard clocks to be perfectly synchronized, the expressions of the two one-way Doppler can be written in the following form

$$E_1(t) = E_{1s}(t) + C_{sc}(t-L) - C_E(t) + T(t) + B(t-L) \\ + A_{sc}(t-L) + EL_{E_1}(t) + P_{E_1}(t) , \quad (1)$$

$$S_1(t) = S_{1s} + C_E(t-L) - C_{sc}(t) + T(t-L) + B(t) \\ + A_E(t-L) + EL_{S_1}(t) + P_{S_1}(t) , \quad (2)$$

where $E_{1s}(t)$, $S_{1s}(t)$ represent the two Doppler perturbations induced by the signal the experiment is attempting to measure. In Eqs.(1,2) we have denoted with $C_E(t)$, $C_{sc}(t)$ the random processes associated with the frequency fluctuations of the clock on Earth and onboard respectively, $B(t)$

the joint effect of the noise from buffeting of the probe by non gravitational forces and from the antenna of the spacecraft, $T(t)$ the joint frequency fluctuations due to the troposphere, ionosphere and ground antenna, $A_E(t)$ the noise of the radio transmitter on the ground, $A_{sc}(t)$ the noise of the radio transmitter on board, $EL_{E_1}(t)$, $EL_{S_1}(t)$, the noise from the electronics on the ground and onboard respectively, and $P_{E_1}(t)$, $P_{S_1}(t)$ the fluctuations on the two links due to the interplanetary plasma. Since the frequency fluctuations induced by the plasma are, to first order, inversely proportional to the square of the radio frequency, by using high frequency radio signals or by monitoring two different radio frequencies transmitted to and from the spacecraft, this noise source can be suppressed to very low levels or entirely removed from the data respectively[14] . In what follows we will assume dual frequency to be used, and disregard the noise effects of the plasma fluctuations in our analysis.

In Eqs. (1,2) it is important to note the characteristic time signatures of the clock noises, $C_E(t)$ and $C_{sc}(t)$, of the probe antenna and spacecraft buffeting noise $B(t)$, of the troposphere, ionosphere, and ground antenna noise $T(t)$, and the transmitters $A_E(t)$, $A_{sc}(t)$. The time signature of the two clock noises, for instance, can be understood by observing that the frequency of the signal received at the ground station at time t contains fluctuations from the onboard clock that were transmitted L seconds earlier and the noise from the ground clock enters with a negative sign at time t due to the heterodyne nature of the Doppler measurement. The time signature of the noises T, $B(t)$, $A_E(t)$, and $A_{sc}(t)$ in Eq. (1,2) can be understood through similar considerations.

Since the major noise sources affecting the two one-way measurements are represented by the fluctuations induced by the Earth troposphere and the mechanical vibrations of the ground station, it has been previously emphasized[5] that it is possible to properly time-shift and linearly combine the one-way Doppler data to cancel these noises. It is easy to see from inspection of Eqs. (1,2) that such a combination is given by the following expression

$$x(t) \equiv S_1(t) - E_1(t - L) \ . \tag{3}$$

Although such a combination removes entirely the noise $T(t)$ from the combined data, it is not guaranteed to achieve maximum signal-to-noise-ratio (SNR) to a given *signal* we are attempting to study. In order to identify the combination of the two one-way Doppler measurements that maximizes the SNR we can proceed mathematically in the following way.

Let us consider the following linear combination $\eta(f)$ of the Fourier transforms of the two one-way measurements, $E_1(t)$, $S_1(t)$

$$\eta(f) \equiv a_1(f, \vec{\lambda}) \; \widetilde{E_1}(f) \; + \; a_2(f, \vec{\lambda}) \; \widetilde{S_1}(f) \; , \tag{4}$$

where the $\{a_i(f, \vec{\lambda})\}_{i=1}^2$ are arbitrary complex functions of the Fourier frequency f, and of a vector $\vec{\lambda}$ containing parameters characterizing the signal and the noises affecting the two Doppler data. For a given choice of the two functions $\{a_i\}_{i=1}^2$, η gives a specific Doppler data combination, and our goal is therefore that of identifying, for a given signal, the two functions $\{a_i\}_{i=1}^2$ that maximize the signal-to-noise ratio[19] , SNR_η^2, of the combination η

$$SNR_\eta^2 = \int_{f_l}^{f_u} \frac{|a_1 \; \widetilde{E}_{1s} + a_2 \; \widetilde{S}_{1s}|^2}{\langle |a_1 \; \widetilde{E}_{1n} + a_2 \; \widetilde{S}_{1n}|^2 \rangle} \; df \; . \tag{5}$$

In equation (5) the subscripts s and n refer to the signal and the noise parts of $(\widetilde{E_1}, \widetilde{S_1})$ respectively, the angle brackets represent noise ensemble averages, and the interval of integration (f_l, f_u) corresponds to the accessible frequency band.

The SNR_η^2 can be regarded as a functional over the space of the two complex functions $\{a_i\}_{i=1}^2$, and the particular set of complex functions that maximize it can of course be derived by solving the associated set of Euler-Lagrange equations. The derivation of the expression of the optimal SNR, $SNR_{\eta_{\rm opt}}^2$, is long but straightforward, and it is equal to (see[20] for details)

$$SNR_{\eta_{\rm opt.}}^2 = \int_{f_l}^{f_u} \mathbf{z}_i^{(s)*} \; (\mathbf{C}^{-1})_{ij} \; \mathbf{z}_j^{(s)} \; df \; . \tag{6}$$

In Eq. (6) the convention of sum over repeated indices is assumed, $\mathbf{z}^{(s)}$ is the vector of the signals, $(\widetilde{E_{1s}}, \widetilde{S_{1s}})$, and $\mathbf{C}$ is the hermitian, non-singular, correlation matrix of the vector random process $\mathbf{z}_n \equiv (\widetilde{E_{1n}}, \widetilde{S_{1n}})$

$$(\mathbf{C})_{rt} \equiv \langle \mathbf{z}_r^{(n)} \mathbf{z}_t^{(n)*} \rangle \; . \tag{7}$$

2.1. *Example application*

As an application of Eq. (6), we have estimated the sensitivity to an ensemble of sinusoidal gravitational wave signals randomly polarized and uniformly distributed over the celestial sphere. The sensitivity is defined as the strength of a sinusoidal gravitational wave required to achieve a signal-to-noise ratio of 1 in a forty-days integration time[6] , and is a function of the Fourier frequency, f. We have chosen the integration time to be equal to forty days since this is the typical duration of Doppler tracking experiments aimed at detecting gravitational waves[6] .

In order to assess the sensitivity gain offered by this experimental configuration over the traditional two-way Doppler tracking technique, we have estimated the sensitivity of a two-way coherent tracking experiment that relies on the same spacecraft and ground configurations, and has an onboard transponder whose noise is characterized by a one-sided power spectral density as given in[5,6] (see[5,6] also for details about noise sources affecting coherent spacecraft Doppler tracking).

As onboard and ground microwave instrumentation has in recent years reached unprecedented frequency stabilities, T has indeed become the major sensitivity limitation of two-way Doppler tracking searches for gravitational radiation[6] . Since the effects from the troposphere can be mitigated by relying on simultaneous measurements performed by a radiometer located in the proximity of the tracking station, our sensitivity analysis reflects the assumption of being able to calibrate out eighty percent of the tropospheric effects (as demonstrated by the CASSINI experiments) from the two one-way and the two-way data.

Figure 2 shows the estimated *optimal* sensitivity (solid-line) obtained by relying on the two one-way measurements, together with that of a two-way coherent tracking experiment (dash-line). In both cases the spacecraft is assumed to be out to a distance of 1 AU. The best sensitivity shown by the optimal combination is in the frequency band $3.0 \times 10^{-4} - 10^{-1}$ Hz, which is of most interest to gravitational wave search experiments. The optimal processing of the two one-way data offers a sensitivity improvement of about a factor of ten over that of a two-way coherent experiment, with an average strain sensitivity of about 2.0×10^{-17} over the frequency band of interest.

3. Conclusions

We have discussed a method for significantly increasing the sensitivity of Doppler tracking radio science experiments. The main result of our analysis shows that by adding an atomic frequency standard and a digital receiver on board the spacecraft it is possible to optimally maximize the SNR of a signal entering into the Doppler data. As an example, we have considered Doppler tracking experiments aimed at the detection of gravitational radiation and shown that a broad-band sensitivity of 2.0×10^{-17} in the milliHertz band can be achieved. This is about a factor of ten better than that obtainable by coherent two-way Doppler tracking searches.

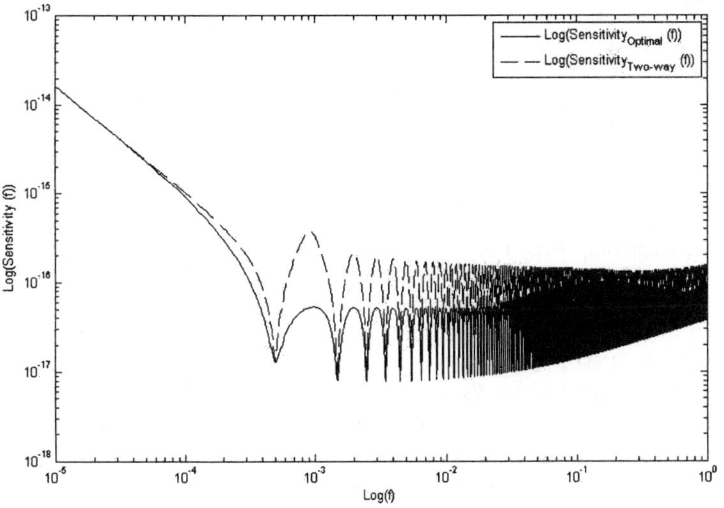

Fig. 2. Sensitivities of the optimal combination (solid-line), and of two-way Doppler tracking (dash-line), to a randomly polarized and uniformly distributed stochastic ensemble of sinusoidal gravitational wave signals. The sensitivity is expressed as a function of the frequency and represents the equivalent sinusoidal strain required to produce a signal-to-noise ration of 1. The spacecraft is assumed to be out to a distance of 1 AU, and an eighty percent troposperic noise calibration is assumed to be applied to the data. The two sensitivity curves reflect the noise spectral levels and shapes, their transfer functions, and the gravitational wave transfer functions described in[5,6] .

We have just started to quantify the sensitivity improvements the technique highlighted in this article will bring to all other radio science experiments. The results of our analysis will be presented in a forthcoming paper.

Acknowledgements

This research was performed at the Jet Propulsion Laboratory, California Institute of Technology, under contract with the National Aeronautics and Space Administration.

References

1. A. Kliore, J.D. Anderson, J.W. Armstrong, S.W. Asmar, C.L. Hamilton, N.J. Rappaport, H.D. Wahlquist, R. Ambrosini, F.M. Flasar, R.G. French, L. Iess, E.A. Marouf, & A.F. Nagy, *Space Sci. Rev.*, **115**, 170, (2004).

2. R.F.C. Vessot. In: *Proceedings of the XXVIIIth RENCONTRE DE MORIOND*, eds. J. Tran Thanh Van, T. Damour, E. Hinds, and J. Wilkerson (Editions FRONTIERS, p.471-489, 1993).
3. B. Bertotti, L. Iess, & P. Tortora, *Nature*, **425**, 374376, (2003).
4. F.B. Estabrook and H.D. Wahlquist, *Gen. Relativ. Gravit.* **6**, 439 (1975).
5. M. Tinto, *Phys. Rev. D* **53**, 5354, (1996).
6. J.W. Armstrong, *Living Reviews in Relativity* **9**, 1, (2006).
7. S.W. Asmar, J.W. Armstrong, L. Iess, & P. Tortora, *Radio Sci.*, **40**, (2005).
8. J.W. Armstrong, F.B. Estabrook, S.W. Asmar, L. Iess, & P. Tortora, *Radio Sci.*, **43**, RS3010, doi:10.1029/2007RS003766 28 June 2008.
9. J.D. Prestage, & G.L. Weaver. In: *Proceedings of the IEEE*, **95**, 11, 2235-2247 (2007).
10. G.J. Dick, J.D. Prestage, C.A. Greenhall, & L. Maleki. In: *Proceedings of the 22nd annual Precise Time and Time Interval (PTTI) Applications and Planning Meeting*, The U.S. Naval Observatory publication. Available at: http://tycho.usno.navy.mil/ptti/1990/Vol
11. *The Radio Atmospheric Sounding and Scattering Instrument (RASSI): JPL proposal for constructing am onboard radio science digital receiver.* (JPL internal publication, 2008).
12. J.W. Armstrong, R. Woo, and F.B. Estabrook, *Ap.J.*, **230**,570 (1979).
13. M. Tinto, *Radio Sci. 37*, **3**, 1045, doi:10.1029/2001RS002473, (2002).
14. B. Bertotti, G. Comoretto, & L. Iess, *Astron. Astrophys.*, **269**, 608616, (1993).
15. J.W. Armstrong, L. Iess, P. Tortora, & B. Bertotti, *Astrophys. J.*, **599**, 806813, (2003).
16. D. Kleppner, R.F.C. Vessot, & N.F. Ramsey, *Astrophys. Space Sci.*, **6**, 13-32, (1970).
17. T. Piran, E. Reiter, W.G. Unruh, and R.F.C. Vessot, *Phys. Rev. D*, **34**, 984, (1986).
18. R.F.C. Vessot, & M.W. Levine. In: *A closeup of the Sun*, Eds. M. Neugebauer, & R.W. Davis. (JPL Publication 78-80, NASA, 1978).
19. C.W. Helström, *Statistical Theory of Signal Detection*, (Pergamon Press, London, 1968).
20. T.A. Prince, M. Tinto, S.L. Larson, and J.W. Armstrong *Phys. Rev. D*, **66**, 122002, (2002).

Part II

Frequency & Metrology

QUANTUM METROLOGY WITH LATTICE-CONFINED ULTRACOLD SR ATOMS

A. D. LUDLOW,[1,2] G. K. CAMPBELL,[1] S. BLATT,[1] M. M. BOYD,[1] M. J. MARTIN,[1] T. L. NICHOLSON,[1] M. SWALLOWS,[1] J. W. THOMSEN,[1] T. FORTIER,[2] C. W. OATES,[2] S. A. DIDDAMS,[2] N. D. LEMKE,[2] Z. BARBER,[2] S. G. PORSEV,[1,3] AND JUN YE[1]

[1]*JILA, National Institute of Standards and Technology and University of Colorado Department of Physics, University of Colorado, Boulder, CO 80309-0440, USA*
[2] *Time and Frequency Division, National Institute of Standards and Technology, Boulder, CO 80302, USA*
[3] *Petersburg Nuclear Physics Institute, Gatchina, Leningrad district, 188300, Russia*

Quantum state engineering of ultracold matter and precise control of optical fields have together allowed accurate measurement of light-matter interactions for applications in precision tests of fundamental physics. State-of-the-art lasers maintain optical phase coherence over one second. Optical frequency combs distribute this optical phase coherence across the entire visible and infrared parts of the electromagnetic spectrum, leading to the direct visualization and measurement of light ripples. At the same time, ultracold atoms confined in an optical lattice with zero differential ac Stark shift between two clock states allow us to minimize quantum decoherence while strengthening the clock signal. For ^{87}Sr, we achieve a resonance quality factor >2.4 x 10^{14} on the $^1S_0 - {}^3P_0$ doubly forbidden clock transition at 698 nm [1]. The uncertainty of this new clock has reached 1 x 10^{-16} and its instability approaches 1 x 10^{-15} at 1 s [2]. These developments represent a remarkable convergence of ultracold atoms, laser stabilization, and ultrafast science. Further improvements are still tantalizing, with quantum measurement and precision metrology combining forces to explore the next frontier.

1. Introduction

In the last decade, the performance of frequency standards based on optical atomic transitions has grown considerably. The performance advancements of these systems are derived from their superior resonance quality factors. Currently, the most accurate optical clocks are based on single trapped ions [3], due to the exquisite control that is possible over their electronic and motional quantum states, as demonstrated in both clock and quantum information experiments. On the other hand, large ensembles of neutral atoms enjoy high measurement precision from the simultaneous measurement of each atom in the ensemble, providing a powerful averaging of measurement noise. However, a longstanding challenge for these neutral atom systems is achieving control and measurement at uncertainties similar to that for single, trapped particles. In particular, motional effects have played a prominent role in limiting neutral atom systems, where traditionally ion-based standards have benefited from a strongly confined measurement environment. Recently, properly designed optical traps [4] have made the use of large neutral atom ensembles compatible with the benefits of atomic confinement. This is possible by ensuring that the ac Stark shifts induced on the two clock states by the optical lattice are equal, leading to zero shift of the clock transition [5]. Implementing a carefully designed frequency standard based on atomic strontium in a one dimensional optical lattice, we have demonstrated an optical frequency standard with a systematic uncertainty at the 10^{-16} level [2], smaller than that of the best Cs fountain primary standards. The evaluation was made possible by a remote comparison of the Sr optical lattice frequency standard to the NIST cold Ca optical standard over a phase-stabilized optical fiber network [6].

Through rigorous experimental evaluations of the Sr standard at this level, we highlighted two key systematic shifts that must be overcome to further improve this system. The first is frequency shifts induced by the blackbody radiation. We offer approaches to improving the precision of the known blackbody shift and also reducing its uncertainty. Secondly, high precision characterization of the Sr frequency standard has uncovered surprisingly non-negligible frequency shifts induced by collisions between the fermionic Sr atoms trapped in the optical lattice. By experimentally identifying the underlying collisional mechanism responsible for these shifts, we have shown that this effect can be substantially reduced and controlled, facilitating further progress of the Sr optical lattice standard.

2. Characterization of the Sr optical frequency standard

Our Sr optical frequency standard is based on ultracold neutral ^{87}Sr atoms confined in a one dimensional optical lattice where the differential Stark-shift between the clock states is cancelled [4]. Depending on the operational mode, the Sr sample undergoes state preparation by spin-polarization to one or two nuclear spin states prior to spectroscopic probing. The clock transition from each of the two nuclear spin states is excited one at a time. Spectroscopy of both states allows for measurement and cancellation of some nuclear spin-dependent shifts on the clock transition, such as the first order Zeeman shift and the vector ac Stark shift induced by the optical lattice. Further details of the experimental operation of the Sr optical standard can be found in several references [e.g., 7,2].

To characterize the Sr optical standard located at JILA, we have constructed a remote fiber optic link between JILA and the NIST Boulder Campus [6] with optical carrier phase stabilization. This link permits high precision measurement between the Sr system and other optical standards at NIST. In particular, frequency measurement against a free-space Ca optical standard enabled direct determination of the most critical systematic shifts of the Sr standard at or below 10^{-16} fractional uncertainty [2]. The results of these measurements are summarized in Table 1, presenting the final error budget for the Sr standard. The combined uncertainty is 1.5 x 10^{-16}, which demonstrates the high potential accuracy of this system, with room for further improvement.

Table 1. Systematic frequency corrections and their associated uncertainties for the $^{87}Sr\ ^1S_0 - {}^3P_0$ clock transition.

Contributor	Shift (10^{-16})	Uncertainty(10^{-16})
Lattice Stark (scalar/tensor)	-6.5	0.5
Fourth order lattice	-0.2	0.2
Blackbody Stark	52.1	1.0
Probe ac Stark	0.2	0.1
First-order Zeeman	0.2	0.2
Second-order Zeeman	0.2	0.02
Collision	8.9	0.8
Line pulling	0	0.2
Servo error	0	0.5
Second-order Doppler	0	<<0.01
Other	0	.1
TOTAL	54.9	1.5

The stability of the Sr standard has been measured by making frequency comparisons between the Sr standard at JILA and the Yb standard at NIST [8], both optical lattice clock systems. The results of this measurement are shown in Figure 1, in the form of total deviation. The 1-s instability of 1 x 10^{-15} is given by the free-running stabilities of the interrogation lasers. At longer timescales just above 1000 s, the combined instability of the two standards reaches below 1 x 10^{-16}.

We have also performed a 50-hour-long absolute frequency measurement of the strontium transition referenced to the NIST-F1 Cs fountain standard [7]. This yields a value for the Sr clock transition frequency with a fractional uncertainty of 8.6 x 10^{-16}, limited by the NIST H-maser and the Cs fountain standard used in the measurement. This represents our fifth, and the most accurate, measurement of the ^{87}Sr clock frequency. It is also one of the most accurate absolute frequency measurements of an optical standard to date. Combining these measurements with those of other groups (Paris, Tokyo), a measurement record of the Sr clock transition is established as the most agreed upon optical standard and it has been used to more tightly constrain fundamental constant variations due to gravitational coupling [9].

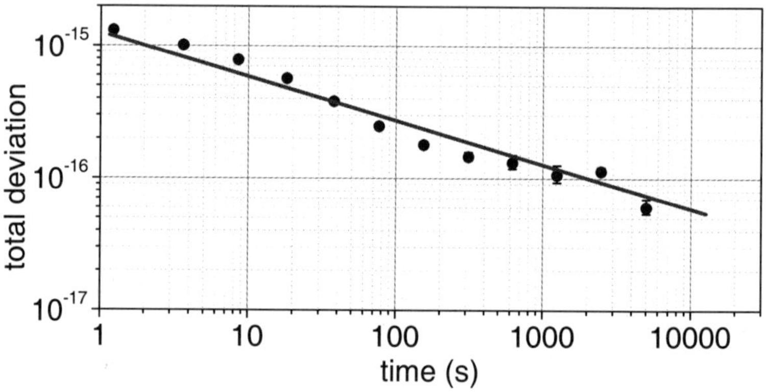

Figure 1. Instability between Sr and Yb optical lattice clocks as measured by the total deviation, in units of fractional frequency.

We have identified two main systematic uncertainties limiting the overall Sr clock accuracy. The first is the uncertainty in the blackbody-radiation-induced frequency shift and the second is the density-dependent shift of the clock transition. We now discuss each of these two mechanisms in more detail, emphasizing the further reduction of their contributions to the overall uncertainty.

3. Blackbody-induced shift

The fractional frequency shift caused by the blackbody radiation is proportional to the differential static polarizability of the two clock states, $5s^2\ ^1S_0$ and 5s5p

$^3P_0$. Because of its dominant coupling to the well-characterized 5s5p $^1P_1$ state, the $5s^2$ $^1S_0$ ground state has a static polarizability that has been determined with an uncertainty of ~ 0.1%. Meanwhile, the static polarizability of the 5s5p $^3P_0$ state is known only at the 1% level. The $^3P_0$ state therefore dominates the uncertainty in the differential static polarizability. This yields a 7 x 10^{-17} uncertainty in the blackbody shift of the Sr clock transition [10]. By reducing the uncertainty of the $^3P_0$ state polarizability to a similar 0.1% level as for the $^1S_0$ state, the possibility is opened for nearly an order of magnitude reduction in the blackbody shift uncertainty, down to 1 x 10^{-17}.

We have identified four low-lying even-parity intermediate states that have a combined contribution to the total static polarizability of the $^3P_0$ state at the level of 90%. When the combined contributions from these four states are determined from experimental data at 0.1% accuracy and the contributions of all other discrete and continuum states are known at the level of 1-2% from calculations, the final 0.1% uncertainty for the polarizability of the $^3P_0$ state will be achieved. The residual contributions from the states outside the four dominant ones have indeed been calculated at the percent level, leaving an urgent need for high precision measurements of contributions from the four dominant states [11]. Their dipole coupling strength to the $^3P_0$ clock state must be determined. Some constraints on their values are already available because of the strong contributions these couplings make to the dynamic polarizability. For example, at the magic wavelength near 813 nm, the dynamic polarizability of $^3P_0$ is identical to the well known value of that of $^1S_0$. Also, measurements of the ac Stark shift at the clock transition frequency have been made [2], helping to determine the dynamic polarizability of $^3P_0$ at 698 nm. When these types of measurements are combined with others, like direct decay rate measurements from the intermediate states to $^3P_0$, the individual contributions can be fully constrained. Reference [11] provides more detailed discussions on this theory and the corresponding experimental planning.

Another important effort to reducing the blackbody shift uncertainty is to bathe the atoms in a more well-defined and controlled blackbody environment than what has been typically accomplished in optical lattice systems. This reduces primarily to a task of maintaining careful temperature homogeneity of the blackbodies surrounding the atomic sample. Cryogenic shields have been used in trapped ion experiments to reduce the blackbody shift and thus its uncertainty. An alternative approach [2] is to load the atoms into a blackbody cavity prior to clock spectroscopy, similar in spirit to those used in infrared radiation metrology. By limiting optical access to small solid-angle viewports for the lattice and clock lasers, the effective emissivity inside a carefully designed blackbody chamber can reach very nearly unity. Then with a careful, homogenous temperature control over the small blackbody chamber, the radiation bath surrounding the atoms can be measured and controlled to enable direct determination of the blackbody shift as the chamber temperature is varied. This direct measurement will then further constrain the differential static polarizability. While such an approach incurs experimental complexity in loading atoms into and controlling the blackbody chamber, it avoids many complexities associated with cryogenic operations. Note that this approach is particularly feasible for atoms held by laser beams, where no trap electrodes are present that can introduce uncontrolled heat loads into the atom environment [12].

4. Collision induced shift

An interesting discovery in our push for ever increasing accuracy of the ^{87}Sr standard is the collision-induced frequency shift on the clock transition. The result is somewhat unexpected, as these atoms are fermions prepared at ultracold temperatures. As the Pauli exclusion principle dictates that identical fermions avoid interactions at ultracold temperatures, interatomic collisions should be strongly suppressed. However, with confinement in the 1-D optical lattice, atoms in different transverse motional states could experience slightly different Rabi rates of excitation. Therefore, at a particular lattice site, while each atom evolves coherently, the atoms evolve to non-identical superpositions of the clock states, at which point they are able to collide. The Rabi excitation pulse is on the order of 100 ms, while the transverse trap oscillation frequency in our lattice is about 450 Hz, giving ample opportunity for collisions. As shown in Figure 2, we have measured this density-dependent collisional shift and its dependence on the excited state fraction, for different atomic sample temperatures. The shift becomes larger at warmer temperatures as the atoms occupy more (transverse) motional states and the Rabi excitation inhomogeneity becomes more pronounced. By misaligning the clock probe beam relative to the strong confinement axis of the lattice, we can further enhance the excitation inhomogeneity, as the effect increases along the weaker, transverse confinement axes. In this case, we also observe an increase in the observed shift [13]. Collision theories incorporating the excitation-inhomogeneity-induced s-wave collisions between fermionic atoms have confirmed these experimental observations.

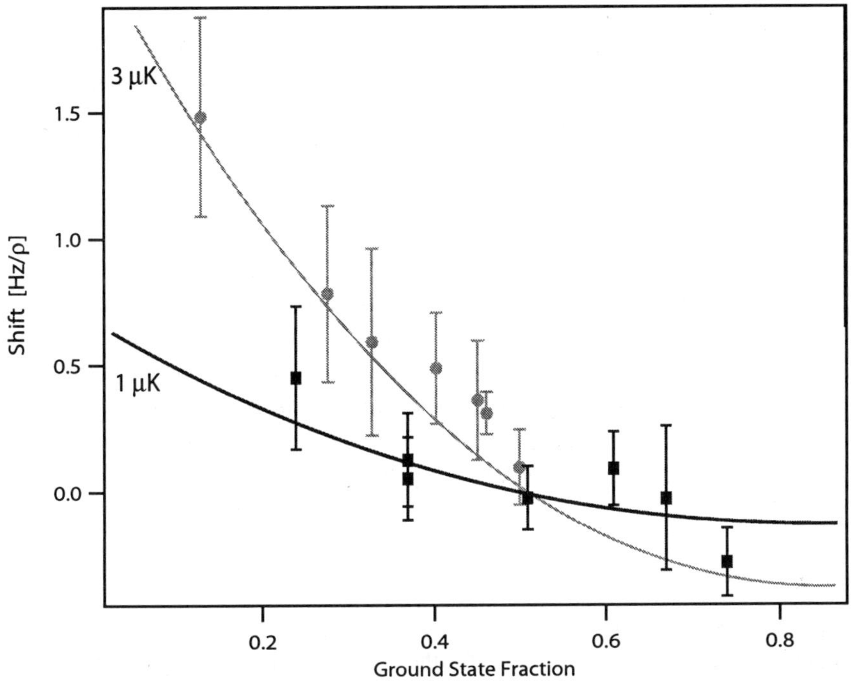

Figure 2. Collisional frequency shift at a given density, as a function of ground state fraction after spectroscopy. The solid circles present the observed shifts for atoms at 3 μK, while the solid squares are for atoms that have been cooled to 1 μK. The solid lines represent theory calculations that are based on excitation-inhomogeneity-induced s-wave collisions between ^{87}Sr atoms during the clock excitation sequence. Note the zero crossing in the shift near the 50-50 mixture of ground and excited state atoms. ρ is the atomic density of $10^{11}/cm^3$.

Significant to the clock operation, we have experimentally observed that if we operate the clock transition at near the 50% excited state fraction (see Figure 2), the collision shift reduces to zero. Since the excitation fraction can be straightforwardly controlled at or better than the few percent level, we can further reduce the collision shift uncertainty beyond the value reported above. Therefore, with a proper design, the 1-D optical lattice clock system continues to offer increased performance. Furthermore, moving to higher dimensional lattices has two possible advantages from the perspective of collisional effects. The first is that by going to higher dimensions, the number of lattice sites increases, giving fewer atoms per lattice site and thus reducing collisions. Second, with strong confinement along all dimensions, the atoms are excited more uniformly. As the excited atoms remain identical, the collision mechanism we have identified is further suppressed. We anticipate that atomic collisions will not be a problem in these optical lattice clocks. As we advance the clock inaccuracy below 1 x 10^{-17}, resonant dipole interactions between neighboring atoms and the related frequency shift might become an interesting interaction to study [14].

5. Conclusions

In our earlier studies we identified two main systematic uncertainties that limited the accuracy of our ^{87}Sr optical clock at the 1 x 10^{-16} level [2]. Since then we have performed a theoretical study of the blackbody radiation-induced frequency shifts and a detailed experimental measurement of density-dependent collisional frequency shifts. These studies now show that both systematic uncertainties have potential to be reduced by an order of magnitude or more.

Acknowledgments

We thank S. Foreman and N. Poli for the optical standard comparison; S. Jefferts, T. Heavner, and T. Parker for the absolute frequency measurement; and P. Julienne, P. Naidon, K. Gibble for discussions on collision-related frequency shifts. We thank funding support from NSF, NIST, DARPA, and ONR.

References

1. M. M. Boyd, T. Zelevinsky, A. D. Ludlow, S. M. Foreman, S. Blatt, T. Ido, and J. Ye, "Optical atomic coherence at the one second time scale," Science **314**, 1430 – 1433 (2006).
2. A. D. Ludlow, T. Zelevinsky, G. K. Campbell, S. Blatt, M. M. Boyd, M. H. de Miranda, M. J. Martin, J. W. Thomsen, S. M. Foreman, J. Ye, T. M. Fortier, J. E. Stalnaker, S. A. Diddams, Y. Le Coq, Z. W. Barber, N. Poli, N. D. Lemke, K. M. Beck, and C. W. Oates, "Evaluation of a Sr lattice clock at 1x10^{-16} via remote optical comparison with a Ca clock," Science **319**,1805-808 (2008).
3. T. Rosenband, D. B. Hume, P. O. Schmidt, C. W. Chou, A. Brusch, L. Lorini, W. H. Oskay, R. E. Drullinger, T. M. Fortier, J. E. Stalnaker, S. A. Diddams, W. C. Swann, N. R. Newbury, W. M. Itano, D. J. Wineland, and J. C. Bergquist, "Frequency ratio of Al+ and Hg+ single-ion optical clocks; Metrology at the 17th decimal place," Science **319**,1808-1812 (2008).

4. J. Ye, H. J. Kimble, and H. Katori, "Quantum state engineering and precision metrology using state-insensitive light traps," Science **320**, 1734-1738 (2008).
5. H. Katori, M. Takamoto, V. G. Pal'chikov, V. D. Ovsiannikov, "Ultrastable Optical Clock with Neutral Atoms in an Engineered Light Shift Trap," Phys. Rev. Lett. **91**, 173005 (2003).
6. S. M. Foreman, A. D. Ludlow, M. H. G. de Miranda, J. Stalnaker, S. A. Diddams, and J. Ye, "Coherent optical phase transfer over a 32-km fiber with 1-s instability at 10^{-17}," Phys. Rev. Lett. **99**, 153601 (2007).
7. G. K. Campbell, A. D. Ludlow, S. Blatt, J. W. Thomsen, M. J. Martin, M. H. de Miranda, T. Zelevinsky, M. M. Boyd, J. Ye, S. A. Diddams, T. P. Heavner, T. E. Parker, and S. R. Jefferts, "Absolute frequency measurement of the ^{87}Sr optical clock transition," Metrologia **45**, 539 – 548 (2008).
8. N. Poli, Z. W. Barber, N. D. Lemke, C. W. Oates, L. S. Ma, J. E. Stalnaker, T. M. Fortier, S. A. Diddams, L. Hollberg, J. C. Bergquist, A. Brusch, S. Jefferts, T. Heavner, and T. Parker, "Frequency evaluation of the doubly forbidden $^1S_0$ - $^3P_0$ transition in bosonic ^{174}Yb," Phys. Rev. A **77**, 050501(R) (2008).
9. S. Blatt, A. D. Ludlow, G. K. Campbell, J. W. Thomsen, T. Zelevinsky, M. M. Boyd, J. Ye, X. Baillard, M. Fouché, R. Le Targat, A. Brusch, P. Lemonde, M. Takamoto, F.-L. Hong, H. Katori, V. V. Flambaum, "New limits on coupling of fundamental constants to gravity using ^{87}Sr optical lattice clocks," Phys. Rev. Lett. **100**, 140801 (2008).
10. S. G. Porsev and A. Derevianko, "Multipolar theory of blackbody radiation shift of atomic energy levels and its implications for optical lattice clocks," Phys. Rev. A **74**, 020502 (2006).
11. S. G. Porsev, A. D. Ludlow, M. M. Boyd, and J. Ye, "Determination of Sr properties for a high accuracy optical clock," Phys. Rev. A **78**, 032508 (2008).
12. M. Chwalla, J. Benhelm, K. Kim, G. Kirchmair, T. Monz, M. Riebe, P. Schindler, A. S. Villar, W. Haensel, C. F. Roos, R. Blatt, M. Abgrall, G. Santarelli, G. D. Rovera, and Ph. Laurent, arXiv:0806.1414 (2008).
13. G. K. Campbell, et al., to be published (2008).
14. D. E. Chang, J. Ye, and M. D. Lukin, "Controlling dipole-dipole frequency shifts in a lattice-based optical atomic clock," Phys. Rev. A **69**, 023810 (2004).

LNE-SYRTE CLOCK ENSEMBLE: NEW ^{87}Rb HYPERFINE FREQUENCY MEASUREMENT - SPECTROSCOPY OF ^{199}Hg AND ^{201}Hg OPTICAL CLOCK TRANSITION

M. PETERSEN, J. MILLO, D. V. MAGALHÃES, C. MANDACHE, S. T. DAWKINS, R. CHICIREANU, Y. LECOQ, J. GUÉNA, F. CHAPELET, P. ROSENBUSCH, P. LAURENT, M. ABGRALL, G. D. ROVERA, G. SANTARELLI, A. CLAIRON AND *S. BIZE

LNE-SYRTE, Observatoire de Paris
61 avenue de l'Observatoire, 75014 Paris, France
**E-mail: sebastien.bize@obspm.fr*

M. E. TOBAR

University of Western Australia, School of Physics
35 Stirling Highway
Crawley, Western Australia

This article describes recent frequency measurements performed with the LNE-SYRTE atomic clock ensemble. First, we report a new determination of the rubidium 87 hyperfine frequency, performed using Rb and Cs atomic fountains. The measurement has a statistical uncertainty below 10^{-16} and an overall uncertainty of 1.1 parts in 10^{15}. Second, we report the first laser spectroscopy of the $^1S_0$–$^3P_0$ optical clock transition at 265.6 nm in laser-cooled fermionic isotopes of neutral mercury. The absolute frequency is measured with a fractional uncertainty of ~ 5 parts in 10^{12} which improves upon previous indirect determinations by more than 4 orders of magnitude, an important step toward the realization of an optical lattice clock.

Keywords: Atomic clocks; Time and Frequency; Laser cooling of atoms

1. Introduction

Over the years, LNE-SYRTE has developed an ensemble of high performance atomic clocks. This ensemble comprises three atomic fountains, FO1, FO2 and FOM. All three use cesium and therefore qualify as primary frequency standards. FOM is a transportable fountain which has been used several times for absolute frequency measurements external to LNE-SYRTE laboratories.[1–3] FO2 is a dual fountain which can also operate with Rb. The LNE-SYRTE atomic clock ensemble also includes a Sr optical lattice clock,[4,5] with a second one under development. Finally, LNE-SYRTE has started developing an optical lattice clock based on neutral mercury. In order to connect these atomic clocks to each other, an ultra-low phase noise reference signal is generated, based on a cryogenic sapphire oscillator (CSO) weakly locked to a hydrogen maser.[6] This original configuration allows for unmatched stability for frequency measurements performed with atomic fountains.[7–9] The results of two measurements made with this clock ensemble are reported in this article.

2. New measurement of the rubidium 87 hyperfine frequency

In this section, a new measurement of the rubidium 87 hyperfine frequency is reported. Following our previous measurements in 2004,[7,8] many subsystems of the rubidium part of the FO2 fountain (FO2-Rb) have been modified or replaced. A 2 dimensional magneto-optic trap (2D-MOT) has been implemented as a source of cold isotopically selected ^{87}Rb atoms for the optical molasses. A new laser system has been constructed, based on a new extended cavity laser diode design with much improved reliability[10] and robust laser stabilization scheme using beat notes against a reference laser and frequency-to-voltage converter (rather locking several laser to satured absorption cells independently with the conventional FM modulation technique). Dichroic collimators have been implemented to allow for simultaneous operation with Rb and Cs. A new microwave synthesizer based on methods recently developed for Cs is now used.[11,12] In general, methods first used with Cs to evaluate and control systematic shifts are now applied to FO2-Rb. As a result of these modifications, FO2-Rb now has a similar level of performance to these achieved with Cs fountains. The best observed fractional frequency instability is $4.5 \times 10^{-14}\tau^{-1/2}$. The total uncertainty of 4.4×10^{-16} is also similar to the best value achieved in Cs, as shown in table 1.

Following these improvements, a new Rb-Cs frequency comparison has been performed lasting from November 2007 to January 2008. In this comparison, FOM was used as the primary standard. It is worth mentioning that comparisons between FO2-Cs and FOM from April 2007 to September 2007, and then again since March 2008 with FO2-Cs loaded with a new 2D-MOT, indicate that the fractional frequency difference between the two Cs fountains is not more than 5×10^{-16}. During the Rb-Cs comparison,

Table 1. FO2-Rb systematic uncertainties. Fractional frequency corrections and fractional frequency uncertainties are in parts per 10^{16}.

Effect	Correction	Uncertainty
Quadratic Zeeman effect	-3472	0.2
Blackbody radiation	120.6	1.6
Cold collisions and cavity pulling	0	2.5
First order Doppler effect	0	2
Microwave leaks	0	< 0.01
Others (micro-wave recoil, Background gas collisions, Ramsey & Rabi pulling,etc)	0	2.5
Total		4.4

the CSO based reference is slowly phase-locked to the 40 889 hydrogen maser (BIPM numbering) with a time constant of $\sim$ 1000 s. Both FOM and FO2-Cs are measuring this common reference. The data of each fountain are averaged over intervals of 100 s, and statistics on the differences are calculated on synchronous data samples. Figure 1 shows the fractional frequency instability of the reference measured by the two fountains during the comparison (black and red), and the fractional frequency instability of the difference between them (blue).

Above 10^5 s, the black and red curves are simply showing the drift of the maser ($\sim -2.4 \times 10^{-15}$ per days in this particular instance). The fountain frequency difference is free of the common mode fluctuations of the reference and averages down as white noise with an instability of 1.3×10^{-13} at 1 s. With a set of 48 000 synchronous data samples of 100 s between Modified Julian Dates (MJD) 54 420 and 54 495, equivalent to 56 effective days of comparison, a statistical resolution of 6×10^{-17} is obtained in the comparison. The mean absolute frequency of the FO2-Rb standard deduced from this measurement is

$$\nu(\mathrm{FO2-Rb})(2007) = 6\ 834\ 682\ 610.904\ 309(8)\ \mathrm{Hz} \qquad (1)$$

with an error dominated by the total systematic uncertainty of table 1 combined with the FOM uncertainty of 9 parts in 10^{16}.

Figure 2 summarizes measurements of the ^{87}Rb hyperfine frequency made at LNE-SYRTE over the years. The 2002 measurement, adopted by BIPM in 2004 as the recommended value for the secondary representation of the SI second, is chosen as the reference ($\nu_{\mathrm{Rb}}(\mathrm{CCTF}) =$ 6 834 682 610.904 324 Hz). The uncertainties include both the statisti-

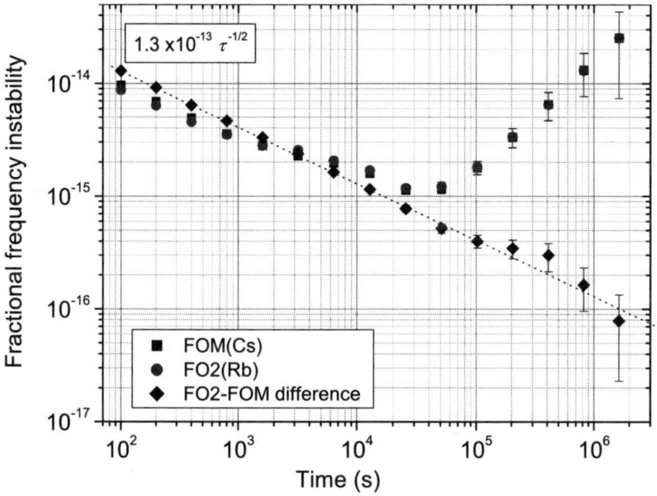

Fig. 1. Allan standard deviation of the fractional frequency fluctuations of maser 40 889 filtered by the CSO as measured by FO2-Rb and FOM (black and red). Allan standard deviation of the fractional frequency difference between these two measurements (blue). Statistics is performed on synchronous data samples of 100 s.

cal and systematic contributions involved in each campaign. A weighted least square fit to these data with a constant gives a goodness of fit (probability for the χ^2 to be larger than the actual χ^2 of data points, assuming a normal distribution) of $Q = 0.7$ indicating good consistency of the data. Considering that, over time, major changes have been done on all clocks involved in this series of measurements, this comes as a strong validation of methods and practices that have been used to evaluate the clock accuracies. By means of a weighted linear least square fit, we find that these data constrain a putative variation of $\nu_{\mathrm{Rb}}/\nu_{\mathrm{Cs}}$ at the level of $d\ln(\nu_{\mathrm{Rb}}/\nu_{\mathrm{Cs}})/dt = (-3.2 \pm 2.3) \times 10^{-16}\ \mathrm{yr}^{-1}$. In terms of fundamental constants, this translates into $d\ln(\alpha^{-0.49} g_{\mathrm{Rb}}/g_{\mathrm{Cs}})/dt = (-3.2 \pm 2.3) \times 10^{-16}\ \mathrm{yr}^{-1}$ and $d\ln(\alpha^{0.49}[m_q/\Lambda_{QCD}]^{0.025})/dt = (3.2 \pm 2.3) \times 10^{-16}\ \mathrm{yr}^{-1}$.[13]

3. Measurement of the $^1S_0$–$^3P_0$ optical clock transition in mercury

In this section, we report our work toward the development of an optical lattice clock using neutral mercury. Neutral mercury has a low sensitivity to blackbody radiation[14–16] whilst retaining all other desirable features for an optical lattice clock. It has the potential to achieve uncertainty in the low

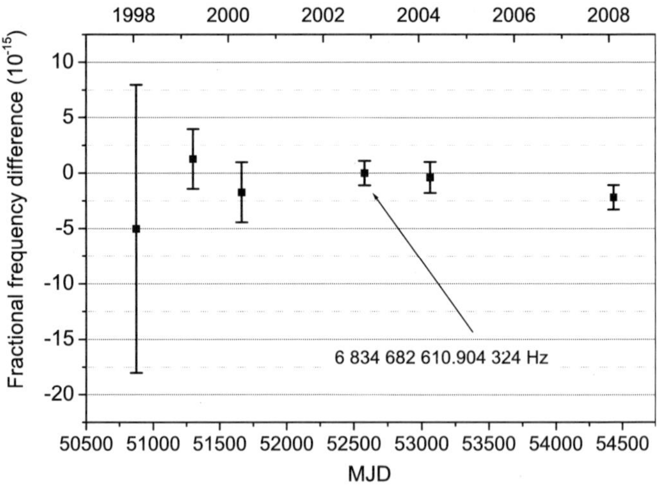

Fig. 2. Measurements of the ^{87}Rb hyperfine frequency performed at LNE-SYRTE since 1998. The 2002 measurement recommended by the CCFT in 2004 is chosen as the reference. The uncertainties include both the statistical and systematic contributions involved in each measurement campaign.

10^{-18} and therefore to compete with the best single ion optical clocks.[17] Laser cooling of neutral mercury has been pursued and achieved only recently[16,18] due to the challenging requirement of deep-UV laser sources.

Mercury has an alkaline-earth like electronic structure similar to those of strontium or ytterbium. In our experiment, laser cooling of mercury is achieved using the $^1S_0$–$^3P_1$ transition at 253.7 nm with a natural linewidth of 1.3 MHz. Cooling light is provided by quadrupling a Yb:YAG thin disk laser. Up to 800 mW of CW power have been generated. In practice, the system is set to generate 100 to 150 mW in order to avoid the rapid degradation of some optical components observed with higher output power. The frequency of the cooling laser is stabilized to the saturated absorption feature observed in a room temperature mercury vapor cell, leading to a linewidth below 100 kHz.[18]

In the experiment, a 2D-MOT is used to generate a slow atom beam which in turn is used to load a conventional MOT. The 2D-MOT is formed at the intersection of two orthogonal pairs of σ^+–σ^- polarized retro-reflected beams with a total power of $\sim$ 50 mW. The longitudinal and transverse diameters at $1/e^2$ are $\sim$ 10 mm and $\sim$ 8 mm. The intersection overlaps with the center of a 2-dimensional quadrupole magnetic field with a gradient of 0.2 mT.mm^{-1} generated by 4 rectangularly shaped coils located outside the vacuum chamber. A 1.5 mm diameter and 10 mm long hole in the 2D-MOT chamber allows the beam of slow atoms confined at the center of the quadrupole field through. The slow atom beam is directed toward the center of the MOT, 70 mm away from the output of the 2D-MOT. The MOT is generated at the intersection of three orthogonal pairs of retro-reflected $\sigma^+ - \sigma^-$ polarized laser beams with a diameter of $\sim$ 6.6 mm and a power of $\sim$ 15 mW each. Coils located outside the vacuum chamber generate the magnetic quadrupole field with a gradient of 0.15 mT.mm^{-1} along the strong axis. At a red detuning of -5.5 MHz corresponding to $-4.3\,\Gamma$, this system is capturing $\sim 5 \times 10^6$ atoms of the most abundant ^{202}Hg isotope, with a loading time constant of 2.3 s. ^{199}Hg and ^{201}Hg fermionic isotopes typically show slightly reduced atom numbers.

Laser light for probing the $^1S_0$–$^3P_0$ clock transition is generated by frequency quadrupling the 1062.5 nm output of a distributed feedback (DFB) laser diode. Up to 7 mW are generated at the transition wavelength of 265.6 nm. The DFB laser diode is injection-locked to an ultra-stable laser source based on an Yb-doped DFB fiber laser stabilized to an ultra-stable Fabry-Perot cavity. A fraction of the ultra stable light is used to stabilize an optical frequency comb generated by a Ti:Sa femtosecond laser whose repetition rate is measured against the LNE-SYRTE flywheel oscillator,[6] which is monitored by several primary fountain frequency standards.[18]

Spectroscopy of the clock transition is performed according to the scheme proposed and demonstrated with strontium[19] and further used with ytterbium.[20] An up-going, linearly polarized, retro-reflected probe beam at 265.6 nm is sent through the MOT. Using a mechanical shutter, the 2D-MOT and MOT beams are switched on for an adjustable duration of 20 to 50 ms then off for 5 ms. During the 5 ms, the probe light is pulsed once with adjustable delay and duration using an acousto-optical modulator. This cycle is repeated continuously. The atoms are recaptured after the 5 ms since the atomic cloud falls under gravity by only 122 μm and expands by only 296 μm at the highest observed temperature of 84 μK. The MOT reaches a steady state determined by the loading rate during the MOT on time and the losses induced by the excitation to the long-lived, untrapped $^3P_0$ state. The average fluorescence of the MOT is measured while stepping the probe frequency.

For a probe diameter of 280 μm at $1/e^2$ and a power of 2 mW, typical spectra have a Doppler broadened linewidth ranging from 360 kHz to 550 kHz full-width at half maximum (depending on the MOT parameters including the quality of beam alignment) with MOT depletion close to 50%. The lowest observed linewidth corresponds to a temperature of 36 μK, which is close to the Doppler temperature of 31 μK related to the natural linewidth of the cooling transition. On top of the Doppler profile, we have observed the Doppler-free recoil doublet,[21] whose components are shifted by $\pm\nu_{\text{recoil}}$ with respect to the atomic transition frequency ν, ν_{recoil} being the recoil frequency of the transition, equal to $\nu \times (h\nu/2mc^2)$ or 14.2 kHz, where c is the speed of light, h is Planck's constant and m is the atomic mass. The recoil doublet observed for the two fermionic isotopes ^{199}Hg and ^{201}Hg is shown in figure 3.

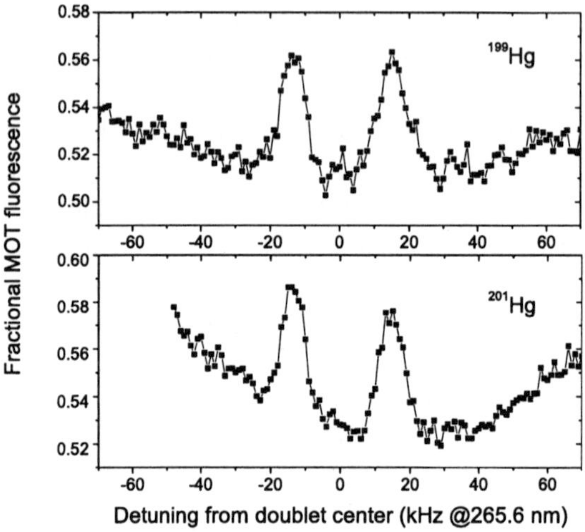

Fig. 3. Recoil doublet observed in the ^{199}Hg (top) and ^{201}Hg (bottom) spectra when a retro-reflected probe beam is used. Data are taken at a rate of one point per second. Spectra have been averaged 4 times.

To determine the transition frequency, we take the center of the recoil doublet. An uncertainty equal to the half-width at half-maximum of the Doppler-free recoil feature is conservatively assigned to this determination. This amounts to a 4.5 kHz uncertainty for both ^{199}Hg and ^{201}Hg. The dominant systematic shift is due to stray magnetic field which adds uncertainties of 3.3 kHz for ^{199}Hg and 2.8 kHz for ^{201}Hg. The measured frequencies are $\nu(^{199}\text{Hg}) = 1\,128\,575\,290\,808.4 \pm 5.6\,\text{kHz}$ for ^{199}Hg and $\nu(^{201}\text{Hg}) = 1\,128\,569\,561\,139.6 \pm 5.3\,\text{kHz}$ for ^{201}Hg.[22] In fractional terms, the uncertainty is $\sim 5 \times 10^{-12}$ which improves upon previous indirect determinations[23–25] by more than 4 orders of magnitude. This completes an important step toward the implementation of the optical lattice clock scheme with neutral mercury.

Acknowledgments

The authors would like to acknowledge support from SYRTE technical services. SYRTE is UMR CNRS 8630 and is associated with UPMC. This work is partly funded by IFRAF and by CNES.

References

1. M. Fischer, N. Kolachevsky, M. Zimmermann, R. Holzwarth, T. Udem, T. W. Hänsch, M. Abgrall, J. Grunert, I. Maksimovic, S. Bize, H. Marion, F. P. D. Santos, P. Lemonde, G. Santarelli, P. Laurent, A. Clairon, C. Salomon, M. Haas, U. D. Jentschura and C. H. Keitel, *Phys. Rev. Lett.* **92**, p. 230802 (2004).
2. M. Niering, R. Holzwarth, J. Reichert, P. Pokasov, T. Udem, M. Weitz, T. W. Hänsch, P. Lemonde, G. Santarelli, M. Abgrall, P. Laurent, C. Salomon and A. Clairon, *Phys. Rev. Lett.* **84**, p. 5496 (2000).
3. C. Roos, M. Chwalla, J. Benhelm, G. Kirchmair, T. Monz, M. Riebe, P. Schmidt, A. Villar, K. Kim, W. Hansel, R. Blatt, M. Abgrall, D. Rovera, P. Laurent and G. Santarelli , Measurement of the absolute transition frequency of the $S_{1/2} - D_{5/2}$ transition in $^{40}Ca^+$ and $^{43}Ca^+$, *22nd European Frequency and Time Forum, Toulouse, France*, (2008).
4. X. Baillard, M. Fouché, R. L. Targat, P. G. Westergaard, A. Lecallier, F. Chapelet, M. Abgrall, G. D. Rovera, P. Laurent, P. Rosenbusch, S. Bize, G. Santarelli, A. Clairon, P. Lemonde, G. Grosche, B. Lipphardt and H. Schnatz, *Eur. Phys. J. D* **48**, p. 11 (2008).
5. S. Blatt, A. D. Ludlow, G. K. Campbell, J. W. Thomsen, T. Zelevinsky, M. M. Boyd, J. Ye, X. Baillard, M. Fouché, R. L. Targat, A. Brusch, P. Lemonde, M. Takamoto, F.-L. Hong, H. Katori and V. V. Flambaum, *Phys. Rev. Lett.* **100**, p. 140801 (2008).
6. D. Chambon, S. Bize, M. Lours, F. Narbonneau, H. Marion, A. Clairon, G. Santarelli, A. Luiten and M. Tobar, *Rev. Sci. Instrum.* **76**, p. 094704 (2005).
7. S. Bize, P. Laurent, M. Abgrall, H. Marion, I. Maksimovic, L. Cacciapuoti, J. Grünert, C. Vian, F. Pereira dos Santos, P. Rosenbusch, P. Lemonde, G. Santarelli, P. Wolf, A. Clairon, A. Luiten, M. Tobar and C. Salomon, *C. R. Physique* **5**, p. 829 (2004).
8. S. Bize, P. Laurent, M. Abgrall, H. Marion, I. Maksimovic, L. Cacciapuoti, J. Grünert, C. Vian, F. Pereira dos Santos, P. Rosenbusch, P. Lemonde, G. Santarelli, P. Wolf, A. Clairon, A. Luiten, M. Tobar and C. Salomon, *Journal of Physics B: Atomic, Molecular and Optical Physics* **38**, S449 (2005).
9. C. Vian, P. Rosenbusch, H. Marion, S. Bize, L. Cacciapuoti, S. Zhang, M. Abgrall, D. Chambon, I. Maksimovic, P. Laurent, G. Santarelli, A. Clairon,

A. Luiten, M. Tobar and C. Salomon, *IEEE Trans. Instrum. Meas.* **54**, p. 833 (2005).

10. X. Baillard, A. Gauguet, S. Bize, P. Lemonde, P. Laurent, A. Clairon and P. Rosenbusch, *Opt. Commun.* **266**, p. 609 (2006).
11. G. Santarelli, G. Governatori, M. Lours, D. Chambon, F. Chapelet, S. Bize, M. E. Tobar, T. Potier and A. Clairon, Phase transient measurement at the microradian level for atomic fountain clocksphase transient measurement at the microradian level for atomic fountain clocks, in *Proc. of the 2006 European Frequency and Time Forum, Braunschweig, Germany*, 2006.
12. D. Chambon, M. Lours, F. Chapelet, S. Bize, M. E. Tobar, A. Clairon and G. Santarelli, *IEEE Trans. Ultr. Ferr. Freq. Contr.Vol.54, Iss.4, April 2007* **54**, p. 729 (2007).
13. V. V. Flambaum and A. F. Tedesco, *Phys. Rev. C* **73**, p. 055501 (2006).
14. (2004), V.G. Pal'chikov, private communication (2004) and H. Katori et al., in proc. of the XIX International Conference on Atomic Physics (2004).
15. M. Petersen, D. Magalhães, C. Mandache, O. Acef, A. Clairon and S. Bize, Towards an optical lattice clock based on neutral mercury, in *Proc. of the 2007 EFTF-FCS joint meeting, Geneva, Switzerland*, 2007.
16. H. Hachisu, K. Miyagishi, S. G. Porsev, A. Derevianko, V. D. Ovsiannikov, V. G. Pal'chikov, M. Takamoto and H. Katori, *Phys. Rev. Lett.* **100**, p. 053001 (2008).
17. T. Rosenband, D. B. Hume, P. O. Schmidt, C. W. Chou, A. Brusch, L. Lorini, W. H. Oskay, R. E. Drullinger, T. M. Fortier, J. E. Stalnaker, S. A. Diddams, W. C. Swann, N. R. Newbury, W. M. Itano, D. J. Wineland and J. C. Bergquist, *Science* **319**, p. 1808 (2008).
18. M. Petersen, J. Millo, D. Magalhães, C. Mandache, S. T. Dawkins, R. Chicireanu, Y. Lecoq, O. Acef, G. Santarelli, A. Clairon and S. Bize, Magneto-optical trap of neutral mercury for an optical lattice clock, in *Proc. of the 2008 Frequency Control Symposium, Honolulu, USA and Proc. of the 2008 European Frequency and Time Forum, Toulouse, France*, 2008.
19. I. Courtillot, A. Quessada, R. Kovacich, A. Brusch, D. Kolker, J.-J. Zondy, G. Rovera and P. Lemonde, *Phys. Rev. A* **68**, p. 030501 (2003).
20. C. W. Hoyt, Z. W. Barber, C. W. Oates, T. M. Fortier, S. A. Diddams and L. Hollberg, *Phys. Rev. Lett.* **95**, p. 083003 (2005).
21. J. L. Hall, C. J. Bordé and K. Uehara, *Phys. Rev. Lett.* **37**, 1339 (1976).
22. M. Petersen, R. Chicireanu, S. T. Dawkins, D. V. Magalhães, C. Mandache, Y. L. Coq, A. Clairon and S. Bize, *Physical Review Letters* **101**, p. 183004 (2008).
23. E. B. Saloman, *Journal of Physical and Chemical Reference Data* **35**, 1519 (2006).
24. K. Burns and K. B. Adams, *J. Opt. Soc. Am.* **42**, p. 716 (1952).
25. K. Burns and K. B. Adams, *J. Opt. Soc. Am.* **42**, p. 56 (1952).

PRECISE MEASUREMENTS OF S-WAVE SCATTERING PHASE SHIFTS WITH A JUGGLING ATOMIC CLOCK

STEVEN GENSEMER, RUSSELL HART, ROSS MARTIN, XINYE XU, RONALD LEGERE, AND KURT GIBBLE

Department of Physics, The Pennsylvania State University, 104 Davey Laboratory University Park, PA 16802, USA

SERVAAS KOKKELMANS

Eindhoven University of Technology, P. O. Box 513, 5600MB Eindhoven The Netherlands

We demonstrate an interferometric scattering technique that allows highly precise measurements of s-wave scattering phase shifts. We collide two clouds of cesium atoms in a juggling fountain clock. The atoms in one cloud are prepared in a coherent superposition of the two clock states and the atoms in the other cloud in an F,mF state. After the two clouds collide, we detect the scattered part of the clock atom's wavefunction for which the phase of the clock coherence is shifted by the difference of the s-wave phase shifts. In this way, we unambiguously observe the differences of scattering phase shifts. These phase shifts are independent of the atomic density to lowest order, enabling measurements of scattering phase shifts with clock accuracy. Recently, we have observed the changes in scattering phase shifts as a function of magnetic field over a range where Feshbach resonances may be expected and inelastic scattering channels open and close. Measurements like these will tightly constrain our knowledge of cesium-cesium interactions. With such knowledge, future measurements may place stringent limits on the time variation of fundamental constants, such as the electron-proton mass ratio, by precisely probing phase shifts near a Feshbach resonance.

1. Introduction

We have demonstrated a fundamentally new type of scattering experiment in which we directly observe scattering phase shifts with clock-like accuracy [1]. Our current statistical uncertainty is as good as 6 mrad. This precision allows us to observe small changes in scattering phase shifts, for example as inelastic scattering channels open and close. We describe the experimental technique and demonstrate that we directly observe the scattering phase shift, which scales differently than the cold collision frequency shift [2]. We also use this picture to analyze frequency shifts for fermionic lattice clocks and show our recent measurements of s-wave phase shifts as a function of magnetic field.

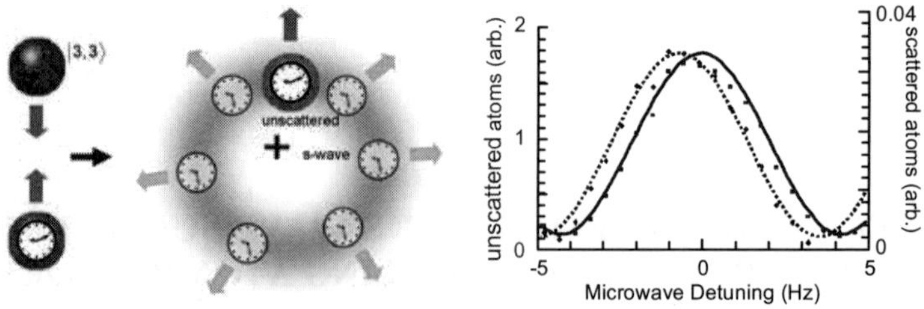

Figure 1. We collide an atom in a coherent superposition of the two cesium clock states with a cesium atom in a pure |F,m⟩ state, such as |3,3⟩. When the clock states scatter, they experience different s-wave phase shifts, shifting the phase of the clock coherence by the difference of the s-wave phase shifts. By detecting only a scattered part of the clock atom's wavefunction, here around 90°, we directly observe the difference of the s-wave phase shifts as a large phase shift of the clock Ramsey fringes (dashed curve). The Ramsey fringe for unscattered clock atoms is the solid curve. Its amplitude is much larger than that for scattered atoms (dashed) since few atoms scatter. Here, the difference of the s-wave scattering phase shifts is $\Phi=\delta_3-\delta_4=-0.35$ for $v_r=8.6$ cm/s.

2. Description of the Experiment

The experiment is schematically illustrated in Fig. 1. We juggle atoms [3] in our cesium fountain clock by launching two clouds of atoms with a short time delay Δt so that the two clouds have a small relative velocity, of order $v_r=g\ \Delta t=10$ cm/s. Both clouds of atoms pass through the microwave clock cavity which puts the atoms in the second cloud into a coherent superposition of the two Cs clock states |Fm⟩=|40⟩ and |30⟩. These atoms collide with the atoms in the first cloud, which are prepared in a pure |Fm⟩ state, such as |44⟩. The spherically incoming s-wave component of the |30⟩ state experiences an s-wave phase shift $\delta_{s,30,Fm}$, or simply δ_3. Similarly, the s-wave component of |40⟩ acquires a phase shift of δ_4. The collision therefore shifts the relative phase of the s-wave scattered clocks states by $\Phi=\delta_3-\delta_4$. After colliding, the atoms return downwardly through the microwave cavity which drives a second $\pi/2$ pulse, converting the phase difference between the clock coherence and the microwave field into a population difference. This gives Ramsey fringes as in Fig. 1b. By detecting only the scattered part of each clock atom's wavefunction, the Ramsey fringes we detect are shifted by $\delta_3-\delta_4$, the difference of the s-wave phase shifts. Thus, we directly observe the difference of s-wave scattering phase shifts with clock-like accuracy.

A key feature of this technique is that the measurement is independent of the atomic density [1]. It is independent of density to 0th order because each scattered clock atom that is detected experiences the two s-wave phase shifts.

The density independence, combined with clock techniques, allows highly precise measurements of the difference of scattering phase shifts which will give unambiguous tests of cesium-cesium interactions and potentially precise tests of fundamental physics.

Highly precise scattering measurements have not been possible because the atomic density n needs to be known to relate the measured number of scattered atoms N during some time interval Δt to the cross section, via $N=n\ v_r\ \sigma\ \Delta t$. To date, the best atomic density measurements do not approach even 1% accuracy [4,5]. Our first measurements had a statistical uncertainty of 7.6 mrad [1] and we expect improvements to 100 μrad accuracy. Our recent measurements have shown a statistical uncertainty of 6 mrad for 20 minutes of data, during which 75% of the time was spent measuring three relatively small backgrounds. With optimal data collection, our current signal-to-noise reduces the averaging time to 5 minutes, or, for 100 μrad, about 2 weeks of averaging.

The phase shift of the scattered atoms' Ramsey fringes is also independent of the interrogation time T between the two Ramsey pulses [1]. This is in contrast to the usual frequency shift in a clock [2] for which Φ increases linearly with the T. As an example, the largest cold-collision frequency shift observed in a Cs clock was 5.5mHz [2], which would be a very small phase shift of a 100 Hz Ramsey fringe (and just a fairly small phase shift of the 1 Hz fringe). Here, the frequency shifts can be very large by comparison; for T=0.127s, the frequency shift is 180 mHz for a relatively small phase shift of $\Phi=-0.14$. Near a Feshbach resonance [6,7], the scattering phase shifts go through π and so the frequency shift of the Ramsey fringes in Fig. 1 will be several Hertz.

The sequence to juggle and obtain Ramsey fringes begins with a vapor cell magneto-optic trap (MOT) repeatedly loading a UHV MOT to collect the first cloud in our juggling fountain [8]. A second cloud is then loaded in the vapor cell MOT, launched and, shortly before it reaches the UHV MOT, the first cloud is launched upwards. Within 1ms, a 9.2 GHz red-detuned moving-frame 3D optical lattice is switched on and, via degenerate Raman sideband cooling, the atoms are cooled to 300-500nK [9] and optically pumped into $|33\rangle$. The second cloud is captured in the UHV MOT, trapped for as little as 1 ms, launched, and then cooled in the optical lattice. The atoms are state-prepared with a series of microwave cavities and pulses of laser beams to purify the state preparation.

3. Brief Theory

We now theoretically describe the scattering of an atom in a coherent superposition of two clock states off of an atom in a pure state $|Fm\rangle$. We relate the familiar cold collision clocks shift and the interferometric scattering to the underlying s-wave phase shifts. In the center of mass frame, we take the wavefunction of the clock atom to be $\psi(\mathbf{r}) = (2v_r)^{-1/2}\,(|30\rangle + i|40\rangle)\exp(ikz)$ after its upward passage through the clock cavity where $v_r = \hbar k/\mu$ and μ is the reduced mass. The clock states acquire s-wave phase shifts as they scatter from atoms in the first cloud, giving:

$$\begin{aligned}\psi(\vec{r}) = {} & \frac{1}{\sqrt{2v_r}}\left[e^{ikz} + e^{i\delta_3}\sin(\delta_3)\frac{e^{ikr}}{kr}\right]|30\rangle \\ & + \frac{i}{\sqrt{2v_r}}\left[e^{ikz} + e^{i\delta_4}\sin(\delta_4)\frac{e^{ikr}}{kr}\right]|40\rangle.\end{aligned} \tag{1}$$

Here δ_3 and δ_4 are the s-wave phase shifts for states $|30\rangle$ & $|40\rangle$ scattering from $|Fm\rangle$.

After scattering, the atoms return back through the microwave cavity where the atoms experience a second $\pi/2$ pulse, but with a variable phase ϕ. Projecting the wavefunction onto the excited state $|40\rangle$, we get $\exp(i\phi/2)\,\langle 30|\psi(r)\rangle + \exp(-i\phi/2)\,\langle 40|\psi(r)\rangle$.

The population of $|40\rangle$ has three contributions. One is from the unscattered part of each clock atom, one from the scattered part, and one from the interference between the scattered and unscattered waves. The unscattered probability current is

$$\vec{j}_{unsc} = \cos^2\left(\phi/2\right)\hat{z}, \tag{2}$$

which represents the usual Ramsey fringes. The interference term is

$$\begin{aligned}\vec{j}_{\rm int} = \frac{\pi}{\rm k^2}\Big(& -2\cos^2\left(\phi/2\right)\left[\sin^2(\delta_3) + \sin^2(\delta_4)\right] \\ & + \frac{1}{2}\sin(\phi)\left[\sin(2\delta_3) - \sin(2\delta_4)\right]\Big)\hat{z},\end{aligned} \tag{3}$$

This term represents the cold collision frequency shift or mean field shift. In the limit of small phase shifts, it reduces to

$$\vec{j}_{\rm int} = \frac{\pi}{\rm k^2}(\delta_3 - \delta_4)\sin(\phi)\,\hat{z}, \tag{4}$$

which is an odd function of ϕ, producing a frequency shift of the Ramsey fringes. This acts as a phase shift of $\delta\Phi = nv_r\Delta t\ 2\pi(\delta_3 - \delta_4)/k^2$ of the Ramsey fringes for atoms detected in the forward direction.

The more interesting term that we detect here is the scattered probability current. By excluding the part of each atom's wavefunction in the forward direction, we detect

$$\vec{j}_{sc} = \frac{\pi}{k^2}\Big[\sin^2(\delta_3) + \sin^2(\delta_4) + 2\sin(\delta_3)\sin(\delta_4)\cos\big(\phi - (\delta_3 - \delta_4)\big)\Big]\hat{r}, \tag{5}$$

Here, the difference of the s-wave scattering phase shifts, $\delta_3 - \delta_4$ is directly observed as a phase shift of the Ramsey fringes. The probability flux in any direction is proportional to the density of scatterers in the first cloud and $a_3^2 + a_4^2$ for small k, where a_i is the s-wave scattering length. If δ_3 and δ_4 are not equal, the fringe contrast is not 100%. But, because the Ramsey fringes represent an interference of the amplitudes of two states, fringes are visible for a wide range of δ_3/δ_4. Even for scattering cross-sections that differ by an order of magnitude, the contrast is more than 60% [10]. Similar expression can be written for higher partial waves, including the interference between different partial waves (e.g. the interference between s & p-waves).

4. Frequency Shifts for Fermion Clocks

Density dependent shifts on the order of the unitary limit are now being seen in lattice clocks based on fermions [11]. Since the temperatures are low, only s-wave scattering is expected and identical fermions cannot have an s-wave interaction. However, if two fermionic atoms experience different Rabi frequencies, there can be an s-wave interaction. In the paradigm of our juggling fountain, here we analyze the collision of two fermionic atoms in two general coherent superpositions $\psi_a(r) = (2v_r)^{-1/2}(\alpha|S\rangle + \beta|P\rangle)\exp[ikz]$, traveling in the forward direction, and, in the backwards direction, $\psi_b(r) = (2v_r)^{-1/2}(\gamma|S\rangle + \delta|P\rangle)\exp[-ikz]$, of the clock states S and P after a Ramsey $\pi/2$ pulse. We take to α and β be real without loss. We first antisymmetrize the total wavefunction of the two atoms, labeled 1 & 2:

$$\psi(\vec{r}_1, \vec{r}_2) = \frac{1}{\sqrt{2}}\big[\psi_a(\vec{r}_1)\psi_b(\vec{r}_2) - \psi_b(\vec{r}_1)\psi_a(\vec{r}_2)\big]. \tag{6}$$

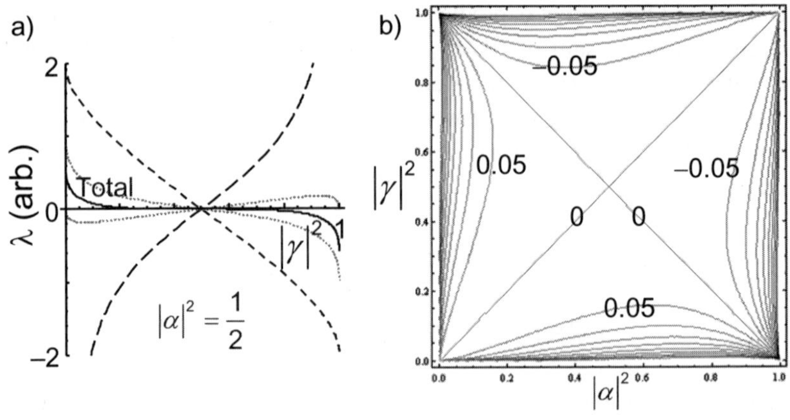

Figure 2. a) s-wave collision shift for two fermions with respective ground state amplitudes α and γ and the same clock phase. The dashed curve represents the shift for the fermion with ground state amplitude $\alpha = 2^{-1/2}$, and the long-dashed that for the other atom. The solid curve represents detecting both equally, giving a well cancelled collision shift. The dotted curves represent a ±30% difference in detection probabilities for the two atoms. b) Contour plot of the total shift on the α, γ plane for equal detection probability. For the shift to be greater than 5% of the unitary limit, one of the two fermions must be nearly in the excited or ground state.

Only the spatially symmetric and internal-state antisymmetric part of this wavefunction can s-wave scatter. Combining Eqs. 1 and 6, with δ_{SP} as the s-wave phase shift, the wavefunction after scattering is

$$\psi^+(\vec{r}_1,\vec{r}_2) = \psi(\vec{r}_1,\vec{r}_2) + (\alpha\delta-\beta\gamma)e^{i\delta_{SP}}\sin(\delta_{SP}) \times \frac{|S\rangle_1|P\rangle_2 - |P\rangle_1|S\rangle_2}{\sqrt{2}} \frac{1}{2v_r}\frac{e^{ik(r_1-r_2)}}{k|r_1-r_2|}\delta(\vec{r}_1+\vec{r}_2). \tag{7}$$

We again apply the second clock $\pi/2$ pulse with a variable phase ϕ. Here, however, we first apply this pulse only in the forward direction and then detect the excited state probability only in the forward direction, tracing over the backwards direction:

$$\vec{j}\bullet\hat{z} = \left[\tfrac{\alpha^2+\beta^2}{2} + \alpha\beta\cos(\phi)\right]\hat{z} + \frac{\pi}{k^2}\Big[(\alpha+\beta)\mathrm{Re}(\alpha\delta-\beta\gamma) \times\sin(2\delta_{SP}) - 2(\alpha+\beta)\mathrm{Im}(\alpha\delta-\beta\gamma)\sin^2(\delta_{SP})\Big]\sin(\phi), \tag{8}$$

where we have neglected terms proportional to $\sin(\delta_{SP})$ and $\sin(\delta_{SP})\cos(\phi)$ since they produce no frequency shift. Here, for γ and δ real, the frequency shift is proportional to $(\alpha+\beta)(\alpha\delta-\beta\gamma)$ divided the fringe contrast $\alpha\beta$. In Fig. 2a, we plot this versus $|\gamma|^2$ for $\alpha=\beta=2^{-1/2}$. For γ near $2^{-1/2}$, the frequency shift is small because the two fermions are nearly identical. The cancellation of the frequency shift is even stronger when we also detect the other particle in the backwards direction. Its frequency shift, by symmetry, is $-(\gamma+\delta)(\alpha\delta-\beta\gamma)/\gamma\delta$ (long dashes in Fig. 2a.). Detecting both atoms, we get a shift proportional to $(\alpha\delta-\beta\gamma)(\alpha+\beta-\gamma-\delta)/(\alpha\beta+\gamma\delta)$ which, plotted in Figs. 2a. (solid) and 2b., is small for a wide range of α and γ. If both atoms aren't detected with the same efficiency, the shift becomes slightly larger as shown by the dotted lines in Fig. 2a for a detection difference of ±30%. It therefore seems that the s-wave frequency shift for a distribution of Rabi frequencies over the sample is small.

It's interesting to note that if the two atoms are excited in different lattice sites, the relative phase of the two clock coherences is different, and of order π. If $\alpha=\beta=\gamma=-\delta$, the above treatment gives a maximal frequency shift. While the tunneling rates are reasonably large for a horizontal lattice[11], when the lattice is vertical, the tunneling between lattice sites is not resonant so one can expect a low tunneling rate [12]. However, for multiple atoms in these lattices, the difference in gravitational energy between lattice sites is small compared to the temperature of the atoms so that tunneling is no longer energetically suppressed. In addition to cold collision shifts, scattering induced tunneling will also lead to large Doppler shifts in the travelling-wave of the clock laser, especially for any atoms that are not in the lowest vibrational state or lattice band. To circumvent these problems, the vertical lattice can be formed with two laser beams at magic wavelength but with an angle between them, as opposed to counter propagating, so that the lattice spacing is equal to the clock wavelength. In such a lattice, all atoms are excited with the same relative phase and there can be no Doppler shifts. This larger spacing also simultaneously allows less tunneling, lower lattice intensity, and reduces other frequency shifts due to the lattice light [13].

5. Feshbach Resonances and Scattering Thresholds

The ability to precisely measure s-wave phase shifts will allow a number of scattering experiments. Several Feshbach resonances for the cesium clock states have been found at magnetic field values below 20 mG by measuring cold collision frequency shifts [14]. Using the cold collision shift is difficult because the phase shifts are so small, especially in comparison to the clock's quadratic Zeeman shift of 427 Hz/G^2. In comparison, for our interferometric technique, as the magnetic field is scanned through a Feshbach resonance, the s-wave phase shift will change by $\pm\pi/2$ so that the frequency shifts can be greater than 1 Hz.

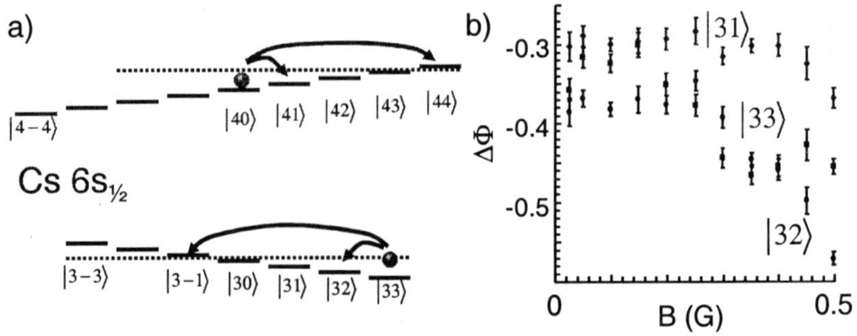

Figure 3. a) Inelastic collisions between Cs hyperfine states |40⟩ & |33⟩ must conserve both z angular momentum (m_F) and energy. At E_c=30 μK and B=0.25 G, the dashed lines show the energy thresholds for Δm=4 changing collisions which are energetically forbidden. b) Preliminary measurements of the difference of s-wave phase shifts ΔΦ for the |3m⟩ state colliding with the clock states as a function of magnetic field at 30 μK. The first and second cloud temperatures are 300 and 500 nK which produces a thermal broadening of ±10 μK. There are a series of thresholds and possibly Feshbach resonances in this range of magnetic fields. Future measurements will extend to higher magnetic fields.

A second effect that we may observe are inelastic threshold effects as a function of magnetic field as depicted in Fig. 3a. Considering the scattering of the |40⟩ state off of |33⟩, all inelastic scattering channels are energetically allowed for small magnetic fields. As the magnetic field increases above 0.23 G, the inelastic collision channel |40⟩|33⟩ → |44⟩|3-1⟩ closes and is no longer energetically permitted. As the field increase further, the |40⟩|33⟩ → |43⟩|30⟩ channel closes, then |40⟩|33⟩ → |42⟩|31⟩, and finally |40⟩|33⟩ → |41⟩|32⟩ closes at 0.89 G. The s-wave phase shift shows a threshold effect as each inelastic channel closes, often accompanied by a change in phase shift that can be 0.1 rad.

Our preliminary measurements of the phase shifts are shown in Fig. 3b for the clock states scattering off of the $|3m_F\rangle$ states. Over a 0.5 G range of magnetic fields, the s-wave phase shifts can change by more than 0.1 rad. These variations could be caused by the opening and closing of inelastic scattering channels or by Feshbach resonances [15]. Measurements are ongoing to extend the range of magnetic fields to greater than 1 G.

6. Conclusions

We have demonstrated a new scattering technique in which we interferometrically observe the difference of s-wave phase shifts. With the high precision that atomic clock techniques enable, we can see small variations in the scattering phase shifts due to Feshbach resonances and the opening and closing of inelastic scattering channels. Precise measurements near Feshbach resonances

[16-18] will highly constrain the interatomic interactions [19] and can also place stringent limits on the time variation of fundamental constants, such as the electron to proton mass ratio [20].

Acknowledgments

We acknowledge stimulating discussions with Boudewijn Verhaar and Eite Tiesinga and financial support from the NSF, NASA, and ONR.

References

1. R. A. Hart, X. Xu, R. Legere, & K. Gibble, Nature **446**, 892 (2007).
2. K. Gibble and S. Chu, Phys. Rev. Lett. **70**, 1771 (1993).
3. R. Legere and K. Gibble, Phys. Rev. Lett. **81**, 5780 (1998).
4. J. R. Ensher, D. S. Jin, M. R. Matthews, C. E. Wieman, and E. A. Cornell, Phys. Rev. Lett. **77**, 4984 (1996).
5. M. T. DePue, C. McCormick, S. L. Winoto, S. Oliver, and D. S. Weiss, Phys. Rev. Lett. **82**, 2262 (1999).
6. H. Feshbach, Ann. Phys. **5**, 357 (1958).
7. E. Tiesinga, B. J. Verhaar, and H. T. C. Stoof, Phys. Rev. A **47**, 4114 (1993).
8. K. Gibble, S. Chang, and R. Legere, Phys. Rev. Lett **75**, 2666 (1995).
9. P. Treutlein, K. Chung, and S. Chu, Phys. Rev. A **63**, 051401 (2001).
10. K. Gibble and B. J. Verhaar, Phys. Rev. A 52, 3370 (1995). See papers in these proceedings by Chris Oates and Jun Ye.
12. P. Lemonde and P. Wolf, Phys. Rev. A. **72**, 033409 (2005)
13. A. V. Taichenachev, V. I. Yudin, V. D. Ovsiannikov, V. G. Pal'chikov, and C. W. Oates, Phys. Rev. Lett. **101**, 193601 (2008).
14. H. Marion, S. Bize, L. Cacciapuoti, D. Chambon, F. Pereira dos Santos, G. Santarelli, P. Wolf, A. Clairon, A. Luiten, M. Tobar, S. Kokkelmans, and C. Salomon, arXiv:physics/0407064, (2006).
15. Eite Tiesinga, private communication (2008).
16. C. Chin, V. Vuletic, A. J. Kerman, and S. Chu, Phys. Rev. Lett. **85**, 2717 (2000).
17. P. J. Leo, C. J. Williams, and P. S. Julienne, Phys. Rev. Lett. **85**, 2721 (2000)
18. C. Chin, V. Vuletic, A. J. Kerman, and S. Chu, E. Tiesinga, P. J. Leo, and C. J. Williams, Phys. Rev. A **70**, 032701 (2004).
19. A. Widera, F. Gerbier, S. Fölling, T. Gericke, O. Mandel and I. Bloch, New J. Phys. **8**, 152 (2006).
20. C. Chin and V.V. Flambaum, Phys. Rev. Lett. **96**, 230801 (2006).

ABSOLUTE FREQUENCY MEASUREMENT OF THE $^{40}Ca^+$ $4s\ ^2S_{1/2} - 3d\ ^2D_{5/2}$ CLOCK TRANSITION

M. Chwalla[†], J. Benhelm[‡], K. Kim[†], G. Kirchmair[‡], T. Monz[†], M. Riebe[†], P. Schindler[†], A. S. Villar[†], W. Hänsel[†], C. F. Roos[‡], and R. Blatt[†,‡,*]

[†] *Institut für Experimentalphysik, Universität Innsbruck*
Technikerstrasse 25, A–6020 Innsbruck, Austria
[*] *E-mail: rainer.blatt@uibk.ac.at*
www.heart-c704.uibk.ac.at
[‡] *Institut für Quantenoptik und Quanteninformation der Österreichischen Akademie der Wissenschaften*
Technikerstrasse 21a, A–6020 Innsbruck, Austria

M. Abgrall, G. Santarelli, G. D. Rovera, and Ph. Laurent
LNE-SYRTE
Observatoire de Paris 61, Avenue de l'Observatoire, 75014 Paris, France

We report on an absolute frequency measurement with a single $^{40}Ca^+$ ion in a linear Paul trap. A frequency comb referenced to the transportable Cs atomic fountain clock of LNE-SYRTE is used to measure the frequency of a laser exciting the $^{40}Ca^+$ $4s\ ^2S_{1/2} - 3d\ ^2D_{5/2}$ electric-quadrupole transition. A transition frequency of $\nu_{Ca^+} = 411\,042\,129\,776\,393.2(1.0)$ Hz was obtained, which corresponds to a fractional uncertainty of 2.4×10^{-15}.

1. Introduction

Fostered by the advent of the optical frequency comb technology, there is strong progress towards the development of optical frequency standards. The best optical frequency standards have recently demonstrated a fractional frequency uncertainty of 10^{-16} or even better for Hg^+, Al^+, and neutral Sr.[1,2] Several other ion candidates have been thoroughly investigated such as Yb^+, In^+, Sr^+,[3–6] but Ca^+[7,8] has not been measured yet. Experiments with single, trapped ions have provided key contributions to the field of optical clocks and the search for a possible time variation of fundamental physical constants in the past 20 years.[9,10]

For an ion clock based on Ca^+ all necessary wavelengths for laser cooling and state manipulation including lasers for photo-ionization can be

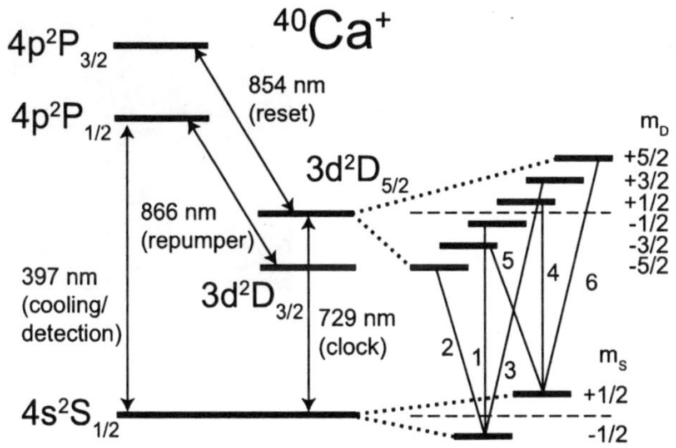

Fig. 1. (a) Partial level scheme of $^{40}Ca^{+}$ showing the $S_{1/2} - D_{5/2}$ clock transition and magnetic sub-levels as well as the transitions used for resetting, repumping, cooling, and detecting the ion.

generated by commercially available solid state lasers. Figure 1 shows the relevant atomic levels of the isotope $^{40}Ca^{+}$. The actual clock transition considered here is the electric quadrupole transition from the $4s\ ^2S_{1/2}$ to the $3d\ ^2D_{5/2}$ level, which has a natural lifetime of 1.17 s.[11] The Zeeman degeneracy is lifted by applying a magnetic field which splits the transition into ten lines. The most important systematic effects like the linear Zeeman and the electric quadrupole shift are canceled by averaging[6] over the six transitions indicated by the numbers.

2. Experimental setup

An overview of our experimental setup is given in Fig. 2 (a). We use a linear Paul trap[13] with four blades separated by 2 mm and two tips of 5 mm separation providing radial and axial confinement. A picture of the trap is shown in Fig. 2 (b). We achieve typical secular trap frequencies of $\omega_r/2\pi$=3.9 MHz in the radial and $\omega_a/2\pi$=1.2 MHz in the axial direction. Single Ca^{+} ions are produced by photo-ionizing a beam of neutral calcium atoms.

A frequency measurement cycle consisted of Doppler cooling on the $S_{1/2} - P_{1/2}$ at a wavelength of 397 nm and repumping at 866 nm to close the cooling cycle if the ion is trapped in the $D_{3/2}$ and at 854 nm to clear out population from the $D_{5/2}$ level. The ion is initialized by optical pumping into $|S_{1/2}, m_S = -1/2\rangle$. After that, laser pulses at 729 nm from a titanium-sapphire laser, which was locked to a high finesse cavity similar in design to

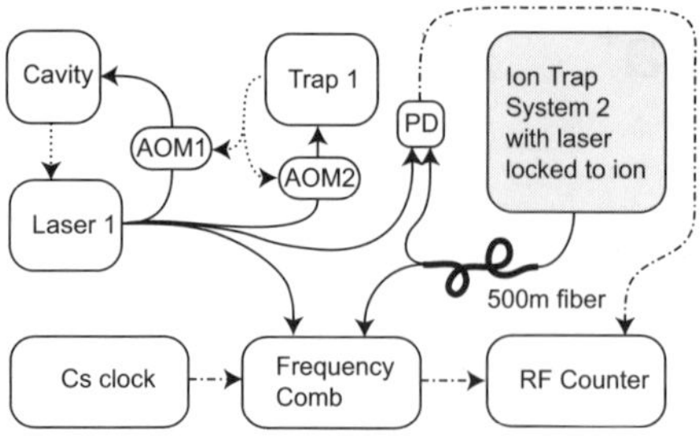

(a)

(b)

Fig. 2. (a) Experiment setup. (b) Ion trap.

the one described in ref.,[16] were used to probe the $S_{1/2} - D_{5/2}$ transition by Ramsey experiments. The laser has a typical linewidth (FWHM) of 50 Hz for time scales exceeding 1 s inferred from an optical beat note with a similar laser system. Long-term drifts are canceled by using feedback from the ion as described below.

Most people involved in this business prefer to use a single pulse excitation scheme. In terms of stability the two methods are almost equivalent.[17] We have chosen the Ramsey method for the following reason: Our initial goal was to perform spectroscopy with entangled states where Ramsey spectroscopy is the only choice. Bell state preparation required individual addressing of the ions for which we used an electro-optical deflector (EOD). Unfortunately, this device induced significant phase errors when switching back and forth between the ions. Therefore, we had to perform the frequency measurement with a single ion and simply kept the measurement strategy. By doing so, we were able to demonstrate that Ramsey's method of separated fields is actually suitable for measurements in the 10^{-15} regime.

The Zeeman multiplet is split into ten components by a constant magnetic field of B =3.087(2) G. In the geometry chosen for the laser polarization, k-vector, and magnetic field[14] only six of those (Δm=0 or $|\Delta m|$=2) could be excited with reasonable coupling strength. To probe transitions starting from $|m_S = +1/2\rangle$, the state was transferred by two π-pulses from $|m_S = -1/2\rangle$. The pulse lengths of the Ramsey pulses were adjusted to $\tau = 50\,\mu$s on all transitions. At a Ramsey probe time of $T_R = 1$ ms, magnetic field fluctuations reduce the contrast to 86% on the most sensitive transition. This is consistent with low-frequency magnetic noise with ΔB_{RMS} =30 μG. The experiments were done asynchronously to the line frequency in order to avoid systematic shifts related to these fluctuations.

State detection of the ion was achieved by electron shelving using a photo multiplier with a discrimination efficiency of well above 99%. This cycle was repeated a hundred times to obtain the mean excitation probability. For determination of the line center of a transition, the phase of the Ramsey pulses was switched by 90° and the experiment was repeated with reversed pulse order. This configuration is free of errors by a systematically incorrectly set phase. The optical phase is set with a precision of better than 0.01° and was checked with the help of an optical beat measurement.

3. The measurement of the $^{40}Ca^+$ $4s\ ^2S_{1/2} - 3d\ ^2D_{5/2}$ transition

The frequency measurement consisted of repeatedly probing the six transitions in the order shown in Fig. 1(b) taking 22 s for each set. The mean excitation probability as well as the radio frequencies of all acousto-optical modulators (AOM) involved were recorded for each setting. In order to calculate the correct frequency deviations the contrast for each transition had to be calibrated and turned out to be constant to better than 1%. The measurement results obtained from transitions 1 and 4 ($|m_S = -1/2\rangle \leftrightarrow |m_D = -1/2\rangle$ and $|m_S = +1/2\rangle \leftrightarrow |m_D = +1/2\rangle$) were frequently used to infer the current laser frequency relative to the ion and the magnetic field strength. The information on the magnetic field was fed back onto the frequency of acousto-optic modulator (AOM2) (see Fig. 2). To compensate for slow drifts of the reference cavity the frequency driving AOM1 was phase continuously stepped according to the measured drift rate. We used a simple model correcting for the two largest systematic shifts, the linear Zeeman and the electric quadrupole shift. Together with a measurement of the quadrupole shift,[15] an evaluation of the whole data set showed that the individual transition frequencies were correctly set with an uncertainty of ±0.4 Hz. The frequency data of the individual transitions together with the Landé g-factor of the $S_{1/2}$ level[19] was used to calculate the g-factor of the $D_{5/2}$ level to be g_D=1.2003340(3).

Part of the probe laser light was sent to a frequency comb for the optical frequency measurement. The frequency comb as well as all radio frequency sources in the experiment were referenced to the transportable Cs fountain atomic clock[18] of LNE-SYRTE which has an accuracy of better than 10^{-15}. All optical fibers involved were stabilized regarding their optical path length. Valid data points for the optical frequency measurement where averaged in the following way: The optical frequency measured by the comb was combined with the AOM frequencies and frequency deviations of six

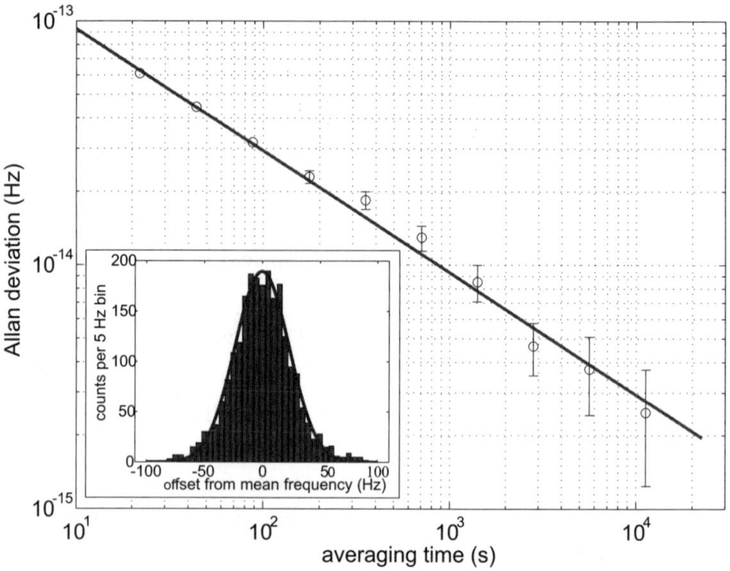

Fig. 3. Frequency instability for the comparison of $^{40}Ca^{+}$ against Cs. The inset shows a histogram of the data.

consecutive Ramsey phase experiments for the transitions in the order 1-6. The second ion trap experiment if operating at the same time could be used to check the validity of the counter results. The optical beat note of the two lasers at 729 nm was compared to a beat note obtained by the difference of the two optical frequencies measured by the comb. It provided an additional means of detecting cycle slips of the counters if the deviation of the two beat signals was larger than 0.5 Hz at times it was available.At the end remaining outliers were removed by rejecting points which were more than 3 σ different from the mean value. The frequency of the $4s\ ^2S_{1/2} - 3d\ ^2D_{5/2}$ transition without correction for systematic shifts could be determined to be 411 042 129 776 395.6(0.5) Hz, limited mainly by the frequency comb stability. The Allan standard deviation for the data is shown in Fig. 3. The solid line represents a fit with $\sigma_y(\tau) = 2.9(1) \times 10^{-13}\ \tau^{-1/2}$. The inset shows a histogram of the deviation from the average frequency which fits a Gaussian distribution with 23(1) Hz standard deviation.

4. Systematic shifts and error budget

Since the linear Zeeman shift, the electric quadrupole shift as well as the AC-Zeeman shift caused by imbalanced currents in the trap electrodes cancel by the chosen measurement scheme, the error of 0.2 Hz is limited by statistics only. The largest remaining shift caused by magnetic fields is due to the second order Zeeman effect. The fine structure splitting of $^{40}Ca^+$ is well known[20] such that the shift could be calculated to be 1.368(1) Hz at a mean magnetic field of 3.087(2) G. A residual magnetic field drift of on average $2(1) \cdot 10^{-8}$ G/s led to a measurement offset of 28(14) mHz because this effect is not completely canceled by averaging over the six transitions without changing their order.

The largest uncertainty in the error budget stems from the AC-Stark shift induced by black body radiation. The ion is exposed to thermal radiation fields emanating from the surrounding vacuum vessel and the ion trap. The latter gets heated up by the applied RF power. Because the tensorial contribution cancels by averaging over all Zeeman transitions, the resulting shifts can be calculated given the static scalar polarizabilities of the $S_{1/2}$ and $D_{5/2}$ states.[21] The actual trap temperature as a function of the applied RF power can only be estimated. With the help of a test setup similar to the ion used trap, we were able to interpolate the temperature. For an input power of 9 W, the trap heats up to about 150±50 °C. Since the trap only covers about one third of the total solid angle while the rest of the vacuum vessel is at 25±2 °C, we assume the black body shift to be 0.9(7) Hz.

All laser sources were switched off during the waiting time of a Ramsey experiment by switching off the RF for the AOMs used to detune the individual lasers. Since this cannot be done perfectly the residual light fields are expected to cause AC Stark shifts. The probe laser leads to an AC-Stark shift $\delta_{probe} = \delta_{AC}/(\frac{\pi T_R}{4\tau} + 1)$ during the Ramsey experiment, where T_R is the probe time and δ_{AC} the light shift during the time τ of the Ramsey pulses. This is mostly caused by non-resonant excitation of dipole transitions.[14] For our parameters, the expected shift was 0.11(2) Hz. The light shift induced by the residual light of the resetting laser at 854 nm could be measured to be -0.005(4) Hz. The shifts by the cooling laser at 397 nm and the repumping laser at 866 nm are estimated by assuming that the residual amount of light is focused tightly onto the ion. For the cooling laser we assume a detuning of half a linewidth where the shift would be maximum. The corresponding shifts and the uncertainties are well below 0.1 Hz. Excess micromotion induced by the trapping field[22] is responsible for second order Doppler and AC-Stark shifts but has been compensated with the help of additional DC electrodes. We achieved modulation indices smaller than 1%, therefore the micromotion-related frequency errors are 0.1 Hz at most. Doppler and gravitational shifts are negligible at this level.

Table 1. An overview of the most important systematic shifts.

Effect	Shift (Hz)	Error (Hz)	Fractional error ($\times 10^{-15}$)
Statistical error	-	0.5	1.2
1st-order Zeeman	0	0.2	0.5
Magnetic field drift	0.03	0.01	0.024
2st-order Zeeman: quantization field	1.368	0.002	0.005
2st-order Zeeman: black body	<0.001	<0.001	<0.002
Electric quadrupole	0	0.2	0.5
AC Stark: probe laser	0.11	0.02	0.05
AC Stark: quenching laser	-0.005	0.004	0.01
AC Stark: repumping laser	0	0.01	0.02
AC Stark: cooling laser	0	0.1	0.2
AC Stark: black body radiation	0.9	0.7	1.7
AC Stark: micromotion	0	0.01	0.02
2nd order Doppler	-0.0007	0.0007	0.002
Graviational shift	0.021	0.004	0.01
Ramsey phase error	0	0.03	0.07
Cs uncertainty	0	<0.4	<1
Total shift	**2.4**	**1.0**	$2.4\ 10^{-15}$

The errors sources described above are listed in Tab. 1. The total shift is 2.4 Hz with an error of 1.0 Hz. Taking into account these systematic effects the corrected value of the $4s\ ^2S_{1/2} - 3d\ ^2D_{5/2}$ transition therefore is $\nu_{Ca^+} = 411\,042\,129\,776\,393.2(1.0)$ Hz, corresponding to a $2.4 \cdot 10^{-15}$ relative uncertainty. To our knowledge this is the most accurate frequency measurement of any Ca^+ transition so far.[23]

5. Conclusion

We have performed an absolute frequency measurement on the $4s\ ^2S_{1/2} - 3d\ ^2D_{5/2}$ clock transition of $^{40}Ca^+$ with the help of an optical frequency comb referenced to a transportable Cs fountain clock. The applied Ramsey method led to a measured frequency of $\nu_{Ca^+} = 411\,042\,129\,776\,393.2(1.0)$ Hz and 2.4×10^{-15} relative uncertainty. It appears realistic that an experiment with $^{40}Ca^+$ especially designed for metrology could lead to an accuracy of better than 10^{-16}–10^{-17}.

Our experiment is mainly dedicated to processing quantum information, a field of research closely related to metrology: a pair of entangled ions can be used for improving the signal-to-noise ratio[24,25] but also for designing states which are immune to certain kinds of environmental perturbations.[12,15] A state like $|S_{1/2}, m_S\rangle\, |S_{1/2}, -m_S\rangle + |D_{5/2}, m_D\rangle\, |D_{5/2}, -m_D\rangle$ which is to first order immune against magnetic field fluctuations could be used for generalized Ramsey experiments with probe times limited only by the probe laser stability and spontaneous decay of the metastable state. Preliminary tests with such states showed significant phase errors caused by the EOD used for individual addressing of the ions. To overcome this problem, high-fidelity entanglement[26] could be generated with a laser beam collectively interacting with the ions, and by combining this wide beam with a second strongly focused laser inducing phase shifts to make the ions distinguishable for carrying out coherent operations on a single ion.

Acknowledgments

This work was supported by the Austrian Science Fund (FWF), the Austrian Academy of Sciences, IARPA, the European network SCALA, and the Institut für Quanteninformation GmbH. The mobile fountain work is supported by CNES, CNRS, LNE and région Ile de France (IFRAF). The SYRTE is a unit associated to CNRS UMR 8630. We thank P. O. Schmidt and H. Häffner for fruitful discussions.

References

1. A. D. Ludlow *et al.*, Science **319**, 1805 (2008).
2. T. Rosenband *et al.*, Science **319**, 1808 (2008).
3. J. Stenger *et al.*, Opt Lett. **26**, 1589 (2001).
4. P. J. Blythe *et al.*, Phys. Rev. A **67**, 020501(R) (2003).
5. J. von Zanthier *et al.*, Opt. Lett. **25**, 1729 (2000).
6. H. S. Margolis *et al.*, Science **306**, 1355 (2004).
7. C. Champenois *et al.*, Phys. Lett. A **331**, 298 (2004).
8. M. Kajita, Y. Li, K. Matsubara, K. Hayasaka, and M. Hosokawa, Phys. Rev. A **72**, 043404 (2005).
9. A. A. Madej and J. E. Bernard, in *Frequency Measurement and Control* (ed. A. N. Luiten) pp.153–195 (Topics in Applied Physics Vol. 79, Springer, Berlin, 2001).
10. S. N. Lea, Rep. Prog. Phys. **70**, 1473 (2007).
11. P. Barton *et al.*, Phys. Rev. A **62**, 032503 (2000).
12. C. F. Roos, M. Chwalla, K. Kim, M. Riebe, and R. Blatt, Nature **443**, 316 (2006).
13. F. Schmidt-Kaler *et al.*, Appl. Phys. B **77**, 789 (2003).
14. H. Häffner *et al.*, Phys. Rev. Lett. **90**, 143602 (2003).
15. M. Chwalla *et al.*, Appl. Phys. B **89**, 483 (2007).
16. M. Notcutt, L.-S. Ma, J. Ye, and J. L. Hall, Opt. Lett. **30**, 1815 (2005).
17. E. Peik and T. Schneiderl and Ch. Tamm, J. Phys. B **39**, 145 (2006)
18. S. Bize *et al.*, J. Phys. B **38**, S449 (2005).
19. G. Tommaseo *et al.*, Eur. Phys. J. D **25**, 113 (2003).
20. R. Yamazaki *et al.*, Phys. Rev. A **77**, 012508 (2008).

21. B. Arora, M. S. Safronova, and C. W. Clark, Phys. Rev. A **76**, 064501 (2007).
22. D. J. Berkeland, J. D. Miller, J. C. Bergquist, W. M. Itano, and D. J. Wineland, J. Appl. Phys. **83**, 5025 (1998).
23. K. Matsubara *et. al.*, Appl. Phys. Express **1**, 067011 (2008).
24. J. J. Bollinger, W. M. Itano, D. J. Wineland, and D. J. Heinzen, Phys. Rev. A **54**, R4649 (1996).
25. D. Leibfried *et al.*, Science **304**, 1476 (2004).
26. J. Benhelm, G. Kirchmair, C. F. Roos, and R. Blatt, Nat. Phys. **4**, 463 (2008).

THE SEMICLASSICAL STOCHASTIC-FIELD/ATOM INTERACTION PROBLEM *

JAMES CAMPARO

Physical Sciences Laboratories, The Aerospace Corporation
PO Box 92957, Los Angeles, CA 90009, USA

No field in nature truly corresponds to the classical monochromatic ideal. Even the "singlemode" laser can be, in certain regards, a poor approximation: the mode has a width that depends on the photon's cavity lifetime, and amplified-spontaneous-emission produces fluctuations in the laser's output. With regard to metrology, the question is twofold: at what level of precision (stability) does a classical field's stochastic nature become relevant in field/matter interactions, and how does the field's stochastic nature manifest itself in measurements. Unfortunately, answers to those questions are only known in a few specific cases, and general intuitive models are lacking. Here, we provide a short overview of the semiclassical stochastic-field/atom interaction, illustrating how the stochastic nature of a classical field can have a significant effect on spectroscopic signals, and how those effects are not always intuitively obvious. In concluding, we argue that simple, general, intuitive models of the stochastic-field/atom interaction are needed, so that researchers can be alerted as to when and how stochastic-field effects may manifest themselves in precision measurements.

1. Introduction

No field in Nature is truly monochromatic: the most ubiquitous light source on the planet, the sun, gives off thermal radiation, and the singlemode laser's emission has a finite linewidth and emits photons in cavity modes other than the main lasing mode. Nevertheless, pick up almost any introductory or advanced quantum text; turn to the section on radiative field/matter interactions, and in nearly all cases the discussion will be restricted to monochromatic fields. This is not to say that treating fields as monochromatic in the semiclassical regime is wrong. Rather, it limits what is perhaps a richer and more varied subject.

Typically, we account for field stochasticity through the convolution picture. Briefly, a stochastic field will always have some "non-delta function" power spectrum, and we describe the atom's response to the stochastic field,

* This work was supported under The Aerospace Corporation's Mission Oriented Investigation and Experimentation program, funded by U.S. Air Force Space and Missile Systems Center under Contract No. FA8802-04-C-0001.

$L(\omega_{atom}-\omega_{field})$ in terms of a convolution between the field spectrum, $P(\omega-\omega_{field})$ and the atom's absorption cross-section, $\sigma(\omega-\omega_{atom})$:

$$L(\omega_{atom} - \omega_{field}) = \int P(\omega - \omega_{field})\sigma(\omega - \omega_{atom})d\omega \,. \tag{1}$$

For example, if we have a positively-skewed field spectrum, as might arise for a field with correlated amplitude and frequency fluctuations, it will produce a negatively-skewed atomic-response function, perhaps a fluorescence lineshape. Moreover, the asymmetry will result in a frequency shift of the atom's response: for a positively-skewed field spectrum we will get a negative frequency shift.

The problem with a convolution picture of the stochastic-field/atom interaction is that it has well-defined limits of applicability. In mathematics, convolutions in frequency arise when we consider the Fourier transform of a quantity that is the product of two temporal functions. In other words, the convolution picture requires the atom's response to the field to be described as the product of an instantaneous absorption cross section and an instantaneous field intensity, which limits it to first-order linear processes. Obviously, the same conclusion is reached when we approach the problem through time-dependent perturbation theory.

2. Stochastic-Field Effects and an Atom's Average Response

2.1. *Multi-photon Excitation*

The first place we might expect the convolution picture to break down is with multiphoton processes: radiative effects that require high intensity and are by definition not first order. If we consider a multimode field, where the modes have uncorrelated phases, then the intensity will show deep fades and extreme spikes all about some average intensity, $\langle I\rangle$, as illustrated in the inset to Fig. 1. The rate of N-photon excitation, and hence the multiphoton signal, S_N, will be equal to a generalized multiphoton cross section, σ_N, and the Nth moment of the field intensity (i.e., $S_N \sim \sigma_N\langle I^N\rangle$) [1]; and for a sufficient number of independent modes ($\sim 10^2$), the central limit theorem implies that the total amplitude of the field will have a normal probability distribution. Thus, $\langle I^N\rangle = N!\langle I\rangle^N$, and we find that an N-photon signal in a stochastic field can be N! times larger than the signal from a monochromatic field of the same average intensity.

This striking prediction was verified in an experiment by Lecompte et al. a number of years ago [2]: a 20 nsec pulsed Nd:Glass laser that could be operated with one to 100 modes was employed for 11-photon ionization of xenon. Essentially, the number of Xe^+ produced was measured as the number of modes in the pulse varied, and this was compared to the number of Xe^+ produced for a single-mode pulse of the same average intensity. The results are shown in Fig. 1: as the number of modes increased, the ionization signal was enhanced by a factor that asymptotically approached $11! = 4\times10^7$... *a 7 orders-of-magnitude increase in ionization efficiency due to the field's stochastic nature.*

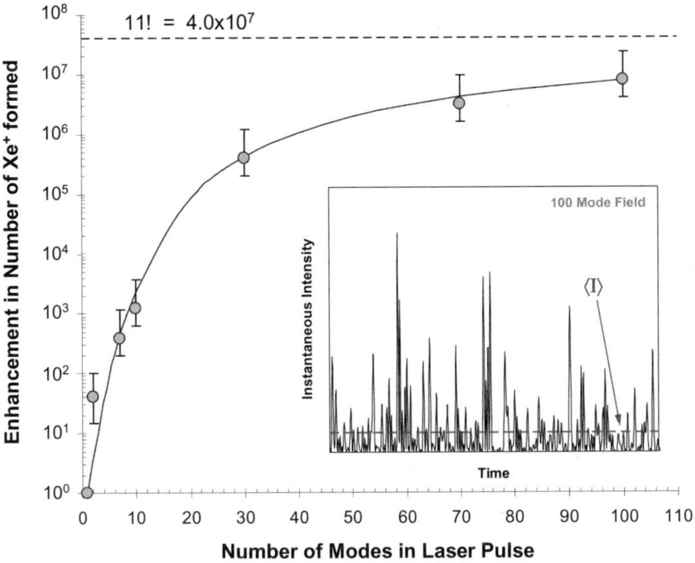

Figure 1: Enhancement of 11-photon ionization of Xe with an amplitude-fluctuating stochastic field; simulated realization of intensity variations shown in inset for illustrative purposes.

2.2. *AC-Stark or Autler-Townes Splitting*

The stochastic nature of a field can also play a profound role in the qualitative nature of the field/atom interaction. As illustrated in Fig. 2, when states $|1\rangle$ and $|2\rangle$ of a three level system are coupled by a strong field, a weak probe tuned around the $|2\rangle \rightarrow |3\rangle$ transition produces a doublet (i.e., the Autler-Townes doublet). The origin of this doublet is best understood through the dressed atom energy level structure of the atom [3]: For $\omega_s < \omega_{21}$, the eigenfunctions of the dressed atom consist of an upper state (with predominant $|2\rangle$ character) and a lower state (with predominant $|1\rangle$ character), so that we can refer to the lower energy resonance as a "sequential" excitation resonance (i.e., $|1\rangle \rightarrow |2\rangle \rightarrow |3\rangle$) and the upper energy resonance as a direct two-photon resonance (i.e., $|1\rangle \rightarrow |3\rangle$). Though the 2-photon process has an "intrinsically" smaller probability of occurrence than the sequential process, the sequential resonance is nonetheless smaller because there are very few atoms in $|2\rangle$ when the laser is off resonance.

The situation is completely different for a field suffering phase fluctuations. In that case, the instantaneous frequency fluctuations of the strong field yield a much larger probability for real transitions into |2⟩ even when the strong field is (on-average) off resonance. As a consequence, the asymmetry of the Autler-Townes doublet reverses, and the sequential excitation resonance becomes larger than the direct 2-photon resonance. This stochastic-field effect was first observed and explained by Hogan et al. [4] and latter studied in detail [5].

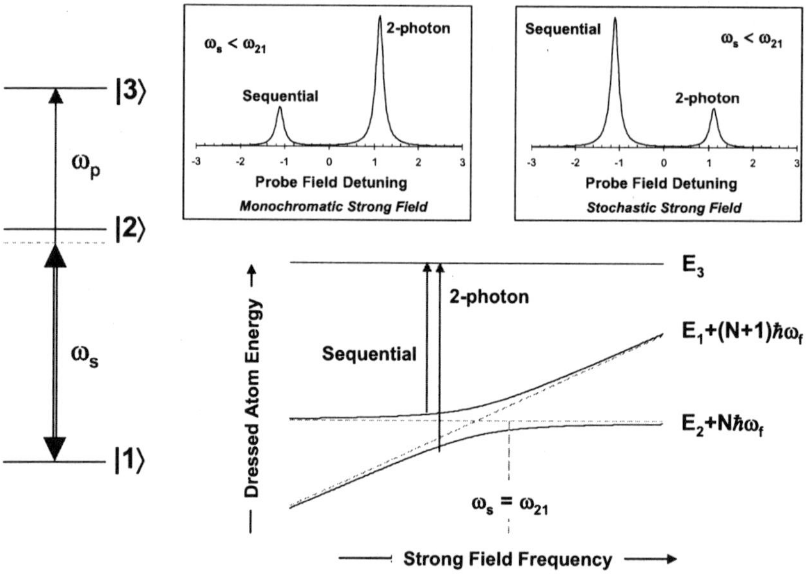

Figure 2: A 3-level system excited by a strong field & weak probe, and the resulting Autler-Townes doublet. The doublet asymmetry reverses when the strong field suffers white frequency fluctuations.

3. Stochastic-Field Effects and an Atom's Dynamic Response

While the mean response of an atom to a stochastic field is important in determining a spectroscopic signal's amplitude and linewidth, it is necessary to remember for precision spectroscopy that stochastic fields will induce stochastic atomic responses, and that these affect signal-to-noise ratios. The importance of the atom's dynamic response to a stochastic field was brought home forcefully to the frequency standards community in the 1990s, during attempts to replace the rf-discharge lamp in Rb vapor-cell clocks with diode lasers [6].

In one experiment [7], the signal amplitude and noise for a commercial lamp-pumped clock were measured. Then, the lamp was removed, replaced with a diode laser ($\Delta\nu_{laser} \sim 50$ MHz), and the signal amplitude and noise remeasured. As anticipated, laser pumping was much more efficient than lamp pumping for clock signal generation, with the clock signal increasing by more than an order of magnitude. (The linewidth of the hyperfine resonance was essentially unchanged.) Surprisingly, however, the noise for the laser-pumped clock signal was more than two orders of magnitude larger, so that the clock's signal-to-noise ratio was actually *poorer* with the laser.

The origin of the increased noise was eventually traced to the influence of laser phase noise on the atomic vapor's absorption cross section [8]. Anytime two quantum states are coupled by a field, a lower-state/upper-state superposition is created, and one can show that fluctuations in laser phase yield fluctuations in the superposition state's expansion coefficients. The expansion coefficients, however, define the atom's absorption cross section. If the coefficients vary randomly, so too does the absorption cross section, and this in turn produces stochastic fluctuations in the transmitted light intensity. Consequently, the absorption process yields laser phase-noise (PM) to transmitted intensity-noise (AM) conversion. This process is intrinsically atomic: it cannot be avoided; it can only be mitigated. Researchers have had success mitigating PM-to-AM conversion either by reducing the magnitude of the laser phase fluctuations (i.e., reducing the laser linewidth) [9], or by making the atoms less sensitive to phase fluctuations through pressure broadening [10].

4. Intuitive Models of the Stochastic-Field/Atom Interaction

4.1. *Rabi Resonances*

All of the isolated examples of the stochastic field's influence on radiative interactions discussed above are exactly that: isolated examples. What is sorely lacking in the area of semiclassical stochastic-field/atom interactions is a cohesive, *intuitive* picture that experimentalists can employ to guide their thinking, identify potential problems in particular experiments, and illuminate areas of research that might broaden our general understanding of radiative interactions. As an example of the type of experiment that may eventually lead to that intuitive picture, investigations of Rabi resonances appear promising.

Rabi resonances are generally described as enhancements in the amplitude of population oscillations when an atom interacts with a time-varying electromagnetic perturbation, and when the perturbation's period matches the Rabi period, $2\pi/\Omega$ [11]. To illustrate how these may lead to a better understanding of the stochastic-field/atom interaction, Camparo et al. [12]

employed a diode laser (Fig. 3) to optically pump a vapor of Rb^{87} atoms in a TE_{011} microwave cavity (contained with 10 torr of N_2). Optical pumping created a population imbalance between the two ground-state hyperfine levels, so that when microwaves (resonant with the 0-0 hyperfine transition) were applied to the sample, atoms returned to the absorbing state with a consequent decrease in the transmitted light intensity, I_{trans}. Thus, the atoms' response to a fluctuating microwave field could be studied by examining I_{trans} with a spectrum analyzer.

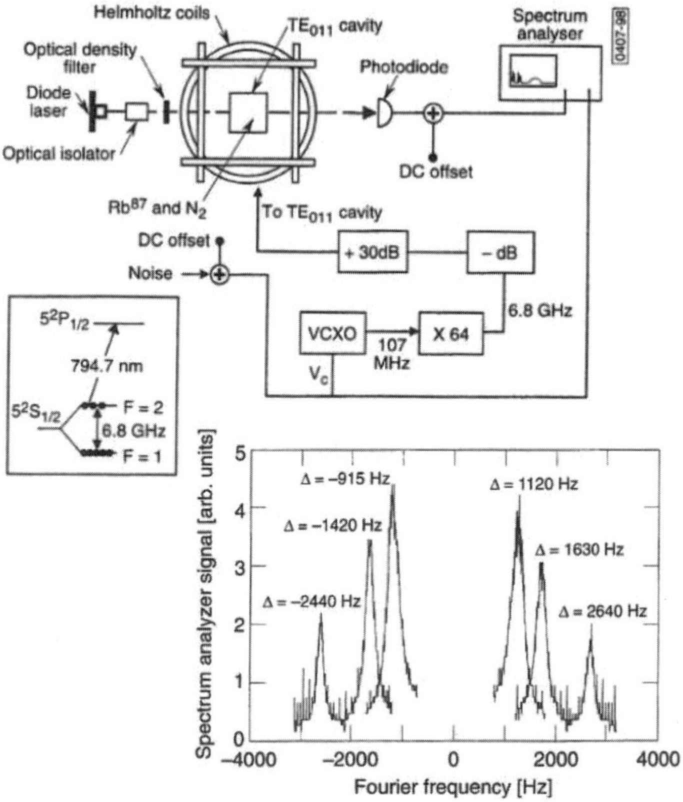

Figure 3: Experiment examining the effect of white frequency fluctuations on a 2-level quantum system: the Rb^{87} 0-0 transition was a surrogate for all two-level systems. Though the frequency fluctuations were white, the atom's response exhibited a "bright line" at Fourier frequencies equal to the Rabi nutational frequency. (VCXO = Voltage-Controlled Crystal-Oscillator).

In the experiment, the Rb^{87} 0-0 hyperfine resonance was used as an illustrative 2-level atom interacting with a stochastic field. Consequently, in addition to the dc voltage, V_c, which controlled the microwave detuning from resonance, Δ, the detuning was randomly modulated by adding a "white" noise voltage to V_c. The atoms' response to the stochastic field was then studied via a spectrum analyzer. Interestingly, the atoms exhibited enhanced "noise" at the Rabi nutational frequency: $[\Omega^2+\Delta^2]^{1/2}$ as shown in the inset of Fig. 3. These results are consistent with a linearized theory of the stochastic-field/atom interaction, which shows that the atom responds to the field like a damped, driven, harmonic oscillator with the Rabi frequency playing the role of the oscillator's resonant frequency [13,14].

4.2. *Bloch-Vector Trajectory Entropy*

In addition to novel experimental investigations of the semiclassical stochastic-field/atom interaction, new theoretical tools will also be necessary if we are to develop an intuitive understanding of the problem. In this vein, Fig. 4 shows theoretical results examining an atom's Bloch-vector trajectory when interacting with a field whose frequency suffers a 2-frequency quasi-periodic variation. In particular, the atom's response was studied in Avan and Cohen-Tannoudji's "instantaneous" frame of Bloch-vector trajectories [15]. This reference frame is essentially the standard rotating frame of resonance problems, but with an additional rotation that keeps the frame's x-axis aligned with the effective field. In the instantaneous frame, trivial aspects of the field's random variations are removed from consideration, and what is left is the atom's intrinsic response to the stochastic field.

The interesting feature of these trajectories is that in general they take on essentially two very specific geometrical patterns: a weak-field pattern and a strong-field pattern. Insight into these trajectories can be gained by calculating a Bloch-vector trajectory entropy, S_B: $S_B = \Sigma p_i \ln(p_i)$, where p_i is the probability of the Bloch-vector trajectory falling within the ith infinitesimal volume element of (instantaneous-frame) Bloch space [16]. The results of this Bloch-vector trajectory entropy analysis are also shown in Fig. 4, where S_B and the amplitude of the atom's population oscillations are plotted as a function of Ω.

Note that in the transition from weak to strong stochastic fields S_B reaches a maximum. Thus, there is what amounts to an "entropy barrier" separating the behavior of Bloch-vector trajectories in weak fields from their behavior in strong fields. (Does this imply that by somehow "clamping" the Bloch vector trajectory's entropy, an atom's stochastic response to a field can be limited?) Moreover, increases in entropy correspond to increases in the amplitude of population variations, so that these amplitude enhancements must come at the expense of what can only be described as randomizations of the instantaneous frame's Bloch-vector trajectory. Therefore, at least with regard to phase-fluctuating/amplitude-stable fields, we may have arrived at a fairly general and relatively simple picture of the stochastic-field/atom interaction, a picture in which the Rabi nutational frequency plays a primary role [13,14].

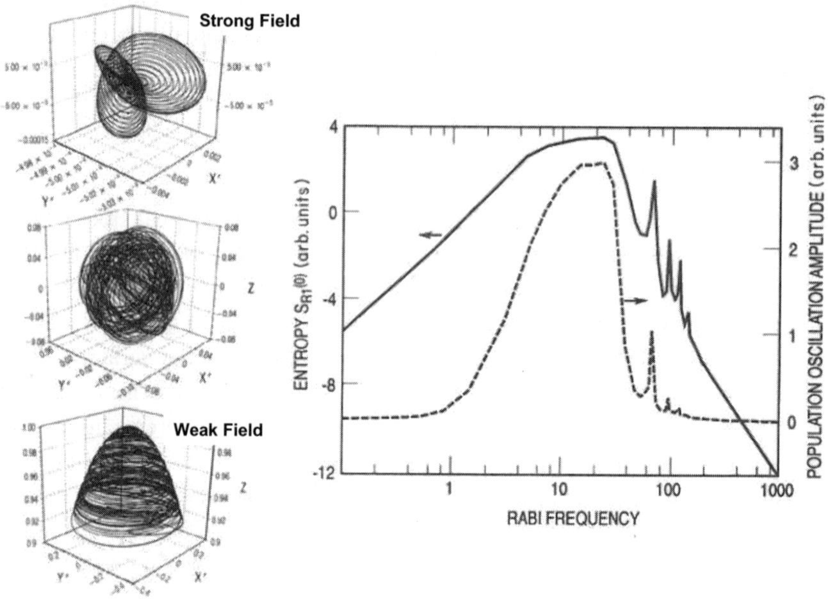

Figure 4: Left - Instantaneous frame Bloch-vector trajectories for an atom interacting with a field suffering 2-frequency quasi-periodic phase variations; Right - Bloch-vector trajectory entropy showing that the transition from weak field to strong field is accompanied by an entropy increase.

5. Summary & Conclusions

In the preceding sections, I have tried to show that the stochastic-field/atom interaction can play an important role in precision spectroscopy, and that stochastic field effects cannot always be adequately captured by a convolution picture of the problem. This is not to say that just because a field is non-monochromatic we must discard our well-worn quantum texts: many lasers come exceeding close to the monochromatic ideal, and not all radiative problems will be sensitive to the field's stochastic nature. Nevertheless, as we push to unprecedented levels of spectroscopic precision, it may be important to keep in mind that the stochastic nature of a field can change both the *quantitative* and *qualitative* aspects of radiative interactions.

Moreover, there is a real need for general intuitive models of the stochastic-field/atom interaction. Experimentalists need to know when and where a "cavalier" treatment of the problem is fair, and if not fair then what effects are anticipated; and they need answers to these questions without launching their own research effort into the stochastic-field/atom interaction problem. At present, it appears that for phase-fluctuating/amplitude-stable fields the Rabi frequency provides a key parameter: the atom responds to the field as if it were a damped driven harmonic oscillator, with its resonant frequency equal to Ω, and Bloch-vector trajectories in the instantaneous frame have easily categorized patterns quantified by the trajectory's entropy. What we don't intuitively understand well is the atom's response to an *amplitude-fluctuating* field. In particular, how do we define a Rabi frequency for such a field that will allow us to categorize the atom's dynamics as non-adiabatic (i.e., weak field) or adiabatic (i.e., strong field)?

References

1. P. Lambropoulos, *Comments At. Mol. Phys.* **20**, 199 (1987).
2. C. Lecompte, G. Mainfray, C. Manus, and F. Sanchez, *Phys. Rev. A* **11**, 1009 (1975).
3. S. Reynaud and C. Cohen-Tannoudji, *J. Physique* **43**, 1021 (1982).
4. P. B. Hogan, S. J. Smith, A. T. Georges, and P. Lambropoulos, *Phys. Rev. Lett.* **41**, 229 (1978).
5. M. W. Hamilton, K. Arnett, S. J. Smith, D. S. Elliott, M. Dziemballa, and P. Zoller, *Phys. Rev. A* **36**, 178 (1987).
6. J. Camparo, Physics Today **60**, 33 (2007).
7. J. Camparo and W. Buell, in *Proc. 1997 IEEE Freq. Control Symp.* (IEEE, Piscataway, NJ, 1997) pp. 253-258.
8. J. Coffer, M. Anderson, and J. Camparo, *Phys. Rev. A* **65**, 033807 (2002).
9. C. Affolderbach, F. Droz, and G. Mileti, *IEEE Trans. Instrum. & Meas.* **55**, 429 (2006).
10. J. Camparo, J. Coffer, and J. Townsend, *J. Opt. Soc. Am. B* **22**, 521 (2005).
11. J. Camparo and R. Frueholz, *Phys. Rev. A* **38**, 6143 (1988).
12 J. Camparo, J. Coffer, and R. Frueholz, *Phys. Rev. A* **58**, 3873 (1998).
13. R. Frueholz and J. Camparo, *Phys. Rev. A* **52**, 472 (1995).
14. R. Frueholz and J. Camparo, *Phys. Rev. A* **54**, 3499 (1996).
15. P. Avan and C. Cohen-Tannoudji, *J. Phys. B: Atom. Molec. Phys.* **10**, 155 (1977).
16. R. Frueholz and J. Camparo, *Phys. Rev. A* **47**, 4404 (1993).

PHASE AND FREQUENCY NOISE METROLOGY

E. RUBIOLA*, V. GIORDANO, K. VOLYANSKIY, L. LARGER

FEMTO-ST Institute (UMR 6174)
CNRS and Université de Franche Comté
Besançon, France
** E-mail: rubiola@femto-st.fr*
** http://rubiola.org*

As a consequence of a general trend in the physics of oscillators and clocks towards optics, phase and frequency metrology is rapidly moving to optics too. Yet, optics is not replacing the traditional radio-frequency (RF) and microwave domains. Instead, it adds tough challenges.

Precision frequency-stability measurements are chiefly based on the measurement of phase noise, which is the main focus of this article. Major progress has been achieved in two main areas. The first is the extreme low-noise measurements, based on the bridge (interferometric) method[1,2] in real time or with sophisticated correlation and averaging techniques.[3,4] The second is the emerging field of microwave photonics, which combines optics and RF/microwaves. This includes the femtosecond laser, the two-way fiber links,[5] the noise measurement systems based on the fiber[6] and the photonic oscillator.[7,8] Besides, the phenomenology of flicker ($1/f$) noise is better understood, though the ultimate reasons are still elusive.

Keywords: Phase noise, Frequency stability, Spectral purity, Oscillator, Microwave photonics.

1. A quick look high-sensitivity measuements

Before getting through the basic principles, let us look at the phase noise spectrum of Figure 1, approximated with the power-law $S_\varphi = \sum_i b_i/f^i$. The device under test is a microwave circulator with the output taken at the isolation port. Isolation results from the destructive interference between two counter-propagating modes in the ferrite bulk. This configuration is the same used in the well known Pound oscillator.[10] The experimentalist familiar with phase noise should notice the following unusual facts.

(1) The low $1/f$ noise of the DUT, $10^{-17}/f$ rad^2/Hz, measured without need of correlation.

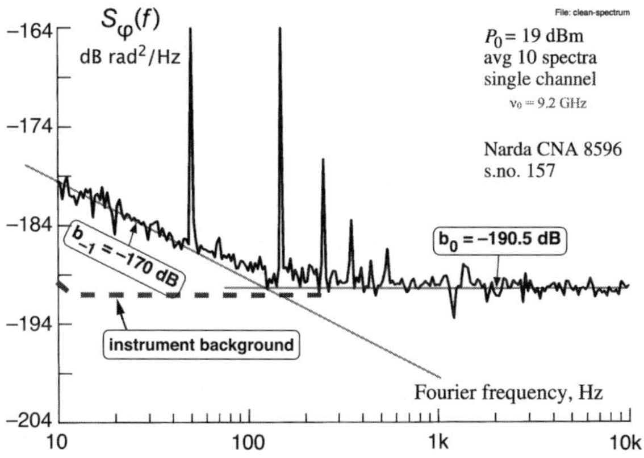

Fig. 1. Phase noise spectrum of a ferrite circulator measured at 9.2 GHz.[9]

(2) The low background noise, $10^{-18}/f + 9{\times}10^{-20}$ rad^2/Hz.
(3) The low level of the residual of the mains, 50 Hz and odd multiples. The even-order harmonics are so small that they are not visible.
(4) The smoothness of the plot, achieved with a small number of averaged samples ($m = 10$).

The background noise deserves more attention. The conversion between spectrum and Allan variance, which is a simple mathematical process, can be done with to any physical quantity. The formula $\sigma^2(\tau) = 2\ln(2)\,h_{-1}$ applies to flicker $S(f) = h_{-1}/f$. Therefore, after converting the phase spectrum $S_\varphi = b_{-1}/f$ into a length-fluctuation spectrum $S_L = h_{-1}/f$, the Allan deviation is $\sigma_L(\tau) = 4.5{\times}10^{-12}$ m. This high stability requires to re-think the mechanical design of electronics. That said, a resolution of parts in 10^{-14} m is common in the field of tunnel and atomic-force microscopy.[11] Also the femtosecond laser owes its success to the unexpected mechanical stability of the optical assembly.

2. Bridge measurements

The instrument, shown in Fig. 2, is based on the carrier suppression by sum of an equal and opposite signal. The null contains only the noise sidebands of the DUT, which are amplified and down-converted to dc by coherent detection.[12] The in-phase signal $x(t)$ is proportional to the normalized-amplitude noise (AM noise) $\alpha(t)$, the quadrature signal $y(t)$ to the PM noise $\varphi(t)$ The following ideas makes this scheme a fortunate choice.

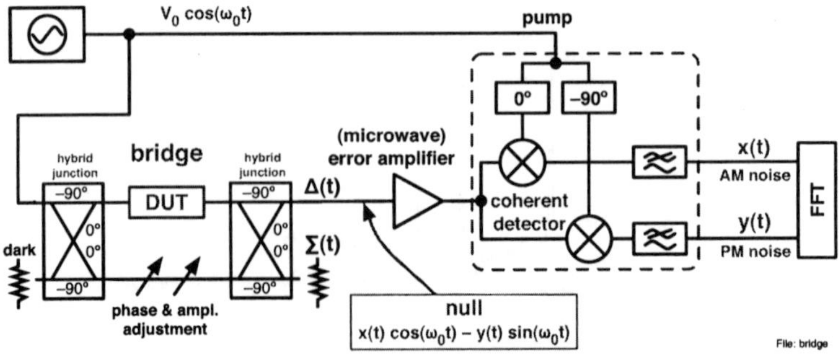

Fig. 2. Bridge (interferometric) method for the measurement of phase noise.

(1) The $1/f$ noise of the passive components the bridge is made of is dramatically reduced by separating coarse and fine the adjustment of the null. The coarse adjustment relies on by-step and fixed-value components. The fine adjustment, though more noisy because of the contact fluctuations, has low weight.
(2) Close-in flicker results from up-conversion of the near-dc $1/f$ noise pumped by the carrier (cf. Sec. 5). Here the error amplifier sees only the fluctuation of the null. Power is too low for the up-conversion process.
(3) The noise floor is improved significantly by microwave amplification of the noise sidebands before the mixer loss.
(4) Microwave amplification of the noise sidebands before detecting reduces the residuals of the mains because the dc electronics is highly sensitive to these fields, the microwave section is not.
(5) The in-phase/quadrature detection is a fully-linear Cartesian process. The dc component of $x(t)$ and $y(t)$ can be used to control the null in closed loop. This is recommended when the experiment lasts more than some half an hour.

The bridge has been used successfully to measure the flicker of ferrites[4,9,13] ($b_{-1} \approx 10^{-17}$ rad^2/Hz, either HF/VHF and microwave carrier) and photodetectors[14] ($b_{-1} \approx 10^{-12}$ rad^2/Hz, 10 GHz modulation). These measurements are out of reach for other methods, the first because of the extremely low noise, the second because of the low microwave power (order of 10 μW).

The photodetector $1/f$ noise is relevant in OEOs (Sect. 6) and in optical-fiber links.[5] In the link, the detector noise can not be removed with the two-way method. The value $b_{-1} = 10^{-12}$ rad^2/Hz at 10 GHz sets the stability limit to $\sigma_x \simeq 1.9{\times}10^{-17}$ s (Allan deviation). In the same condition the fractional-frequency stability of frequency transfer is $\sigma_y \simeq 4.2{\times}10^{-17}/\tau$,

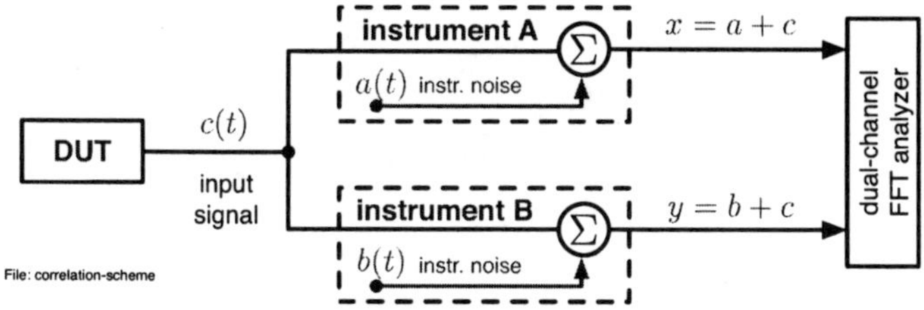

Fig. 3. Basic correlation measurement.

not far from the performance of short-range links. This is seen using $S_y = (f^2/\nu_0^2)\, S_\varphi$ and converting S_y into Allan deviation.

3. Correlation measurements

Two equal instruments measure the same DUT, as shown in Fig. 3. This figure also introduce the main symbols. Notice that $x(t)$ and $y(t)$ are *not* the same thing as in Sec. 2. We denote the Fourier transform with the uppercase of the time-domain function, the complex conjugate with $*$, and the average of m samples with $\langle\,\rangle_m$. Thanks to the Wiener-Kinchine theorem, the cross-spectrum is $S_{yx} = YX^*$. Hence

$$S_{yx} = \langle [B + C] \times [A + C]^* \rangle_m \qquad (1)$$

$$= CC^* + O(1/m)\ . \qquad (2)$$

Thus, S_{yx} converges to the spectrum $S_c = CC^*$ of the DUT, with a residual random term of the order of $1/m$.

In the laboratory practice, it is often convenient to display $|\,\langle \Re\{S_{yx}\}\rangle_m\,|$ because it is the minimally-biased always-positive estimator, assuming that S_c is real. With white noise we use the ergodicity principle to access the ensemble (a part of) by sweeping the frequency. Thus we have access to the average and to the standard deviation of $|\,\langle \Re\{S_{yx}\}\rangle_m\,|$. With small m, the single-channel noise is still not rejected. This means that the term $O(1/m)$ is larger than the DUT noise. In this condition, the deviation of $|\,\langle \Re\{S_{yx}\}\rangle_m\,|$ is almost equal to the average. In logarithmic scale, the spectrum looks as a band of constant thickness that moves downwards on the analyzer display. Increasing m, the term $O(1/m)$ becomes lower than the DUT noise. Thus, the deviation becomes smaller than the average. The track attains a constant value and shrinks. This is shown in Fig. 4, where for $m > 32$ the deviation of $|\,\langle \Re\{S_{yx}\}\rangle_m\,|$ becomes significantly lower than the average.

Interestingly, the fact that the deviation/average ratio that decreases for m larger than a threshold validates the measurement mathematically,

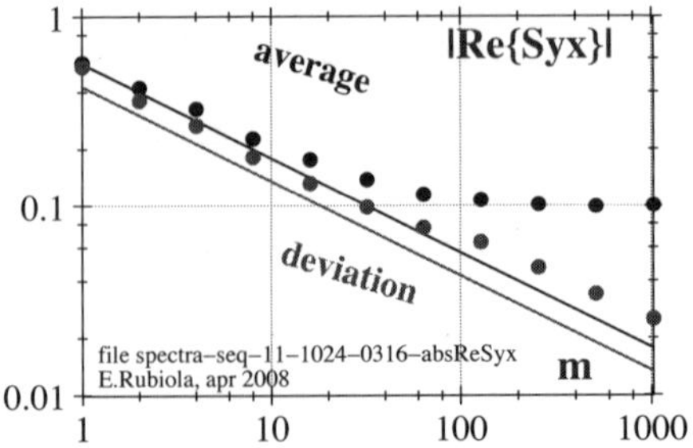

Fig. 4. Convergence of $|\langle \Re\{S_{yx}\}\rangle_m|$ to S_c in the case of the white noise floor. In the example, the DUT noise is 10 dB lower than the single-channel background of the instrument.

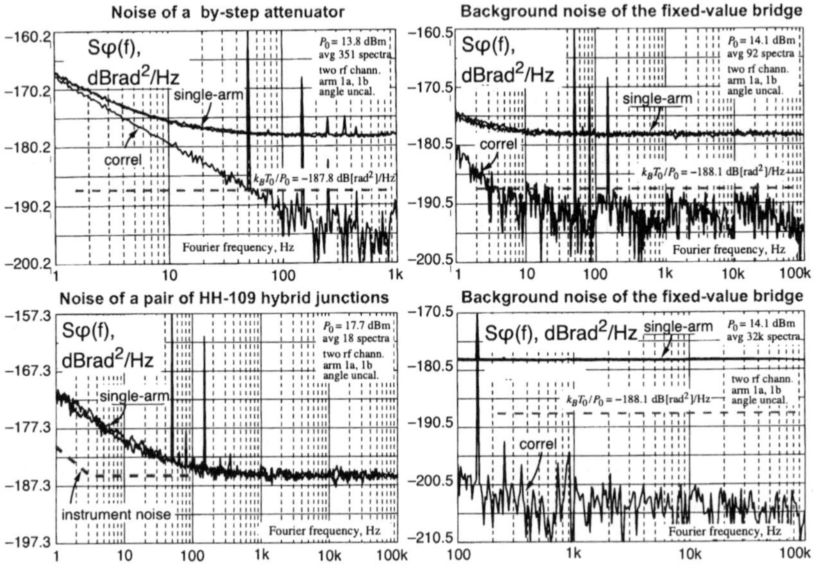

Fig. 5. Measurement of low-phase-noise devices. The background noise of the instrument is measured in the same conditions, but with different frequency span and number of averaged spectra.

without need of checking on the instrument background.

Correlation is an old idea,[15–17] which could be used routinely in phase noise measurements only with the FFT analyzer.[18] The use of correlation on a bridge instrument enables the measurement of some devices that could not be measured before.[4] Figure 5 shows some examples.

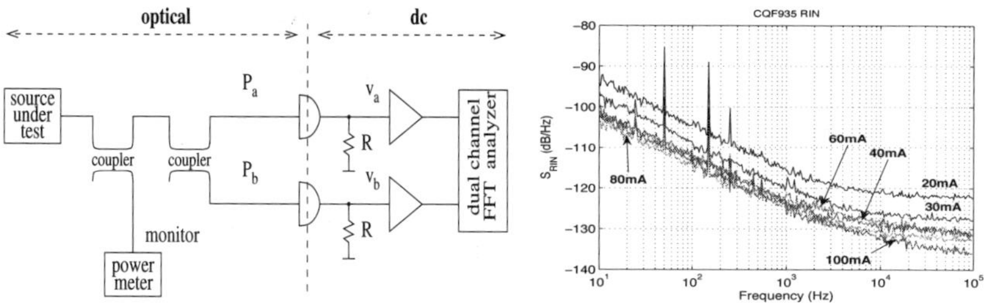

Fig. 6. Measurement of the laser RIN spectrum.

4. Measurement of AM noise and RIN spectra

In the case of PM noise, the single-channel background is usually measured by removing the DUT and by replacing it with a short cable, which is virtually noise free. In the case of AM noise and RIN, the DUT can *not* be removed because the DUT is the only source of signal. Thus, the validation of the instrument requires a reference of suitably low amplitude noise, which is not available in general. Correlation solves the problem because the convergence of the spectrum to the DUT noise can be assessed mathematically through the deviation/average ratio, as explained in Sec. 3. Figure 6 shows an example of laser RIN measurement.

5. Flicker in electronic and optical devices

Though the ultimate reasons for the $1/f$ noise are still elusive, the mechanism of the close-to-the-carrier flicker in RF/microwave devices is simple. Understanding it starts from the simple observation that

- near-dc flicker exists per se, though often made in accessible by an output filter.
- the output spectrum is white in the microwave band,
- close-in noise appears only in the presence of a carrier.

The reason is that close-in noise is brought from dc to the vicinity of the carrier by parametric modulation, which at first approximation is linear (Fig. 7). This simple model accounts for the following experimental facts about the $1/f$ noise spectra [19, Chap. 2].

- The noise is about independent of the carrier power.
- The noise of cascaded devices is the sum of the individual spectra, regardless of the order in which the devices are chained.

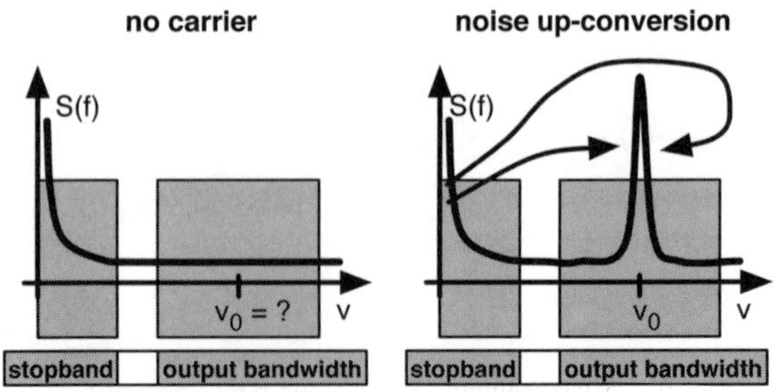

Fig. 7. Flicker up-conversion in electronic devices.

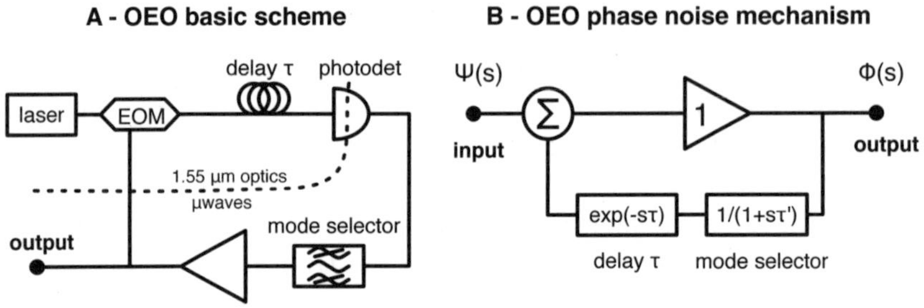

Fig. 8. Basic scheme and phase-noise model in the OEO.

- When equal devices are connected in parallel, the noise is that of one device divided by the number of devices.

The typical value of the flicker parameter b_{-1} in amplifiers is of 10^{-14} rad^2/Hz for bipolar RF units, of 10^{-12} rad^2/Hz for SiGe microwave units, and of $5{\times}10^{-11}$ rad^2/Hz for HBT microwave units; and of 10^{-12} rad^2/Hz for high-speed photodetectors.

6. Noise in OEOs

The opto-electronic oscillator (OEO) is an appealing alternative for high spectral purity microwave oscillators.[7,8] Figure 8 shows the basic scheme of an optical-fiber OEO and its phase noise model [19, Chap. 5]. Microwave oscillation can take place at any frequency multiple of $1/\tau$. A bandpass filter is necessary to select a single frequency. In Fig. 8 B all the signals are the Laplace transform of the *phase fluctuation* of the microwave signal or modulation. Non-linearity necessary in real oscillator for the power not to decay or diverge. Using amplitude and phase as the base representation, all the non-linearity goes in the amplitude, the phase is linear. The underlying

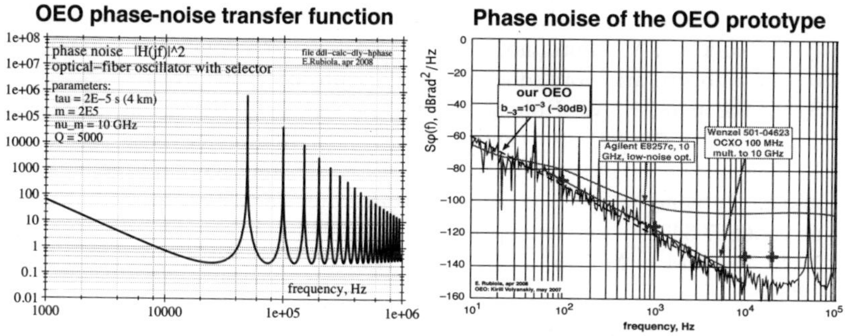

Fig. 9. Examples of OEO phase-noise transfer function and spectrum.

physical fact is that time can not be stretched[a]. As a relevant consequence, phase noise is additive. This eliminates the difficulty inherent in the parametric nature of some noise processes, like flicker. The input $\Psi(s)$ is the phase noise of all the oscillator components. The output $\Phi(s)$ is the phase noise at the oscillator output. The amplifier gain is exactly equal to one because the amplifier repeats the phase of the input. Elementary feedback theory yields to the phase-noise transfer function $\mathrm{H}(s) = 1/[1-\mathrm{B}(s)]$, where $\mathrm{B}(s)$ represents the delay and the low-pass equivalent of the mode selector. The function $|\mathrm{H}(jf)|^2$ is an extension of the Leeson model.[20]

Figure 9 shows an example of transfer function $|\mathrm{H}(jf)|^2$ and of phase noise spectrum with 4 km optical fiber, i.e., with a delay of 20 μs. The peaks of $|\mathrm{H}(jf)|^2$ at n/τ (50 kHz, 100 kHz, etc.) are due to the contiguous microwaves modes kept below the oscillation threshold by the mode selector. The spectrum is measured with two delay lines used as the frequency discriminator, taking the average cross-spectrum to enhance the sensitivity.[6] The $1/f^3$ noise is no more than 2 dB higher than the theoretical value, evaluated by putting the $1/f$ noise of the electronics in the Leeson formula. The phase noise $10^{-3}/f^3$ (frequency flicker) is equivalent to the Allan deviation $\sigma_y = 3.7{\times}10^{-12}$.

References

1. E. N. Ivanov, M. E. Tobar and R. A. Woode, *IEEE Trans. Ultras. Ferroelec. and Freq. Contr.* **45**, 1526(November 1998).
2. E. Rubiola, V. Giordano and J. Groslambert, *Rev. Sci. Instrum.* **70**, 220(January 1999).

[a]This may not be true in the presence of strong non-linearity.

3. E. Rubiola and V. Giordano, *Rev. Sci. Instrum.* **71**, 3085(August 2000).
4. E. Rubiola and V. Giordano, *Rev. Sci. Instrum.* **73**, 2445(June 2002), Also on the web site arxiv.org, document arXiv:physics/0503015v1.
5. S. M. Foreman, K. W. Holman, D. D. Hudson, D. J. Jones and J. Ye, *Rev. Sci. Instrum.* **78**, 021101 1(February 2007), Invited article.
6. K. Volyanskiy, J. Cussey, H. Tavernier, P. Salzenstein, G. Sauvage, L. Larger and E. Rubiola, *J. Opt. Soc. Am. B - Opt. Phys.* **25** (2008), In press. Preprint arXiv:0807.3494v1 [physics.optics] 22 July 2008.
7. X. S. Yao and L. Maleki, *J. Opt. Soc. Am. B - Opt. Phys.* **13**, 1725(August 1996).
8. X. S. Yao, L. Davis and L. Maleki, *J. Ligtwave Technol.* **18**, 73(January 2000).
9. E. Rubiola, Y. Gruson and V. Giordano, *IEEE Trans. Ultras. Ferroelec. and Freq. Contr.* **51**, 957(August 2004).
10. R. V. Pound, *Rev. Sci. Instrum.* **17**, 490(November 1946).
11. D. Sarid, *Scanning Force Microscopy* (Oxford, 1994).
12. A. J. Viterbi, *Principles of Coherent Communication* (McGraw Hill, New York, 1966).
13. R. A. Woode, E. N. Ivanov and M. E. Tobar, *Meas. Sci. Technol.* **9**, 1593(September 1998).
14. E. Rubiola, E. Salik, N. Yu and L. Maleki, *IEEE Trans. Microw. Theory Tech.* **54**, 816(February 2006), Preprint available on arxiv.org, document arXiv:physics/0503022v1, March 2005.
15. R. Hanbury Brown, R. C. Jennison and M. K. Das Gupta, *Nature* **170**, 1061(December 20, 1952).
16. R. F. C. Vessot, R. F. Mueller and J. Vanier, A cross-correlation technique for measuring the short-term properties of stable oscillators, in *Proc. IEEE-NASA Symposium on Short Term Frequency Stability*, (Greenbelt, MD, USA, 1964).
17. F. L. Walls, S. R. Stain, J. E. Gray and D. J. Glaze, Design considerations in state-of-the-art signal processing and phase noise measurement systems, in *Proc. Freq. Control Symp.*, (Atlantic City, NJ, USA, 1976).
18. W. F. Walls, Cross-correlation phase noise measurements, in *Proc. Freq. Control Symp.*, (IEEE, New York, 1992, Hershey, PA, May 27-29 1992).
19. E. Rubiola, *Phase Noise and Frequency Stability in Oscillators* (Cambridge University Press, Cambridge, UK, November 2008).
20. D. B. Leeson, *Proc. IEEE* **54**, 329(February 1966).

OPTICAL SPECTROSCOPY OF ATOMIC HYDROGEN FOR AN IMPROVED DETERMINATION OF THE RYDBERG CONSTANT

J. L. FLOWERS[1,*], P. E. G. BAIRD[2], A. EVANS[2], H. A. KLEIN[1],
C. D. LANGHAM[1], L. MAIRINIAC[1,3], H. S. MARGOLIS[1] and B. R. WALTON[1]

[1] *National Physical Laboratory (NPL), Hampton Road, Teddington, Middlesex TW11 0LW, UK*
[2] *Clarendon Laboratory, University of Oxford, Oxford OX1 3PU, UK*
[3] *ENSMM, 26 Chemin de l'Epitaphe, 25030 Besançon CEDEX, France*
* *E-mail: jeff.flowers@npl.co.uk*

This paper describes the current status of an experiment designed to make absolute frequency measurements of a series of optical two-photon $2S - nS, nD$ transitions in atomic hydrogen. The $2S_{1/2} - 8D_{5/2}$ transition has been observed and a preliminary measurement of its frequency has been made using a femtosecond optical frequency comb. Detailed studies of systematic frequency shifts are in progress. When combined with results from several other experiments worldwide, this work will contribute to an improved determination of the Rydberg constant.

1. Introduction

The Rydberg constant, which is currently known to 6.6 parts in 10^{12}, is determined by comparing measured transition frequencies in hydrogen and deuterium to the theoretical expressions for the energy levels.[1] Although the $1S - 2S$ transition frequency in hydrogen has been measured to 1.4 parts in 10^{14} by laser spectroscopy on a cooled atomic beam,[2] this is not sufficient to reduce the uncertainty in the Rydberg constant because of the uncertainties in quantum electrodynamic (QED) and nuclear size corrections to the energy levels. To separate out these effects requires other transitions to be measured with higher precision than has been achieved to date. At the UK National Physical Laboratory (NPL) an experiment is underway that aims to make new and improved measurements of $2S - nS, nD$ transitions frequencies in atomic hydrogen by two-photon laser spectroscopy on a metastable hydrogen atomic beam.[3]

2. Experimental arrangement

The design of the hydrogen atomic beam is similar to that of Biraben *et al.*[4] Hydrogen atoms are produced in the ground 1S state by molecular dissociation in a radio-frequency discharge source and then excited to the metastable 2S state by electron impact excitation. The 2S state atoms then pass through the atom-laser interaction region before entering the detection region in which an applied electric field mixes the 2S and 2P states. This quenches the metastable state and generates Lyman-α photons at 121.5 nm, which are detected by a pair of photomultiplier tubes. The photomultiplier count rate recorded is proportional to the number of metastable atoms reaching the detection region. The operating parameters of the metastable hydrogen beam have been optimized to give signals of approximately 8000 counts per second in the absence of any atom-laser interactions along the beam path.[3]

The two-photon $2S-nS, nD$ transitions are excited using a Ti:sapphire laser that is stabilized to an ultralow expansion (ULE) cavity with a free spectral range of 500 MHz and a finesse of around 5000. The ULE cavity mirror coatings were chosen to cover the wavelength range 730 – 868 nm, enabling all $2S-nS, nD$ transitions with $n \geq 5$ to be studied. A double-pass acousto-optic modulator is used to provide a tunable frequency offset from a ULE cavity mode close to each transition of interest. A build-up cavity constructed around the metastable atomic beam produces the counter-propagating beams required for the Doppler-free two-photon excitation and increases the excitation rate. The transitions are detected by monitoring the reduction in the number of metastable atoms reaching the far end of the atomic beam as the laser frequency is tuned through the resonance.

For absolute frequency measurements, a small fraction of the Ti:sapphire laser output is sent via an optical fibre link to one of the NPL femtosecond optical frequency combs.[5,6] Compared to previous work,[7] in which the transitions that could be studied were limited by the need for near-coincidences with suitable optical frequency standards, the use of a femtosecond comb makes it straightforward to relate any optical frequency directly to a suitable microwave frequency reference such as a hydrogen maser or caesium fountain primary frequency standard.

3. Present status of experiment

Initial work on this experiment has focussed mainly on the $2S_{1/2}-8D_{5/2}$ two-photon transition, selected because it is the strongest of the $2S-8S, 8D$

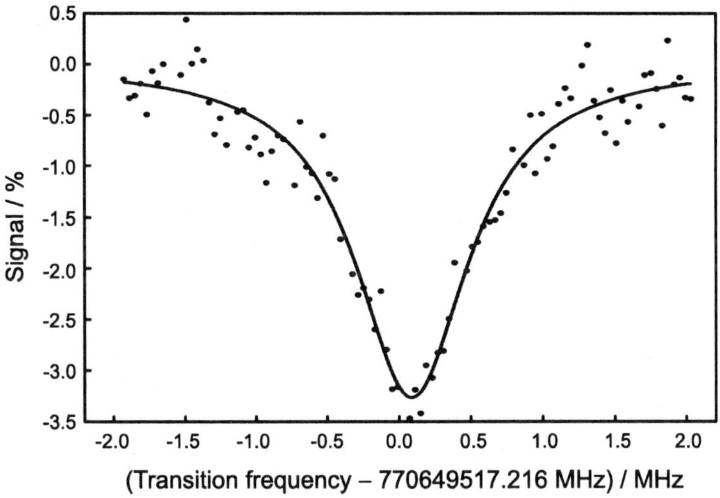

Fig. 1. Laser-induced signal from the $2S_{1/2}-8D_{5/2}$ two-photon transition in atomic hydrogen, shown as a percentage of the off-resonance photomultiplier count rate.

transitions that were studied in earlier experiments.[7] Subsequent to the initial observation of this transition at NPL, several improvements to the apparatus have been implemented. In particular, new photomultiplier tubes with CsI photocathodes have been installed which are significantly less sensitive to scattered light from the Ti:sapphire laser system. To eliminate drifts in the background count rate, active control of the inlet hydrogen gas pressure and the electron gun emission current has been implemented. Finally we have increased the finesse of the buildup cavity around the metastable beam in order to improve the signal-to-noise level of the induced transitions. A typical two-photon resonance signal following these improvements is shown in Fig. 1.

Detailed studies of systematic frequency shifts are in progress, and will benefit from the ability to measure a series of $2S-nS, nD$ transition frequencies under nominally identical conditions. In order to calculate the lineshapes of the two-photon transitions and to correct for the second-order Doppler shift, a precise knowledge of the velocity distribution of the metastable hydrogen atoms is required. To measure this velocity distribution we are using a 656 nm diode laser to observe the Doppler-broadened, single-photon, $2S-3P$ transition using co- and counter-propagating laser beams which are collinear with the metastable beam. Two resonance peaks are observed, the separation of which depends on the beam velocity due to the first-order Doppler effect, and by fitting modelled lineshapes[7] to the experimental data it is possible to derive the velocity distribution.

4. Optical excitation of the 2S state

In the present apparatus the metastable 2S state atoms are created by electron impact excitation, resulting in relatively low yields and a high energy spread. As a consequence, a high laser power is required to induce the two-photon $2S - nS, nD$ transitions and so the limiting systematic effects in this first experiment are expected to be ac Stark shifts and saturation effects.[7] These will vary for different atomic trajectories through the laser beam, leading to both broadening and shifts of the transitions and meaning that detailed modelling of the line profiles will be required.

To reduce these effects, we plan to modify our apparatus to produce the metastable atoms by two-photon optical excitation from the ground state using 243 nm radiation. Optical excitation, in contrast to electron impact excitation, imparts no additional momentum to the atoms and so should generate an intense, well-collimated source of metastable atoms which can be cooled to increase further the atomic density in the beam. This in turn will enable the $2S - nS, nD$ transitions to be excited using lower laser power, thereby reducing the light shifts.

For the optical excitation, we will use a commercial 243 nm laser source which has many similarities to that described by Kolachevsky *et al.*[8] This consists of a 972 nm extended-cavity diode laser and tapered amplifier, the output of which is frequency doubled in a temperature-tuned potassium niobate crystal to generate 486 nm radiation. The output from this first doubling stage is then doubled again in temperature-tuned beta barium borate to produce up to 20 mW of 243 nm radiation. To achieve the requisite long-term frequency stability, the 972 nm laser is locked to a second ULE cavity.

A linear atomic beam geometry is required when optical excitation is used to produce the 2S state because the atoms are no longer deflected by electron collisions. A separate apparatus is therefore being constructed to test the optical excitation efficiency, while electron impact excitation continues to be used for experiments in the primary apparatus. The atomic hydrogen source is similar to that used for $1S - 2S$ spectroscopy experiments[2] and the design incorporates a buildup cavity for the 243 nm radiation to enhance the metastable atom production rate.

5. Conclusion

This experiment is expected to lead to new and improved absolute frequency measurements of a series of $2S - nS, nD$ transitions in atomic hydrogen. It complements experiments being carried out elsewhere, including absolute frequency measurements of the hydrogen $1S - 3S$ transition,[9] spectroscopy of He^+ and highly charged hydrogen-like ions to test QED corrections to energy levels in hydrogenic systems[10,11] and an experiment to measure the proton charge radius by spectroscopy of muonic hydrogen.[12] When combined with the existing high accuracy measurement of the $1S - 2S$ transition in atomic hydrogen, this international effort will lead to significantly reduced uncertainty in the value of the Rydberg constant.

Acknowledgments

This work is supported by the UK Department of Innovation, Universities and Skills as part of the NMS Pathfinder Metrology Programme.

References

1. P. J. Mohr and B. N. Taylor, *Rev. Mod. Phys.* **77**, 1 (2007).
2. T. W. Hänsch *et al.*, *Phil. Trans. Roy. Soc. A* **363**, 2155 (2005).
3. J. L. Flowers, P. E. G. Baird, L. Bougueroua, H. A. Klein and H. S. Margolis, *IEEE Trans. Instrum. Meas.* **56**, 331 (2007).
4. F. Biraben, J. C. Garreau, L. Julien and M. Allegrini, *Rev. Sci. Instrum.* **61**, 1468 (1990).
5. V. Tsatourian *et al.*, these proceedings.
6. B. R. Walton, H. S. Margolis, V. Tsatourian and P. Gill, *IET Optoelectron.* **2**, 182 (2008).
7. B. de Beauvoir *et al.*, *Eur. Phys. J. D* **12**, 61 (2000).
8. N. Kolachevsky, J. Alnis, S. D. Bergeson and T. W. Hänsch, *Phys. Rev. A* **73**, 021801(R) (2006).
9. O. Arnoult, F. Nez, C. Schwob, L. Julien and F. Biraben, *Can. J. Phys.* **83**, 273 (2005).
10. A. Ozawa *et al.*, *Phys. Rev. Lett.* **100**, 253901 (2008).
11. K. Hosaka *et al.*, *Phys. Rev. A* **69**, 011802(R) (2004).
12. R. Pohl *et al.*, *Can. J. Phys.* **83**, 339 (2005).

Part III

Clock Application in Space

RECENT PROGRESS ON THE ACES MISSION

L. CACCIAPUOTI

European Space Agency, Research and Scientific Support Department
Noordwijk ZH, 2200 AG, The Netherlands
E-mail: Luigi.Cacciapuoti@esa.int

C. SALOMON

Laboratoire Kastler Brossel, CNRS, ENS, 24 Rue Lhomond,
75005, Paris, France
E-mail: Christophe.Salomon@lkb.ens.fr

Atomic Clock Ensemble in Space (ACES) is a mission designed to perform accurate tests of Einstein's theory of general relativity and develop applications in time and frequency metrology, global positioning and navigation, geodesy, and remote sensing. Mission concept, scientific objectives, and status of ACES will be discussed and the latest test results presented.

Keywords: Cold-atom clocks; Space clocks; Fundamental physics tests.

1. ACES Mission Overview

Atomic Clock Ensemble in Space (ACES)[1–3] is an ESA mission in fundamental physics based on the performances of a new generation of atomic clocks operated in the microgravity environment of the International Space Station (ISS) and on a network of ultra-stable clocks on ground. The station is orbiting at a mean elevation of 400 km with 90 min. of rotation period and an inclination angle of 51.6°. Transported on orbit by the Japanese transfer vehicle HTV, ACES will be installed at the external payload facility of the Columbus module (Fig. 1) using the ISS robotic arm.

The ACES payload accommodates two atomic clocks: PHARAO, a primary frequency standard based on samples of laser cooled cesium atoms, and the active hydrogen maser SHM. The performances of the two clocks are combined to generate an on-board timescale with the short-term stability of SHM and the long-term stability and accuracy of the cesium clock PHARAO. The on-board comparison of PHARAO and SHM and the distribution of the ACES clock signal are ensured by the Frequency Comparison and Distribution Package (FCDP), while all data handling processes are controlled by the eXternal PayLoad Computer (XPLC). A GNSS receiver installed on the ACES payload and connected to the on-board time scale will provide precise orbit determination of the ACES clocks. One of the main objectives of the ACES mission consists in maintaining a stable and accurate on-board timescale which will be used to perform comparisons of atomic frequency standards. The ACES clock signal will be transferred on ground by a time and frequency transfer link in the microwave domain (MWL). MWL compares the ACES frequency reference to a set of ground clocks, enabling fundamental physics tests and applications in different areas of research.

The planned mission duration is 18 months with a possible extension to 3 years. During the first two months, the functionality of the clocks and of MWL will be tested. Then, a period of 4 months will be devoted to the performance evaluation of the clocks. During this phase, a signal with frequency inaccuracy in the 10^{-15} range will be available to ground users. In microgravity, the linewidth of the atomic resonance of PHARAO will be tuned by two orders of magnitude, down to sub-Hertz values (from 11 Hz to 110 mHz). After the clocks optimization, performances in the 10^{-16} range both for frequency instability and inaccuracy are expected. In the second part of the mission (12 to 30 months), the on-board clocks will be compared to a number of ground atomic clocks operating both in the microwave and optical domain. These measurements will test general relativity and search

Fig. 1. Left: The Columbus module on-board the ISS and the nadir-pointing location of the Columbus External Payload Facility (CEPF), where the ACES payload will be installed. Right: The ACES payload.

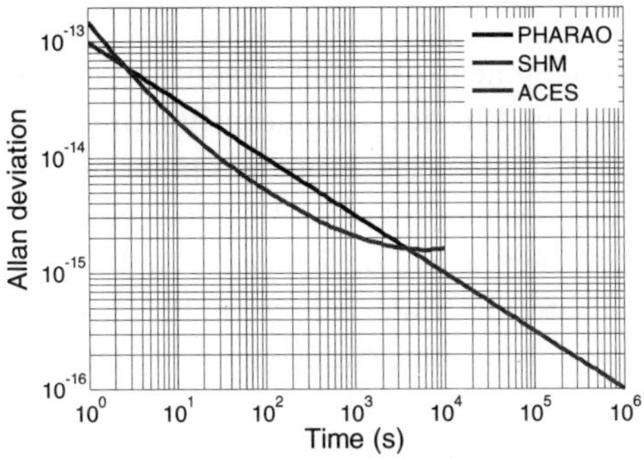

Fig. 2. Fractional frequency instability specified for the PHARAO, SHM, and ACES clock signals in space.

for new physics beyond the Standard Model.

2. Mission Objectives

ACES will conduct first experiments with cold atoms in a freely falling laboratory, it will perform fundamental physics tests to high resolution, and develop applications in different areas of research.

A new generation of clocks reaching frequency instability and inaccuracy of few parts in 10^{16} will be validated by ACES. PHARAO will combine laser cooling techniques and microgravity conditions to significantly increase the interaction time and consequently reduce the linewidth of the clock transition. Improved stability and better control of systematic effects will be demonstrated in the space environment. PHARAO will reach a fractional frequency instability of $1 \cdot 10^{-13} \cdot \tau^{-1/2}$, where τ is the integration time expressed in seconds, and an inaccuracy of few parts in 10^{16}. The reliability offered by active H-masers will be made available for space applications by SHM. SHM will provide a clock signal with fractional frequency instability down to $1.5 \cdot 10^{-15}$ after only 10^4 s of integration time. Two servo-loops will lock together the clock signals of PHARAO and SHM generating an on-board time scale combining the short-term stability of the H-maser with the long-term stability and accuracy of the cesium clock (Fig. 2).

This clock signal will be distributed by MWL. Frequency transfer with time deviation better than 0.3 ps at 300 s, 7 ps at 1 day, and 23 ps at 10 days of integration time will be demonstrated. These performances, surpassing existing techniques (TWSTFT and GPS) by one to two orders of magnitude, will enable common view and non-common view comparisons of ground clocks with 10^{-17} frequency resolution after few days of integration time. The recent development of optical frequency combs[4,5] awarded with the 2005 Nobel Prize in Physics, significantly simplifies the link between optical and microwave frequencies. From this point of view, ACES will take full advantage of the recent progress of optical clocks,[6,7] today reaching instability and inaccuracy levels of few parts in 10^{17}. In addition, ACES will deliver a global atomic time scale with 10^{-16} accuracy, it will allow clocks synchronization at an uncertainty level of 100 ps, and contribute to international atomic time scales (TAI, UTC).

These performances will be used to test general relativity to high accuracy. According to Einstein's theory, identical clocks placed at different positions in stationary gravitational fields experience a frequency shift that, in the frame of the PPN approximation, depends directly on the Newtonian potential at the clock position. The comparison between the ACES on-board clocks and ground-based atomic clocks will measure the frequency variation due to the gravitational red-shift with a 70-fold improvement on previous experiments,[8] testing Einstein's prediction at the 2 ppm uncertainty level. Time variations of fundamental constants will be measured by comparing clocks based on different transitions or different atomic species.[9] Any transition energy can be expressed in terms of the fine structure constant α and the two dimensionless constants m_q/Λ_{QCD} and m_e/Λ_{QCD}, depending on the quark mass m_q, the electron mass m_e and the QCD mass scale Λ_{QCD}.[10,11] ACES will perform crossed comparisons of ground clocks both in the microwave and in the optical domain with a resolution of $1 \cdot 10^{-17}$ after few days of integration time. These comparisons will impose strong and unambiguous constraints on time variations of fundamental constants reaching an uncertainty of $1 \cdot 10^{-17}$/year in case of a 1-year mission duration and $3 \cdot 10^{-18}$/year after three years. ACES will also perform tests of the Local Lorentz Invariance (LLI). According to this principle, the outcome of any local test experiment is independent of the velocity of the freely falling apparatus. In 1997, LLI tests based on the measurement of the round-trip speed of light have been performed by comparing clocks on-board GPS satellites to ground hydrogen masers.[12] In such experiments, LLI violations would appear as variations of the speed of light c with the direction and the relative velocity of the clocks. ACES will perform a similar experiment by measuring relative variations of the light velocity at the 10^{-10} uncertainty level.

Other applications of the ACES clock signal are currently under investigation. ACES will demonstrate a new "relativistic geodesy". This method, based on a precision measurement of the Einstein's gravitational red-shift, will take advantage of the accuracy of ground-based optical clocks to resolve differences in the Earth gravitational potential at the 10 cm level. The GNSS receiver on-board the ACES payload will allow to monitor the GNSS network and develop interesting applications in Earth remote sensing. This includes studies of the oceans surface via GNSS reflectometry measurements and the analysis of the Earth atmosphere with GNSS radio-occultation experiments. At the 10^{-18} level, ground clocks will be limited by the fluctuations of the Earth potential suggesting that future high precision time references will have to be placed in the space environment where these fluctuations are significantly reduced. From this point of view, ACES could pioneer new concepts for global time keeping and positioning based on a reduced set of ultra-stable space clocks in orbit associated with simple transponding satellites.

3. ACES Status

The ACES Mission is presently in C/D phase. All instruments and subsystems are in an advanced status of development with engineering models delivered or in final assembly. The Mission Preliminary Design Review has been successfully concluded in 2007, consolidating the overall mission concept and demonstrating the feasibility of ACES. ACES clocks and subsystems will be soon assembled and tests conducted on the fully integrated system will measure the end-to-end performances. The Mission Critical Design Review, scheduled in mid 2009, will close the engineering model development phase and release the manufacturing of the flight hardware.

3.1. *PHARAO*

PHARAO is a cesium clock based on laser cooled atoms developed by SYRTE, LKB, and CNES. Its concept is very similar to ground based atomic fountains, but with a major difference: PHARAO will be operated under microgravity conditions. Atoms, launched in free flight along the PHARAO tube, cross a resonant cavity where they interact two times with a microwave field tuned on the transition between the two hyperfine levels of the cesium ground state. The interrogation method, based on two spatially separated oscillating fields (Ramsey scheme), allows the detection of an atomic line whose width is inversely proportional to the transit time between the two interaction regions. In a microgravity environment, the velocity of the atoms along the ballistic trajectories is constant and can be continuously changed over almost two orders of magnitude (5-500 cm/s) allowing the detection of atomic signals with sub-Hertz linewidth.

The engineering model of the PHARAO clock has been completed and is presently under test at CNES premises in Toulouse. Design and recent results are presented in Ref. 13. Microwave resonance signals (Ramsey fringes) with a signal-to-noise ratio of ~ 700 have been recorded demonstrating the correct interfacing of PHARAO subsystems and the correct operation of the clock. For a launch velocity of 3.42 m/s, the duration of the free flight between the two Ramsey interaction regions is about 100 ms, corresponding to a typical width of the central fringe of about 5 Hz. In microgravity, the longer interaction times will allow PHARAO to measure linewidths 10 to 50 times narrower. Fig. 3 shows an Allan deviation of the PHARAO clock. A fractional frequency instability of $2.3 \cdot 10^{-13} \cdot \tau^{-1/2}$ between 1 s and 10^4 s can be measured when the PHARAO microwave source is driven by an external cryogenic oscillator. This situation closely approaches the specified stability of $1 \cdot 10^{-13} \cdot \tau^{-1/2}$ only reachable in microgavity environment. The result is in perfect agreement with theoretical predictions based on atom number, resonance width, and cycle duration. When the microwave source is driven by its internal ultra-stable quartz oscillator, the measured stability is $4 \cdot 10^{-13} \cdot \tau^{-1/2}$. This value is set by the phase noise of the quartz oscillator which is sampled by the atoms in the microwave cavity (Dick effect). In space, this effect will be reduced by a factor of ~ 5 because of the

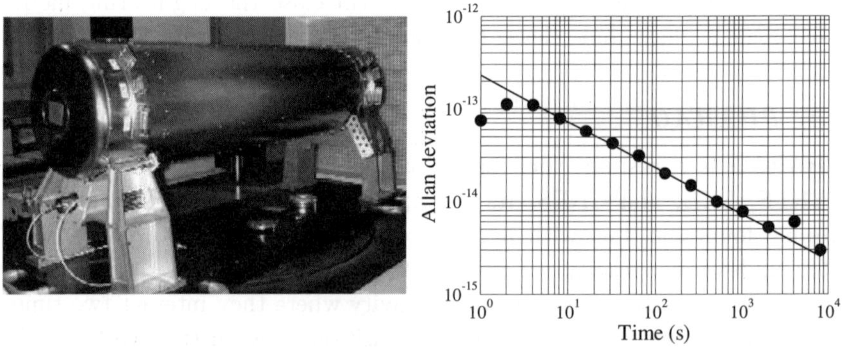

Fig. 3. Left: Engineering model of the PHARAO cesium tube. Fully assembled, the tube has a volume of $990 \times 336 \times 444$ mm^3 and a mass of 44 kg. Right: Allan deviation of the PHARAO clock obtained by operating the PHARAO microwave source with an external cryogenic sapphire oscillator. The solid line corresponds to a fractional frequency instability of $2.3 \cdot 10^{-13} \cdot \tau^{-1/2}$

much narrower resonance width. Flight software and performance tests on the PHARAO clock have been recently completed. In the coming months, the long-term stability and accuracy of the clock will be evaluated.

3.2. *SHM*

Because of their simplicity and reliability, hydrogen masers are used in a large variety of applications. Passive and active masers are expected to be key instruments in future space missions, satellite positioning systems, and high-resolution VLBI (Very Long Baseline Interferometry) experiments. The clock operates on the hyperfine transition of atomic hydrogen at 1.420405751 GHz. H_2 molecules are dissociated in a plasma discharge and the resulting beam of H atoms is state selected and sent in a storage bulb. The bulb is surrounded by a full-size aluminium cavity tuned to the atomic resonance that sustains the microwave field resulting from the stimulated emission of H atoms on the clock transition.

Developed by SPECTRATIME under ESA coordination, SHM provides ACES with a stable fly-wheel oscillator. The main challenge of SHM is represented by the low mass and volume figures required by the space clock with respect to ground H-masers. SHM has a mass budget of 42 kg and a volume of $390 \times 390 \times 590$ mm^3. A dedicated Automatic Cavity Tuning (ACT) system has been designed to maintain the microwave cavity tuned to the resonance. The ACT system continuously injects two tones symmetrically placed around the H-maser signal. The two tones are coherently detected and the unbalance between their power levels is used to close a feedback loop acting on a varistor which steers the resonance frequency of the microwave cavity and stabilizes it against thermal drifts. This method will allow SHM to reach fractional frequency instabilities down to $1.5 \cdot 10^{-15}$ at 10^4 s of integration time. SHM engineering model will be completed and tested by the end of 2009.

3.3. *FCDP*

The Frequency Comparison and Distribution Package (FCDP) is the central node of the ACES payload. Developed by ASTRIUM and TIMETECH under ESA coordination, FCDP is the on-board hardware which compares the signals delivered by the two space clocks, measures and optimizes the performances of the ACES frequency reference, and finally distributes it to the microwave link.

The engineering model of FCDP has been completed and tested. The

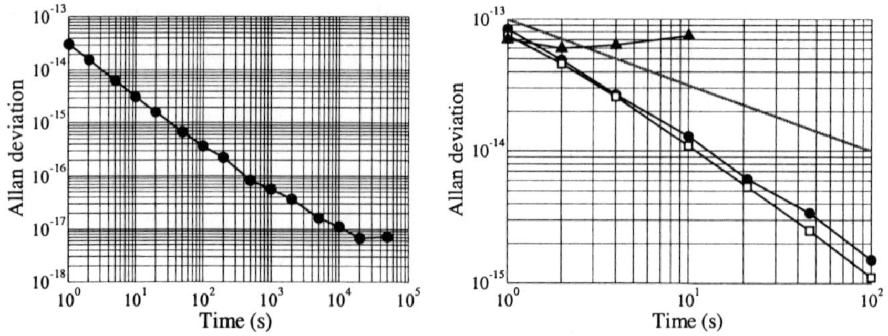

Fig. 4. Left: Allan deviation of the noise introduced by FCDP in the comparison of the two on-board clocks. Right: Allan deviation of the PHARAO microwave source phase-locked to the cryogenic sapphire oscillator and measured against the cryogenic sapphire oscillator itself (circles). Measurements are compared to the simulated behavior (squares), to the Allan deviation of the free-running microwave source of PHARAO (triangles), and to the expected performances of the PHARAO clock (solid line).

noise introduced by FCDP on the distributed clock signal rapidly averages down entering the 10^{-18} regime already after 10^4 s of integration time. Fig. 4 shows the Allan deviation of the noise floor of FCDP phase comparator. The curve decreases as the inverse of the integration time, dropping below $1 \cdot 10^{-17}$ after 10^4 s. In addition, a specific test has been conducted at CNES premises in Toulouse to validate the performances of the FCDP phase-locked loop which stabilizes the local oscillator of PHARAO on the clock signal generated by SHM. The microwave source of PHARAO has been phase-locked via FCDP to the clock signal provided by a cryogenic sapphire oscillator (CSO). The excellent short-term stability of the CSO is crucial for correctly identifying the noise contribution of FCDP when the servo-loop is closed. Fig. 9 shows the Allan deviation plot of the PHARAO microwave source phase-locked to the cryogenic sapphire oscillator and measured against the CSO itself (circles). For integration times longer than the servo-loop time constant, the Allan deviation of the noise contribution introduced by the short-term servo-loop rapidly decreases, closely following the simulated behavior (squares).

3.4. *MWL*

The ACES clock signal distributed by FCDP is finally transmitted to ground stations by the ACES microwave link. MWL is developed by ASTRIUM, KAYSER-THREDE, and TIMETECH under ESA coordination.

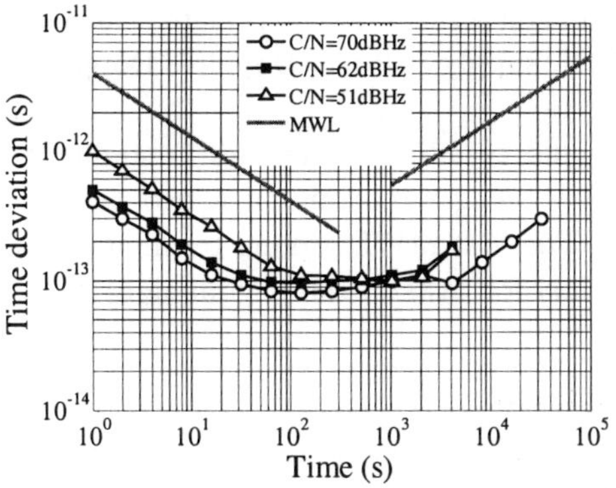

Fig. 5. Stability of the carrier phase, expressed in time deviation, for different signal to noise density ratios (C/N). Measurements are compared to MWL system requirements.

The proposed MWL concept is an upgraded version of the Vessot two-way technique used for the GP-A experiment in 1976[8] and the PRARE geodesy instrument. The system operates continuously with a carrier frequency in the Ku-band. The high carrier frequencies of the up and down links (13.5 GHz and 14.7 GHz respectively) allow for a noticeable reduction of the ionospheric delay. A third frequency in the S-band (2.2 GHz) is used to determine the Total Electron Content (TEC) and correct for the ionospere time delay. A PN-code modulation (100 Mchip/s) of the carrier removes the phase ambiguity between successive comparison sessions separated by large dead times. The system is designed for multiple access capability, allowing up to 4 simultaneous ground users distinguished by the different PN-codes and Doppler shifts.

The engineering model of the flight segment electronic unit has been completed and tested. MWL long-term stability is ensured by the continuous calibration of the receiver channels provided by a built-in test-loop translator. For shorter durations ($\sim$ 300 s), time stability is driven by the noise performance of the Ku transmitter and receiver and of the DLL (Delay-Locked Loop) boards. The 100 MHz chip rate allows to reach a time stability better than 2 ps already with code measurements. Carrier phase stability is shown in Fig. 5, where time deviations down to 70 fs at about 100 s are reported. For longer durations, time deviation remains well below the 1 ps level even in the worst conditions of signal to noise density ratio (C/N). The thermal sensitivity of the system has been measured and used to calibrate the MWL phase comparisons against temperature variations.

The sensitivity against a series of key parameters such as clock input power, received signal C/N, supply voltage, Doppler, Doppler rate, etc. has been characterized.

The MWL ground terminal electronics is similar to the MWL flight hardware, symmetry being important in a two-way system to reduce instrumental errors. The first MWL ground terminal, presently under development, will be ready for tests in the beginning of 2009.

3.5. *Ground Segment*

The ACES Users Support and Operations Center (USOC) will operate the payload via the Columbus control center, it will remotely control the MWL ground terminals, receive telemetry and scientific data, and finally analyze, archive and distribute them to the scientific community. The architecture of the ACES ground segment has been designed. The physical interfaces towards the Columbus control center, the MWL ground terminals, the ground clocks, and the ACES service providers (precise orbit determination, data analysis, technical and scientific support) have been defined. A mission operations concept is already available. It includes mission planning, execution, data analysis and archiving, scientific and engineering support needed during the complete life-cycle of ACES.

4. Acknowledgments

The authors express their warm thanks for their valuable contribution to the ACES science team, to the ACES project team at ESA, ASTRIUM, KAYSER-THREDE, and TIMETECH, to the PHARAO team at LNE-SYRTE and CNES, to the SHM team at SPECTRATIME. This work is supported by ESA and CNES.

References

1. C. Salomon *et al.*, *Comptes Rendus de l'Acadmie des Sciences - Series IV - Physics* **2**, 1313(November 2001).
2. L. Cacciapuoti *et al.*, *Nuclear Physics B Proceedings Supplements* **166**, 303(April 2007).
3. C. Salomon, L. Cacciapuoti and N. Dimarcq, *International Journal of Modern Physics D* **16**, 2511(December 2007).
4. R. Holzwarth *et al.*, *Physical Review Letters* **85**, 2264(September 2000).
5. S. A. Diddams *et al.*, *Physical Review Letters* **84**, 5102(May 2000).
6. T. Rosenband *et al.*, *Science* **319**, 1808(March 2008).
7. A. D. Ludlow *et al.*, *Science* **319**, 1805(March 2008).

8. R. F. C. Vessot *et al.*, *Physical Review Letters* **45**, 2081(December 1980).
9. T. M. Fortier *et al.*, *Physical Review Letters* **98**, 070801(February 2007).
10. V. V. Flambaum *et al.*, *Physical Review D* **69**, 115006(June 2004).
11. V. V. Flambaum and A. F. Tedesco, *Physical Review C* **73**, 055501(May 2006).
12. P. Wolf and G. Petit, *Physical Review A* **56**, 4405(December 1997).
13. P. Laurent *et al.*, *Applied Physics B: Lasers and Optics* **84**, 683(September 2006).

THE SAGAS MISSION

PETER WOLF

LNE-SYRTE, Observatoire de Paris, CNRS, UPMC,
61 av. de l'Observatoire, 75014 Paris, France

THE SAGAS COLLABORATION

See [1,2] for a complete list of names and affiliations

We summarize the scientific and technological aspects of the SAGAS (Search for Anomalous Gravitation using Atomic Sensors) project, submitted to ESA in June 2007 in response to the Cosmic Vision 2015-2025 call for proposals. The proposed mission aims at flying highly sensitive atomic sensors (optical clock, cold atom accelerometer, optical link) on a Solar System escape trajectory in the 2020 to 2030 time-frame. SAGAS has numerous science objectives in fundamental physics and Solar System science, for example tests of general relativity and the exploration of the Kuiper belt. The combination of highly sensitive atomic sensors and of the laser link well adapted for large distances will allow measurements with unprecedented accuracy and on scales never reached before. We present the proposed mission with particular emphasis on the measurements and payload technology.

1. Introduction

The SAGAS mission will study all aspects of large scale gravitational phenomena in the Solar System using quantum technology, with science objectives in fundamental physics and Solar System exploration. The large spectrum of science objectives makes SAGAS a unique combination of exploration and science, with a strong basis in both programs. The involved large distances (up to 53 AU) and corresponding large variations of gravitational potential combined with the high sensitivity of SAGAS instruments serve both purposes equally well. For this reason, SAGAS brings together traditionally distant scientific communities ranging from atomic physics through experimental gravitation to planetology and Solar System science.

The payload will include an optical atomic clock optimised for long term performance, an absolute accelerometer based on atom interferometry and a laser link for ranging, frequency comparison and communication. The complementary instruments will allow highly sensitive measurements of all aspects of gravitation via the different effects of gravity on clocks, light, and the free fall of test bodies,

thus effectively providing a detailed gravitational map of the outer Solar System whilst testing all aspects of gravitation theory to unprecedented levels.
SAGAS technology takes advantage of the important heritage from cold atom technology used in PHARAO and laser link technology designed for LISA (Lasers Interferometric Space Antenna). It will provide an excellent opportunity to develop those technologies for general use, including development of the ground segment (Deep Space Network telescopes and optical clocks), that will allow such technologies to be used in many other mission configurations for precise timing, navigation and broadband data transfer throughout the Solar System.
The trajectory design is driven by the available launcher options in the Cosmic-Vision frame. With the largest launcher available in that frame (Ariane5 ECA) the optimum is a ΔV-EGA trajectory with an Earth gravity assist 2 years after launch and a Jupiter gravity assist another 1.5 years later. Hyperbolic escape velocity is reached after Jupiter and the SAGAS spacecraft reaches 50 AU in a hyperbolic coast. With the chosen trajectory SAGAS will reach a heliocentric distance of 38.9 AU over the nominal mission duration (15 years) and 53.3 AU over the extended mission (20 years). With other existing, more powerful launcher options (AtlasV, DeltaIV, Proton) the first Earth gravity assist can be omitted, leading to a two year reduction in travel time. However, those launchers were not available in the Cosmic Vision frame.
In this short contribution we only briefly discuss the science objectives, with some more detail on the payload technology. A much more detailed description of the complete mission can be found in [1].

2. Science Objectives

SAGAS will carry out a large number of tests of fundamental physics, and gravitation in particular, at scales only attainable in a deep space experiment. The unique combination of onboard instruments will allow 2 to 5 orders of magnitude improvement on many tests of special and general relativity, as well as a detailed exploration of a possible anomalous scale dependence of gravitation. It will also provide detailed information on the Kuiper belt mass distribution and determine the mass of Kuiper belt objects and possibly discover new ones. During the transits, the mass and mass distribution of the Jupiter system will be measured with unprecedented accuracy.
It is interesting to note that the sensitivity of SAGAS for most science objectives can be significantly enhanced when making optimal use of the asynchronous two-way optical link in the post-processing of the data, as discussed in detail in

[4]. The science objectives are summarised in table 1. They are investigated in more detail in [1,2].

Table 1. Science objectives of SAGAS in Fundamental Physics and Solar System science. (GR: General Relativity, PA: Pioneer Anomaly, IS: Ives-Stilwell, M_E: Earth mass).

Science Objective	Expected Result	Comments
Test of Universal Redshift	1×10^{-9} of GR prediction	10^5 gain on present
Null Redshift Test	1×10^{-9} of GR prediction	10^3 gain
Test of Lorentz Invariance	3×10^{-9} to 5×10^{-11} (IS or "time dilation" test)	10^2 to 10^4 gain fct. of trajectory
PPN test	$\delta(\gamma) \leq 2\times10^{-7}$	10^2 gain may be improved by orbit modelling
Large Scale Gravity	- Fill exp. data gap for scale dependent modif. of GR - Identify and measure PA to < 1% per year of data	Different observation types and large range of distances will allow detailed "map" of large scale gravity
Variation of Fund. Const.	$\delta\alpha/\alpha \leq (2\times10^{-9})\ \delta(GM/rc^2)$	250-fold gain on present
Upper limit on Grav. Waves	$h \leq 10^{-18}$ @ 10^{-5} to 10^{-3} Hz	Integration over one year
Kuiper Belt (KB) Total Mass	$\delta M_{KB} \leq 0.03\ M_E$	Dep. on mass distribution and correlation of clock meas.
KB Mass Distribution	Discriminate between different common candidates	Will contribute significantly to solution of the "KB mass deficit" problem
Individual KB Objects (KBOs)	Measure M_{KBO} at ≈ 10%	Depending on distance of closest approach
Planetary Gravity	-Jupiter Gravity at $\leq 10^{-10}$ -Study Jupiter and its moons	10^2 gain on present for Jupiter idem for other planet in case of 2nd fly-by
Technology Developement	Develops S/C and ground segment technologies for wide use in future missions (interplanetary timing, navigation, broadband communication,...)	

3. Payload

The SAGAS payload is composed of two instruments (accelerometer and optical clock) and the optical link which is used as a scientific instrument (frequency comparison and Doppler measurements), and also for data transfer. A particular feature of the SAGAS payload is the synergy between the different payload parts which significantly simplifies the overall design and reduces cost and technology development: The accelerometer also provides the absolute calibration of the quartz Ultra Stable Oscillator (USO) with respect to the Cs atoms, which in turn is used to generate all on-board RF signals used in the optical clock, the link and the time-tagging of observations. The optical clock and optical link share the same frequency, thereby avoiding use of a femtosecond frequency comb and allowing common use of some of the laser sources, with the clock providing a highly narrow and stable laser for the link. In short, the combination of on board instruments is not only optimal and complementary for the science objectives, but also close to ideal for the simplicity and coherence of the technology.

3.1. *Cold Atom Accelerometer*

To reach the scientific objectives of SAGAS, accurate measurements of accelerations along three orthogonal axes are required. The design payload and its characteristics arise from the development of the fields of cold atom physics and atom interferometry. Key technologies are identical to those already developed within the ACES project for the PHARAO payload. Compared to PHARAO, the SAGAS accelerometer shares identical key technologies and similar payload architecture and subsystems.

The accelerometer is based on the use of cold atoms and Raman transitions for the manipulation of the atomic wave-packets. The atoms are Cs atoms, which can be easily cooled in an optical molasses using all solid-state semi-conductor diode lasers. The sensitivity depends only on the wave vector **k** of the Raman lasers and the square of the time between pulses T [5]. Each cycle (cooling, state preparation, interrogation sequence, detection) lasts 3 s, for an interrogation time T = 1 s with about $3\text{x}10^5$ detected atoms. The three axes of acceleration are successively measured within the same vacuum tube using three orthogonal pairs of Raman lasers.

The sensitivity to acceleration per cycle is given by $1/(kT^2.\text{SNR})$, where SNR is the signal to noise ratio of the measurement. Different sources of noise limit the SNR: atom shot noise, phase noise between Raman lasers, acceleration noises. The latter is due essentially to S/C vibrations stemming from unbalance of the attitude control fly wheels and was estimated using Bepi-Colombo mission studies.

It turns out that the limiting effect is the residual phase noise between Raman lasers due to the imperfections of the reference frequency at 9.2 GHz. By using the spectral noise density of the PHARAO flight model frequency reference (Quartz USO), we obtain a SNR=200 per shot (3s total) leading to a sensitivity limit of $3\text{x}10^{-10}$ ms^{-2} per shot. As the sensitivity improves with the square root of the measurement time, the required sensitivity ($5\text{x}10^{-12}$ ms^{-2}) is obtained after only 3 h per axis, which gives 9 h for the three axes.

To achieve the requirement for SAGAS, the accelerometer has to be accurate at a level of $5x10^{-12}$ ms^{-2}, *i.e.* all acceleration biases need to be controlled at that level. Compared to ground experiments, the absence of gravity, and therefore of average velocity of the atomic cloud, allows reduction of most of the systematics. Moreover, some of the error sources decrease with increasing interaction time *T*, as they depend on the interaction with the Raman laser. These considerations make possible the extrapolation from ground performances to the needs for SAGAS. Main sources of systematic error are: magnetic field effects, one- and two-photon light shifts, laser wave-front distortions, S/C vibrations and self-gravity. The main limiting effect on accuracy comes from laser wave-front distortions. Although the effect can be calibrated *in situ* by changing the interaction time and/or the temperature of the atomic sample, more detailed ground studies should be carried out to confirm that the effect can be controlled at the required $5x10^{-12}$ ms^{-2} level.
As all the elements needed to realize a micro-wave atomic clock at moderate accuracy (10^{-12} in relative frequency) are present in the payload, one can measure periodically the drift of the reference quartz oscillator, which is used to generate all radio or micro-wave frequencies for the accelerometer, the optical clock, the optical link and the on-board time scale. The atomic source and detection system are the same as for the accelerometer and the interrogation may be done by the micro-wave antenna used for the state preparation. Sensitivity of 10^{-12} per shot and the required 10^{-12} accuracy can easily be achieved.

3.2. *Optical Trapped Ion clock*

The clock component of the overall mission scenario is based on the use of an on-board optical frequency standard with frequency stability $\leq 10^{-17}$ for 10 day integration times (i.e. an Allan variance of $\sigma_y(\tau) \leq 1x10^{-14}\ \tau^{-1/2}$), where clock frequency data is downlinked to Earth by means of a high power link laser. This required performance is one order of magnitude better than the PHARAO clock and requires the development of a new generation of space clocks. Instead of the microwave frequencies used in PHARAO, the SAGAS clock uses optical frequencies. Ground based optical clocks are evolving at a significantly fast rate, with one or two single ion clocks already demonstrating instabilities ~ $4x10^{-17}$ at 10^4 s and $2x10^{-17}$ accuracy in realising the unperturbed ion frequency [6]. There is good reason to expect similar performance for a range of other optical clock systems, which is extendable to below the target specification at longer times.
Currently, there are two distinct high accuracy optical atomic clock architectures, one based on a single cold trapped ion approach, and one on cold neutral atoms

held within an optical lattice. The single ion architecture presently represents a higher technology readiness level than is the case with neutral atom lattice clocks. In addition neutral atom clocks exhibit larger mass, volume and power consumption when compared to the ion clock arrangement. As a result, only the single ion clock is considered for this SAGAS proposal.

Within this ion clock arena, there are a number of possible ion species and isotopes which have been developed with state-of-the art characteristics: $^{199}Hg^+$, $^{27}Al^+$, $^{115}In^+$, $^{40}Ca^+$, $^{171}Yb^+$ quadrupole, $^{171}Yb^+$ octupole, $^{88}Sr^+$. For all species the quantum-limited theoretical stabilities surpass the stability requirement of 10^{-17} @ 10 days. The experimentally observed stabilities are typically a factor 3 to 5 above this limit, but still within the 10^{-17} @ 10 days specification.

Technological complexity in a space environment (generation of deep UV wavelengths, trap technology, UV optics and fibres, alignment tolerances,...) excludes a number of the above mentioned species, leaving three possibilities: $^{40}Ca^+$ (729 nm), $^{171}Yb^+$ quadrupole (436 nm), and $^{88}Sr^+$ (674 nm). These show little difference in sensitivity to perturbing effects, essentially the effect of blackbody radiation and electric fields from the trap. As a result the choice is based on the availability of the required laser technology and the present state of ground systems. Taking together the currently available power levels for both clock and cooler, the need for only one single pass doubling stage, clock laser linewidth and the clock accuracy already achieved, it is considered that the $^{88}Sr^+$ 674 nm clock represents the most feasible ion clock option at this time. It is acknowledged that the technology underpinning all three options has the capability to evolve in the near to medium term, and this should be a core component of technology refinement activity during early stages of the L class mission preparations.

The $^{88}Sr^+$ ion clock system design comprises a number of sub-components, including:

(i) The primary physics package, an RF end-cap trap for ionising and confining a single $^{88}Sr^+$ ion within an ultra-high vacuum chamber pumped by a small ion pump and non-evaporable getter pump.

(ii) A laser platform providing Doppler cooling of the ion and auxiliary lasers for repumping and clear-out.

(iii) The 674 nm clock laser frequency FM-stabilised to a very high finesse ultra-low-expansion (ULE) cavity mounted on a temperature-stabilised and evacuated platform. The clock laser is directly linked to the laser link (DOLL), see section 3.3.

(iv) A monitoring and control processor driving the clock sequence (cooling-probing-detection). The processor also monitors frequency and amplitude data

necessary to determine normal laser and ion operational conditions and initiate resetting and recovery algorithms where necessary, and laser unit failure.
(v) A redundancy level of 2 or 3 units for both cooling, clock and high power link laser, plus a redundancy level of 3 units for the repumper and clear-out DFB lasers. All redundancy units for each wavelength to be fibre multiplexed as standard, allowing redundant unit activation on determination of prior unit failure mode.
The progress in the accuracy of single ion optical clocks has been rapid and it is foreseeable that within the next few years the level of 1 part in 10^{17} will be demonstrated in several ground based systems, including Sr^+. The $^{88}Sr^+$ ion clock systematic frequency dependencies are shown in [1,2] in the form of the expected uncertainty budget. The two dominant contributions are the linear Zeeman shift and the black body radiation shift. They will be controlled at the 10^{-17} level using Zeeman pair averaging and magnetic shielding for the former and ± 1 K temperature control for the latter. Thus, overall uncertainty of $\sim 10^{-17}$ is achievable within the spacecraft environment, provided external magnetic field and temperature variations are sufficiently low and adequately controlled.
Reloading of the trap will be necessary over a mission period of 15 years. A miniature dispenser, containing a few mg of Sr and a hot wire as an electron source are needed. Automatic reloading of the trap will induce an interruption of clock operation for a few minutes. Charging problems of the trap and vacuum system will be avoided by illumination with ultraviolet LEDs. The discharge procedure will also be used periodically to eliminate charges from high energy radiation
Clock payload requirements are developed by extrapolation from existing ground-based clock arrangements, together with attention to existing laser and opto-electronic hardware and reference to the existing ACES/PHARAO microwave clock payload. This includes a redundancy level of three lasers for both the clock laser ECDL and the high power 844 nm cooling laser and single pass doubler. The use of available DFB lasers for the 1092 nm repumper laser and 1033 nm clear-out laser will also allow a redundancy level of at least three per wavelength.

3.3. *Deep space Optical Laser Link (DOLL)*

We propose an original optical link concept for SAGAS that takes full advantage of the particular technology available on-board and the synergy between payload components (narrow, stable and accurate laser from the clock, accurate microwave from the accelerometer) whilst being specially tailored to achieve the required science objectives. Concerning the technology, particular emphasis was paid to making maximum use of existing developments (ACES/PHARAO diode laser technology, LISA telescope technology, SLR/LLR ground stations,). DOLL features in particular:

- continuous wave laser operation in both directions (two-way system).
- heterodyne detection schemes on-board and on ground.
- high data transfer rate with simultaneous science measurements.
- asynchronous operation allowing optimal combination of on-board and ground measurements.
- large stray light rejection from heterodyne detection and due to the possibility of a large controlled frequency offset between the up and down link using the accurate on-board microwave.

More generally, we believe that DOLL will not only find its use in SAGAS, but also contribute significantly toward the development of general optical, high accuracy interplanetary navigation and broadband communication.

DOLL is based on continuous two-way laser signals exchanged between the ground station and the S/C, with independent heterodyne detection of the incoming signal at both ends (no transponder scheme). The fundamental measurement is the frequency difference between the local oscillator and the incoming signal. This measurement is particularly adapted to SAGAS because of the availability of the very narrow clock laser ($\approx$ 10 Hz linewidth) on board and on the ground, stabilized to the atomic transitions. The on-board and ground data are combined in post treatment and analysed in order to extract the science observables: Doppler ($\rightarrow$ velocity difference), clock frequency difference, ranging. This allows asynchronous operation, *i.e.* the combination of ground and onboard measurements (sum or difference) with a delay that can be freely chosen thereby optimising the signal to noise ratio individually for any given science objective (see [4] for details).

To avoid complexity onboard the spacecraft, the signals emitted on the ground will be offset in frequency to largely compensate for the Doppler frequency shift (up to 70 GHz), so the frequency received at the S/C is close to the nominal clock frequency, allowing direct heterodyne beat with the local clock laser. Additionally, the up and down signals will be linearly polarized with orthogonal polarizations (for stray light rejection), which implies variable polarization direction for the ground station telescope.

Data is encoded onto the laser signals of the up and down link by modulating the phase of the reference microwave signal. Data is decoded by using a tracking oscillator phase locked on the incoming signal. The corresponding achievable data rate is about 3000 bps at 30 AU with a negligible probability of error bits or cycle slips. The encoded signals also allow rough ranging and timing at the ≈3 m (10 ns) level.

Note that for orbit determination the Doppler and ranging observables are, in principle, redundant. For SAGAS the precise orbit is obtained from the Doppler, whilst the ranging (orders of magnitude less precise) serves only to improve initial conditions for the integration.

The main subsystems of the DOLL space segment are the telescope and the optical bench providing the laser source (see [1,2] for details). The present baseline is to use a telescope design similar to that of the LISA mission (40 cm aperture), adapted for the SAGAS wavelength (674 nm) and including a high definition CCD camera (similar to COROT or LISA).

The ground segment for DOLL will consist of several (3 to 6, with a minimum of 3 to ensure permanent coverage) laser tracking stations. Present satellite and lunar laser ranging stations are well adapted for this purpose, but will require upgrades concerning the optical set-up (wavelength, polarization control, pointing accuracy, adaptive optics) and need to be equipped with high performance optical clocks

The main error sources affecting the performance of DOLL are related to power and pointing issues, stray light and effects of the Earth's atmosphere.

Assuming 1 W at emission, at a distance of 30 AU, a total attenuation of ≈ 0.14 from losses in the atmosphere, the telescope, and from a pointing error of 0.3" leaves about 22000 photons/s at the photodiode, which corresponds to negligible shot noise of the heterodyne detection.

Stray light issues are strongly reduced by the narrow filter of the heterodyne detection PLL (bandwidth ≈ 1 kHz). Particular care has to be taken to reduce in-band stray light from the outgoing signal. This is achieved by offsetting the outgoing frequency by about δf = 1 to 10 GHz using a PLL driven by the local RF signal. The main limiting effect on the performance of DOLL will be fluctuations due to the Earth's atmosphere. At high frequency (around the 1 kHz bandwidth of the detection PLL) these can lead to frequent cycle slips and/or loss of lock. At low frequency atmospheric effects lead to phase fluctuations that can affect directly the observables and corresponding science measurements. These effects are discussed in detail in [1,2], showing that they are only significant at short integration times ($< 10^3$ s), but below the clock or accelerometer noise at longer integration times. Additionally atmospheric noise can be further reduced when using appropriate combinations of the up and down link [4].

The main DOLL critical technology issues are the 0.3" pointing requirement and the laser reliability. They are discussed in more detail in [1,2]. Much of the

DOLL technology is based on heritage from PHARAO (ECDL, AOM, RF synthesis, Quartz,...), COROT (CCD, pointing) and on LISA technology developments (telescope, optics,...). That ensures a relatively high TRL (Technology Readiness Level) for most of the sub-systems, which is however contingent on the reliability and the development of the high power laser source at 674 nm.

Finally, we note, that the development of DOLL as well as the optical clocks for the DOLL ground stations fits well into current technology drives towards the use of optical clocks and optical communication in NASA and ESA interplanetary missions and DSN. Experimental work in a terrestrial environment (ground and low Earth orbit) is underway in order to demonstrate the feasibility of DOLL and experimentally quantify perturbing effects, in particular due to the atmosphere.

4. Conclusion

We have presented a brief description of the scientific aspects of the SAGAS project, with a little more emphasis on the payload technology. This work is based on the original mission proposal submitted to ESA in June 2007 in response to the Cosmic Vision 2015-2025 call for proposals. Complete detail can be found in [1,2]. The outcome of the ESA selection procedure is now known, and unfortunately SAGAS has not been deemed a priority. However, independently of that outcome, we believe that it is worth pursuing the investigations initiated by the Cosmic Vision call, in order to further study the scientific and technological implications of this type of mission. In that respect current R&D is oriented towards development of atomic sensors and optical clocks for space applications, as well as the realization of coherent optical links for deep space navigation, clock comparisons and broad-band data transfer.

References

1. P. Wolf, et al., arXiv : 0711.0304, (2007).
2. P. Wolf, et al., *Experimental Astronomy*, http://dx.doi.org/10.1007/s10686-008-9118-5, (2008).
3. Ph. Laurent, et al., *Appl. Phys.* **B84**, 684-690, (2006).
4. S. Reynaud, et al., Phys. Rev. **D77**, 122003, (2008). arXiv:0801.2896.
5. A. Peters et al, Metrologia **38**, 25-61, (2001).
6. T. Rosenband, et al., Science **319**, 1808 (2008)

SMALL MERCURY MICROWAVE ION CLOCK FOR NAVIGATION AND RADIO-SCIENCE*

JOHN D. PRESTAGE, SANG CHUNG, ROBERT THOMPSON, PAUL MACNEAL, THANH LE

Jet Propulsion Lab, California Institute of Technology, 4800 Oak Grove Dr. Pasadena, CA, 91109, USA

We have recently completed a breadboard ion-clock physics package based on Hg ions shuttled between a quadrupole and a 16-pole rf trap. With this architecture we have demonstrated short-term stability ~1-2x10^{-13} at 1 second, averaging to 10^{-15} at 1 day. This development shows that H-maser quality stabilities can be produced in a small clock package, comparable in size to an ultra-stable quartz oscillator required for holding 1-2x10^{-13} at 1 second. This performance was obtained in a sealed vacuum configuration where only a getter pump was used to maintain vacuum. Because the tube is sealed, the Hg source and Neon buffer gas are held indefinitely, for the life of the tube. There is no consumption of Hg in this system unlike in a Cs beam tube where lifetime is often limited by Cs depletion. The vacuum tube containing the traps has now been under sealed vacuum conditions for over three years with no measurable degradation of ion trapping lifetimes or clock short-term performance. A 3-liter prototype physics package now in development will be described.

1. Introduction

Small atomic clocks operating in deep space could be used for many planetary science investigations. Ultra-stable quartz crystal oscillators (USO's) have flown on many missions for planetary atmosphere investigations during occultation[1]. A small space-qualified atomic frequency standard with stability as good as 10^{-15} over several hours averaging interval would enable one-way deep space navigations, where Doppler data is accumulated in a down-link only fashion. Currently, deep space navigation is implemented by measuring the Doppler frequency shift of a 2-way link from a ground station to a spacecraft (s/c) and the coherent return link. Typically, these links are maintained for 7-8 hours per s/c track, requiring full use of a 34-meter antenna in the Deep Space Network (DSN) for the time the s/c is sufficiently above the horizon. With a suitable on-board atomic clock, the navigational Doppler beacon could be

* This research was carried out at the Jet Propulsion Laboratory, California Institute of Technology, under a contract with the National Aeronautics and Space Administration.

generated at the s/c so that many ground radio telescope facilities could be used for navigation since no specialized high power uplink would be required. Additionally, when more than one s/c orbit the same planet, they can be tracked simultaneously with one earth based antenna. Multiple s/c tracking by a single antenna can reduce antenna usage and mission navigation costs.

2. Physics Package Status

A breadboard physics package has been operating for a few years to investigate short-term clock stability, operational lifetime, and in-vacuum operation. A second generation physics package is now under construction and a 3d drawing is shown in Figure 1. The mechanical package shown is designed to withstand generic random vibration levels of over 10 g_{rms}. A modal analysis was made of the structure using more than 3000 nodal mesh points to determine resonance frequencies of various mechanical subassemblies. All frequencies are 200 Hz and higher, adequate for many launch requirements. Various electronics circuit boards can be designed to conform to the spaces available within this structure. The functional layout is described in reference 1.

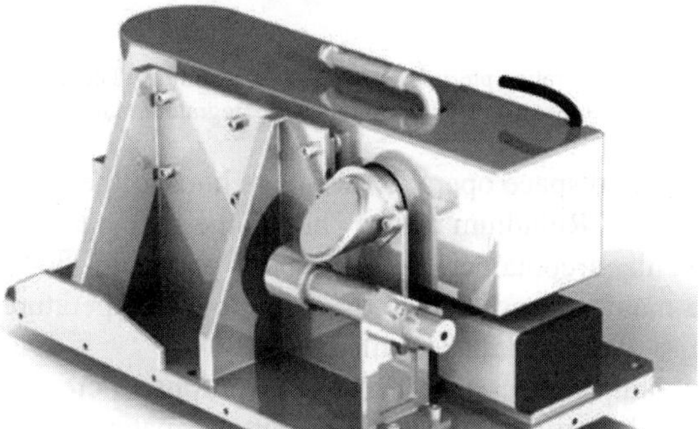

Figure 1. The ~ 3 liter physics package with baseplate, photomultiplier tube and vacuum tube mounting through the two layer magnetic shield. An outer magnetic shield is not shown.

3. Lamp Lifetime Strategies

The ion clock relies on optical pumping to prepare and to detect the $^{199}Hg^+$ microwave 40.5 GHz transition. The light source is a fused silica bulb ~ 12 mm diameter by ~ 30 mm in length filled to ~ 1 Torr Argon buffer gas with a quantity of metallic ^{202}Hg. The lamp is excited at ~ 200 MHz inside a resonant LC tank circuit, with the bulb placed inside a 2 turn coil to inductively excite a discharge inside. Typically 10W is dissipated within the exciter circuit and bulb for clock operation. Under these and similar operating conditions lamps have functioned well for 5 years or more.

Figure 2. The second-generation physics package under assembly. The package volume is ~ 3 liters. Various electronic boards are also being developed to integrate into the full instrument.

For in-vacuum space operation we have housed the lamp bulb in a package similar to the GPS Rubidium clock. That is, the bulb and exciter coil are placed inside an ovenized container, with the bulb potted in an Indium thermal heat sink base to maintain and control the cold point temperature and consequently vapor pressure of Hg inside the sealed bulb.

Mercury lamps are used in many applications to create UV light for water purification and sterilization. UV transmitting materials become opaque over time when exposed to high fluxes of UV light and possible chemical interactions with Hg vapor inside, as hundreds of watts of power are used to generate either 254 nm or 185 nm light. This degradation is similar to the slow darkening seen in our Hg clock application, though the rate of darkening is accelerated because higher power and temperature are typically used in UV light purification/ sterilization applications.

Alternate materials have been investigated which may be more resistant to chemical reactions with Hg and also more UV resistant. The most successful of these is Sapphire as described by Hereaus[2] where a sapphire coating inside of the fused silica bulb acts as a barrier to chemical reactions where HgO is formed near the surface of the SiO_2 bulb. Figure 3 show a dramatic decrease in darkening of the glass when sapphire is applied. Other materials and a description of the process is given in reference 3.

Sapphire bulb material has been proposed for use in GPS Rubidium. Some test Rubidium bulbs were built[4] and showed behavior similar to conventional glass bulbs.

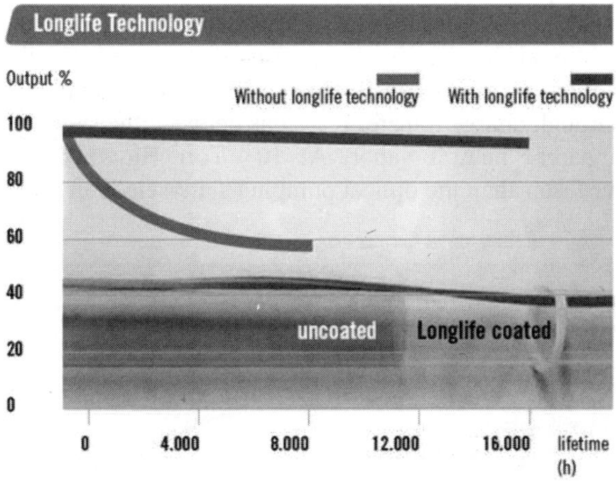

Graph of operating life of an uncoated amalgam lamp and a Heraeus amalgam lamp (254nm, 300 W) with Longlife coating.

Figure 3. Improvements to the lifetime of fused silica Hg ultraviolet lamps have recently been achieved. See reference 2.

4. Simplifying Ion Clocks for Space and Commercial Use

A major simplification and reduction of the clock instrument has come about by eliminating active vacuum pumps from the physics vacuum tube. Adapting fabrication methods of modern vacuum tube amplifiers, the ion tube is constructed from materials to withstand bake-out to ~ 400 °C, well above the operating temperature for many glass-to-metal seals for microwave and UV optical windows. Titanium-to-alumina and titanium-to-sapphire joints are a practical solution since these joints can be baked to 400 °C, UV grade sapphire windows can transmit light at 194 nm, and titanium is sufficiently non-magnetic and will not degrade the high-Q ion clock transitions. Following bake-out, the tube is then sealed-off with getter pump elements inside.

The best indication of vacuum quality inside the tube is the length of time ions are maintained within the trap. Hg ions can be lost from the trap via charge transfer reactions with residual background gases. Data measured nearly a year following the tube seal-off from the pump stand shows trap time up to 1000 hours (figure 4), several hundred times improvement over "open" systems bake-able only to 200 °C with vacuum supplied by a mechanical turbo-pump. The sealed tube approach increases system reliability, improves vacuum quality, and eliminates mass and power required for active pumping.

In normal operation, the sealed vacuum tube does not consume Hg, an advantage since a potential life-limiting mechanism is eliminated. The tube was charged with ^{199}Hg vapor from a small appendage HgO oven. After filling to adequate operational levels, the oven was de-powered. If the HgO oven remained on following tube seal-off, Hg vapor would build up extinguishing the clock signal through charge transfer reactions between the state-selected trapped ions and the parent neutral vapor. At 10^{-8} Torr Hg vapor pressure, these reactions occur faster than the optical pumping state-selection process.

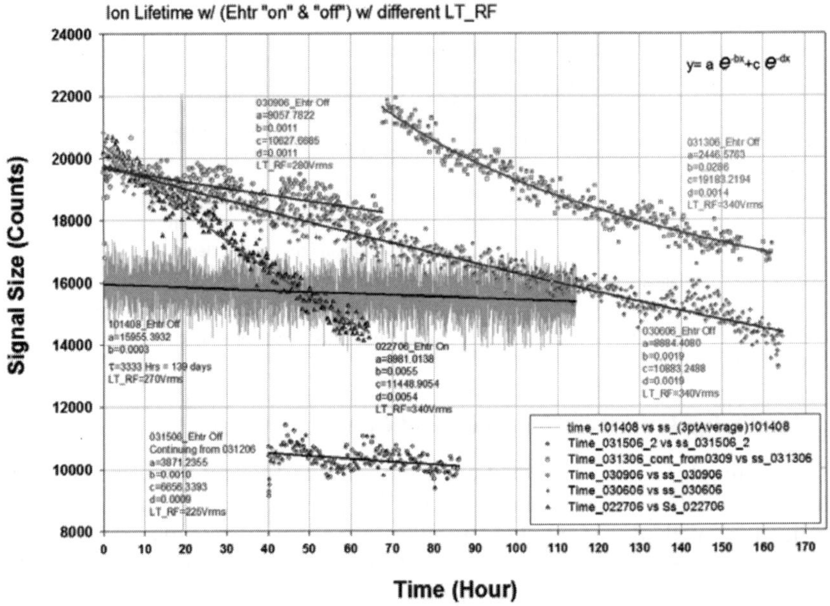

Figure 4. Ion lifetime within the quadrupole loading trap. The slow decay of the ion fluorescence signal is found to depend on the rf level of the confining field. The blue curve was measured nearly three years after all other plotted decay curves.

4. Novel Line acquisition methods

The ion clock resonance signal shown in Figure 5 is derived from the classic Rabi single microwave square-wave pulse[5]. The resonance linewidth is determined by the duration, T, of the pulse. The resulting frequency linewidth is 0.799/T ~ 0.27 Hz for typical ion-clock measurements, corresponding to a 3 second microwave pulse.

To measure the $^{199}Hg^{+}$ clock reference frequency, the output of a tracking synthesizer is typically stepped to either side of the resonance line center, to the points of steepest frequency slope. The sequence of measurements yields a sequence of numbers, C(1), C(2), C(3),...C(n), each the number of photon pulses accumulated in some time interval, usually 2 seconds or less. Such successive fluorescence measurements are made on opposite sides of line center. The simplest proportional loop adjusts center frequency of modulation according to

$$\Delta f \propto R_{loop} Lw \frac{C(1)+C(3)-2C(2)}{2\pi\, SignalSize} \tag{1}$$

in order to "track" the line center. In this expression, *Lw* is the linewidth (.799/T for the Rabi interrogation) and *SignalSize* is the height of the resonance signal above background light level as shown in Figure 5. The sequence of frequency corrections, Δf are summed so that the tracking synthesizer approaches the line center approximately exponentially. The "second difference" *C(1)+C(3)-2C(2)* is used so that linear drifts in the UV light source output will not cause a frequency offset.

Signal size depends on several variables including the microwave power so that occasionally the sequence of frequency measurements is interrupted to measure the signal size via fluorescence measurements on line center and several linewidths away from line center, typically 1 Hz or more. Another method of determining the line-center is based upon the analytical expression for the Rabi line shape[5],

$$P = \frac{(2b)^2}{(\omega-\omega_0)^2+(2b)^2}\sin^2\left(\frac{T}{2}\sqrt{(\omega-\omega_0)^2+(2b)^2}\right)$$

$$= \frac{\sin^2\left(\frac{\pi}{2}\sqrt{1+y^2}\right)}{1+y^2} \equiv Rabi(y) \quad \text{where} \tag{2}$$

$$y^2 = \frac{(\omega-\omega_0)^2}{(2b)^2} = \left(2T(\nu-\nu_0)\right)^2 \tag{3}$$

and we have assumed power is optimized, $bT = \frac{\pi}{2}$

The steepest point for use in frequency discrimination occur at $y \cong \pm\, 0.761$, while the half-max points occur at $y \cong \pm\, 0.799$.

In practice, the Rabi resonance line shape is a very good model for the measured signal as shown in Figure 5. The full measurement yields a background light level, Bck, a signal-size, Amp, and a line-center frequency, ε. The total light collected vs. frequency is thus described by three parameters as

$$Light = Bck + Amp * Rabi(y - \varepsilon) \tag{4}$$

A method for measuring frequency stability of the resonance line would be to make 3 frequency measures as shown in Figure 5, and invert the equation

$$Light = Bck + Amp * Rabi(y - \varepsilon) \tag{5}$$

to extract values for *Bck*, *Amp*, and ε.

This method allows frequency tracking of the line-center, collection of engineering data of both lamp light output and signal size. The signal size measurement can be used to continuously optimize microwave power levels, a useful measurement for long term autonomous clock operation. This line acquisition method is complete in three measurements unlike the proportional method described earlier which only approaches line-center exponentially.

In practice, three measurements, C(1), C(2) and C(3), are made at different frequencies y_1, y_2, and y_3. The following 3 equations are then inverted.

$$\begin{aligned} C(1) &= Bck + Amp * Rabi(y_1 - \varepsilon) \\ C(2) &= Bck + Amp * Rabi(y_2 - \varepsilon) \\ C(3) &= Bck + Amp * Rabi(y_3 - \varepsilon) \end{aligned} \tag{6}$$

Three values for y are shown in Figure 5, $y \cong \pm 0.761$ and $y = 0$. To monitor the line-center frequency, the 3 most recent measurements are used as described above to determine line-center frequency, background light level and signal size. Every measurement is followed by an inversion of the three equations and yields a new set of parameter estimates. Figure 6 shows simulated frequency tracking where center frequency is changed from $\boldsymbol{y} = 0$ to $\boldsymbol{y} = -0.75$ at step 10 of the 100 step sequence. This simulation is based on an average signal-to-noise ratio of ~ 32. The response time is very fast compared to the proportional methods. The step size of this instantaneous line-center change can be too large for the tracking to follow, in part because far from line-center, there may not be a unique inverse solution to the 3 equations at the heart of the method.

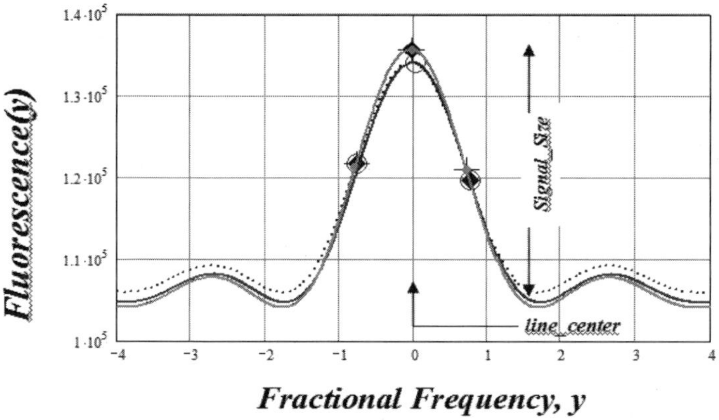

Figure 5. The theoretical Rabi line-shape with a constant background light level Bck = 105,000 counts and Signal Size = Amp = 30,000. A signal-to-noise level of ~ 30 has been simulated, and 3 successive solutions are shown.

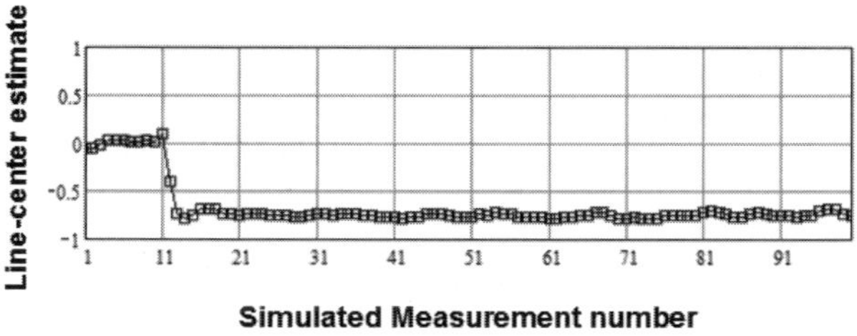

Figure 6. The response of the tracking algorithm is shown for an instantaneous change in line center (at step 10) by y = 0.75.

This non-uniqueness can also cause problems as the Signal-to-Noise ratio diminishes. Because this method separates the influences of background light, signal size and line-center, a linear drift in the light level from the lamp does not seem to force a frequency pulling of the estimated line center. Figure 7 shows a simulated large light level drift imposed over the simulated frequency tracking noise data, with no evidence of frequency pulling.

We have tested this algorithm in laboratory measurements where $^{199}Hg^+$ ion frequency was tracked using an H-maser local oscillator. The magnetic shields for the ion clock were removed so that ambient magnetic fields would change the clock frequency to test the algorithm's ability to follow changes. The Allan deviation of the resulting sequence of line-center frequency estimates is shown in Figure 8.

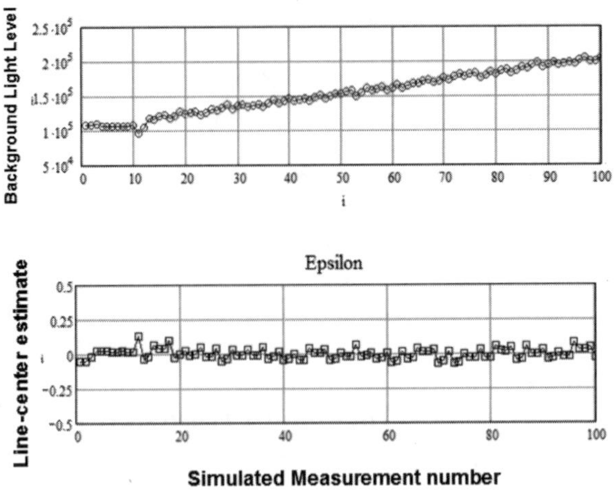

Figure 7. The non-linear curve-fit tracking algorithm does not show a frequency shift when the light-source drifts in its output level. In the above simulated line tracking, a large linear drift was imposed on the light level starting at measurement 10.

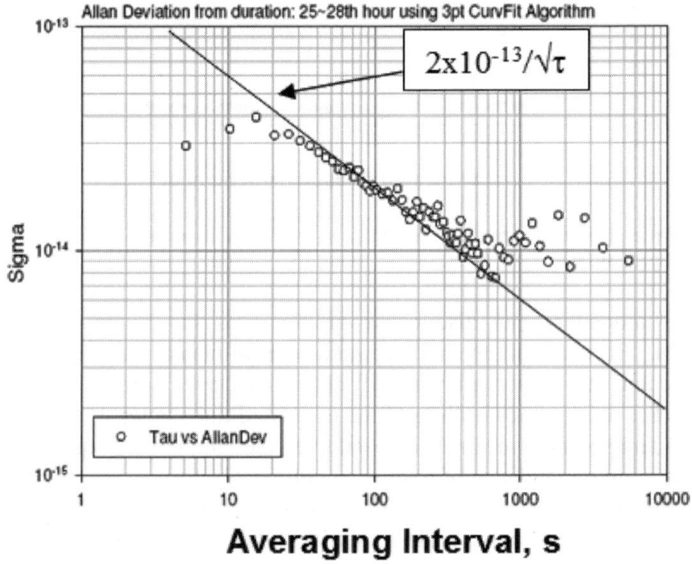

Figure 8. In-lab frequency tracking of the 199Hg+ clock transition using the non-linear curve-fit tracking algorithm as outlined in this paper. During this measurement the ion clock magnetic shields were removed resulting in frequency instabilities beyond 1000 seconds averaging interval.

Acknowledgments

This work was carried out by the Jet Propulsion Laboratory, California Institute of Technology, under contract with the National Aeronautics and Space Administration.

References

1. John D. Prestage and Gregory Weaver, "Atomic Clocks and Oscillators for Deep-Space Navigation and Radio Science", *Proc. IEEE*, vol. 95, No. 11, November 2007. pp. 2235-2247.
2. See *Heraeus Website*, http://www.heraeus-noblelight.com/fileadmin/ user_upload/PDF/disinfection/Lampe_eng.pdf
3. A V Krasnochub and A I Vasiliev, "Qualitative model of the operating mechanism of the protective coating for low pressure Hg lamps", *J. Phys. D: Appl. Phys.* 39 (2006) pp. 1378–1383.
4. Thomas C. English and Ernst Jechart, "Development of a Sapphire Lamp for Use in Satellite-Borne Atomic Rubidium Clocks", Proc. *35th Ann. Frequency Control Symposium, Ft. Monmouth*, NJ, 1981, pp. 637-645.
5. Norman F. Ramsey, "Molecular Beams", Oxford press, 1985, p. 118.

ASTRO-COMB: REVOLUTIONIZING PRECISION SPECTROSCOPY IN ASTROPHYSICS

C. E. Cramer, C.-h. Li, A. G. Glenday, D. F. Phillips, G. Furesz, D. Sasselov, A. Szentgyorgyi, and R.L. Walsworth

Harvard-Smithsonian Center for Astrophysics, Cambridge, MA, 02138, USA

A. Benedick and F. X. Kärtner

Research Laboratory for Electronics, Massachusetts Institute of Technology, Cambridge, MA, 02139, USA

Precision Doppler spectroscopy in astrophysics is limited by the stability of calibration sources, as well as the number, width and distribution of available calibration lines. Broadband frequency combs generated from mode-locked femtosecond lasers have revolutionized precision laboratory spectroscopy. We present a laser comb with up to 40 GHz line-spacing, generated from a 1 GHz repetition-rate "source comb" and a Fabry-Perot filtering cavity. The line spacing can be optimized for use with any high-resolution astronomical spectrograph. Our prototype, which we have recently tested with the Tillinghast Reflector Echelle Spectrograph on Mt. Hopkins, provides a 100-nm band of calibration light centered on 850 nm, in the center of the a region difficult to calibrate with atomic sources. The astro-comb should allow more than an order-of-magnitude improvement in sensitivity to changes in Doppler-shifts and cosmological redshifts, with significant impact on many fields, including stellar seismology, cosmology, and the search for extrasolar earths.

Keywords: laser frequency combs; radial velocity measurements; spectrograph calibration

1. Precision Spectroscopy in Astrophysics

High-precision Doppler-shift/redshift spectroscopy ushered in a new era in astronomy and astrophysics at the end of the 20th century. In particular, the search for extrasolar planets has benefited tremendously from high precision measurements of stellar spectra. Extending the reach of precision astronomical spectroscopy into the near-infrared will allow astronomers to address previously unanswerable questions about low mass stars, including whether they can host habitable planets. Improving the precision by one or two orders of magnitude will not only bring about the possibility of finding earth-like planets in solar systems like our own, but will also open the door to a new era of precision cosmology.

Radial-velocity searches for extrasolar planets use the periodic Doppler shift of stellar lines to constrain the mass and orbital parameters of exoplanets. These oscillatory radial velocities result from the reflex motion of the host star around the barycenter of the exo-solar system. The state-of-the-art High-Accuracy Radial-velocity Planet Searcher (HARPS) spectrograph at the ESO La Silla telescope has recently detected a stellar source with radial velocity varying by 60 cm/sec over timescales of a few months.[1] 60 cm/s precision is sufficient to find a 5-earth-mass planet around a sun-like star. To find an earth-mass planet in an earth-like orbit, a precision of 5 cm/s is necessary.

The astro-comb, will help to extend current observational techniques to search for earth-like planets in the habitable zones of stars less massive than the sun. The Doppler shift for a given planetary mass is larger when the host star is less massive, bringing the signal within reach of existing astronomical spectrographs. Jupiter orbiting in the habitable zone of a typical M dwarf star of 0.25 times the sun's mass would produce a radial velocity signal of 20 m/s, and earth a few m/s. Low-mass stars are brightest in the near infrared, where radial velocity calibration is most difficult. The development of a good calibrator from 700-1000 nm may allow detection of terrestrial planets in the habitable zone.[2]

Cosmology will also benefit from accurate, precise radial velocity measurements. A direct measurement of the temporal variation of the redshift of astronomical sources over a wide redshift range, known as the Sandage–Loeb test,[3,4] would be an unique probe of the nature of dark energy and allow us to determine the total matter density of the universe.[5] The Sandage–Loeb test requires measurement of tiny redshift changes (about 6 cm/sec) over very long periods of time (10 years) through observation of faint sources. The test will become feasible with the next generation of extremely large telescopes provided other systematic effects at the cm/s level can be overcome.[6]

A careful study of systematic effects is required to push radial velocity measurements to a precision of 1 cm/s. The astro-comb is a prototype near-infrared wavelength calibrator that promises to achieve the fractional stability and accuracy of 3×10^{-11} necessary to measure 1 cm/sec radial velocity variation of astronomical objects.

2. Calibrating Radial Velocity Measurements

For a spectrograph to detect exo-earths or perform a Sandage–Loeb test, its calibrator must be able to provide cm/s precision reproducibly over timescales of a week up to several decades. Calibrators currently in use are inadequate. While iodine vapor absorption cells[7] and Thorium-Argon (ThAr) cathode lamps[8] have been used to achieve m/s precision, both have inherent limitations. An iodine absorption cell placed in the telescope's optical path produces a set of narrow, finely-spaced absorption features superimposed on the stellar spectrum between 520 and 600 nm. While the calibration takes place in the same light path as the science data, roughly half the incident light is blocked and some stellar features are masked. Further, since the radial velocity precision is a function of the number of spectral features observed simultaneously, the iodine cells cannot exploit the full spectrum of stars or other astronomical observables. While ThAr lamps calibrate a broader passband than iodine cells, their utility in the red is compromised by strong, broad Ar features which typically saturate detector pixels before even moderate signal to noise is achieved in the thorium features. Finally, the intrinsic stability of the ThAr lamp is limited: Ar lines are known to be variable, individual Th lines are only stable at 1 m/s, and the lamps have finite lifetimes, exhibiting subtle changes with age. Spectrograph calibration is a crucial hurdle in measuring radial velocities with precision below 1 m/s.

The ideal wavelength calibrator for precision radial velocity measurements would be stable and reproducible over long timescales, and have a large number of lines with comparable amplitude uniformly distributed across the spectral range of interest. Laser frequency combs have been suggested as a near-ideal wavelength calibrator.[9,10] The spectrum of a laser frequency comb consists of millions of narrow emission lines spaced by the repetition rate, f_{rep} of the pulse train emitted by a mode-locked femtosecond laser. The absolute frequency f of the mth comb line is determined by f_{rep} and the system's carrier-envelope offset frequency, f_{ceo}: $f = f_{ceo} + m \times f_{rep}$, where m is an integer. Both f_{rep} and f_{ceo} can be synchronized with radio-frequency oscillators referenced to atomic clocks. Referenced to GPS, comb-lines have long-term fractional stability and accuracy below 10^{-12}. The spectrum of a typical mode-locked Titanium:sapphire laser spans 600 nm to 1.2 μm and through frequency doubling it is possible to cover the full CCD bandwidth. Therefore, frequency combs have the essential properties of an ideal spectrograph wavelength calibrator.

The chief impediment to using combs for astronomical spectrograph

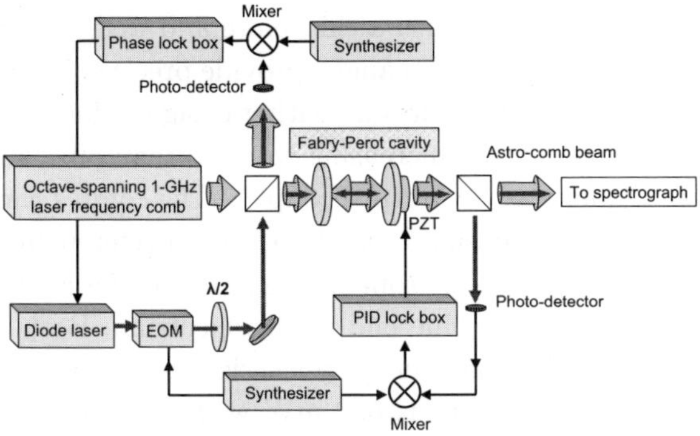

Fig. 1. Schematic diagram of astro-comb. A stabilized 1 GHz source comb generated by a mode-locked femtosecond laser passes through a Fabry-Perot cavity that filters out unwanted comb lines, increasing the line spacing by up to 40 GHz. The Fabry-Perot cavity is locked to a diode laser which is in turn phase locked to the source comb. For the demonstration spectra shown in Fig. 2, the Fabry-Perot output is coupled into an optical spectrum analyzer.

calibration is that the comb spectrum is too dense. High resolution spectrographs, designed to collect light from faint sources, typically have a resolving power, R, between 30,000 and 150,000, corresponding to a resolution between 2 and 10 GHz. A 1 GHz repetition rate produces lines too closely spaced for the spectrograph to resolve. We have therefore augmented a source comb with a stable broadband Fabry-Perot cavity to increase the comb line spacing up to 40 GHz over a range of 100 nm to create an "astro-comb" suitable for use with astronomical spectrographs[11] shown schematically in Fig. 1. Here, we both describe the astro-comb and present preliminary data from a test with an astronomical spectrograph.

3. The Astro-Comb

As described above, the astro-comb consists of a 1 GHz source comb coupled to a Fabry-Perot filtering cavity. Our astro-comb spans 100 nm centered on 850 nm, and has a line spacing that can be tuned from 10 to 40 GHz. The source comb is an octave-spanning optical frequency comb with a 1 GHz repetition rate generated by a mode-locked Ti:sapphire femtosecond laser. This ring-cavity laser uses a set of broadband double-chirped mirror (DCM) pairs, optimized Kerr-lens mode locking and an optimized output coupler.[12] The DCMs, along with a thin plate and wedge of BaF_2 inserted into the laser cavity at Brewster's angle, provide precise dispersion compensation. The BaF_2 plate, which has a coating designed for 4% transmission in the center of the spectrum increasing to more than 50% above 1050 nm and below 650 nm, acts as the output coupler. The comb's carrier envelope offset frequency, f_{ceo} is measured via the 1f-2f self-referencing beatnote, in which frequency-doubled light from the red end of the comb spectrum is compared to light from the blue end of the spectrum. Both f_{rep} and f_{ceo} are stabilized to low-noise frequency synthesizers, referenced to an atomic clock. The source comb output is 600 mW with a spectrum of 600 nm to 1,200 nm, shown in Fig. 2 together with the corresponding astro-comb output spectrum. Once mode-locked, the source comb is robust and operates continuously for many hours. Through comparison to a hydrogen maser, we have determined the frequency stability of the source comb characterized by f_{rep} and f_{ceo} to be better than 10^{-12} over timescales ranging from seconds to hours.

The stabilized source comb light passes through a Fabry-Perot cavity (FP) that increases the line spacing by filtering out unwanted comb lines. While the source comb lines are spaced by f_{rep}, astro-comb lines are spaced by the FP cavity free spectral range (FSR). A typical astro-comb FSR is 37 GHz, and we have measured the finesse of the FP to be 250, a value consistent with the theoretical limit estimated from the mirror reflectivity and Fresnel losses. The cavity's FSR can easily be tuned, allowing the astro-comb to be optimized for use with different spectrographs. We lock the FP cavity to the source comb via an injected diode laser signal, which in turn is phase locked to the source comb (Fig. 1).

We create an astro-comb by adjusting the FSR to an integer multiple of

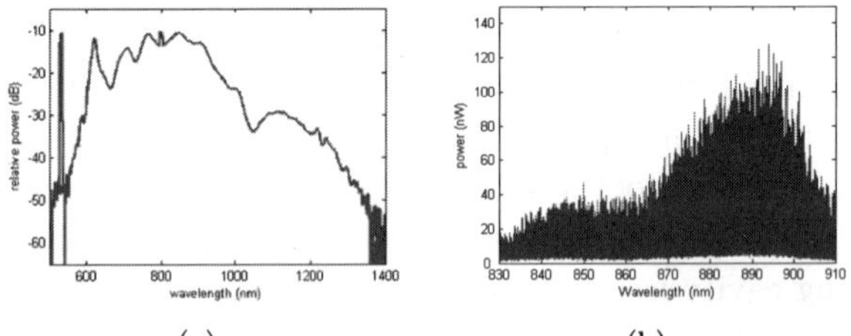

(a) (b)

Fig. 2. (a) Source comb output spectrum. (b) Astro-comb output spectrum, tuned to have a line spacing of 37 GHz and a bandwidth of 100 nm.

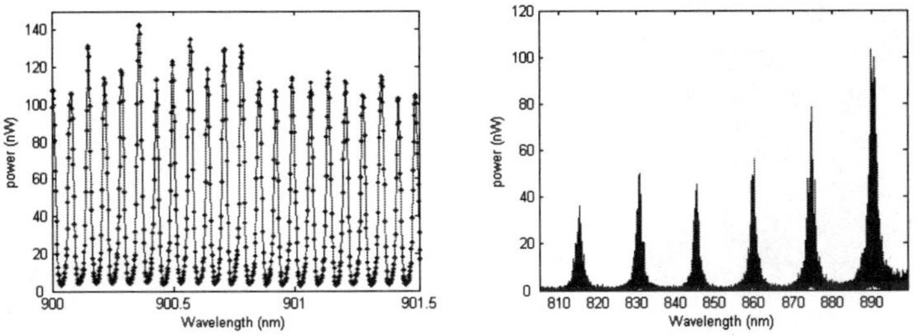

(a) (b)

Fig. 3. (a) Individual lines in the astro-comb output spectrum shown in Fig. 2b. The width of the lines is set by the 8 GHZ resolution of the OSA. (b) Vernier-like pattern that results from a mismatch between comb line spacing and FP FSR.

f_{rep}. We record the astro-comb spectrum by directing the output through a single-mode fiber to an optical spectrum analyzer (Ando 6317: resolution = 8 GHz; reproducibility = 2 GHz). Fig. 2b shows the astro-comb output spectrum spanning a 100 nm bandwidth, with 37 GHz line-spacing and 10–100 nW in each comb line. Individual filtered comb lines are shown in Fig. 3a The amplitude variation across the spectrum is primarily that of the source comb, but there is an additional small variation due to a relative shift between FP transmission resonances and source comb lines caused by the group delay dispersion (GDD) of the FP mirrors. The mirrors used in the plane-parallel FP have 99% reflectivity and a group delay dispersion (GDD) of <50 fs^2 in the range of 755 nm to 920 nm. If the ratio between the FSR and f_{rep} is not an integer, the span of the filtered comb lines is narrower and groups of filtered comb lines appear repeatedly within the bandwidth of the mirrors (Fig. 3b). The single-sided suppression of extraneous comb lines by the FP is more than 25 dB, consistent with the measured FP finesse of 250.

By the end of 2008, we will have used the astro-comb to calibrate astronomical measurements made with the Tillinghast Reflection Echelle Spectrograph (TRES) at Fred Lawrence Whipple Observatory. TRES, with $R = 63,000$ and a passband of 385-910 nm, is designed to measure the spectra of objects that are bright in the near-infrared, making it an ideal testbed for the astro-comb. A calibration image illustrating the difference between the astro-comb and a ThAr lamp is shown in Fig. 4a. To create this calibration frame, a filter blocking light to the red of 750 nm was placed in front of the ThAr lamp and a filter blocking light to the blue of 750

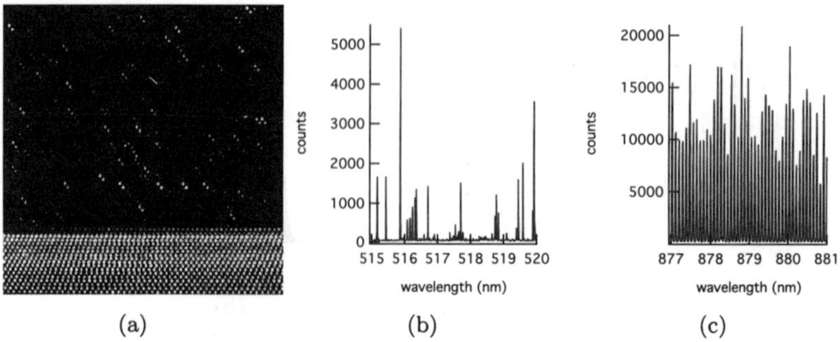

Fig. 4. Preliminary calibration data from the TRES spectrograph using ThAr below 750 nm and the astro-comb above 750 nm. For clarity, only an illustrative section of the CCD image and extracted spectra are shown. (a) CCD image of the cross-dispersed spectrum. Wavelength increases from top to bottom and left to right; the red end of the spectrum is in the lower right-hand corner of the image. (b) Typical ThAr spectrum extracted from (a). (c) Typical astro-comb spectrum also extracted from (a).

nm was placed at the astro-comb's output. Astro-comb light was fed into the spectrograph along the same optical path as the ThAr lamp, and the CCD was exposed to both calibration sources for 180 s. The top portion of Fig. 4a contains the cross-dispersed ThAr spectrum while the lower half contains light from the astro-comb. Fig. 4b and 4c show extracted portions of the ThAr and astro-comb spectra.

4. Astro-comb Wavelength Calibration

We estimate the performance of the astro-comb as a wavelength calibrator by considering astro-comb accuracy and stability. The 1 GHz source comb has a short-term fractional stability of 10^{-12} when f_{rep} and f_{ceo} are locked to a high-quality local RF oscillator. Referencing the local oscillator to GPS results in precision and long-term stability better than 10^{-12}. Despite the inherent stability of the source comb, changes in the FP cavity could change the apparent position of line centers seen by the spectrograph. A conservative upper limit on the stability of the diode laser and FP cavity locks is half the diode linewidth, or 500 kHz. The corresponding change in extraneous line suppression is $\sim$0.2%. The demonstrated suppression factor of 4×10^{-3} leads to an upper limit on astro-comb line drift of $4 \times 10^{-3} \times 0.2\% \times 1\text{GHz} = 8$ kHz. Converting this to a systematic error in an astrophysical radial velocity measurement, we find an upper limit of $(8\ \text{kHz}/377\ \text{THz}) \times (3 \times 10^{10}\ \text{cm/s}) = 0.6$ cm/s. Astro-comb accuracy and stability are therefore well matched to detecting exo-earths and measuring the expansion rate of the universe. The use of GPS as the absolute reference allows the comparison of measurements made at different observatories and separated by long time intervals as will be necessary for the detection of planets with long orbital periods and the Sandage–Loeb test. Radial velocity measurements of M dwarf stars require an order of magnitude less precision.

We now estimate how well the astro-comb will calibrate TRES and the HARPS-NEF spectrograph that is being built to detect earth-like planets around stars in the Northern sky. The spectrum recorded by a spectrograph CCD is a convolution of the comb line profile with the spectrograph instrumental profile, with the latter typically dominant because it is much broader. Assuming the full width at half maximum (FWHM) of a line profile incident on the CCD is sampled by n pixels, the error of the line position is estimated by[14]

$$\Delta v_{single} = A \times \mathrm{FWHM}/(\mathrm{SNR} \times \mathrm{n}^{1/2}) \tag{1}$$

where SNR is the signal-to-noise ratio of the peak pixel, and A is an order-unity factor that depends on the comb spectral profile. In one calibration exposure, the total velocity precision available from the entire spectrograph depends on the total number, N, of comb lines included in the calibration:

$$\Delta v = \Delta v_{single}/\mathrm{N}^{1/2} = A \times \mathrm{FWHM}/(\mathrm{SNR} \times (\mathrm{nN})^{1/2}) \tag{2}$$

Following the analysis of Murphy,[9] the greatest velocity precision will be obtained from an astro-comb wavelength calibrator when the line spacing is 2.5 resolution elements of the spectrograph. For TRES ($R \approx 60,000$) we choose a comb line spacing of 15 GHz, which will result in approximately 500 comb lines in each ~15 nm echelle order covered by comb light. The FWHM of the comb lines convolved with the instrumental response is 0.01 nm (5 km/s), sampled by n = 4 pixels. Taking A = 0.41[9] and SNR= 100, we can achieve a velocity precision of 45 cm/s through an astro-comb calibration of one echelle order of TRES in the absence of other systematic effects. The HARPS-NEF spectrograph, with $R = 120,000$ should attain a theoretical sensitivity below 1 cm/s.

5. Conclusion

We have constructed and tested a laser frequency comb suitable for the calibration of narrow-band high resolution astronomical spectrographs in the near-infrared. Initial testing of the astro-comb with a high-resolution astronomical spectrograph are underway. Future work will generalize the technique to span the entire range of broadband spectrographs. The astro-comb's precision, stability and wavelength coverage will remove calibration from the list of significant systematic errors in precision radial velocity measurements. In practice, the precision of Doppler-shift/redshift measurements will also be affected by drift in the spectrograph, imperfect telescope pointing, telescope instability and astronomical light-source fluctuations. Nonetheless, by potentially enabling velocity-shift precision of ~1 cm/sec, broad wavelength coverage, and reproducibility over many years and between telescopes, astro-combs should revolutionize astrophysical spectroscopy.

Acknowledgements

The authors are grateful to Dave Latham, Cullen Blake and Avi Loeb for useful discussions, and would like to thank Emilio Falco and the FLWO staff for their assistance in bringing the astro-comb to Mt. Hopkins. This work was supported by the Smithsonian Institution, the Harvard University Origins of Life Initiative, NSF-ATI and DARPA.

References

1. S. Udry, *et al.*, A.&A. **469**, L43-L47 (2007).
2. J. C. Tarter, *et al.*, Astrobiology **7**, 30-65 (2007).
3. A. Sandage, Ap. J. **136**, 319-333 (1962).
4. A. Loeb, Ap. J. **499**, L111-L114 (1998).
5. P. S. Corasaniti, D. Huterer, and A. Melchiorri, Phys. Rev. D **75**, 062001 (2007).
6. J. Liske, *et al.*, Mon. Not. R. Astron. Soc. **386**, 1192-1218 (2008).
7. R. P. Butler, *et al.*, PASP **108**, 500-509 (1996).
8. C. Lovis, *et al.*, Proc. SPIE **6269**, 62690P23 (2006).
9. M. T. Murphy, *et al.*, Mon. Not. R. Astron. Soc. **380**, 839-847 (2007).
10. P. O. Schmidt, S. Kimeswenger, H. U. Kaeufl, The 2007 ESO Instrument Calibration Workshop Proceedings, Garching, Germany, Jan. 23-26, 2007.
11. C.-h. Li, *et al.*, Nature **452**, 610-612 (2008).
12. A. Benedick, J. Birge, O. D. Mücke, M. Sander, F. X. Kärtner, In: *Digest of European Conference on Lasers and Electro-Optics CLEO/Europe 2007*, Munich, Germany, 11-27 June, 2007.
13. K. L. Corwin, Z.-T. Lu, C. F. Hand, R. J. Epstein, and C. E. Weiman, Appl. Opt. **37**, 3295-3298 (1998).
14. J. W. Brault, Mikrochimica Acta **93**, 215-227 (1987).

HIGH FREQUENCY VERY LONG BASELINE INTERFEROMETRY: FREQUENCY STANDARDS AND IMAGING AN EVENT HORIZON

Sheperd Doeleman*

Haystack Observatory, MIT,
Westford, MA 01886, USA
**E-mail: sdoeleman@haystack.mit.edu*

VLBI is a powerful astronomical technique that allows data from widely separated radio telescopes to be combined into interferometric arrays that can image the sky with exceptional angular resolution. At the heart of a VLBI station is a highly stable frequency reference that locks heterodyne receivers and enables the phase of radiation from a cosmic source to be preserved. Subsequent correlation of data streams from two VLBI stations provides fourier components of sky brightness, and with many antennas, an accurate image of the radio sky can be reconstructed with an angular resolution equal to the ratio of observing wavelength to antenna separation. At the highest frequencies and longest baselines, angular resolutions approach 30 micro arcseconds, over a thousand times finer than the Hubble Space Telescope. With such resolution, a number of astrophysical phenomena can be studied in detail, including the Super Massive Black Hole candidate, SgrA* (Sagittarius A "star"), at the center of the Milky Way Galaxy. To maintain phase coherence for observing wavelengths shortward of 1.3mm requires frequency standards to be stable over 1-10 second intervals, which is a coherence time set by atmospheric turbulence. Historically, Hydrogen masers have been used as VLBI references, but for short wavelength observations new frequency sources are available that outperform masers over the integration times of interest. I will discuss the stability requirements for VLBI and describe how future VLBI experiments can image the accretion disk and Event Horizon of the black hole at the Galactic Center.

1. Frequency Standards for VLBI

The resolving power of Very Long Baseline Interferometry (VLBI) has led to numerous important scientific advances, from the study of superluminal jets and ultraluminous infrared mergers in extragalactic objects, to Galactic micro-quasars and the central massive black hole in the Milky Way. As with connected-element arrays, VLBI relies on the fundamental ability of radio band receivers to accurately preserve the phase of detected cosmic radiation. Subsequent correlation of signals from different radio antennas measure complex-valued Fourier components of the radio sky, providing the data necessary to synthesize images. For connected-element arrays, a common local oscillator (LO), derived from a station frequency reference, is used for all antenna receivers. On VLBI baselines longer than ~100km, sharing a common frequency reference is not feasible (though transmission of LO signals for VLBI with satellites has been demonstrated[1]), and each VLBI site must use a separate frequency reference. This practice imposes a strict requirement on the stability of the frequency standards used for VLBI. The central role for this frequency reference is shown in a schematic view of the VLBI signal path in Figure 1. Frequency standards are typically

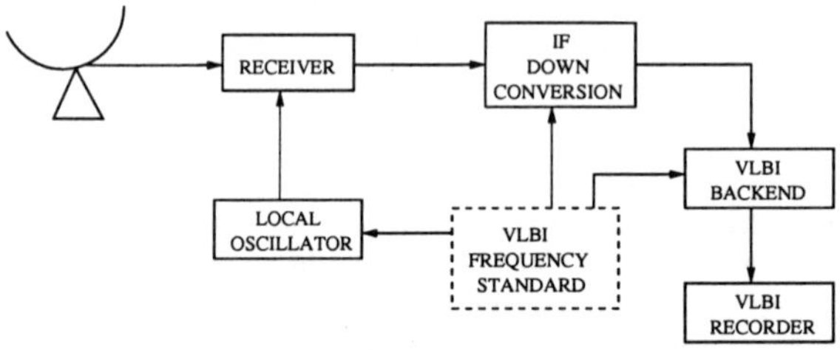

Fig. 1. Schematic view of the role of the VLBI frequency standard. The frequency standard typically provides a number of outputs at 5 or 10 MHz which are used to a) lock an LO synthesis chain for use in a mixing receiver, b) lock synthesizers in an Intermediate Frequency (IF) downconversion system used to place desired sky frequencies in the VLBI backend band, and c) provide a reference for the digital sampling clock in the VLBI backend.

characterized by the two-sample Allan variance, $\sigma_y^2(\tau)$, which describes the stability of the frequency source on different integration time scales. The time (τ) over which a VLBI baseline can coherently integrate depends on the stability of the frequency standards used at each antenna roughly as:

$$\omega\tau\sigma_y(\tau) < 1 \tag{1}$$

where ω is the observing frequency in radians/sec, $\sigma_y(\tau)$ is the square root of the Allan variance (the Allan standard deviation), and the limit keeps rms phase fluctuations below 1 radian. The signal loss on a two station VLBI baseline due to instability in the frequency standards is then approximately

$$\text{Loss} = 1 - e^{-(\omega\tau\sigma_y(\tau))^2} \tag{2}$$

Thus, to limit coherence losses to $< 5\%$ over 100 seconds, a Rubidium frequency standard with fractional stability of $\sigma_y(100s) = 3 \times 10^{-13}$ would be useful for VLBI observing frequencies below about 1GHz. Hydrogen masers, which excite coherent emission of the hyperfine 1420 MHz transition, can have exceptional stability on integration times from 10 to 10,000 seconds and are used almost without exception as VLBI frequency references worldwide. With good H-Masers reaching $\sigma_y(10s) = 1.5 \times 10^{-14}$, VLBI observations below $\sim$ 230 GHz can be carried out with coherence losses limited to $< 5\%$. Rogers & Moran give exact expressions for VLBI coherence loss.[2]

In addition to instability of frequency standards, the other major sources of coherence loss in VLBI correlation signals are propagation effects through the atmosphere. Below $\sim$ 1 GHz, phase fluctuations in the received cosmic signals caused by ionospheric turbulence dominate, and can limit VLBI coherence times to $\sim$ 90 sec or less.[3] Above $\sim$ 5 GHz, ionospheric effects, which decrease in magnitude as λ^2, are relatively unimportant, and it is not uncommon to integrate VLBI signals for up to 10 minutes at 5 GHz. At higher frequencies, turbulence in the troposphere begins to limit VLBI coherence times. The coherence limit of $\sim$ 8$-$10 seconds seen in the 129 GHz VLBI data shown in Figure 2 is not uncommon at this frequency in fair weather conditions. Though the troposphere limits high frequency VLBI

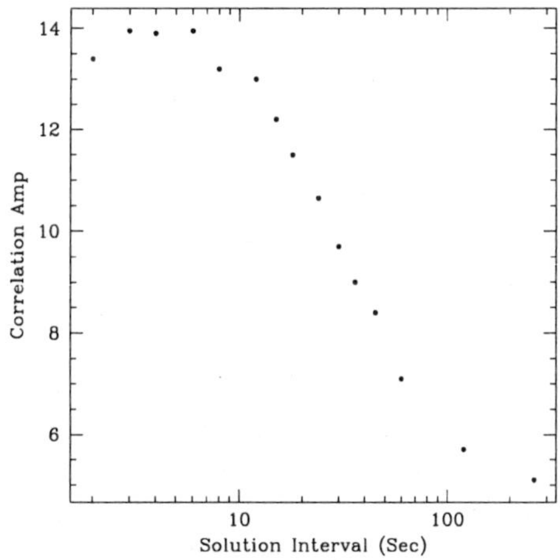

Fig. 2. Coherence losses (decrease in correlation amplitude) on a 160 km VLBI baseline observing at 129 GHz. Integration of the VLBI signal on a range of time scales shows that below $\sim$ 10 seconds, the coherence losses imposed by tropospheric turbulence fall below $\sim$ 5%. At this frequency, contribution to coherence losses from instability in the frequency standards (H-Masers) will be $< 5\%$.

coherence times, high altitude sites often experience very good weather conditions. At the ALMA site in Chile, the measured coherence time of the atmosphere is > 10 seconds 60% of the time at 230 GHz and 45% of the time at 345 GHz.[4] In terms of Allan standard deviation, this corresponds to $\sigma_y(10s) < 2 \times 10^{-14}$ over 40% of the time. For this reason, VLBI at these frequencies requires frequency standards whose Allan Standard Deviation is better than 2×10^{-14} at 10 seconds. While the best H-Masers have Allan variances that lead to coherence losses $< 5\%$ at 230 GHz, for H-Masers with $\sigma_y(10s) = 5 \times 10^{-14}$ the losses exceed 15%. At 345 GHz and 450 GHz, the coherence losses due solely to H-Maser frequency standards (with $\sigma_y(10s) = 5 \times 10^{-14}$) can exceed 24% and 40% respectively.

There are two options to procuring frequency standards that meet these stringent requirements. The first is to characterize existing H-Masers and improve their performance. This involves replacing the existing internal quartz crystal oscillators, which lock to the 1420 MHz maser line and are typically of older vintage, with new very low phase noise crystals. For high frequency VLBI work, one also runs the H-masers with a high hydrogen flux to increase the signal-to-noise ratio of the maser line, and this puts additional strain on the vac-ion pumps used to maintain maser vacum, thus requiring more maintenance. These modifications can be made, and recent 230 GHz VLBI experiments with Hydrogen masers running at high flow rates and with installation of modern quartz oscillators have been carried out.

A second option is to explore alternative frequency standards. Cryogenic oscillators can now routinely achieve $\sigma_y(\tau)$ an order of magnitude better than H-Masers at $\tau = 10$sec. These new standards are promising and may be ideal for high frequency VLBI, but will require testing in VLBI observations as well as technical adaptation for use a VLBI station reference.

2. A Cryogenic Sapphire Oscillator for VLBI

It has been known for some time[5] that cooled sapphire dielectric resonators can be used to produce extremely stable frequency references. They are well studied and used by metrology and atomic physics groups in precision tests of modern physics, including violations of Lorentz invariance[6] and time evolution of the fine structure constant.[7] Modern cooled sapphire oscillators (CSO) achieve $\sigma_y(\tau) < 6 \times 10^{-16}$ for $10s < \tau < 100s$, at least an order of magnitude better than H-Masers over the same range.[8] This highly stable performance relies on keeping the sapphire oscillator at a turning point in a frequency-temperature curve near 7 K. Liquid Helium in long-hold

(30 day) dewars is the cooling system of choice; CSOs cooled with pulse tube systems have also been constructed, and while they do not have the stability of dewar-cooled oscillator, they are still more stable than the best H-masers by a factor of ~ 2.[9,10] Early CSO frequency standards were used successfully in 2.3 GHz VLBI observations,[11] but H-Maser stability was sufficient for VLBI observing frequencies at the time, and H-Masers had better stability than early CSOs for $\tau > 100$s.

The increased stability of modern CSOs now make them attractive frequency standards for high frequency VLBI. To adapt them for use as VLBI station references requires constructing a low phase noise loop that will lock a GPS-conditioned 10 MHz quartz oscillator to the CSO resonant output near 11.2 GHz. This loop must be carefully designed to retain the stability characteristics of the CSO, and must use the highest quality quartz crystals to ensure low phase noise on the output reference signal. Figure 3 shows the effects of frequency standard stability on VLBI experiment coherence when using an H-maser and a CSO. For integration times less than 100s, the superior stability of the CSO results in little coherence loss at observing frequencies up to ~ 345 GHz, while an H-maser will incur more coherence loss at these short integration times.

3. The Super Massive Black Hole at the Galactic Center

At a distance of ~ 8 kPc [[12]], SgrA*, the compact radio, near infrared and X-ray source at the Galactic Center, is thought to mark the position of a super-massive black hole of mass $\sim 4 \times 10^6\ M_\odot$.[13,14] Proper motions of SgrA*[15] confirm that it traces a significant amount of the mass that is inferred by stellar motions and orbits. Due to its proximity, SgrA* is the only galactic nucleus that can be studied with VLBI on sub-AU linear scales ($R_{\rm sch} \sim 10\mu$as ~ 0.1AU). The ionized ISM, however, scatter-broadens images of SgrA* with a λ^2 dependence, and VLBI at the highest frequencies is the only available means to set important limits on intrinsic structures near the event horizon. VLBI at 43 GHz and 86 GHz has detected evidence for intrinsic structure of SgrA*, but these observations remain dominated by scattering effects, and the intrinsic sizes at these wavelengths (set by the optical depth of the emission) are much larger than the apparent size of the event horizon.[16,17] Evidence from light curves of SgrA* flares from the radio to Xray[18–20] implicate structures on smaller ($\sim 5 - 15R_{\rm sch}$) scales. Only at frequencies above 230 GHz does the scattering size become smaller than the VLBI array resolution allowing direct measurement of intrinsic structure on these scales corresponding to the innermost accretion region.

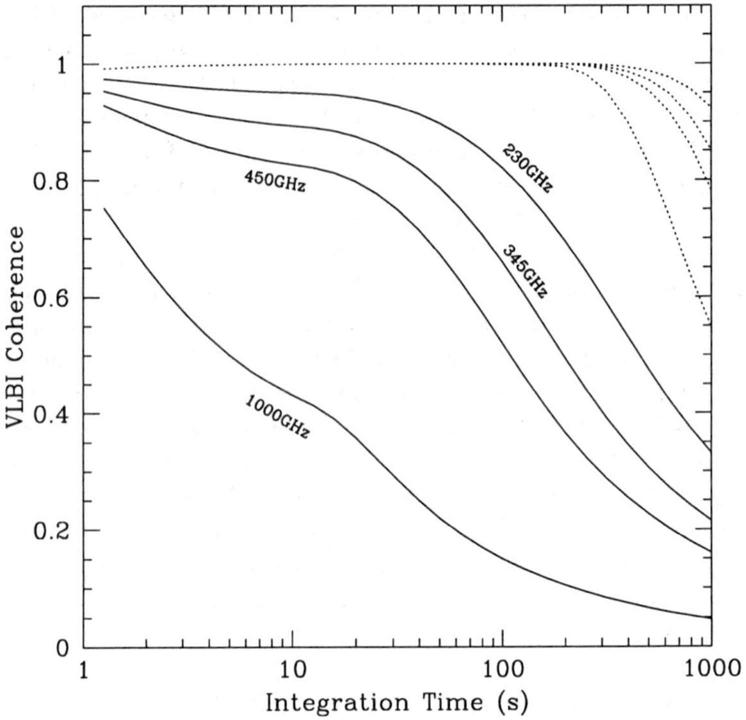

Fig. 3. Coherence of VLBI data when both Cryogenic Sapphire Oscillators and Hydrogen Masers are used as frequency standards. Coherence is calculated by numerically integrating exact expressions.[2] The solid lines show coherence if two H-masers are used at each station in a VLBI baseline, and the dotted lines show coherence when CSO standards are used. At the highest VLBI frequencies, use of a CSO is preferable to reduce coherence losses to negligible levels. For the H-maser, a $\sigma_y(1sec) = 3 \times 10^{-13}$ and a $\sigma_y(10sec) = 3 \times 10^{-14}$ were assumed, with $\sigma_y(\tau) \propto \tau^{-0.5}$ for longer integration times. For the CSO, $\sigma_y(1sec) = 2 \times 10^{-15}$, $\sigma_y(10sec) = 7 \times 10^{-16}$ and remains at this level until $\tau = 200$ sec, and then $\sigma_y(\tau) \propto \tau^{0.5}$ for longer integration times.

In April 2007, SgrA* and several quasar calibrators were observed over two consecutive days at a frequency of 230 GHz with a three station VLBI array.[21] The array included the James Clerk Maxwell Telescope (JCMT) on Mauna Kea, the Arizona Radio Observatory Submillimeter Telescope (ARO/SMT) on Mt Graham in Arizona, and one 10m dish of the Coordinated Array for Research in Millimeter-wave Astronomy (CARMA) in California. Projected baseline lengths on SgrA* ranged from $500 \times 10^6\lambda$

on the shortest baseline to $3500 \times 10^6\lambda$ on the longest. The interferometric fringe spacing on the longest baseline was ~ 55 μarcseconds, corresponding to a linear size of about the orbit of Mercury.

SgrA* was detected on both the ARO/SMT-JCMT and ARO/SMT-CARMA baselines. No detections were found on the CARMA-JCMT baseline, which is attributable to the lower sensitivity of that baseline compared with the others. A circular Gaussian model was fit to the VLBI data (shown in Figure 4). The weighted least-squares best-fit model has a total flux density of 2.4 ± 0.5Jy and full width at half maximum (FWHM) of 43 (+14,-8) μas where errors are 3σ. On the assumption of a Gaussian profile, the intrinsic size of Sgr A* can be extracted from our measurement assuming that the scatter broadening due to the ISM adds in quadrature with the intrinsic size. At a wavelength of 1.3 mm the scattering size extrapolated from previous longer-wavelength VLBI[22] is $\sim 22\mu$as. Removing the scattering effects results in a 3σ range for the intrinsic size of Sgr A* equal to 37 (+16,−10) μas. The 3σ intrinsic size upper limit at 1.3 mm, combined with a lower limit to the mass of Sgr A* of $4 \times 10^5 M_\odot$ from measured proper motions yields a lower limit for the mass density of $9.3 \times 10^{22} M_\odot \mathrm{pc}^{-3}$. This density lower limit and central mass would rule out most alternatives to a black hole for Sgr A* because other concentrations of matter would have collapsed or evaporated on timescales that are short compared with the age of the Milky Way.[23] It should be noted, however, that while structure on $4R_{\mathrm{sch}}$ scales is present in SgrA*, models other than the circular Gaussian can be fit to the data. This is illustrated by the dotted line in Figure 4, which shows the expected flux density as a function of baseline length for a uniform circular annulus with inner diameter 35μas and outer diameter 80μas that has been scatter broadened by the ISM. Future higher-sensitivity observations will distinguish between these two models by allowing detections of SgrA* on the CARMA-JCMT baseline, which is now represented in Figure 4 only as an upper limit.

Because of gravitational lensing effects due to the extreme gravity near the assumed black hole, radiation emitted from near the event horizon of a non-spinning black hole will have an apparent size of $3\sqrt{3}R_{\mathrm{sch}}$. For SgrA*, this expected diameter is $5.2R_{\mathrm{sch}} \simeq 52\mu$as, which differs by 3σ from the size derived from a Gaussian model. Even if the black hole is maximally spinning (a=1), the diameter of the event horizon in the equatorial plane ($\sim 45\mu$as) would still exceed the estimated size. This suggests that SgrA* is not an optically thick emission region that symmetrically enfolds the black hole. Rather, it is likely due either to emission from a jet or from

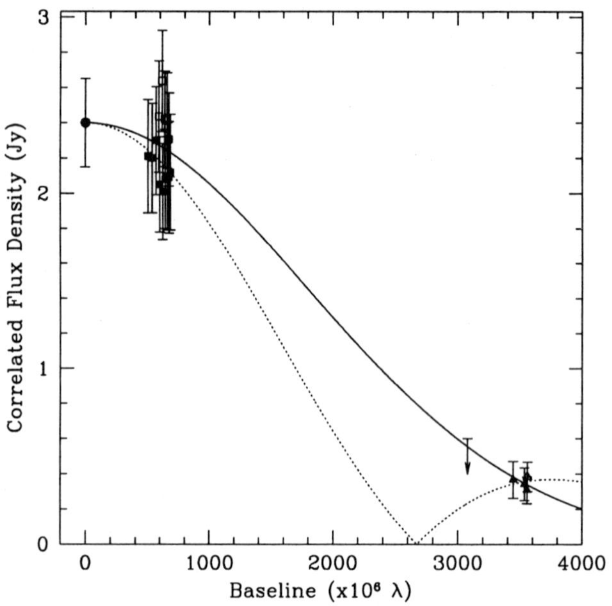

Fig. 4. Shown are the correlated flux density data on the ARO/SMT-CARMA and ARO/SMT-JCMT baselines plotted against projected baseline length (errors are 1σ). Squares show ARO/SMT-CARMA baseline data and triangles show ARO/SMT-JCMT data, with open symbols for 10 April and filled symbols for 11 April. The solid line shows the weighted least-squares best fit to a circular Gaussian brightness distribution, with FWHM size of 43.0 μas. The dotted line shows a uniform thick-ring model with an inner diameter of 35 μas and an outer diameter of 80 μas convolved with scattering effects due to the interstellar medium. The total flux density measurement made with the CARMA array over both days of observing is shown as a filled circle. An upper limit for flux density of 0.6 Jy, derived from non-detection on the JCMT-CARMA baselines, is represented with an arrow near a baseline length of $3075 \times 10^6 \lambda$.

the approaching (and therefore Doppler enhanced) side of an accretion disk that is inclined to our line of sight.[24–26] Either scenario results in emission that is offset from the black hole position. This marks the first time that astronomical observations of any kind have directly constrained the spatial relationship between SgrA* and the black hole. Future VLBI observations with more antennas will be able to search for time variable structures that are predicted as localized 'hot-spots' of emission within the accretion disk in-spiral to the black hole event horizon.[27] At observing frequencies above 230 GHz, the frequency references discussed here will enable VLBI observations with negligible coherence losses.

Acknowledgments

VLBI at mm/submm wavelengths would not be possible without the dedicated support of staff and scientists as all participating facilities. VLBI work at the MIT Haystack Observatory is supported through grants from the National Science Foundation.

References

1. Levy G S et al 1986 *Science* **234** 187
2. Rogers A E E & Moran J M 1981 *IEEE Trans. Inst. Meas.* **IM-30** 283
3. Chaterjee S 1999 *VLBA Memo Series* **22**
4. Holdaway M A 1997 *MMA Memo Series* **169**
5. Jones S K, Blair D G, Buckingham M J 1988 *Electron. Lett.* **8** 321
6. Wolf P et al 2004 *Phys. Rev. D*, **70**, 051902
7. Marion H et al 2003 *Phys. Rev. Lett.* **90** 150801
8. Hartnett J G, Locke C R, Ivanov E N, Tobar M E & Stanwix P L 2006 *App. Phys. Lett.* **89** 203513
9. Wang R T & Dick G J 2003 *Proceedings of the 2003 IEEE International Frequency Control Symposium* 371
10. Wang R T, Calhoun M D, Kirk A, Diener W A, Dick G J & Tjoelker R L 2005 *Proceedings of the 2005 IEEE International Frequency Control Symposium* 919.
11. Costa M E, Blair D G, Buckingham M J, Giles A J, Jones S K, Luiten A N, Turner P J, Young A C, Hong P & Mann A G 1992 *Meas. Sci. Technol.* **3** 718
12. Reid M 1993 *ARA&A* **31** 345
13. Schödel R et al 2002 *Nature* **419** 694
14. Ghez A M et al 2005 *ApJ* **620** 744
15. Reid M J & Brunthaler A 2004 *ApJ* **616** 872
16. Bower G C, Falcke H, Herrnstein R M, Zhao J H, Goss W M & Backer D C 2004 *Science* **304** 704
17. Shen Z Q, Lo K Y, Liang M -C, Ho P T P & Zhao J-H 2005 *Nature* **438** 62
18. Eckart A et al 2006 *A&A* **450** 535
19. Yusef-Zadeh F et al *ApJ* **644** 198
20. Marrone D et al *ApJ* **682** 373
21. Doeleman S. S et al 2008 *Nature* **455** 78
22. Bower G C et al 2006, *ApJ* **648** L127
23. Maoz E 1998 *ApJ* **494** L181
24. Falcke H, Melia F & Algol E 2000 *ApJ* **528** L13
25. Noble S C, Leung, P K, Gammie, C F & Book, L G 2007 *Class. Quant. Gravity* **24** S259
26. Broderick A E & Loeb A 2006b *MNRAS* **367** 905
27. Doeleman S S, Fish V L, Broderick A E, Loeb A & Rogers A E E *Preprint* astro-ph/0809.3424

OPTICALLY-PUMPED SPACE CESIUM CLOCK FOR GALILEO: RESULTS OF THE BREADBOARD*

R. RUFFIEUX, P. BERTHOUD†, M. HALDIMANN†, S. LECOMTE
Centre Suisse d'Electronique et de Microtechnique SA, Neuchâtel, Switzerland

V. HERMANN, M. GAZARD
Thales Electron Devices, Vélizy, France

R. BARILLET, S. GUERANDEL, E. DE CLERCQ, C. AUDOIN
SYRTE – Observatoire de Paris, Paris, France

Thales Electron Devices is leading a French-Swiss industries and research institutes consortium aiming at developing an Optically-pumped Space Cesium Clock for Galileo. The technical objective of this development is to demonstrate a frequency stability of $1\times10^{-12}\ \tau^{-1/2}$ compatible with an operational lifetime of twelve years. The present clock demonstrator combines the simplest and best technologies available, among those the single optical wavelength scheme and the dark fringe Ramsey cavity. Presently a clock operational signal-to-noise ratio of 21'300 $Hz^{1/2}$ has been recorded for a Cs oven heated at 100°C. The clock frequency stability has been measured to be $2.3\times10^{-12}\ \tau^{-1/2}$. The current limitations and future improvements are discussed.

1. Introduction

The on-board clocks baseline for Galileo, the European Global Navigation Satellite System, is based on both Passive Hydrogen Maser (PHM) and Rubidium Atomic Frequency Standards (RAFS) technologies [1]. While RAFS is a very compact clock (3.4 kg, 2.4 liters), the PHM is bulkier (18 kg, 26 liters) but provides a 5-fold improvement in short-term frequency stability, below $1\times10^{-12}\ \tau^{-1/2}$ from 1 s to 10'000 s. Based on past developments [2, 3], the optically-pumped Cesium beam frequency standard has been proposed as an excellent alternative to the existing clock technologies. While significantly less bulky than the PHM, it has been devised to be almost as stable as the PHM in the short-term and even better in the long-term.

* This work is supported by the European Spatial Agency (ESA).
† Now at Oscilloquartz SA, Neuchâtel, Switzerland

In the frame of a "Feasibility study of cesium clock technology for Galileo", which is supported by the European Space Agency, Thales Electron Devices, France (TED), has set up a French-Swiss consortium [4]: TED is in charge of the project management and of the development of specific technologies for the atomic resonator, CSEM develops and tests the complete demonstrator, SYRTE supports this development mainly for clock and metrology aspects, and Oerlikon Space Zurich, Switzerland (OSZ), advices for opto-electronics aspects. This feasibility demonstration activity has started in June 2006. The objective is to demonstrate a frequency stability of $1\times10^{-12}\ \tau^{-1/2}$ with industrially and space-feasible technologies along with a lifetime of 12 years.

In these proceedings, we report on the results of the demonstrator for an "Optically-pumped Cesium Space Clock" (OSCC). First we describe its experimental setup, then we present the measured results, and finally we analyze the noise budget of the clock. In the conclusion, we present solutions for possible clock performance improvement.

2. Experimental Setup

The realized atomic resonator breadboard fully takes into account constraints on volume, mass, and space-compatible industrial technologies. It also combines simple, innovative, and performing technologies like single optical wavelength scheme, homogeneous magnetic C-field, and dark fringe Ramsey cavity that are compatible with near-future space qualification. Less emphasis has been yet put in the laser & optics sub-systems that have been developed at the laboratory level using free-space propagating laser beams generated by a single DFB laser diode. High and low level electronics have also been developed at the laboratory level and include a fully digital electronics frequency lock loop of the quartz local oscillator.

A detailed description of the experimental setup can be found in P. Berthoud *et al.* [5].

3. Experimental results

Once the OSCC demonstrator has been integrated, the clock has been operated with parameters very similar to those reported in P. Berthoud *et al.* [5]. By scanning the RF field frequency over the central Ramsey fringe, the dark fringe gives a minimal fluorescence photo-current at resonance (300 pA), which mainly arises from remaining spurious light (the fraction of unpumped atoms being negligible). The peak-valley amplitude of this Ramsey fringe is 920 pA.

The full width at half maximum is 750 Hz, giving an atomic quality factor of $Q_{at} = 1.22 \cdot 10^7$. The detection photo-current noise density has been measured at an RF frequency detuning of 350 Hz, which corresponds about to the RF modulation depth. The average photo-current noise density is 43 fA/Hz$^{1/2}$. Combined with the above mentioned Ramsey signal, the operational signal-to-noise ratio is 21'300 Hz$^{1/2}$ with signal and noise measured at half-height of the central Ramsey fringe (corresponding to the clock operational conditions).

3.1. *Measured Frequency Stability and Noise Budget*

By operating the cesium atomic clock with the above-mentioned parameters, a frequency stability of $2.3 \times 10^{-12}\ \tau^{-1/2}$ measured against an active hydrogen maser has been obtained:

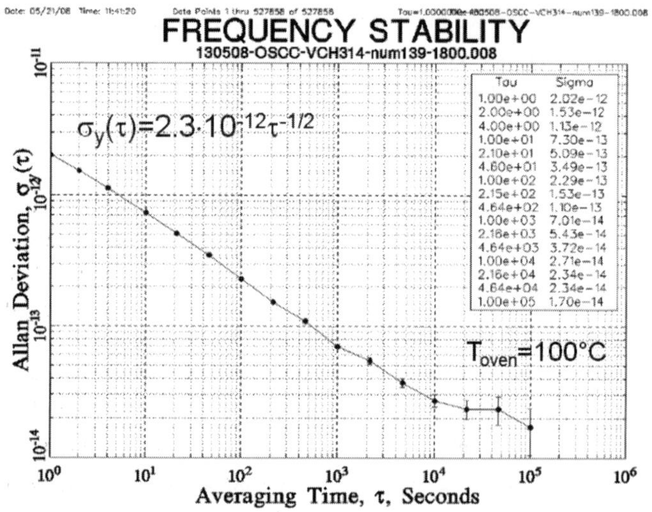

Figure 1: Measured clock relative frequency stability at an oven temperature of 100°C.

For averaging times over 10'000 s, the clock frequency stability (Figure 1) reaches a floor. This is due to the absence of servo loops for the microwave power, the light power, and the magnetic field. We plan to add these loops in a near future and expect to reach a frequency stability floor well below 10^{-14}.

For further clock improvement the clock signal noise has to be analyzed. For this purpose we use the formalism developed in [6].

With the same notations as in [6] and assuming that the operating point is at half maximum of the resonance curve, we obtain a very useful equation for the power spectral density (PSD) of the electronic flux fluctuations at the output of the photo-detector:

$$S_{\Delta i} = eI_{PV}F^{*}(\beta\gamma\eta) + 2eI_{SL} + 4kT/R_{sf} + e^{2}S_{extra} \quad (1)$$

if the amplitude and frequency noises of the laser and microwave signals induce negligible effects [7, 8]. I_{PV} and I_{SL} are respectively the peak-to-valley detected current and the spurious light detected current. In the right-hand side:

- the first term describes the fundamental noises; it includes the flux of atoms involved in the transition at the operating point, the effective noise figure F^* and the amplification factor $\beta\gamma\eta$ from atom to photo-electron;
- the second term is the effect of the spurious light which is supposed to be shot noise limited;
- the third term is the effect of the resistive components in the photo-detector apparatus;
- the fourth term S_{extra} is the effect of additional noises independent of the laser and of the microwave signals.

With the current experimental parameters the noise budget can be expressed as the following:

Table 1. Clock signal current spectral density (CSD) analysis and relative frequency stabilities @ 1 s. Unit for noise CSD: $fA/Hz^{1/2}$. Noise values measured at the Fourier frequency corresponding to the modulation frequency of the clock interrogation microwave frequency. N_{limit} and σ_{limit} stand for the reachable levels if the extra noise contribution (N_{extra}) is eliminated. Oven temperature: 100°C.

Noise analysis						Freq. Stab. @ 1s	
$N_{total, measured}$	$N_{fundamental}$	$N_{spurious\ light}$	$N_{photodiode}$	N_{extra}	N_{limit}	$\sigma_{measured}$	σ_{limit}
43	17	11	10	37	22	$2.3\text{x}10^{-12}$	$1.2\text{x}10^{-12}$

Table 1 summarizes the noise CSD budget, the measured frequency stability, and the limit frequency stability with an oven temperature of 100°C. N_{limit} and σ_{limit} are the calculated values if the extra noise is eliminated. This also shows that it is eventually feasible to operate the clock at an oven temperature compatible with a lifetime of 12 years and to reach the frequency stability goal of $1.2\times10^{-12}\ \tau^{-1/2}$.

3.2. *Clock Environmental Sensitivities*

Different clock environmental sensitivities have been measured:

- Light shift coefficient: $2.8\cdot10^{-13}\ mW^{-1}$
- Transverse and longitudinal magnetic field coefficients: $< 2\cdot10^{-14}\ G^{-1}$
- RF power coefficient: $2.2\cdot10^{-13}\ dB^{-1}$

•Atomic resonator temperature coefficient: $5 \cdot 10^{-14}$ K^{-1}
As already mentioned, to improve the clock long-term stability, additional servo loops must be added according to the above measured sensitivities.

4. Conclusions

In these proceedings, we have reported on the latest development for an "Optically-pumped Space Cesium Clock for Galileo". The architecture of this compact Cs beam frequency standard takes advantage on two key features, a single optical wavelength scheme and a dark fringe Ramsey cavity.

By operating the atomic resonator with a narrow band DFB laser tuned on the Cs D2:44' transition at 852 nm and an oven temperature of 100°C, a frequency stability of $2.3\times10^{-12}\ \tau^{-1/2}$ has been measured. Long-term frequency stability will be improved with additional servo loops. A detailed clock signal noise budget analysis has been performed and it turns out that the clock can reach a frequency stability of $1.2\times10^{-12}\ \tau^{-1/2}$ if spurious technical noise sources can be eliminated.

References

[1] F. Droz *et al.*, Proc. of the 20th European Time and Frequency Forum, 2006, Braunschweig, Germany
[2] S. Guérandel *et al.*, Proc. of the 16th European Time and Frequency Forum, 2002, St-Petersburg, Russia
[3] S. Lecomte *et al.*, Proc. of the 2007 IEEE International Frequency Control Symposium jointly with the 21st European Time and Frequency Forum, pp. 1127-1131, Geneva, Switzerland, May 2007
[4] V. Hermann *et al.*, Proc. of the 2007 IEEE International Frequency Control Symposium jointly with the 21st European Time and Frequency Forum, pp. 77-80, Geneva, Switzerland, May 2007
[5] P. Berthoud *et al.*, Proc. of the 2008 Proc. of the 22nd European Time and Frequency Forum, Toulouse, France, April 2008
[6] S. Guérandel *et al.*, Proc. of the 2007 IEEE International Frequency Control Symposium jointly with the 21st European Time and Frequency Forum, pp. 1050-1055, Geneva, Switzerland, May 2007
[7] C. Audoin, V. Candelier and N. Dimarcq, IEEE Trans. Instr. Meas., Vol. 40, N°2, April 1991, pp. 121-125
[8] G. Lucas-Leclin, P. Cerez, and N. Dimarcq, J. Phys. B, Vol. 32, 1999, pp.327-340

Part IV

Optical Clocks I: Lattice Clocks

OPTICAL LATTICE CLOCK: SEVEN YEARS OF PROGRESS AND NEXT STEPS

HIDETOSHI KATORI, MASAO TAKAMOTO, AND TOMOYA AKATSUKA

Department of Applied Physics, Graduate School of Engineering, The University of Tokyo, Bunkyo-ku, 113-8656 Tokyo, Japan.

We discuss configurations of optical lattice clocks from the viewpoint of lattice geometries and quantum statistics of interrogated atoms. This leads to two promising schemes: a one-dimensional lattice loaded with spin-polarized fermions and a three-dimensional lattice loaded with bosons. These two optical lattice clocks based on fermionic ^{87}Sr and bosonic ^{88}Sr were demonstrated by measuring the frequency stability as well as their isotope shift. In addition, we address the ongoing experimental challenges for optical lattice clocks.

1. Introduction

An idea for an optical lattice clock[1] was proposed 7 years ago at the last Symposium on Frequency Standards and Metrology, held in St. Andrews. After early experiments that demonstrated a "magic wavelength"[2] and its narrow spectrum,[3] development of optical lattice clocks soon became an active competition involving strong research groups that introduced new ideas[4, 5] and elaborate laser techniques.[6] To date, these endeavors have reduced the clocks' uncertainty down to 1×10^{-15} and even below,[7, 8] making optical lattice clocks a promising candidate for future atomic clocks, in addition to traditional single ion clocks. For the ^{87}Sr-based lattice clock, frequency measurements have been performed by the JILA,[7] SYRTE,[9] and Tokyo-NMIJ[10] groups, which showed excellent agreement at 6×10^{-16}, comparable to the uncertainty of International Atomic Time.

Looking at fractional uncertainties of 1×10^{-16} and below, collisional frequency shift, blackbody radiation (BBR) shift,[11] and hyperpolarizability effects,[12] all of which depend on interrogated atomic elements and experimental configurations, are becoming major concerns. In this proceeding, we first discuss the optimal design of lattice geometries in view of the quantum statistics and related spins of interrogated atoms.[13] This leads to two promising configurations for the clock: a one-dimensional (1D) lattice loaded with spin-polarized fermions[14] and a 3D lattice loaded with bosons. We present a frequency comparison of these two optical lattice clocks using fermionic ^{87}Sr and bosonic

^{88}Sr. Such a comparison will offer an important step in ascertaining the clocks' uncertainty beyond the Cs limit. As for the latter two issues, i.e., the BBR and lattice-laser-related uncertainties, we discuss prospects for a cryogenic clock, a Hg-based optical lattice clock,[15] and a "blue-detuned" magic wavelength.

2. Optical lattice clocks with bosons and fermions and their frequency comparison

2.1. *One-dimensional optical lattice with fermions*

In 1D optical lattice clocks, the application of spin-polarized fermions may minimize the collisional frequency shift by utilizing their quantum-statistical properties: The anti-symmetrization of the wavefunction yields a two-particle correlation function at zero distance to be $g^{(2)}(0) = 0$, and hence reduces collisional frequency shifts. The application of spin-polarized fermions for precision spectroscopy had been discussed for Cs clocks[16] and experimentally demonstrated in an RF transition with ultracold ^6Li atoms.[17] However, its application to optical transitions is not obvious because the order of magnitude of shorter transition wavelength makes the Lamb–Dicke condition more severe. To guarantee the identity of fermions during the excitation process, the Bloch vector of each atom should evolve uniformly, which requires careful removal of inhomogeneous phase shifts among atoms that arise from the motional effects and the excitation field.

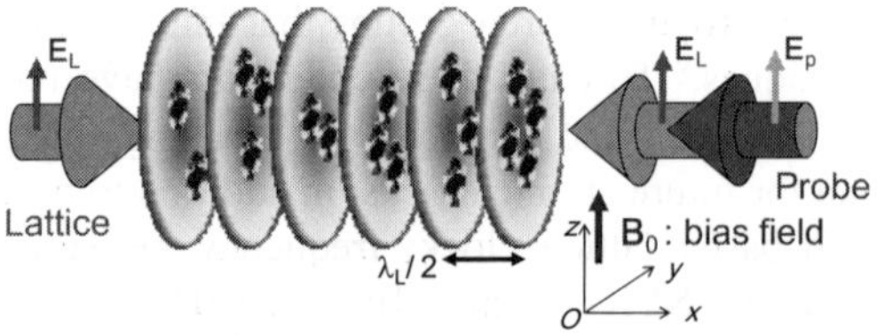

Figure 1. Experimental configuration for 1D optical lattice clock with spin-polarized fermions that are indicated by upward-pointing arrows.

Figure 1 shows the schematic diagram for a "spin-polarized" 1D optical lattice clock,[14] where upward arrows correspond to spin-polarized fermionic atoms. Figure 2 demonstrates the Rabi oscillations of atoms on the $^1S_0\,(F = m_F = 9/2)$ □ $^3P_0\,(F = m_F = 9/2)$ clock transition. The first Rabi π-pulse excitation at 2 ms showed 98% excitation of atoms, which indicates nearly uniform excitation of spin-polarized fermionic systems necessary for the Pauli blocking of collisions.[17] Collision shifts would therefore be suppressed if the p- and higher odd partial wave collisions do not contribute. Although such collision shifts have not been detected at our measurement uncertainties of about 1×10^{-15}, collision shift measurement with reduced uncertainty will provide insights into the decoherence mechanism of fermionic systems and the contribution of higher partial wave collisions in ultracold atomic systems. In Fig. 2(a), a relatively rapid decay of the envelope of Rabi oscillation is observed, which is attributed to the interference of n-dependent Rabi frequency $\Omega_{n,n}$. Collapses and revivals up to 100 ms were observed, as shown in Fig. 2(b). The oscillation period of the envelope function is roughly given by $\tau_B = 2\pi / \left|\Omega_{1,1} - \Omega_{0,0}\right|$, which is the beat note of the Rabi frequencies for atoms in the $n_x = 0$ and $n_x = 1$ vibrational states. A detailed study of the dynamics of Rabi oscillations is reported in Ref. 18.

We note that a metrologically important feature of the 1D optical lattice is that the light field polarization is spatially uniform and well defined. This allows us to cancel out the vector light shift[14] by alternately interrogating the two transition frequencies $f_\pm$ of the $^1S_0\,(F = 9/2,\ m_F = \pm 9/2)$–$^3P_0\,(F = 9/2,\ m_F = \pm 9/2)$ clock transitions to deduce the averaged value as $f_0 = (f_+ + f_-)/2$. This averaging protocol also allows us to cancel out the first-order Zeeman shift.[19] Therefore, we can simulate virtual spin-0 atoms by applying this protocol.

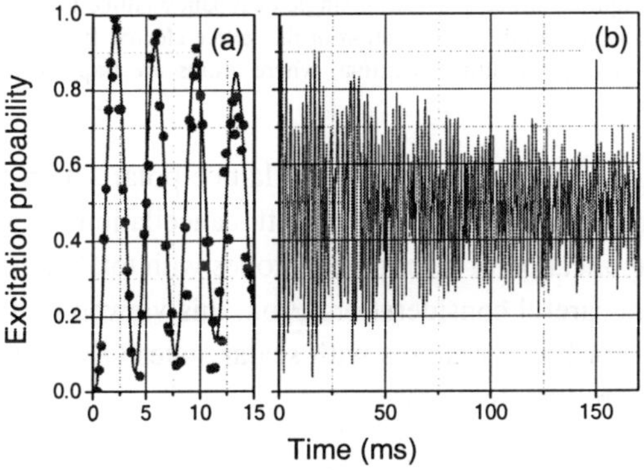

Figure 2. Rabi oscillations of spin-polarized ^{87}Sr atoms on the $^1S_0\,(F = m_F = 9/2)$–$^3P_0\,(F = m_F = 9/2)$ transition in a 1D optical lattice. (a) For the first Rabi excitation at 2 ms, more than 98% of the atoms were excited, demonstrating a high degree of spin polarization. (b) Collapses and revivals were observed up to 100 ms.

2.2. *Three-dimensional optical lattice*

To suppress atomic collisions and confine atoms in the Lamb–Dicke regime, as proposed initially,[1] it seems straightforward to apply 3D optical lattices with less than a single atom in each lattice site, as shown in Fig. 3(a). Technically, it is more challenging to realize 3D optical lattices, considering their stability in position as well as local polarization, than 1D ones. One of the solutions is to apply "folded optical lattices,"[20] where the 3D optical lattice consists of a single standing wave of light. In this configuration, local polarization of lattice sites may remain unchanged, as the two orthogonal electric field vectors oscillate in phase at the local lattice site. Even though the local polarization can be set to be linear in this lattice, the direction of the electric field vector spatially changes from one site to the next by nearly 90°, as shown in Fig. 3(b).

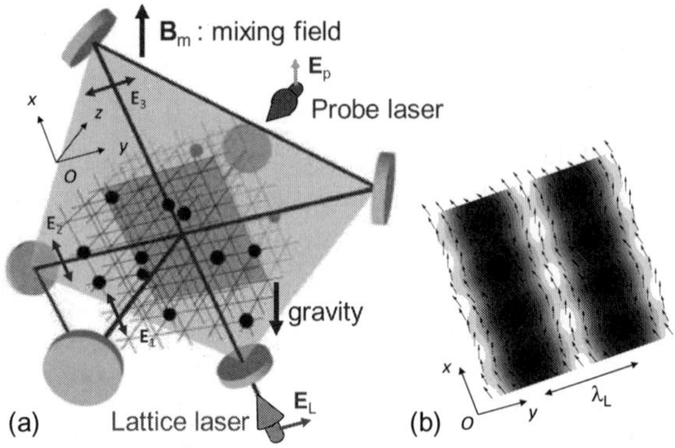

Figure 3. (a) A 3D optical lattice provides a single-occupancy lattice to suppress the bunching of bosonic atoms. (b) The amplitude and direction of the lattice electric field vector $\mathbf{E}_L$ corresponding to the shaded area in (a). Amplitude maxima, where atoms are trapped, are indicated by light regions.

Considering the local character of 3D lattice polarization that makes it less effective to perform the vector light shift cancellation technique mentioned previously, the application of fermionic isotopes with total angular momentum of $F \neq 0$ will require careful consideration. From this viewpoint, the 3D lattice will be more suitable for bosonic atoms with scalar states ($J = 0$) since the coupling to light polarization is a minimum.

2.3. *Frequency comparison of optical lattice clocks with bosons and fermions*

We have built two optical lattice clocks in different vacuum chambers: one consisted of a 1D optical lattice with spin-polarized fermionic ^{87}Sr, and the other consisted of a phase-stable 3D optical lattice with bosonic ^{88}Sr. While hyperfine mixing gave a transition moment for the clock transition in ^{87}Sr, an external magnetic field of about 2.34 mT with a probe laser intensity of $I_p = 400$ mW/cm^2 was applied to ^{88}Sr to introduce a Rabi frequency of 9 Hz on the clock transition by mixing the $^3P_0$ state with $J = 1$ states.[21]

Figure 4 shows a schematic of our experiment. Optical lattice clocks with ^{87}Sr and ^{88}Sr atoms were alternately prepared and interrogated by clock lasers detuned by the isotope shift of about 62 MHz. The clock lasers were then servo-locked to the respective transition frequencies of f_{87} and f_{88}, and the frequency difference $f_{88} - f_{87}$ was recorded as a time series. The Allan deviation evaluated by the beat note of these two "independent" clocks reached 5×10^{-16} for an averaging time of 2,000 s. After the careful elimination of systematic uncertainties, in particular, for the ^{88}Sr lattice clock that was seriously perturbed by mixing fields, we deduced the isotope shifts on the clock transitions to be $f_{88} - f_{87} = 62{,}188{,}138.4(1.3)$ Hz.[13]

In this experiment, the 1D and 3D clock comparison consists of 6 successive measurement cycles, which include 4 cycles to measure both sides of the Rabi excitation line shapes for the outermost Zeeman components $f_{\pm}$ of ^{87}Sr and 2 cycles for ^{88}Sr, corresponding to a total measurement time of 6 s. If a clock laser with its coherence time sufficiently longer than 0.5 s, which is comparable to the time required for atom loading, is available, seamless interrogation of the clock transitions will be possible, i.e., while interrogating one of the clocks, the other lattices could be loaded with atoms. These attempts will lead to lattice clocks with significantly improved stability.

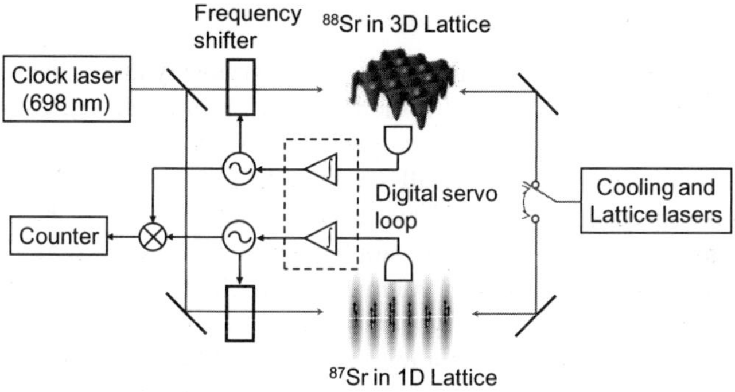

Figure 4. Two optical lattice clocks with different isotopes and lattice configurations were operated sequentially, and their beat note was evaluated.

3. Future directions and outlook

3.1. *Remaining issues to be explored in Sr lattice clocks*

3.1.1. *Minimally destructive or QND measurements of atoms to speed up the clock cycle time.*

Currently, the stability of our optical lattice clock is typically $3 \times 10^{-14}\,\tau^{-1/2}$ in comparing 1D and 3D clocks, as described in the previous section. The stability is severely limited by the Dick effect,[22] which is responsible for the small duty ratio of the clock measurements. The actual interrogation time of 60 ms was less than one-tenth of the entire cycle time of 1 s, most of which was spent on cooling and capturing atoms. This time-consuming reloading process may be avoided if minimally destructive or quantum non-demolition (QND) schemes[23, 24, 25] are applied to the state detection of the clock transition, as the schemes prevent atoms from being heated out of the lattice trap and allow the reuse of trapped atoms, as in the case of single-ion based clocks. These improvements will allow us to explore the fractional uncertainties of 10^{-17} in a reasonable averaging time of 100–1,000 s and reveal newer aspects that have not been discussed seriously, as described below.

3.1.2. *Blackbody shift: Cryogenic lattice clock*

In a few 10^{-17} uncertainties, BBR shifts[11, 12] will seriously affect the uncertainty budget of Sr-based lattice clocks. As the BBR shift rapidly decreases as $1/T^4$ for a surrounding temperature T, a cryogenic environment, even at the liquid nitrogen temperature of $T = 77$ K, will reduce BBR shifts to 10 mHz.[12] Therefore, the corresponding uncertainty will have an affect only at 10^{-18} or below. Such a cryogenic environment will be readily applicable to lattice clocks, as a cryogenic region, a few cubic millimeters in volume, is sufficient for their operation. For transferring atoms into the tiny cryogenic volume, a moving optical lattice that was used to transfer Sr atoms into a microstructured Stark chip[26] in our previous experiment will be directly applied. The experiment is now in progress.

3.2. *Exploring new atomic elements*

Most of the uncertainties discussed above such as collision shifts, BBR shifts, and hyperpolarizability effects depend on each atomic parameter; therefore, they can be improved by appropriate choice of atomic elements.[27] On the other hand, the potential high precision of the lattice clock scheme and its wide applicability to other atom species will make optical lattice clocks useful tools in investigating possible variations in the fine structure constant.[28] Assuming $\Delta\alpha/\alpha = 10^{-16}$, at which level per year astrophysical determinations gave controversial results, the fractional change of the clock frequency $\delta\nu/\nu_0$ of Sr-, Yb-, and Hg-based lattice clocks is expected to be 6.2×10^{-18}, 3.1×10^{-17}, and 8.1×10^{-17}, respectively.[29] Heavier atoms such as Yb and Hg tend to show higher sensitivity, as the relativistic corrections depend on the nuclear charge as Z^2, while a Sr-based optical lattice clock may be used as an anchor to detect the fractional frequency change of a Yb- or Hg-based optical lattice clock.

As a first step in this direction, we have recently demonstrated laser cooling and trapping of Hg atoms for the first time.[15] Hg is the heaviest nonradioactive atom trapped so far, and enables sensitive atomic searches for "new physics" beyond the standard model, including α-variation. The systematic accuracy of a Hg-based optical lattice clock is theoretically evaluated to be better than 10^{-18}, because the BBR shifts are more than one order of magnitude smaller than those of Sr or Yb. Highly accurate and stable Hg-based clocks will provide a new avenue for research of optical lattice clocks and the time variation of the fine-structure constant.

3.3. *A "blue-detuned" magic wavelength: Prospects for quantum metrology*

It is natural to think about the possibility of finding the magic wavelength for a blue-detuned dipole trap[30, 31] that confines atoms in the intensity minima of the electric field, as in the case of ions in Paul traps. These traps would effectively moderate the influence of higher-order light shifts. For Sr, such wavelengths will be found on the blue side of the $5s^2\ {}^1S_0$–$5s5p\ {}^1P_1$ transition at 461 nm. The far-off-resonance condition is more difficult to satisfy because the magic wavelength can be found close to the transitions arising from the $5s5p\ {}^3P_0$ state, as indicated in Fig. 5(a).

One such wavelength is found at $\lambda_L \sim 390$ nm on the blue side of the $5s5p\ {}^3P_0$–$5s6d\ {}^3D_1$ transition at 394 nm. For this magic wavelength, the laser intensity of $I_L = 10$ kW/cm^2 gives a reasonable trap depth of about 200 kHz. The effective light intensity that atoms experience will be about one-tenth of the maximum intensity, as the atoms are trapped near the node of the standing wave. Assuming $I_L = 1$ kW/cm^2, the photon scattering rate is calculated to be 0.4 /s and $d\nu_{ac}/d\nu_L \sim 1.9 \times 10^{-9}$. Figure 5(b) shows the differential light shift measured for a probe laser intensity of 4.5 W/cm^2 scanned around 390 nm. This measurement was performed for ^{87}Sr atoms trapped in the 1D magic wavelength optical lattice operated at $\lambda = 813.4$ nm. The "blue" magic wavelength was determined to be 389.889 (9) nm.

With the magic wavelength located at violet wavelengths, the closer interatomic separation of only 195 nm will raise the intriguing possibility of controlling atomic interactions with mid-infrared dipole transitions. For example, in the presence of laser radiation tuned to the $5s5p\ ^3P_0$–$5s4d\ ^3D_1$ transition at $\lambda_d = 2.6\ \mu m$, two neighboring atoms in this optical lattice will undergo a strong dipole–dipole interaction, which may serve as an interesting tool to investigate atom entanglement for quantum information processing (QIP) and quantum metrology.

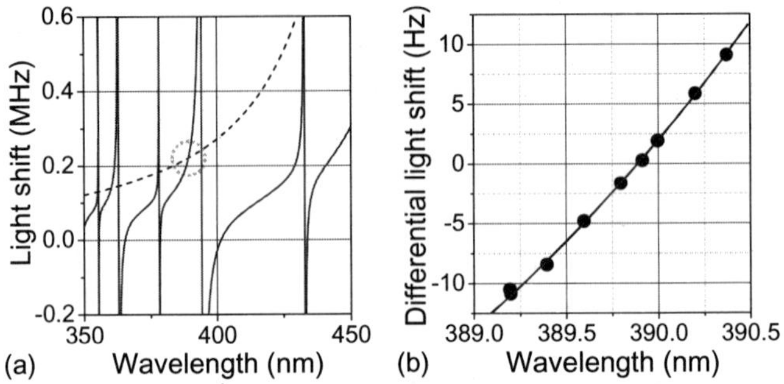

Figure 5. (a) Light shift for the $^1S_0$ state (dashed blue line) and $^3P_0$ state (solid red line) calculated for a lattice laser intensity of 10 kW/cm^2. On the higher frequency side with respect to the $^1S_0$–$^1P_1$ transition at 461 nm, several blue-detuned wavelengths such as that at 390 nm and 360 nm are found. (b) The differential light shift measured around 389.9 nm with a light intensity of 4.5 W/cm^2.

Acknowledgment

This work is supported by Core Research for Evolutional Science and Technology (CREST) and Strategic Information and Communications R&D Promotion Programme (SCOPE).

References

1. H. Katori, in the 6th Symposium on Frequency Standards and Metrology, edited by P. Gill (World Scientific, Singapore, 2002), pp. 323.
2. M. Takamoto, and H. Katori, Phys. Rev. Lett. **91**, 223001 (2003).
3. M. Takamoto, F. L. Hong, R. Higashi, and H. Katori, Nature **435**, 321 (2005).
4. Z. W. Barber et al., Phys. Rev. Lett. **96**, 083002 (2006).
5. A. Brusch, R. Le Targat, X. Baillard, M. Fouch, and P. Lemonde, Phys. Rev. Lett. **96**, 103003 (2006).
6. M. M. Boyd et al., Science **314**, 1430 (2006).
7. S. Blatt et al., Phys. Rev. Lett. **100**, 140801 (2008).
8. Z. W. Barber et al., Phys. Rev. Lett. **100**, 103002 (2008).
9. X. Baillard et al., Eur. Phys. J. D **48**, 11 (2008).

10. F.-L. Hong et al., http://arXiv.org/abs/0811.1816 (2008).
11. S. G. Porsev, and A. Derevianko, Phys. Rev. A **74**, 020502 (2006).
12. H. Katori, M. Takamoto, V. G. Pal'chikov, and V. D. Ovsiannikov, Phys. Rev. Lett. **91**, 173005 (2003).
13. T. Akatsuka, M. Takamoto, and H. Katori, Nat. Phys. advanced online publication, doi: 10.1038/nphys1108 (2008).
14. M. Takamoto et al., J. Phys. Soc. Jpn. **75**, 104302 (2006).
15. H. Hachisu et al., Phys. Rev. Lett. **100**, 053001 (2008).
16. K. Gibble, and B. J. Verhaar, Phys. Rev. A **52**, 3370 (1995).
17. S. Gupta et al., Science **300**, 1723 (2003).
18. M. Takamoto, and H. Katori, To appear in J. Phys. Soc. Jpn. **78** (2009).
19. J. E. Bernard, L. Marmet, and A. A. Madej, Opt. Commun. **150**, 170 (1998).
20. A. Rauschenbeutel, H. Schadwinkel, V. Gomer, and D. Meschede, Opt. Commun. **148**, 45 (1998).
21. A. V. Taichenachev et al., Phys. Rev. Lett. **96**, 083001 (2006).
22. G. Santarelli et al., IEEE Trans. Ultrason. Ferroelectr. Freq. Control **45**, 887 (1998).
23. R. Long, A. K. Tuchman, and M. A. Kasevich, Opt. Lett. **32**, 2502 (2007).
24. S. Chaudhury, G. A. Smith, K. Schulz, and P. S. Jessen, Phys. Rev. Lett. **96**, 043001 (2006).
25. P. G. Petrov, D. Oblak, C. L. G. Alzar, N. Kjaergaard, and E. S. Polzik, Phys. Rev. A **75**, 033803 (2007).
26. T. Kishimoto et al., Phys. Rev. Lett. **96**, 123001 (2006).
27. V. D. Ovsiannikov, V. G. Pal'chikov, H. Katori, and M. Takamoto, Quantum Electronics **36**, 3 (2006).
28. J.-P. Uzan, Rev. Mod. Phys. **75**, 403 (2003).
29. E. J. Angstmann, V. A. Dzuba, and V. V. Flambaum, Phys. Rev. A **70**, 014102 (2004).
30. N. Davidson, H. Jin Lee, C. S. Adams, M. Kasevich, and S. Chu, Phys. Rev. Lett. **74**, 1311 (1995).
31. T. Mueller-Seydlitz et al., Phys. Rev. Lett. **78**, 1038 (1997).

THE Yb OPTICAL LATTICE CLOCK

N. D. LEMKE, A. D. LUDLOW, Z. W. BARBER*, N. POLI†, C. W. HOYT‡, L. S. MA§, J. E. STALNAKER¶, C. W. OATES*, L. HOLLBERG, J. C. BERGQUIST, A. BRUSCH, T. M. FORTIER, S. A. DIDDAMS, T. HEAVNER, S. JEFFERTS, AND T. PARKER

National Institute of Standard and Technology,
Boulder, CO 80305 USA
** e-mail: oates@boulder.nist.gov*

We describe the development and latest results of an optical lattice clock based on neutral Yb atoms, including investigations based on both even and odd isotopes. We report a fractional frequency uncertainty below 10^{-15} for ^{171}Yb.

Keywords: Optical clocks; Optical frequency standards; Optical lattices

1. Introduction

At the previous Symposium on Frequency Standards and Metrology, H. Katori proposed using $^1S_0 \rightarrow ^3P_0$ transitions in alkaline earth like atoms to make high performance optical clocks based on large numbers of neutral atoms tightly confined in an optical lattice.[1] This proposal was based on the use of Sr atoms, as were the initial experiments.[2–4] In 2004 Porsev et al. proposed the use of the analogous transition in Yb at 578 nm,[5] and we have been developing lattice clocks based on Yb over the past four years. While there are considerable similarities between the Sr and Yb atomic systems, there are also some important differences. Most significantly, Yb has an

*Present address: Montana State University Spectrum Lab P.O. Box 173510 Bozeman, MT 59717-3510 USA

†Present address: LENS and Dipartimento di Fisica, Universita di Firenze, INFN - sezione di Firenze - 50019 Sesto Fiorentino, Italy

‡Present address: Bethel University, St. Paul, MN 55112 USA

§Present address: State Key Laboratory of Precision Spectroscopy, East China Normal University, Shanghai, China

¶Present address: Department of Physics and Astronomy, Oberlin College, Oberlin, Ohio 44074, USA abundant isotope (^{171}Yb) that has nuclear spin = 1/2 and correspondingly simple energy level structure and optical pumping. Due to its 1/2 spin, ^{171}Yb has zero tensor polarizability, which could make it more straightforward to use in three-dimensional lattices, for which the sites cannot have uniform polarization.

In the course of this work we implemented a technique, magnetic field-induced spectroscopy (MIS), that enables excitation of the $^1S_0 \rightarrow {}^3P_0$ clock transition, which is otherwise completely forbidden in even isotopes.[6,7] This technique uses an external magnetic field to provide the required level mixing that occurs naturally in odd isotopes. In this way we can access the conceptually simpler bosonic isotopes, which may be especially attractive for clocks based on atoms confined in three dimensions. However with bosons there are two additional (and potentially significant) frequency shifts to worry about: a second-order Zeeman shift associated with the magnetic bias field, and a light shift that results from the increased probe light intensity required to excite the effectively weaker transition.

We summarize results based on the most abundant Yb isotope, ^{174}Yb (nuclear spin = 0) as well as more recent results on ^{171}Yb. In addition to uncertainty budgets for the two systems, we include absolute frequency measurements made via an octave-spanning femtosecond-laser frequency comb against other microwave and optical standards.

2. Apparatus

The ^{174}Yb apparatus (see Figure 1) has been described in detail elsewhere;[7–9] here we summarize the relevant details and modifications used for the study of ^{171}Yb. In order to load atoms into the lattice it is necessary to first cool the atoms. We achieve this with two stages of laser cooling: the first uses the strong $^1S_0 \rightarrow {}^1P_1$ transition at 399 nm to load atoms into a magneto-optic trap (MOT) from a thermal beam, while the second uses the intercombination line at 556 nm to cool the atoms from 5 mK to about 50 μK, thus requiring optical lattices 5 to 10 times deeper than those used for Sr atoms. After the two stages of laser cooling there are approximately 4 x 10^5 atoms in the trapping region, of which about 10^4 are transferred into the pancake-like potential wells (10 to 20 atoms per well) of the 1-D optical lattice. The horizontally oriented lattice is formed by retroreflecting a tightly focused laser beam (waist = 30 μm, power = 1 W) through the MOT region. We tune the wavelength of the lattice light to its"magic" value for the Yb clock transition, near 759 nm, where the ground and excited clock states experience equal lattice light shifts.

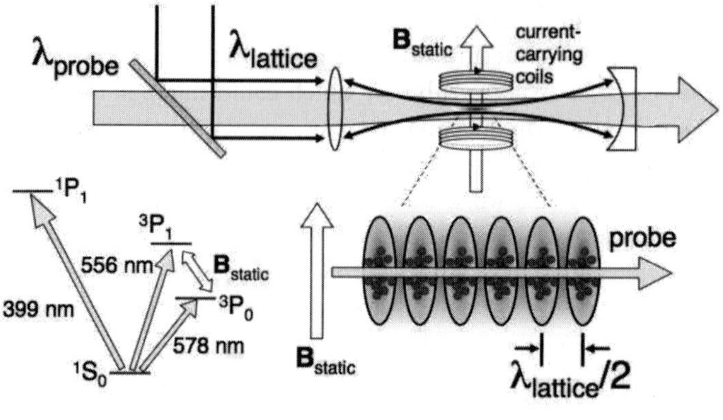

Fig. 1. Schematic of the Yb lattice clock apparatus. After two stages of laser cooling, the atoms are confined in a 1-D optical lattice whose wavelength is tuned near its magic value of 759.4 nm. Prestabilized light at 578.42 nm is used to probe the clock transition. For ^{174}Yb a magnetic bias field assists the probe light in excitation of the transition.

For spectroscopy with ^{174}Yb, we first turn off all lasers (except for the lattice) and the MOT magnetic field gradient. We then turn on the probe laser light at 578 nm (that propagates collinearly with the lattice light) and a magnetic bias field of about 1 mT (10 G). The 578 nm light is produced through sum-frequency generation in a periodically poled waveguide of a Nd:YAG source at 1.319 μm and a fiber laser at 1.03 μm. The frequency is pre-stabilized with a vertically mounted, high finesse cavity. With this system we have resolved linewidths as narrow as 4 Hz (see Figure 2) and achieved a single-shot signal-to-noise ratio of more than 15 by using a re-pump laser at 1.39 μm to normalize shot-to-shot atom number fluctuations.

With ^{171}Yb we do not need a bias field to enable excitation of the transition, since mixing due to the non-zero nuclear magnetic moment yields a natural linewidth of 10 mHz for the clock transition.[5] However, we use a weak magnetic field and a 10 ms period of optical pumping before the spectroscopy to spin polarize the sample (typical sample polarization $>$ 98 %). This is accomplished with light at 556 nm tuned to the appropriate resonance (the clock resonances are split by a 0.5 mT magnetic field). We then alternate 578 nm measurement cycles between the $m = \pm 1/2$ states in the presence of a weak bias field (0.1 mT) to split the two resonances by 500 Hz. Locking to the average of this value enables us to suppress stray magnetic field effects and vector light shifts.[10]

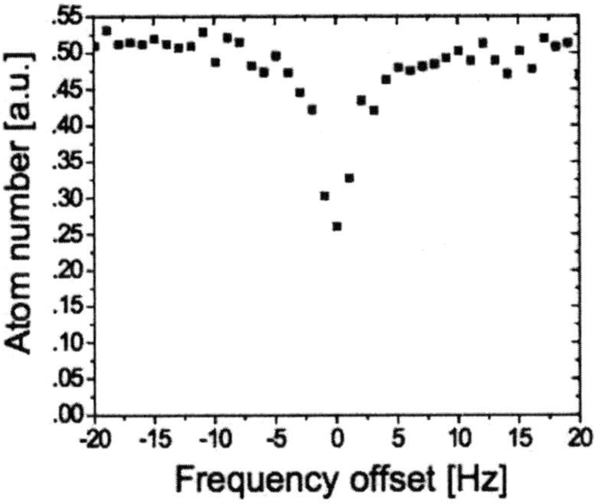

Fig. 2. Spectrum of the magnetic field-induced $^1S_0 \leftrightarrow {}^3P_0$ resonance in ^{174}Yb with the lattice laser near the magic wavelength. Each point represents a ~540 ms experimental cycle. A Gaussian fit yields a Fourier transform-limited FWHM of ≈ 4 Hz. The vertical scale is not normalized to full atom depletion, but the observed contrast is about 50 %.

In order to evaluate the performance of the Yb clock we usually measure against a Ca optical clock at 657 nm that resides in our laboratory.[11] We span the 62 THz gap between the clock lasers with a femtosecond-laser frequency comb.[12] We also lock the frequency of the optical lattice to the comb to make sure it remains fixed at the magic wavelength. The Ca clock lacks the small absolute uncertainty of the lattice clock, but has good short-term stability and provides a robust flywheel against which we can evaluate the various Yb clock shifts. By alternating conditions for a given parameter we can evaluate its effects at the sub-10^{-15} level in a few minutes (Figure 3 shows the Allan deviation of the beat note between the Ca and Yb lasers connected by the frequency comb).

3. Systematic uncertainties for ^{171}Yb and ^{174}Yb

3.1. *Key frequency shifts for the Yb lattice clock*

With the system described in the previous section we have evaluated the systematic effects for both ^{171}Yb and ^{174}Yb. In addition to the common effects such as blackbody radiation shifts and lattice wavelength shifts, each has shifts that are peculiar to the isotope, which we consider first.

Use of the even isotope greatly reduces atom orientation-related effects such as the vector light shift or the first-order Zeeman effect.[1,5] However we

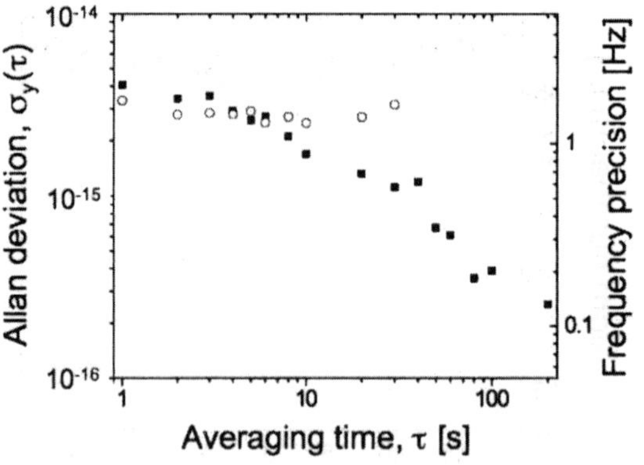

Fig. 3. Fractional frequency instability of the beat note between light locked to the Yb clock transition and (i) light locked to the Ca reference cavity (open circles) and (ii) light locked to the Ca clock transition (closed circles).

need to evaluate carefully the second-order Zeeman effect associated with magnetic bias field and the probe light-induced Stark shift. For magnetic field-induced spectroscopy these shifts are inextricably connected with the strength of excitation; they are typically about the same size as the spectroscopic resolution, so they can be reduced only by working at higher resolution.[6] Ultimately, however, the resolution is limited by the prestabilization of the laser, so clearly clocks based on the MIS method will benefit considerably from better reference cavities.

In order to reduce the uncertainty associated with these two effects, we have evaluated the frequency shifts for a range of values for the magnetic field and probe intensity. We have the ability to turn these "knobs" to values considerably larger than our typical operating values, so we can achieve small uncertainties for these effects. In Figure 4 we show plots used to evaluate these effects. From these we can extrapolate to zero-field values, but we still have to consider how well we can hold these values and how stray magnetic fields will contribute to the uncertainty.

For ^{171}Yb, the protocol that alternates (and averages) between spin polarized m $= \pm 1/2$ states suppresses atom orientation-related shifts.[10] However, since the vector light shift may be important for multidimensional lattices, we made a preliminary measurement of this effect. We varied the polarization of our 1-D lattice from linear to circular and found a frequency shift (equal and opposite for our two states) of 28 Hz. This value is smaller than the estimate by Porsev et al.,[5] but still large enough to warrant con-

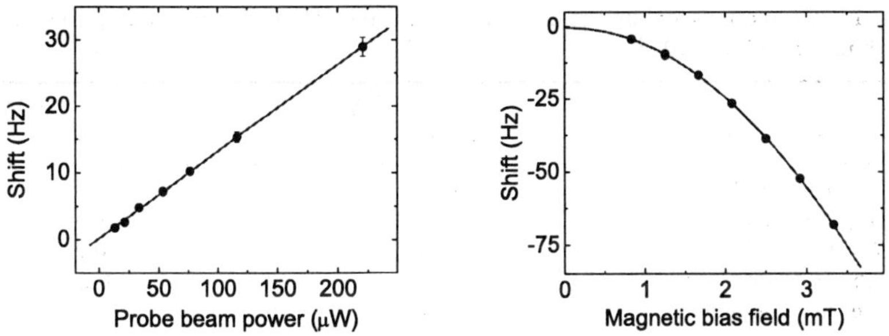

Fig. 4. Measurements (circles) of the clock frequency vs. probe intensity (left) and magnetic bias field (right) for ^{174}Yb. From the fit on the right (solid line) we determine the second-order Zeeman coefficient (β = -6.12(10) Hz/mT^2), which agrees well with theoretical estimates.[5,9]

sideration in the design of three-dimensional lattices.

3.2. *Uncertainty budgets for ^{171}Yb and ^{174}Yb*

Table 1 shows the total uncertainty budgets for the frequency standards based on the two isotopes. The total uncertainty for ^{174}Yb is about four times that of ^{171}Yb. This is a result of more precise evaluations of some of the common effects but also reflects the larger values for the magnetic field and probe light uncertainties for the even isotope. Effects due to the lattice (magic wavelength uncertainty, hyperpolarizability) should have similar values for the two isotopes. The hyperpolarizability was originally of considerable concern for Yb-based lattice clocks because of a possible coincidence between the magic wavelength and two-photon transitions from the excited state to higher-lying states. Indeed the magic wavelength turned out to be close to one of these resonances, but an evaluation of these effects has shown them to be small.[8] Interestingly, the collision effects have turned out to be larger for the fermionic isotope, although this is probably a result of the residual atomic temperature that prevents the fermions from being excited in an indistinguishable way. Most of the effects listed have uncertainties that can be reduced well below the 10^{-16} level by simply performing more precise measurements of the type already described. A possible exception is the shift due to ambient blackbody radiation (BBR). The uncertainty listed

Table 1. Corrections and uncertainties for ^{174}Yb and ^{171}Yb (fractional frequency shift relative to 578 nm in units of 10^{-16})

	^{174}Yb		^{171}Yb	
Systematic	Corr.	Unc.	Corr.	Unc.
Latt. Pol.	0.0	5.7	0.0	2.9
Hyperpolarizability	3.5	0.7	3.5	0.7
2nd Order Zeeman	-346	3.8	-0.2	0.0
Probe Light Shift	116	3.8	0.2	0.0
Density dependent	-1.9	9.6	-19	1.0
Blackbody shift	-25	2.5	-25	2.5
AOM Chirp	0.0	0.2	0.0	0.2
Residual Servo Error	4.8	0.2	0.0	0.0
Total Uncertainty ($\times 10^{-16}$)		15		3.2

for this effect is due primarily to the uncertainty in the theoretically calculated blackbody coefficient[13] used to extrapolate the transition frequency to that at zero kelvin. Experimentally derived values will be most likely be needed to reduce this uncertainty further. Alternatively, if we defined the transition frequency at a temperature of 300 K rather than 0 K, then the BBR shift uncertainty would depend mainly on the uncertainty of the ambient temperature, which would yield a fractional uncertainty of about 3×10^{-17} K^{-1} at 300 K.

3.3. *Frequency measurements for ^{171}Yb and ^{174}Yb*

We measured the absolute frequency of the clock transition for both isotopes through use of a mode-locked femtosecond-laser frequency comb[12] that connects the Yb lattice standard to other microwave and optical standards at NIST. In Figure 5 we show independent measurements of the ^{174}Yb clock frequency[9] against a Cs-fountain calibrated maser at NIST[14] and the Hg^+ optical clock.[15] Because these measurements were taken independently on different days over several months, they form a closed loop (in conjunction with Hg^+-Cs frequency measurements[15]) between the three standards and show good agreement at the 1 Hz level. Due to the much higher stability of the optical vs. optical measurements, the time required for a single measurement was minutes rather than the 24-hour period needed to average down the maser noise.

We have more recently measured the ^{171}Yb clock frequency against masers calibrated by the NIST Cs fountain and derived a preliminary absolute frequency of 519 295 836 590 863.2(0.8) Hz. The fractional frequency uncertainty of 1.5 x 10^{-15} resulted primarily from uncertainties in the comparison itself.

4. Conclusions and future prospects

Thus far, research with the Yb optical lattice clock has demonstrated its viability as a high performance optical frequency standard. The issue of hyperpolarizability has been resolved, and for the most part the frequency shifts are well understood. Measurements indicate that performance at a level below 10^{-16} seems accessible even with the present 1-D lattice system. To go further, the Yb lattice clock now faces the same issues that exist for other lattice clocks. First of all, if we want to evaluate systematic effects efficiently at the 10^{-17} level and beyond, we need to improve the stability of existing lattice clocks, which is currently limited by the cavities used to prestabilize the probe laser frequency. New cavity designs may be required for the neutral atom clocks to fulfill their potential. As an additional benefit, working at higher resolution will considerably reduce the shifts for the bosonic lattice clock. Second, future lattice clocks will most likely confine atoms in two or three dimensions (underfilled, to remove collision effects) and will use considerably shallower lattices (to minimize lattice shifts).[16]

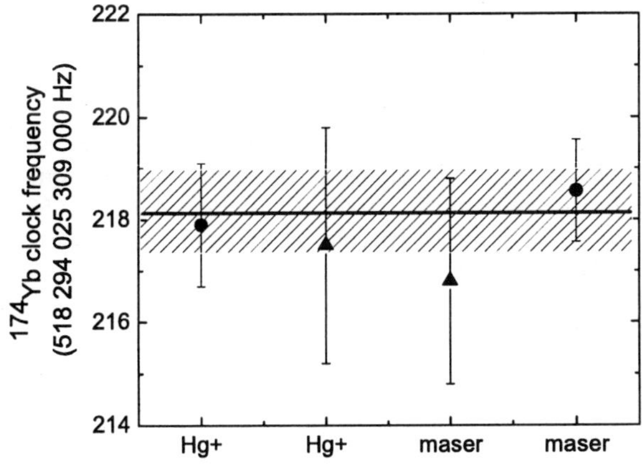

Fig. 5. Independent measurements of the ^{174}Yb clock frequency vs. the Hg^+ optical clock and a maser calibrated by the NIST Cs fountain. The resultant value is 518 294 025 309 217.8(0.9) Hz.

However, we then will need to develop improved laser cooling techniques for Yb (e.g., sideband cooling in the lattice). Finally we will need to address key issues such as the BBR shift, fluctuating gravitational potentials, and the challenges of high-precision clock comparisons over long distances. Comparisons of the ratios of the Yb and Sr optical lattice clocks paired in different locations seem particularly feasible and provide one way to bypass the frequency-transfer limitations, while still providing stringent tests of clock performance.

References

1. H. Katori, M. Takamoto, V. G. Pal'chikov and V. D. Ovsiannikov, *Phys. Rev. Lett.* **91**, p. 173005 (2003).
2. M. Takamoto and H. Katori, *Phys. Rev. Lett.* **91**, p. 223001 (2003).
3. R. Le Targat, X. Baillard, M. Fouché, A. Brusch, O. Tcherbakoff, G. D. Rovera and P. Lemonde, *Phys. Rev. Lett.* **97**, p. 130801 (2006).
4. A. D. Ludlow, T. Zelevinsky, G. K. Campbell, S. Blatt, M. M. Boyd, M. H. G. de Miranda, M. J. Martin, J. Thomsen, S. Foreman, J. Ye, T. Fortier, J. Stalnaker, S. Diddams, Y. L. Coq, Z. Barber, N. Poli, N. Lemke, K. Beck and C. W. Oates, *Science* **319**, p. 1805 (2008).
5. S. G. Porsev, A. Derevianko and E. N. Fortson, *Phys. Rev. A* **69**, p. 021403(R) (2004).
6. A. V. Taichenachev, V. I. Yudin, C. W. Oates, C. W. Hoyt, Z. W. Barber and L. Hollberg, *Phys. Rev. Lett.* **96**, p. 083001 (2006).
7. Z. W. Barber, C. W. Hoyt, C. W. Oates, A. V. Taichenachev, V. I. Yudin and L. Hollberg, *Phys. Rev. Lett.* **96**, p. 083002 (2006).
8. Z. W. Barber, J. E. Stalnaker, N. D. Lemke, N. Poli, C. W. Oates, T. M. Fortier, S. A. Diddams, L. Hollberg, C. W. Hoyt, A. V. Taichenachev and V. I. Yudin, *Phys. Rev. Lett.* **100**, p. 103002 (2008).
9. N. Poli, Z. W. Barber, N. D. Lemke, C. W. Oates, L. S. Ma, J. E. Stalnaker, T. M. Fortier, S. A. Diddams, L. Hollberg, J. C. Bergquist, A. Brusch, S. Jefferts, T. Heavner and T. Parker, *Phys. Rev. A* **77**, p. 050501R (2008).
10. M. Takamoto, F. L. Hong, R. Higashi, M. Imae and H. Katori, *J. Phys. Soc. Jpn.* **75**, p. 104302 (2006).
11. C. W. Oates, F. Bondu, R. W. Fox and L. Hollberg, *The European Physical Journal D* **7**, 449 (1999).
12. T. M. Fortier, A. Bartels and S. A. Diddams, *Opt. Lett.* **31**, p. 1011(April 2006).
13. S. G. Porsev and A. Derevianko, *Phys. Rev. A* **74**, p. 020502R (2006).
14. T. P. Heavner, S. R. Jefferts, E. A. Donley, J. H. Shirley and T. E. Parker, *Metrologia* **42**, p. 411 (2005).
15. W. H. Oskay, S. A. Diddams, E. A. Donley, T. M. Fortier, T. P. Heavner, L. Hollberg, W. M. Itano, S. R. Jefferts, M. J. Delaney, K. Kim, F. Levi, T. E. Parker and J. C. Bergquist, *Phys. Rev. Lett.* **97**, p. 020801 (2006).
16. P. Lemonde and P. Wolf, *Phys. Rev. A* **72**, p. 033409 (2005).

OPTICAL LATTICE CLOCK WITH Sr ATOMS

P. G. WESTERGAARD, A. LECALLIER, J. LODEWYCK AND P. LEMONDE*

LNE-SYRTE, Observatoire de Paris, CNRS, UPMC
61 Avenue de l'observatoire, 75014 Paris, France
**E-mail: pierre.lemonde@obspm.fr*
http://lne-syrte.obspm.fr/

We review the recent progress at LNE-SYRTE on an optical lattice clock using Sr atoms. We discuss the required lattice depth to cancel motional effects in such a clock and show that a relatively small lattice depth is sufficient provided the lattice is oriented vertically. We also demonstrate an accurate frequency measurement of the Sr clock transition with accuracy in the low 10^{-15} range. Finally, we describe a new non-destructive method to detect the atoms in the clock and discuss the expected associated improvement in the frequency stability of the clock.

Keywords: Optical lattice clocks; Strontium

1. Introduction

The idea of using cold atoms confined in an optical lattice for an optical clock was proposed by H. Katori at the sixth symposium on frequency standards and metrology.[1] Since then, this idea has become an experimental reality. We review here the contribution of LNE-SYRTE to these developments. A first point is a study of the depth of the required lattice depth to cancel motional effects in such a clock. We show that a relatively small lattice depth (about 10 recoil energies) is sufficient provided the lattice is oriented vertically. In that case gravity lifts the degeneracy between adjacent potential wells strongly inhibiting tunneling induced motion. In turn, offering the possibility to operate the lattice at small lattice depths both minimizes the requirements in terms of lattice induced shift rejection and opens the possibility to operate lattice clocks with atoms of small dynamic polarizabilities like Hg. In a second point, we present recent experimental developments. We demonstrate a frequency measurement of the Sr clock transition with accuracy in the low 10^{-15} range. Finally, we describe a new non-destructive method to detect the atoms and discuss the expected associated gain in terms of frequency stability.

2. Motional Effects and Required Lattice Depth

Confinement in an optical lattice is a new and specific feature of lattice clocks as compared to the previous generation of clocks with neutral atoms. In early papers like Ref. [1] it is implicitly assumed that each microtrap can be treated separately as a quadratic potential in which case the situation is similar to the case of trapped ion case and fully understood.[2] With an accuracy goal in the $10^{-17} - 10^{-18}$ range in mind however, this is correct at very high trapping depths only.[3] The natural energy unit for the trap dynamics is the recoil energy associated to the absorption or emission of a lattice laser photon, $E_r = \frac{\hbar^2 k_L^2}{2m_a}$ with k_L the wave vector of the laser and m_a the atomic mass. For Sr and the above accuracy goal the trap depth corresponding to the independent trap limit is typically $U_0 = 100\,E_r$ in a strictly periodic potential.

For a number of reasons however, one would like to work with traps as shallow as possible. Firstly, the residual shift of the clock transition by the trapping light decreases with U_0. The first order perturbation is canceled by tuning to the magic wavelength except for a small tensorial effect which depends on the hyperfine structure of the atom. Hyperpolarizability terms may be much more problematic depending on possible coincidences between two photon resonances and the magic wavelength.[4,5] The associated shift scales as U_0^2. Another potential problem is the frequency shift associated to non-resonant M1 and E2 couplings.[6] Due to the different spatial dependence of these contributions with respect to the dominant E1 coupling, they lead to a change in the curvature of each potential well and subsequently to a frequency shift scaling as $U_0^{1/2}$. All these shifts would be minimized by a reduction of U_0 and their evaluation would be greatly improved if one could vary this parameter over a broader range. Secondly, for some of the possible atoms for a lattice clock, such as Hg,[7,8] lattice induced two-photon ionization can occur possibly limiting the achievable resonance width and leading to a frequency shift. Finally, technical aspects like the required laser power at the magic wavelength can be greatly relaxed if one uses shallow traps. This can make the experiment feasible or not if the magic wavelength is in a region of the spectrum where no readily available high power laser exists, such as in the case of Hg. For this atom, reaching $100\,E_r$ would require a peak intensity of $500\,\mathrm{kW/cm^2}$ at 340 nm.

When considering shallow traps, the independent trap limit no longer holds, and one cannot neglect tunneling of the atoms from one site of the lattice to another. This leads to a delocalization of the atoms and to a band structure in their energy spectrum. Depending on the initial state of the atoms in the lattice, one faces a broadening and/or a shift of the atomic transition of the order of the width of the populated energy bands. For Sr, a trapping depth of the order of 100 E_r corresponds to a lower band width in the mHz range.

The deep reason for such a large required value of U_0 is that site-to-site tunneling is a resonant process in a periodic lattice. A much lower U_0 can be used provided tunneling is made non-resonant by lifting the degeneracy between adjacent sites. This can be done by adding an acceleration to the lattice, leading to the well-known Wannier-Stark ladder of localized states.[9,10] The experimental realization of such a scheme is straightforward: the lattice should be directed vertically, using gravity as the additional acceleration. In this configuration, trap depths down to $U_0 \sim 5 - 10\,E_r$ are sufficient for the above accuracy goal.[3]

3. Accuracy Evaluation

3.1. *Clock Features*

3.1.1. *Atomic Source*

Atoms initially come from an oven at 550 °C in which a Sr vapor is produced. A set of microtubes forms the output of the oven and the transverse velocity selection of a Sr atomic beam which then freely propagates towards the capture region at an average velocity of 500 m/s. The atomic beam is slowed down by a Zeeman slower of 25 cm length.[11] The atomic average velocity when they enter the capture region is 25 m/s. Atoms are then captured in a magneto-optical trap (MOT). The blue MOT laser light at 461 nm is generated from an infrared extended cavity laser diode-semiconductor power amplifier system of output power 450 mW at 921 nm. A PPKTP crystal in a resonant Fabry-Pérot ring cavity is then used to generate the second harmonic at the desired wavelength.[12] About 220 mW of blue laser light are available.

The lattice laser light is generated by a Ti-Sapphire laser of output power 1 W at the magic wavelength. A resonant Fabry-Perot cavity enables to enhance the power seen by the atoms up to 12.5 W. The beam waist at the MOT location is 89 μm, corresponding to a maximal lattice depth of 1400 E_r or 200 μK. The length of the Fabry-Pérot cavity is actively stabilized to the laser frequency via a Hänsch-Couillaud scheme which presents the advantage to not require any modulation of the cavity length or laser frequency. An active stabilization scheme of this servo-loop offset is used to take full advantage of the available power. Finally, the intra-cavity power is also actively stabilized.

During the MOT phase, the lattice is set at its maximal depth and two drain lasers aligned to the center the lattice and tuned to the $^1S_0 - {}^3P_1$ (at 689 nm) and $^3P_1 - {}^3S_1$ (at 688 nm) transitions are switched on. Atoms crossing these beams are pumped to $^3S_1$ which decays back to ^3P states. These atoms end up trapped into the two metastable $^3P_0$ and $^3P_2$ states. If their motional energy is smaller than the lattice depth they are captured and kept in the these states. This process is cumulative since the MOT dynamics keeps bringing atoms to the lattice center. The drain laser beam waist is about 50 μm. With powers of 3 μW (689 nm) and 30 μW (688 nm) the capture rate is about 10^5 per second. Capture ends by bringing the confined atoms back to the ground states. The MOT and drain laser beams are switched off and two repumping beams tuned to $^3P_0 - {}^3S_1$ (at 679 nm) and $^3P_2 - {}^3S_1$ (at 707 nm) switched on for about 1 ms.

After the capture phase, the atomic temperature is close to the depth of the lattice, i.e. 200 μK. For a number of reasons, we further cool the atoms down to the ground state of the lattice. The main reason is that we want to be able to reduce the lattice depth for the interrogation phase. This is achieved by sideband cooling along the lattice direction and Doppler cooling along the other directions, both using the narrow $^1S_0 - {}^3P_1$ transition at 689 nm. More than 95 % of the atoms are brought to the axial vibrational ground state with this technique, corresponding to a temperature of a few μK. Finally atoms need to be pumped to the extreme Zeeman substates for controlling the magnetic field in the experiment. This is done using again the $^1S_0 - {}^3P_1$ $(F = 9/2)$ transition and a bias magnetic field of 30 μT.

3.1.2. *Interrogation and detection*

After the optical pumping phase, the lattice depth is adjusted to its desired value for the interrogation. The magnetic field is also adjusted to about 100 μT. An ultra-stable laser pre-stabilized to a reference Fabry-Pérot cavity is used to probe the clock transition. The beam waist is about 200 μm. Typical interrogations of 20 ms duration with an intensity of a few μW/cm^2 are performed using a π polarisation. Finally, the population of both clock states is measured using a sequence of 461 nm light pulses. A first pulse of 2 ms duration gives a fluorescence signal proportional to the number of atoms in the ground state $^1S_0$. Atoms in $^3P_0$ are then pumped back to the ground state and detected similarly. Atomic resonances are probed alternately on both sides allowing the generation of an error signal proportional to the frequency difference between the clock laser and the atomic frequency. This is in turn used in a digital servo-loop to stabilize the clock laser to the atomic transition.

3.1.3. *Resonance and frequency stability*

The usual configuration of the ^{87}Sr clock is an interrogation time of 20 ms and a magnetic field amplitude $B = 87\,\mu$T. In this configuration, resonances of 30 Hz width are observed. At this value of the magnetic field, adjacent Zeeman transitions are separated by 96 Hz and are therefore perfectly resolved.

The present frequency stability of the clock is limited by the stability of the clock laser. The Allan deviation of the comparison between the Sr optical lattice clock and the FO2 Cs fountain [13] is $\sigma_y(\tau) = 6 \times 10^{-14}\tau^{-1/2}$, with roughly equal contributions from the Sr clock and from the fountain. After 1 hour of averaging time, the frequency noise of the measurements averages down to 10^{-15}, corresponding to a 0.5 Hz frequency resolution.

3.2. *Evaluation of Systematic Effects*

3.2.1. *Zeeman effect*

Interrogating symmetrical Zeeman components in principle cancels the first order Zeeman effect. One then expects the clock frequency to only depend on the quadratic Zeeman effect. In Fig. 1 the measured clock frequency is plotted as a function of the bias field, together with the expected dependence (continuous curve). No residual first order dependence is visible in these data. For a field of $B = 87\,\mu$T, the quadratic Zeeman shift is $\Delta_Z = -0,15$ Hz.

3.2.2. *Lattice Shift*

Figure 1 shows the present level of control of the lattice induced shift, representing the clock frequency with sub-Hz resolution as a function of U_0, for U_0 ranging from $50\,E_r$ to $500\,E_r$. The light shift of individual clock states ranges up to 1.8 MHz while the clock transition frequency shift is not resolved and reaches at $U_0 = 500\,E_r$, $\Delta_L = 0.5(0.5)$ Hz. If one scales the effect linearly down to $U_0 = 10\,E_r$, the effect reduces to $2(2)\times10^{-17}$, already very close to the ultimate expected accuracy for the Sr lattice clock. The frequency shift due to the atomic hyperpolarizability with its quadratic

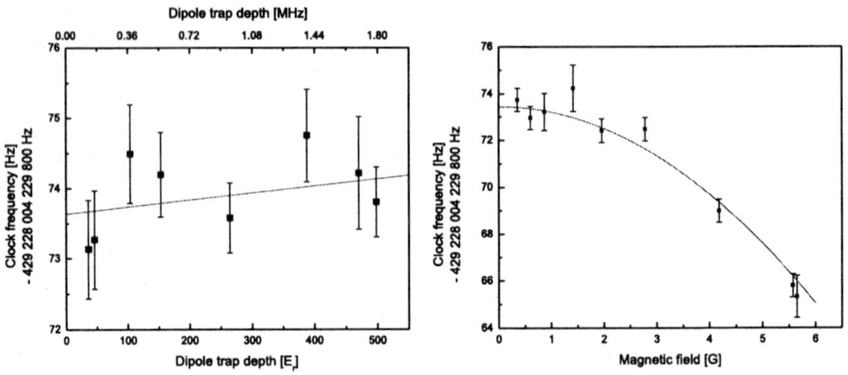

Fig. 1. Left: Clock frequency as a function of the bias magnetic field. Dots: measurements. Red curve: expected quadratic dependence. Right: Clock frequency versus U_0 expressed in recoil energy units (lower scale) and in frequency units (upper scale).

dependence on the lattice depth was also measured.[4] It turns out that scaled down to a lattice depth of $10\,E_r$, this frequency shift is $-1(1)\times 10^{-18}$ and therefore constitutes no real limitation.

3.2.3. *Other Effects and Accuracy Budget*

A number of other possible systematic effects were investigated. An important one is the line pulling due to the various stray resonances. Those include transverse sidebands due to the confinement of the atoms in the trap, and Zeeman transitions associated to the other m_F states. An upper limit to this effect was estimated taking into account the height, the width and the detuning of these other resonances, leading to an upper limit of 0.5 Hz or 1.1×10^{-15} in fractional units. Collisions between cold atoms have also been studied.[14] The atomic density was varied by adjusting the duration of the loading phase. No resolved frequency shift was observed and an uncertainty of 0.1 Hz was attributed to this effect. Finally the blackbody radiation shift was estimated by measuring the vacuum chamber temperature with accurate Pt resistance temperature detectors and using the theoretical BBR sensitivity.[15,16] At the temperature $T = 301(1)$ K, the frequency shift is $\Delta_{\rm BBR} = -2.39(5)$ Hz with roughly equal contributions to the error bar of the uncertainties of the temperature measurement and of the theoretical calculation. The overall accuracy budget of the clock is reported in Table 1. It gathers all the above discussed effects as well as a technical effect to which we attributed a conservative uncertainty of 1 Hz. As a matter of fact, one of the series of measurements that were performed exhibits a non Gaussian distribution with a dispersion of about 1 Hz. Though the other series do not show this behavior we conservatively added this line to the uncertainty budget. The frequency of the Sr clock transition is then 429228004229873.6(1.1) Hz, with a fractional accuracy of 2.6×10^{-15}. This value is in perfect agreement with the measurements performed at JILA [17] and Tokyo university.[18]

Table 1. Accuracy budget of the ^{87}Sr lattice clock.

Effect	Correction (Hz)	uncertainty (Hz)	Fractional uncertainty (10^{-15})
Zeeman	0.15	0.1	0.2
Light shift	0.1	0.1	0.2
Pulling	0	0.5	1.1
Collisions	0	0.1	0.2
Blackbody radiation shift	2.39	0.05	0.1
Technical effect	0	1	2.3
accuracy FO2	0	0.2	0.4
Total	**2.64**	**1.1**	**2.6**

4. Non-Destructive Detection

The key to further improvement of the clock performance will be the optimization of its frequency stability. The main limitation to the stability remains the clock laser frequency noise associated to the Dick effect.[19] This degradation will be reduced by improving the clock duty-cycle. This is illustrated in Figure 2 where the clock Allan deviation (relative to the clock laser free running Allan deviation) is plotted as a function of the duty cycle in the case of white frequency noise of the clock laser. Currently, about 90 % of the cycle time is spent accumulating the atoms in the dipole trap, while these atoms are lost in the detection scheme using fluorescence. Getting rid of this loss would enable recycling the atoms and then reduce the dead time, possibly down to about 10 %. The associated gain in frequency stability would be more than an order of magnitude. We have developed a new detection scheme which allows to get rid of this loss.

The principle of this new scheme is to detect a phase shift induced on a laser beam by the atoms.[20] We probe the atoms with a weak laser beam resonant with the $^1S_0 - {}^1P_1$ transition, modulated at 90 MHz by an electro-optic phase modulator in a carrier suppression regime. The dispersion prop-

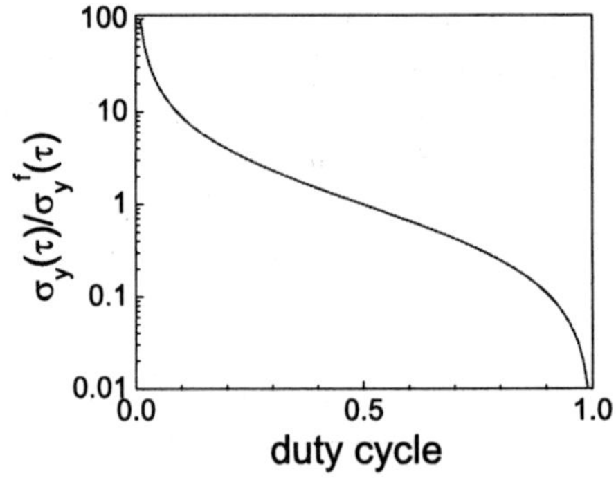

Fig. 2. Clock Allan deviation $\sigma_y(\tau)$ (relative to the clock laser free running Allan deviation $\sigma_y^f(\tau)$) as a function of the duty cycle in the case of white frequency noise of the clock laser.

erties of the atomic medium result in a phase shift of the laser beam, with an opposite sign for positive and negative modulation sidebands. This phase shift is subsequently measured by an interferometric setup using a strong local oscillator allowing a shot noise limited homodyne detection.

With this detection scheme, we were able to measure the number of captured atoms in the dipole trap, as well as the transition probability with a signal to noise ratio up to 100 per cycle. More than 95 % of the atoms are kept in the dipole trap. This new scheme, in addition to an improved clock laser under construction should lead to an optimized stability. Finally, this new detection scheme is capable of generating atomic ensemble spin squeezing, ultimately enabling to overcome the quantum projection noise limit in the clock.[21]

5. Conclusion

Within a few years, it has been possible with optical lattice clocks to come close to the performance of atomic fountains. Surpassing an accuracy of 10^{-16} now seems a realistic short term goal.[22] In the case of Sr lattice clocks, the expected limitation will probably be the blackbody radiation shift at a level close to 10^{-17}. With Hg atoms however, for which the sensitivity to black body radiation is almost 20 times smaller, another order of magnitude could in principle be gained. Several utilizations of such high performance clocks can be thought of. A competitive test of the stability of the fine structure constant was recently performed by combining the measurements of the Sr transition in several laboratories.[23] A further step could be the development of space projects. One could for instance test the laws ruling gravity at the solar system scale, such as proposed in the SAGAS project.[24]

Acknowledgments

The research leading to these results has received funding from CNES, ESA and the European Community's Seventh Framework Programme, ERA-NET Plus, under Grant Agreement No. 217257.

References

1. H. Katori, in *Sixth Symposium on frequency standards and metrology*, ed. P. Gill (World Scientific, Singapore, 2002).
2. D. Leibfried, R. Blatt, C. Monroe and D. Wineland, *Rev. Mod. Phys.* **75**, p. 281 (2003).
3. P. Lemonde and P. Wolf, *Phys. Rev. A* **72**, p. 033409 (2005).
4. A. Brusch, R. Le Targat, X. Baillard, M. Fouché and P. Lemonde, *Phys. Rev. Lett.* **96**, p. 103003 (2006).
5. Z. W. Barber *et al.*, *Phys. Rev. Lett.* **100**, p. 103002 (2008).
6. A. V. Taichenachev *et al.*, *Phys. Rev. Lett.* **101**, p. 193601 (2008).
7. H. Hachisu *et al.*, *Phys. Rev. Lett.* **100**, p. 053001 (2008).
8. M. Petersen *et al.*, *Phys. Rev. Lett.* **101**, p. 183004 (2008).
9. G. Nenciu, *Rev. Mod. Phys.* **63**, p. 91 (1991).
10. M. Glück, A. R. Kolovsky and H. J. Korsch, *Phys. Rep.* **366**, p. 103 (2002).
11. I. Courtillot *et al.*, *Opt. Lett.* **28**, p. 468 (2003).
12. R. Le Targat, J.-J. Zondy and P. Lemonde, *Opt. Comm.* **247**, p. 471 (2005).
13. S. Bize *et al.*, *J. Phys. B: At. Mol. Opt. Phys.* **38**, S449 (2005).
14. R. Le Targat *et al.*, *Phys. Rev. Lett.* **97**, p. 130801 (2006).
15. S. G. Porsev and A. Derevianko, *Phys. Rev. A* **74**, p. 020502 (2006).
16. S. G. Porsev, A. D. Ludlow, M. M. Boyd and J. Ye, *Phys. Rev. A* **78**, p. 032508 (2008).
17. G. K. Campbell *et al.*, *Metrologia* **45**, p. 539 (2008).
18. F.-L. Hong *et al.*, *arXiv:0811.1816* (2008).
19. A. Quessada *et al.*, *J. Opt. B : Quantum Semiclassical Opt.* **5**, p. S150 (2003).
20. P. J. Windpassinger *et al.*, *Phys. Rev. Lett.* **100**, p. 103601 (2008).
21. J. Appel *et al.*, *arXiv:0810.3545* (2008).
22. A. D. Ludlow *et al.*, *Science* **319**, 1805 (2008).
23. S. Blatt *et al.*, *Phys. Rev. Lett.* **100**, p. 140801 (2008).
24. P. Wolf *et al.*, *Exp. Astron.*, in press.

DEVELOPMENT OF AN OPTICAL CLOCK BASED ON NEUTRAL STRONTIUM ATOMS HELD IN A LATTICE TRAP

E. A. CURTIS, Y. B. OVCHINNIKOV, I. R. HILL [1], G. P. BARWOOD, AND P. GILL

National Physical Laboratory, Hampton Road
Teddington, Middlesex TW11 0LW, UK

At the National Physical Laboratory we are developing a neutral atom optical clock based on strontium atoms held in an optical lattice during the clock frequency measurement. The innovation of lattice-type clocks has greatly reduced velocity-related systematic frequency shifts compared to ballistic expansion neutral atom clocks. However, with predicted fractional uncertainties as low as 10^{-18}, many previously unexplored systematic effects become relevant, especially in dealing with shifts due to blackbody radiation. We are designing a compact, diode-laser-based system, with a view to future transportable systems, and which includes a novel permanent-magnet-based Zeeman slower and a blackbody measurement chamber.

1. Introduction

The $^1S_0 - {}^3P_0$ transition in neutral ^{87}Sr is one of the potential candidates for the redefinition of the SI second and is being pursued in a number of laboratories worldwide. Further frequency measurements need to be made and compared between various laboratories, and in-depth studies of systematic effects need to be performed in order to fully understand these optical clocks at the level necessary for redefinition of the second. At the National Physical Laboratory we are designing and implementing such a neutral strontium lattice clock, focusing on building a compact experimental system that will facilitate the study of various systematic effects, especially those due to blackbody radiation.

2. Experimental system

We are designing an experimental system with a compact footprint. Strontium clocks benefit from the fact that all the transition wavelengths we would like to excite can be reached using diode-laser technology. We intend to study the clock transition in both ^{87}Sr (linewidth ~ few mHz) and ^{88}Sr, using magnetic field mixing [1] to access the clock transition in the latter. The clock laser system is diode-laser based and uses a vertically oriented, vibrationally

---

[1] Also at Blackett Laboratory, Imperial College, Prince Consort Road, London SW7 2AZ, UK

insensitive cavity, with the aim of reaching a linewidth of < 1 Hz. Our vacuum system measures just over 1 m in length, half that in width, and less than 30 cm at its tallest point. It includes an oven, a Zeeman slower, a 2D transfer MOT chamber, and a final small-scale lattice chamber in which the clock transition will be measured (see Figure 1). The oven consists of a heater-wire-wrapped crucible to hold the strontium and a microchannel nozzle to help collimate the beam. Both can be heated to > 600 °C, and the nozzle is kept hotter than the crucible at all times to avoid blockages.

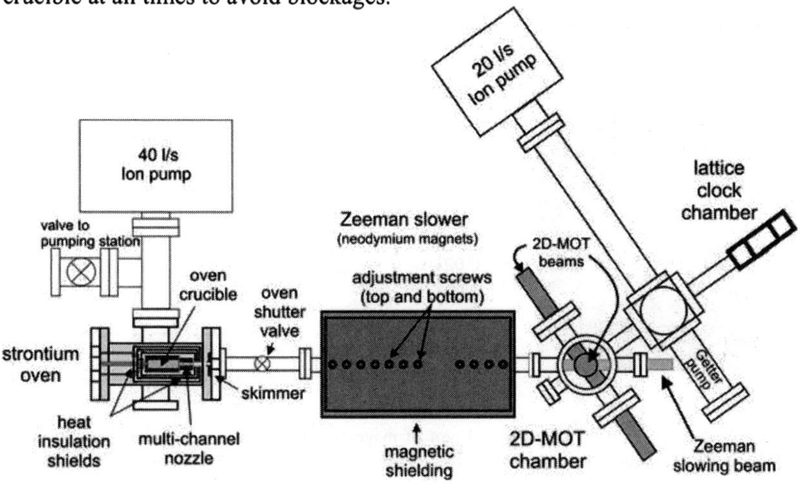

Figure 1. Experimental layout of the compact vacuum system for the strontium lattice clock.

The first-stage cooling transition radiation is provided by a Toptica TA-SHG, which includes a diode laser and tapered amplifier, and uses second harmonic generation in a $KNbO_3$ crystal to produce up to 450 mW of light at 461 nm. This laser is stabilized by locking its IR diode output to a scannable Fabry-Perot low-finesse cavity. An external cavity diode laser at 689 nm is stabilized via the Pound-Drever-Hall technique to a high-finesse ULE cavity for second-stage cooling, which further reduces the temperature of the atoms for more efficient loading into the optical lattice trap. The lattice laser system is under development.

2.1. *Permanent magnet Zeeman slower*

Experiments involving strontium usually rely on a Zeeman-slowed atomic beam to load a magneto-optical trap (MOT) as a first stage of cooling and confinement before transferring these atoms into the lattice trap. One key innovation in the miniaturization of our system is the development of a novel type of Zeeman slower based on permanent magnets [2], as shown in Figure 2. By using magnets rather than current-carrying coils as a source for the Zeeman slower's transverse magnetic field, we eliminate the need for (and noise from) power supplies and water cooling. Additionally, the Zeeman slower can be removed from the vacuum system for adjustment without breaking vacuum and its field can be fine-tuned in situ. Time-of-flight measurements are now underway with our prototype 25 cm Zeeman slower and we expect to be able to reduce its conservatively designed length in the future.

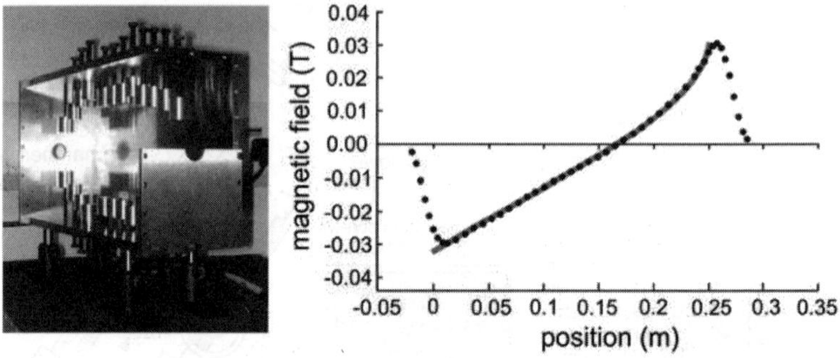

Figure 2. Picture of Zeeman slower and plot of magnetic field vs. position along the axis of the slower: solid line is theoretical calculation, points are measurements taken from real Zeeman slower.

2.2. *Blackbody radiation shift measurement chamber*

Due to the introduction of engineered light shift trapping and the progress in reducing many systematic effects over the last few years, at present the largest systematic uncertainties in the strontium lattice clock are due to the limited knowledge and control of the frequency shift due to blackbody radiation (BBR). There are two major sources of BBR in our experiment: the oven that is the atomic source, and the room temperature vacuum chamber in which the atoms are held during the frequency measurements. We will avoid the oven radiation by transferring our trapped atoms out of the line of sight of the oven using a 2D MOT. This process will allow us to move the atoms into a small chamber specifically designed to study the BBR shift due to the non-zero temperature of the vacuum apparatus.

Theoretical calculations [3] give this frequency shift $\delta\nu_{Sr}$ in Hz in strontium as a function of environmental temperature, T, as:

$$\delta\nu_{Sr} = -2.354\ (T/300)^4. \qquad (1)$$

The difficulty in making an accurate experimental assessment of the true magnitude of the shift is due to the control needed over the temperature of the environment in the vicinity of the lattice-trapped atoms. One solution would be to transfer the atoms to a cryogenic environment for the clock interrogation cycle, greatly reducing the magnitude of the shift (note the T^4 dependence in Eq. 1), but, more importantly, potentially reducing the uncertainty in the measurement. In contrast, at NPL we are designing a BBR measurement chamber that can be enclosed in an oven to raise the temperature of that section of the vacuum system by more than 150 °C. At its maximum this should induce a frequency shift up to about 10 Hz, which should be measurable by our system. The difficulties in the development of this chamber will be in assessing the temperature gradients and inhomogeneities in the trap enclosure, as well as its radiative properties, and designing the chamber such that there is sufficient optical access for the cooling, lattice, and clock beams.

It is also possible to make an indirect measurement of the BBR shift by making a measurement of the static or dc Stark shift and then inferring the ac Stark shift from that measurement. This has been done experimentally in Cs fountain clocks by measuring the clock frequency when the atoms have been subject to a uniform dc field produced by a pair of parallel plates [4]. This measurement has not yet been performed on strontium atoms and we intend to build this capability into the lattice clock BBR measurement chamber.

2.3. *Sub-Hz linewidth laser development*

In order to take advantage of the extremely narrow linewidths of the strontium clock transitions, we are developing a pair of a diode laser systems (grating stabilized external cavity diode lasers), stabilized via the Pound-Drever-Hall technique to high-finesse, Fabry-Perot cavities, with a goal of sub-Hz linewidth. A variety of new cavity designs have come into play in an effort to minimize the effects of vibrations on the cavity transmission [5,6], including the vibrationally insensitive, vertically mounted ULE (ultra-low expansion material) cavities used in our experiment. In an effort to maintain our goal of a small experimental footprint while ensuring a vibrationally stable and low-drift environment, the cavity vacuum cans are designed as shown in Figure 3.

Each vertical cavity is enclosed in two layers of temperature-stabilized aluminum, with Peltiers both inside and outside used to cool the chamber to the temperature of the zero crossing of thermal expansion for the system. The support system for the cavities is a set of three Teflon rods held in a ULE ring, which itself is supported by Viton balls sitting in indentations in the base of the

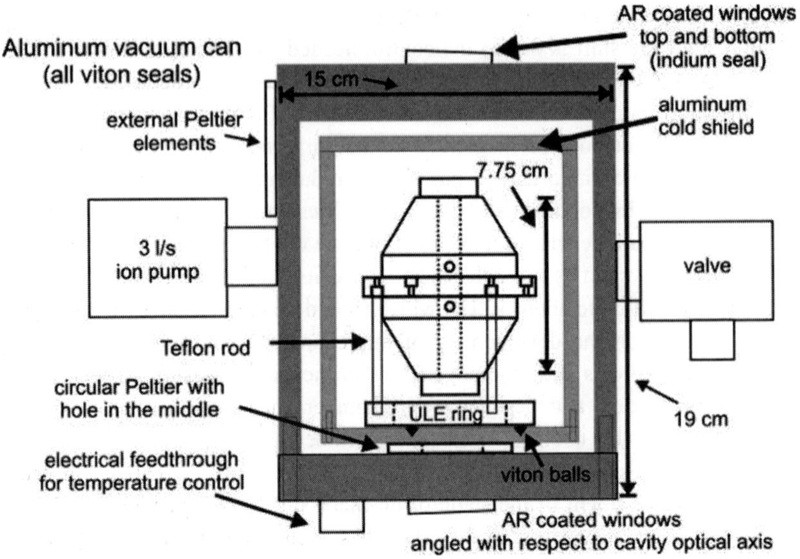

Figure 3. Temperature stabilized vacuum enclosure for vertical cavity.

inner cold shield. The cavity will reside on an active vibration isolation platform in a vibrationally isolated room at NPL. When complete, a measurement of the beat frequency between the two laser outputs or between one laser and a frequency comb will be made to assess and reduce the realized laser linewidth, before attempting to probe the clock transition.

Acknowledgments

This work is supported by the UK Department for Innovation, Universities & Skills NMS Pathfinder Metrology Programme, and through the European Metrology Research Programme Joint Research Project "Optical clocks for a new definition of the second."

References

1. A.V. Taichenachev *et al.*, *Phys. Rev. Lett.* **96**, 083001 (2006).
2. Y. B. Ovchinnikov, *Opt. Commun.* **276**, 261 (2007).
3. S. G. Porsev and A. Derevianko, *Phys. Rev. A* **74**, 020502R (2006).
4. E. Simon, P. Laurent, and A. Clairon, *Phys. Rev. A* **57**, 436 (1998).
5. M. Notcutt *et al.*, *Opt. Lett.* **30**, 1815 (2005).
6. S. A. Webster *et al., Phys. Rev. A* **75**, 011801(R) (2007).

DECOHERENCE AND LOSSES BY COLLISIONS IN A ^{88}Sr LATTICE CLOCK

J.S.R. Vellore Winfred, Ch. Lisdat*, T. Middelmann, T. Legero, F. Riehle and U. Sterr

Physikalisch-Technische Bundesanstalt
Bundesallee 100, 38116 Braunschweig, Germany
**E-mail: christian.lisdat@ptb.de*

We report the observation and interpretation of collision-induced perturbations in a ^{88}Sr lattice clock. Losses are observed in the collision channels $^1S_0{+}^3P_0$ and $^3P_0{+}^3P_0$. Furthermore, we observe broadening and shift of the clock transition by collisions.

Keywords: collisions, decoherence, optical lattice clocks

1. Introductions

Optical clocks with single ions[1] and neutral atoms[2] have recently surpassed Cs fountain clocks in instability and uncertainty. With neutral atom clocks employing large atom numbers at moderate or high density, collisional effects must be investigated. Currently, most often the fermionic ^{87}Sr is used because it promises a strong suppression of collision-induced shifts and losses for indistinguishable collision partners. Collision induced shifts are observed nevertheless due to inhomogeneous excitation.[3] This suppression is not present for the bosonic ^{88}Sr since s-wave collisions are allowed. However, ^{88}Sr offers larger atom numbers due its higher natural abundance, and therefore possibly better signal-to-noise ratio and higher stability. The precise limitations due to collisions for clock operation with ^{88}Sr have not been investigated in detail so far.

2. Measurements with Sr

Strontium atoms are cooled to microkelvin temperatures using a multi-stage cooling process. In the first cooling stage, about $4 \cdot 10^7$ atoms are captured within 200 ms from a Zeeman-slowed atomic beam and are cooled to 2 mK in a magneto optical trap (MOT) operating at the broad $^1S_0{-}^1P_1$

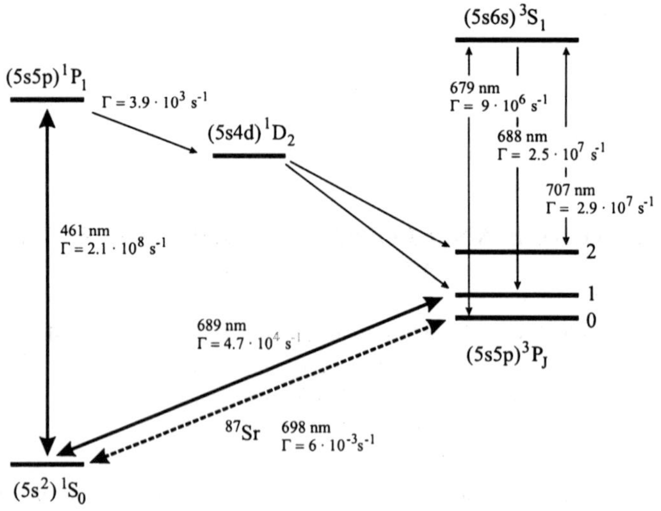

Fig. 1. Simplified electronic level structure of atomic strontium. Γ denotes the spontaneous emission rate.

transition at 461 nm (see Fig. 1). The second and third cooling stage utilize the spin forbidden $^1S_0$–$^3P_1$ transition at 689 nm. First, this cooling laser is spectrally broadened for 70 ms and then single frequency cooling is applied for 70 ms, leading to $8 \cdot 10^6$ atoms at a temperature of 3 μK.

During the cooling process the atomic cloud is superimposed with a 1D optical lattice operated at the 813 nm magic wavelength of the strontium clock transition. About 10^6 atoms at 3 μK are transferred into this optical lattice. At a laser power $P = 300$ mW and a beam waist radius $w_0 = 32$ μm (corresponding to a peak intensity $I = 75$ kW/cm^2) a trap depth $U_0 = 52$ μK is reached. For the dipole trap a Ti:sapphire laser pumped by a 10 W Verdi is used. The output beam is coupled into a polarization-maintaining optical fiber and can be blocked by a mechanical shutter. The output beam passes through polarization optics before being focused on the center of the atom cloud. The horizontally directed trap beam is linearly polarized with its polarization oriented perpendicular to the axis of the MOT magnetic field and perpendicular to gravity. We have verified the beam parameters of the dipole trap obtained with a beam analysis camera by the observation of the breathing and sloshing modes in the dipole trap. The observed frequencies confirm the expected waist radii of the beams. The corresponding trap parameters are also consistent with the observed axial and radial sidebands of the clock transition $^1S_0$–$^3P_0$ when interrogated in

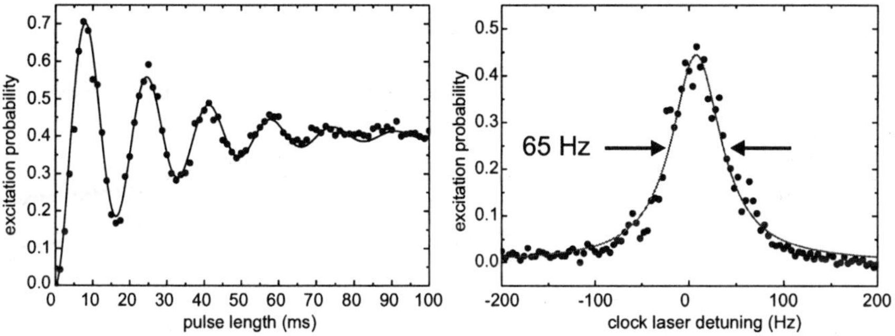

Fig. 2. Rabi oscillations (left-hand-side; 3 mT magnetic field; 83 W/cm^2 peak intensity) observed by variation of the excitation pulse length and a spectrum of the clock transition (right-hand-side) recorded at similar clock laser intensity and a homogeneous magnetic field of 0.75 mT by tuning the clock laser. The excitation pulse length was 35 ms. The solid lines are fits of damped cosine function and a Lorentzian, respectively.

the lattice.

A dc magnetic field of up to 3 mT is used to enable the clock transition in ^{88}Sr at 698 nm by field-induced state-mixing.[4,5] The clock laser is described in a contribution by Legero *et al.* in these proceedings.[6] At maximum a clock laser intensity of 125 W/cm^2 can be applied. The transition is excited by pulses of lengths between 3 ms and 500 ms. The population in the state $^1S_0$ is detected by fluorescence of a blue MOT-phase. Atoms in the $^3P_0$ state are detected in the same way after repumping to the ground state via the $^3S_1$ and $^3P_1$ states (Fig. 1) using laser diodes at 678 nm and 707 nm. We have observed Rabi oscillations and linewidths of down to 65 Hz (see Fig. 2).

3. Results

We observe different density dependent perturbations of the excitation on the clock transition in the optical lattice. With $^1S_0$ atoms only, we find a trap lifetime of 7.5 s to 8 s. When the upper clock state is populated, inelastic collisions shorten the trap lifetime. Collisions between atoms in the $^3P_0$ state can be distinguished from collisions $^1S_0$+$^3P_0$ by removing or leaving the ground state atoms in the lattice for a variable hold time. We observe very different decay characteristics (Fig. 3). The experimental results of excitation spectra and lifetime measurements can be successfully modeled by integration of coupled set of differential equations and we find collision loss rates for $^3P_0$+$^3P_0$ which are compatible with the ones reported

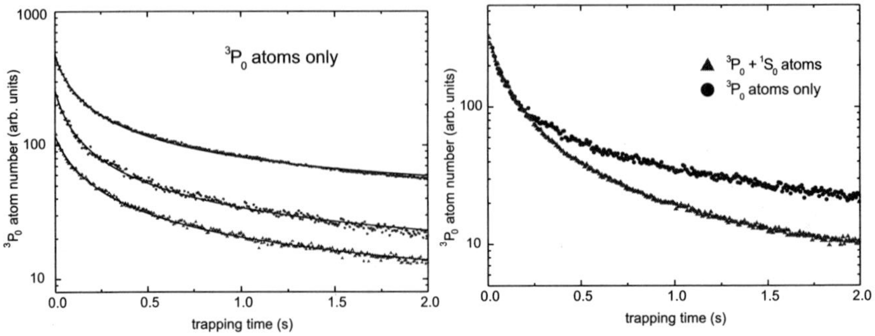

Fig. 3. Losses of $^3P_0$ atoms for different initial atom numbers if only excited state atoms are trapped (left-hand-side). The solid curves are fits to extract the loss rates. Right-hand-side: Comparison of the time evolution of $^3P_0$ atoms with and without $^1S_0$ atoms in the trap. The trace with excited state only atoms is slightly offset in time to correct for a small difference in the initial atom numbers and thus to visualize the initially parallel decay.

by Traverso *et al.*[7]

In addition to the atom loss, line broadening is observed which also depends on the atom density (Fig. 4). The observed spectra can be modeled with a density matrix formalism including the inelastic losses and a density dependent dephasing rate. The broadening of the transition due to the inelastic collisions is not sufficient to describe the observed linewidths. Beyond the discussed collision induced decoherence processes, we observe a density dependent shift of the clock transition, which will be subject of

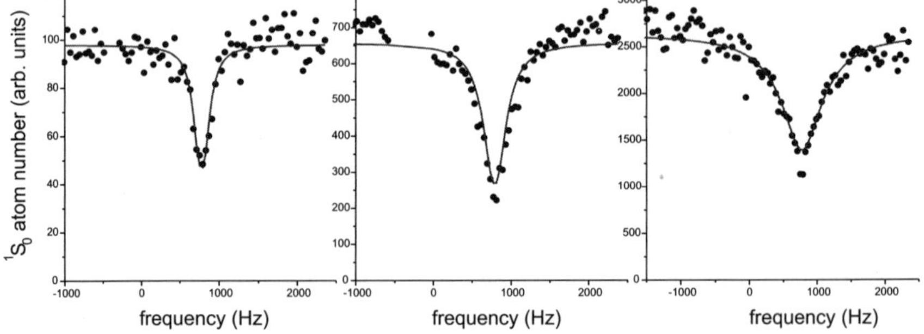

Fig. 4. Broadening of the clock transition by collisions. The density rises from the left to the right graph. The solid line is a fit according to the model which includes inelastic losses and dephasing collisions. The frequency scale is given with respect to an arbitrary offset.

further investigations.

4. Conclusion and outlook

We have observed density dependent losses of ^{88}Sr atoms trapped in an optical lattice. Two loss channels were identified, namely via collisions $^1S_0+^3P_0$ and $^3P_0+^3P_0$. The clock transition $^1S_0-^3P_0$ is broadened and shifted by collisions. Losses and broadening can be modeled using the density matrix formalism and the resulting rate constants will be used to optimize the parameters of a ^{88}Sr lattice clock.

Acknowledements

We acknowledge support from the Deutsche Forschungsgemeinschaft under SFB 407 and the Centre for Quantum Engineering and Space-Time Research (QUEST), from ESA and DLR.

References

1. T. Rosenband, D. B. Hume, P. O. Schmidt, C. W. Chou, A. Brusch, L. Lorini, W. H. Oskay, R. E. Drullinger, T. M. Fortier, J. E. Stalnaker, S. A. Diddams, W. C. Swann, N. R. Newbury, W. M. Itano, D. J. Wineland and J. C. Bergquist, *Science* **319**, 1808 (2008).
2. A. D. Ludlow, T. Zelevinsky, G. K. Campbell, S. Blatt, M. M. Boyd, M. H. G. de Miranda, M. J. Martin, J. W. Thomsen, S. M. Foreman, J. Ye, T. M. Fortier, J. E. Stalnaker, S. A. Diddams, Y. L. Coq, Z. W. Barber, N. Poli, N. D. Lemke, K. M. Beck and C. W. Oates, *Science* **319**, 1805 (2008).
3. S. Gupta, Z. Hadzibabic, M. W. Zwierlein, C. A. Stan, K. Dieckmann, C. H. Schunck, E. G. M. van Kempen, B. J. Verhaar and W. Ketterle, *Science* **300**, 1723 (2003).
4. A. V. Taichenachev, V. I. Yudin, C. W. Oates, C. W. Hoyt, Z. W. Barber and L. Hollberg, *Phys. Rev. Lett.* **96**, 083001 (2006).
5. X. Baillard, M. Fouché, R. L. Targat, P. G. Westergaard, A. Lecallier, Y. L. Coq, G. D. Rovera, S. Bize and P. Lemonde, *Opt. Lett.* **32**, 1812 (2007).
6. T. Legero, C. Lisdat, J. S. R. Vellore Winfred, H. Schnatz, G. Grosche, F. Riehle and U. Sterr: "Clock laser system for strontium lattice clock" in these proceedings.
7. A. Traverso, R. Chakraborty, Y. N. M. de Escobar, P. G. Mickelson, S. B. Nagel, M. Yan and T. C. Killian, arxiv:0809.0936 (2008).

LATTICE YB OPTICAL CLOCK AND CRYOGENIC Cs FOUNTAIN AT INRIM*

FILIPPO LEVI, DAVIDE CALONICO, LUCA LORINI,
CLAUDIO CALOSSO, ELIO K. BERTACCO AND ALDO GODONE

Istituto Nazionale di Ricerca Metrologica, Str. delle Cacce 91
Torino, 10132, Italy

GIOVANNI A. COSTANZO

Politecnico di Torino,Dipartimento di Elettronica, C.so Duca degli Abruzzi 24
Torino,10129, Italy

In this paper we present the current status in the realization of a new generation of high accuracy frequency standards at INRIM, whose relative frequency uncertainty is expected to achieve and possibly pass the 10^{-16} range. Currently we are developing a nitrogen cooled cryogenic Cs fountain and a dipole trap neutral Yb optical clock. This development will allow us to make an absolute measurement of the Yb clock frequency against a primary frequency reference in the low 10^{-16} range.

1. Introduction

Since 2003 INRIM is operating IT-CsF1 [1], a laser cooled Cs fountain primary frequency standard. The ultimate accuracy currently reached by this standard is 5×10^{-16}; the main limiting uncertainty sources being the blackbody radiation shift, the spin exchange (or atom density) shift and last but not least the microwave leakage shift. IT-CsF1 was used to provide accuracy evaluation of TAI 20 times during the last 5 years. Aiming to improve the present accuracy and measurement capabilities, INRIM has started the development of a cryogenic fountain, designed to take account the experience acquired with IT-CsF1, and of an Yb optical clock, recognized as one of the most interesting atomic system for the next generation of high accuracy standards.

2. The cryogenic fountain

The new Cs fountain, developed in tight cooperation with NIST time and frequency division [2], will be operated at liquid nitrogen temperature. The

* This work is supported by "Compagnia di San Paolo" and "Regione Piemonte".

whole interaction region cavity and drift tube, as well as the C-field solenoid, and the two innermost magnetic shields are placed inside a cryostat; a third magnetic shield is placed outside it. The whole structure is under rough vacuum in order to guarantee thermal insulation. The optical system is designed and engineered aiming to guarantee long and stable operation times.

2.1. *Accuracy perspective of F2*

Since the launching geometry will be (1,1,1), there is no need of an upper window and the lower entrance in the cavity body is a long tube with a rough surface, that prohibits low angle total reflection. The interaction region is to good approximation a true blackbody radiator. In preliminary experiments we were able to cool the fountain structure down to 90 K, measuring a resonance temperature of the cavity at 95 K. The overall temperature gradient in the drift region is 3 K. In these conditions the blackbody bias will be $2\times x10^{-16}$ with an uncertainty in the low 10^{-17} range.

The second major source of uncertainty in F1 is the density shift. In F1 the density shift uncertainty is composed of two terms, a type A uncertainty as the result of differential measurements performed during the whole measurement campaign, and a type B due to possible non-linearity in the zero density extrapolation. In F2 several improvements are implemented: i) better design and control of the optical system and as a consequence of the shape of the loaded atomic sample ii) direct molasses trapping and launching, that will result in a density reduction and, iii) strong reduction of the Type B uncertainty. Furthermore if we will need to further reduce the spin exchange shift F2 is predisposed for multi-ball operation [3], a techniques that allows to reduce the density shift by at least a factor 5 without stability degradation. The overall density shift is estimated to be of few units of 10^{-16}.

The microwave leakage will also be reduced in the new fountain because i) the required power supplied from the synthesizer is reduced by more than 10dB, ii) the only possible leak source below the Ramsey cavity is much more far away than in F1. The state selection cavity will be completely switched off during the Ramsey interaction and detection, as in present F1 operations.

The leakage from the Ramsey cavity in the drift tube can be modeled similarly to what was done in F1 and gives no significant contribution to the uncertainty. [4].

Hence, the overall uncertainty of F2 is expected to approach $1{\sim}2\times10^{-16}$.

3. The Yb optical clock

Among many atoms candidates to be used for the realization of high accuracy lattice optical clocks, Yb has some interesting features that triggered our choice:

- Several isotopes available, both fermions and bosons.
- Two stage laser cooling and MOT with dipole transition $^1S_0 \rightarrow ^1P_1$ (399nm) and intercombination line $^1S_0 \rightarrow ^3P_1$ (556nm).
- Easiness of the second laser cooling stage because of its relatively large transition probability.
- $^1S_0 \rightarrow ^3P_0$ clock transition accessible in bosons (magnetic field induced) and in fermions with low nuclear magnetic number (^{171}Yb, I=1/2).
- "Light shift free" wavelength for dipole trap radiation at 759nm (peak of Ti:S laser).
- No need of any repumping light.

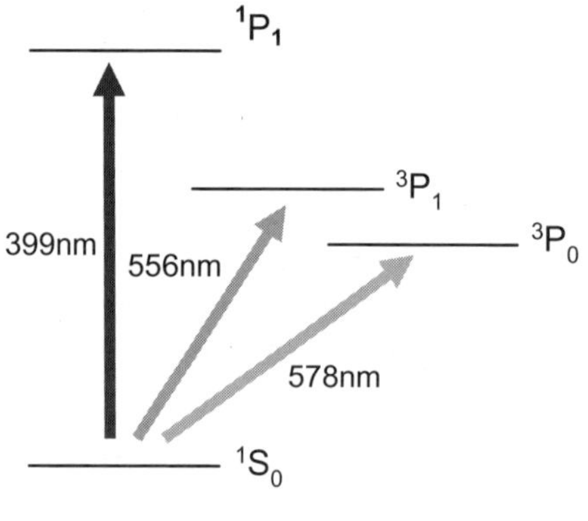

Figure 1. Yb cooling and clock transitions.

The Yb source for the MOT loading is realized with an oven and a multi-channel nozzle made by a micro-tube bundle (1 cm × 200 μm), allowing for a bright beam with some 20 mrad diveregence. A second identical beam is realized for laser locking.

Direct MOT loading is achieved without the need of Zeeman slower as well as of a slowing beam. This latter possibility will be exploited at a later time if a stronger and faster MOT loading is needed.

3.1. *Experimental results*

The blue radiation at 399 nm is synthesized by second harmonic generation in a PPKTP non linear crystal. IR radiation from Ti:Sapphire laser is enhanced in a build-up ring cavity_and_blue light up to 70 mW has been obtained from a ~550 mW IR laser beam. An 800 nm high power diode laser will soon replace the Ti:S laser, that will be used to create the dipole trap at 759 nm.

The green light at 556 nm is synthesized by second harmonic generation of an Yb:doped fiber laser. Up to 10 mW are generated by 300 mW IR radiation in a single-pass PPKTP non linear crystal.

Up to now we were able to perform a complete spectroscopy of the Yb beam both in the green and in the blue and to lock the lasers on atomic beam signals. We were able as well to load and characterize the Yb MOT in the blue for the various Yb isotopes.

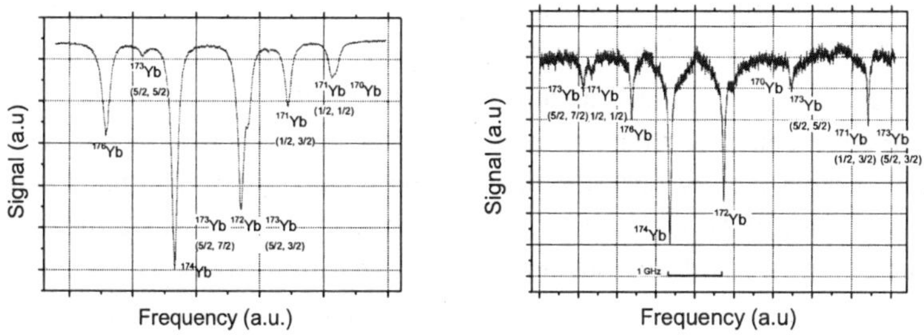

Figure 2. Spectroscopy of the Yb beam at 399 nm and at 557nm.

Up to 10^4 atoms are loaded in the 399 nm MOT, operating with la laser intensity of 6 mW/cm^2 , magnetic linear gradient of 48 G/cm, a detuning with respect to resonance transition of –22 MHz and the Yb oven at 350 °C. In this conditions, the trap loading time has been measured to be ~700 ms.

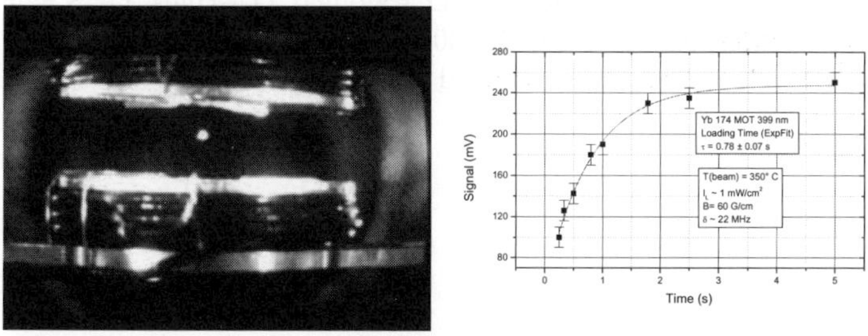

Figure 3. Picture of the Yb MOT loaded with blue light and its loading transient.

We are also developing the system to generate the yellow clock radiation at 578nm; this wavelength will be realized by sum frequency generation in a PPMgO:$LiNbO_3$ waveguide, using a Nd:YAG laser operating at 1319nm and a Yb:doped fiber laser operating at 1030 nm.
Yellow light will be stabilized on a ULE cavity whose finesse is expected to exceed 200000. The cavity has a notched design [5] to minimize its sensitivity to acoustic vibrations. The cavity thermal control system has been tested using a dummy cavity, and temperature measurements show a thermal stability better than 1 μK at 1 s integration time. In this measurement, the temperature was around 23 °C. As the zero crossing of the TCE is expected to be in the range 20-30 °C with 95% probability, this result is suitable for a laser frequency stabilization at $<10^{-15}$ at 1 s in terms of Allan deviation

4. Conclusions and perspectives

INRIM efforts to develop high accuracy standards and to improve the measurements capabilities of the laboratory results in two major research projects. The cryogenic fountain, for the realization of the SI second with a relative frequency accuracy of $1\text{x}10^{-16}$, is now in advanced development and we hope to have it fully operative by the end of next year. The realization of an optical lattice clock with Yb atoms is in progress: the physical structure and the optical bench has been developed, and present activity is focused on the generation of an ultracold atomic sample for clock transition spectroscopy.

References

1. F Levi, D Calonico, L Lorini and A Godone, *Metrologia* 43 (2006) 545–555
2. see also S. Jefferts paper in this proceedings
3. F. Levi, A. Godone, L. Lorini, S.R. Jefferts, T.P. Heavner, and C. Calosso, *Proc. 2001 Freq. Stand. Metrology Symp.* 466-468 (2001)
4. J.H.; Shirley, F. Levi, T.P Heavner, D. Calonico, Dai-Hyuk Yu, S.R Jefferts, IEEE Transactions on UFFC 53,12, 2006
5. Design developed by Till Rosenband of NIST

Part V

Optical Clocks II: Ion Clocks

^{171}YB$^+$ SINGLE-ION OPTICAL FREQUENCY STANDARDS

CHR. TAMM, B. LIPPHARDT, T.E. MEHLSTÄUBLER, M. OKHAPKIN,
I. SHERSTOV, B. STEIN, E. PEIK

Physikalisch-Technische Bundesanstalt,
Bundesallee 100, 38116 Braunschweig, Germany

Singly ionized ytterbium offers a number of possibilities to realize frequency standards in the optical frequency range. Two narrow-linewidth reference transitions, one at a wavelength 436 nm and the other at 467 nm, are of particular interest for practical implementations of single-ion optical clocks. The frequency ratio of these transitions is expected to be a very sensitive probe for the constancy of the fine structure constant. We outline our related experimental work and in particular the status of the 436 nm (688 THz) single-ion optical frequency standard developed at PTB.

1. Energy level system of Yb$^+$

A single laser-cooled ion confined in a radiofrequency trap is a nearly ideal reference for an optical frequency standard. Yb$^+$ has an alkali-like strong $^2S_{1/2}$ - $^2P_{1/2}$ resonance transition in the near-ultraviolet range permitting efficient cooling to the Doppler limit. At variance with other alkali-like ions like Ca$^+$, Sr$^+$, and Hg$^+$, Yb$^+$ has a number of low-lying energy levels originating from excitation of the 4f electron shell (see Fig. 1). This allows for a technically convenient laser cooling scheme [1] and provides unique possibilities for optical frequency standards. The extremely small natural linewidth of the $^2S_{1/2}$ - $^2F_{7/2}$ electric-octupole transition at a wavelength of 467 nm poses particular technical challenges but is attractive because it allows to realize an optical frequency standard whose stability is limited only by the frequency instability of the laser that probes the transition [2]. This system has been investigated extensively at NPL [3].

At first sight it might appear that the $^2S_{1/2}$ - $^2D_{3/2}$ electric-quadrupole transition of Yb$^+$ at 436 nm can not be used as the reference transition of an optical frequency standard because the $^2D_{3/2}$ level is rapidly populated by spontaneous decay during laser cooling which prevents the detection of individual quantum jumps to this level. However, the hyperfine structure of ^{171}Yb$^+$ makes it possible to decouple the $^2D_{3/2}$(F=2) hyperfine sublevel from the

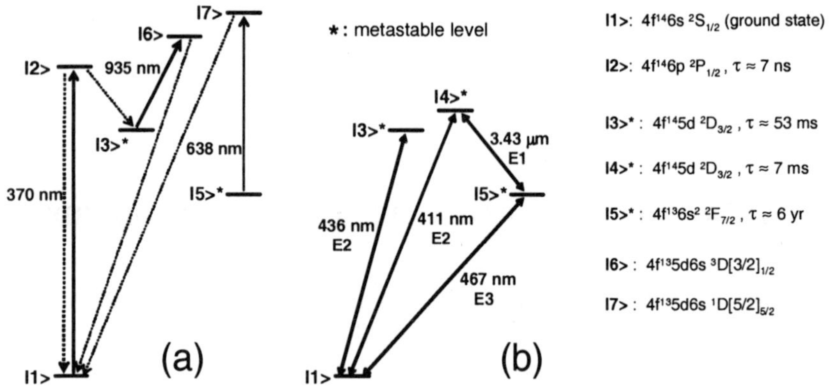

Figure 1. Section of the energy level system of Yb^+, showing (a) the employed laser cooling and repumping scheme and (b) narrow-linewidth transitions between low-lying metastable levels and the ground state that can serve as frequency standard reference transitions. The electronic configurations of the levels and experimentally relevant natural lifetimes are indicated on the right.

laser cooling excitation cycle [4]. An optical frequency standard based on the $^2S_{1/2}(F=0)$ - $^2D_{3/2}(F=2)$ transition of $^{171}Yb^+$ has been developed at PTB [5,6]. Together with transitions in $^{199}Hg^+$, $^{88}Sr^+$, ^{87}Sr, and ^{87}Rb, the frequency of the 436 nm $^{171}Yb^+$ single-ion standard has been recognized as a secondary representation of the SI second [7].

Atomic-structure calculations indicate that the frequency of the $^2S_{1/2}$ - $^2F_{7/2}$ transition in Yb^+ shows a uniquely strong dependence on the value of the fine structure constant α and that there is a large difference in the α-sensitivities of the $^2S_{1/2}$ - $^2F_{7/2}$ and $^2S_{1/2}$ - $^2D_{3/2}$ frequencies [8]. The ratio of the transition frequencies $R = \nu(S - F)/\nu(S - D)$ is expected to vary with α according to $\delta R/R \approx -7.0 \cdot \delta\alpha/\alpha$. This sensitivity is larger than for any other combination of reference frequencies of atoms or ions which have been considered so far for optical frequency standards. The high sensitivity and the possibility to use a single trap setup to perform the Yb^+ transition frequency ratio measurement is a strong motivation for the further development of both the Yb^+ 467 nm and the Yb^+ 436 nm standards.

2. Experimental work at PTB

Presently our experimental setup comprises two $^{171}Yb^+$ ion traps, four extended-cavity diode lasers which are commonly used for both traps, and a femtosecond

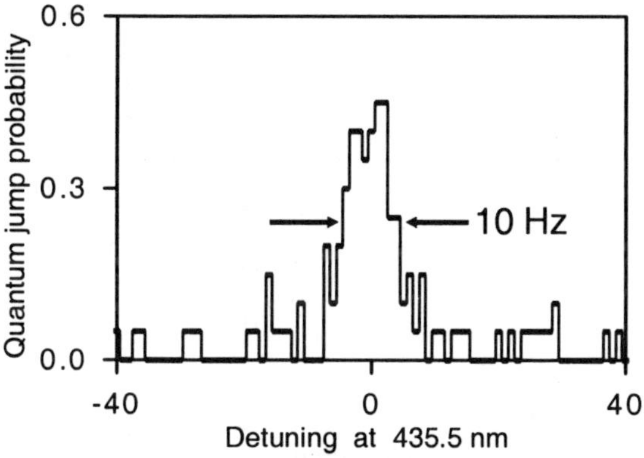

Figure 2. Absorption spectrum of the 436 nm reference transition of a trapped $^{171}Yb^{+}$ ion. Each data point corresponds to an average over 20 measurement cycles. During each cycle, the probe light is applied as a π-pulse of 90 ms duration.

fiber laser frequency comb generator which permits absolute optical frequency measurements using a local caesium fountain reference. The output of two of the diode lasers is frequency doubled for generation of the 370 nm cooling and 436 nm probe radiation; the other two lasers produce repumping light at 935 nm and at 638 nm. The probe laser frequency is stabilized by locking it to a high-finesse ULE cavity mounted on a passive vibration isolation platform. Figure 2 shows that the probe laser frequency stability is sufficient to resolve the 436 nm reference transition of a trapped $^{171}Yb^{+}$ ion with an essentially Fourier-limited linewidth of 10 Hz.

In order to directly compare two 436 nm $^{171}Yb^{+}$ frequency standards, the two ion traps were operated simultaneously and two independent digital servo systems were used to lock the probe beam frequencies to the $^2S_{1/2}(F=0)$ - $^2D_{3/2}(F=2)$ reference transitions of the trapped ions. The frequency difference was observed with and without applying external perturbations to one of the standards. These experiments enabled measurements of the electrical quadrupole moment of the $^2D_{3/2}$ state and of the scalar and tensorial quadratic dc Stark shift coefficients of the reference transition [9]. Without external perturbations, one expects that the largest systematic frequency shift is caused by the quadrupole shift resulting from the gradient of the static electric stray field in the trap. Figure 3 shows measurements of the frequency offset between the two traps without applied perturbations. The offset was determined for various orientations of the static magnetic field which would lead to varying quadrupole and tensorial Stark shifts. The scatter of the data points in Fig. 3

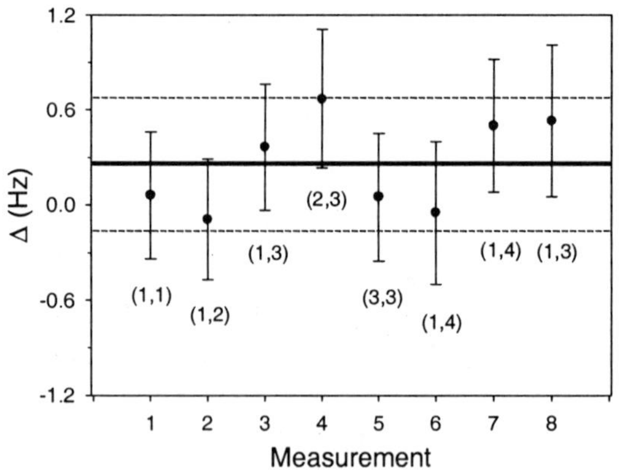

Figure 3. Frequency difference between two 436 nm $^{171}Yb^+$ standards in the absence of external perturbations. The data points are labeled according to the combination of magnetic field orientations used (trap 1, trap 2). Orientations 1, 2, 3 are approximately orthogonal to each other while orientation 4 is perpendicular to the trap symmetry axis. The data were taken within two days and are displayed in temporal order. The error bars denote 1σ statistical uncertainty. The solid line is the weighted average of the data points and the dashed lines mark the average statistical uncertainty.

does not significantly deviate from a random distribution. Thus there is no evidence of magnetic-field-orientation dependent shifts that are larger than the statistical uncertainty of the individual measurements of approximately 0.4 Hz (relative uncertainty $6 \cdot 10^{-16}$) .

By numerical simulations we investigated how the stability of a single-ion optical frequency standard can be optimized if realistic experimental conditions such as dead times in the interrogation sequence are taken into account. The dynamic behaviour of the servo algorithms employed in the frequency comparison experiments was analyzed in order to quantify the effects of fluctuations and drift of the probe laser frequency [10]. In the trap comparison experiments, an instability of the frequency difference (Allan deviation) of $\sigma_y(\tau) \approx 1.5 \cdot 10^{-14} (\tau/s)^{-1/2}$ is observed if the reference transition is resolved with a Fourier-limited linewidth of 30 Hz in a measurement cycle time of 90 ms. This result is in good agreement with our numerical simulations if quantum projection noise is assumed as the only noise source.

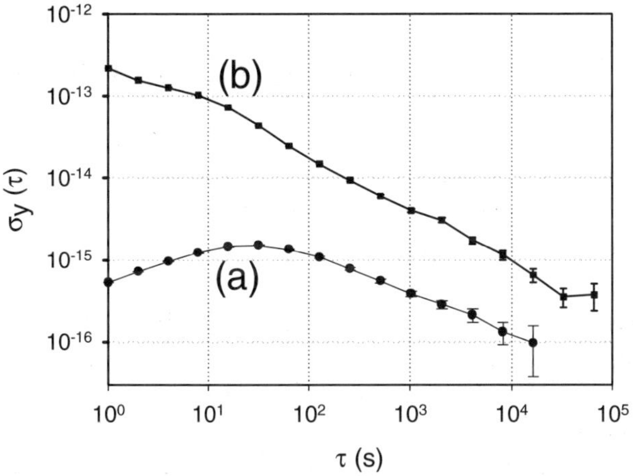

Figure 4: Allan deviation $\sigma_y(\tau)$ as a function of averaging time τ in recent long-time measurements using the 436 nm single-ion $^{171}Yb^+$ standard. (a): instability of the servo frequency difference in an operation mode where two digital servo control loops operate alternately in order to determine systematic frequency shifts. (b): absolute frequency measurement relative to a caesium fountain standard (see text).

Recent improvements of the 436 nm standard concerned in particular an optimized and comprehensive software control of the measurement sequence and the capability of unattended continuous long-term operation. As a result, the stray-field induced quadrupole shift and other systematic shifts can now be evaluated without using an external reference, and with a statistical uncertainty of $\sigma_y(10^4\ s) \approx 1 \cdot 10^{-16}$ (see Fig. 4). Figure 4 also shows the instability of a recent absolute frequency measurement relative to the PTB caesium fountain CSF1. Here the measurement instability is essentially determined by the instability of the fountain standard. As a result of the long measurement time of approximately 120 h, the Allan deviation is reduced to $\sigma_y(6.5 \cdot 10^4\ s) \approx 4 \cdot 10^{-16}$, well below the present systematic uncertainty estimates of both standards.

Acknowledgments

This work was partly supported by Deutsche Forschungsgemeinschaft in SFB 407 and by grant RFP1-06-08 from the Foundational Questions Institute (fqxi.org). We gratefully acknowlegdge cooperation with S. Weyers in frequency measurements relative to CSF1.

References

1. A.S. Bell, P. Gill, H.A. Klein, A.P. Levick, Chr. Tamm, D. Schnier, *Phys. Rev.* **A44**, R20 (1991).
2. R. Blatt, R. Casdorff, V. Enders, W. Neuhauser, P.E. Toschek, in *Frequency Standards and Metrology*, Proceedings of the Fourth Symposium, Ancona, Italy, 1988, ed. A. DeMarchi (Springer, Berlin, 1989), p. 306.
3. P.J. Blythe et al., *Phys. Rev.* **A67**, 020501(R) (2003); K. Hosaka et al., *IEEE Trans. Instrum. Meas.* **54**, 759 (2005).
4. Chr. Tamm, D. Engelke, in *Frequency Standards and Metrology*, Proceedings of the Fifth Symposium, Woods Hole, USA, 1995, ed. J.C. Bergquist (World Scientific, Singapore 1996), p. 283.
5. Chr. Tamm, D. Engelke, V. Bühner, *Phys. Rev.* **A61**, 053405 (2000).
6. Chr. Tamm, B. Lipphardt, H. Schnatz, R. Wynands, S. Weyers, T. Schneider, E. Peik, *IEEE Trans. Instrum. Meas.* **56**, 601 (2007).
7. www.bipm.org/utils/common/pdf/CCTF17.pdf
8. V.V. Flambaum, V.A. Dzuba, *arXiv:*0805.0462v2 (2008).
9. T. Schneider, E. Peik, Chr. Tamm, *Phys. Rev. Lett.* **94**, 230801 (2005).
10. E. Peik, T. Schneider, Chr. Tamm, *J. Phys. B: At. Mol. Opt. Phys.* **39**, 145 (2006).

AN OPTICAL CLOCK BASED ON A SINGLE TRAPPED $^{88}Sr^+$ ION

H. S. MARGOLIS[1,*], G. P. BARWOOD[1], G. HUANG[1], H. A. KLEIN[1], S. N. LEA[1], G. MARRA[1], V. TSATOURIAN[1,2], B. R. WALTON[1] and P. GILL[1]

[1]*National Physical Laboratory (NPL), Teddington, Middlesex TW11 0LW, UK*
[2]*Heriot-Watt University, Riccarton, Edinburgh EH14 4AS, UK*
[*]*E-mail: helen.margolis@npl.co.uk*

The 5s $^2S_{1/2}-4d\,^2D_{5/2}$ electric quadrupole transition in a single trapped and laser-cooled $^{88}Sr^+$ ion provides the reference for a highly stable and accurate optical clock. This paper describes recent progress on the $^{88}Sr^+$ optical clocks under development at NPL. An ultrastable laser at 674 nm is used to excite the clock transition and a fractional frequency resolution of 2×10^{-14} has been achieved, corresponding to a linewidth of 9 Hz. A relative frequency instability of 2×10^{-15} at 5000 s has been observed in the comparison between two similar systems and the fractional uncertainty in the absolute frequency of the clock transition is 3.8×10^{-15}. Improvements to the experimental arrangement are in progress and are expected to lead to a $^{88}Sr^+$ optical clock with stability and reproducibility exceeding that of the primary caesium standard.

1. Introduction

An important class of optical atomic clocks are those based on forbidden transitions in single laser-cooled ions stored in rf quadrupole traps.[1–5] Since the confining electric field vanishes at the centre of the trap, the single cooled ion closely approximates the spectroscopic ideal of a particle at rest in a perturbation-free environment and is therefore an excellent option for a clock of very high accuracy.

The reference transition for the $^{88}Sr^+$ single ion optical clock is the 5s $^2S_{1/2}-4d\,^2D_{5/2}$ electric quadrupole transition at 445 THz (674 nm). This system, which is being studied both at the UK National Physical Laboratory (NPL)[4,6] and at the National Research Council (NRC) in Canada,[5,7] has the advantage of a narrow clock transition (natural linewidth 0.4 Hz) combined with cooling and probe laser wavelengths that can be produced using commercially available diode laser technology. It has also recently been adopted as a secondary representation of the second.[8]

2. Experimental arrangement

A single $^{88}Sr^+$ ion is produced by electron impact ionization of strontium atoms evaporated from an oven and confined in an endcap trap.[9] The trapping field is provided by an rf voltage at a frequency of 17–18 MHz applied to the inner endcap electrodes, which have a separation of approximately 0.6 mm. Three pairs of mutually orthogonal current-carrying coils around the trap are used to control the dc magnetic field experienced by the ion, whilst a mu-metal shield reduces the effect of ac magnetic fields. The typical applied dc magnetic field is around 1 μT and causes the clock transition to split into ten Zeeman components.

The ion is laser-cooled into the Lamb-Dicke regime on the 5s $^2S_{1/2}$–5p $^2P_{1/2}$ transition at 422 nm (Fig. 1). Since the 5p $^2P_{1/2}$ state can also decay to the metastable 4d $^2D_{3/2}$ state, an additional repumper laser at 1092 nm is required to maintain effective Doppler cooling. The 422 nm cooling radiation is generated by single pass frequency doubling of 844 nm radiation from an extended-cavity diode laser in periodically poled KTP, whilst the 1092 nm radiation is provided by a DFB laser. Fluorescence from the 422 nm cooling transition is detected using a photomultiplier tube and the micromotion of the ion is monitored and minimized in three dimensions using the rf photon correlation technique.[10]

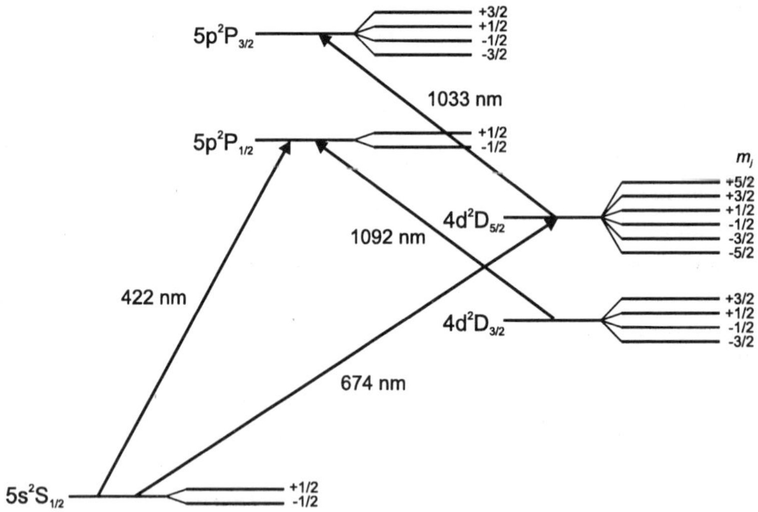

Fig. 1. Low-lying energy levels of $^{88}Sr^+$, showing the transitions relevant for operation of the optical clock. (Not to scale.)

The 5s $^2S_{1/2} - 4d\,^2D_{5/2}$ optical clock transition is probed using a 674 nm extended-cavity diode laser which is stabilized to a high-finesse ultra-low-expansion (ULE) cavity using the Pound-Drever-Hall technique and shifted into resonance with the clock transition using a single-pass acousto-optic modulator (AOM). The $^{88}Sr^+$ ion is interrogated using a computer-controlled sequence of operations which alternates between cooling the ion and probing the clock transition, with the weak clock transition being observed using the electron shelving technique. During the probe laser pulses, the cooling and repumping lasers are blocked to prevent ac Stark shifts of the 674 nm clock transition. To minimize the dead-time in each interrogation cycle, a 1033 nm extended-cavity diode laser has recently been introduced to drive the 4d $^2D_{5/2} - 5p\,^2P_{3/2}$ transition and hence return the ion rapidly to the cooling cycle.

The centre frequency of the 5s $^2S_{1/2} - 4d\,^2D_{5/2}$ Zeeman structure is determined by using a four-point servo scheme with a typical cycle time of 15 − 20 s to probe a pair of Zeeman components which are symmetrically located around line centre. The corrections applied to the AOM drive frequency have two parts. The first, primary, servo is a simple proportional gain control as described by Bernard *et al.*[11] The secondary servo is used to compensate for the drift rate of the ULE cavity and differs from that reported previously.[6] The modified scheme is similar to that used by Peik *et al.*[12] and works by adjusting the assumed drift rate in such a way as to maintain zero quantum jump rate imbalance between the interrogation frequencies on either side of each Zeeman component. Absolute frequency measurements of the clock transition frequency are performed by simultanously recording the AOM offset frequency and measuring the frequency of the light locked to the ULE cavity using a femtosecond optical frequency comb.

3. 674 nm probe laser system

To optimize the performance of the $^{88}Sr^+$ optical frequency standard so that its stability is close to the quantum projection noise limit set by the 0.4 s lifetime of the 4d $^2D_{5/2}$ level, the 674 nm probe laser linewidth must be reduced to the sub-hertz level for timescales of a few seconds.

The probe laser performance can be assessed by heterodyne beat frequency comparison between two independently stabilized 674 nm laser systems, with the linear component of the ULE cavity drift compensated by mixing the beat with the output from a ramped function generator. The beat linewidth observed in this way is a function of the measurement time, increasing from 2 Hz at 3 s to 6 Hz at 30 s. The Allan deviation of the beat frequency (without linear drift compensation) was observed to have a minimum instability of about 2.5 parts in 10^{15} at 1 s. When one of these lasers is used to interrogate a single Zeeman component of the clock transition in $^{88}Sr^+$ with a probe pulse length of 100 ms, a Fourier-transform-limited linewidth of 9 Hz is observed,[13] with some further improvements possible by using longer interrogation times.

These results were achieved using standard cylindrical high finesse reference cavities which were carefully isolated from acoustic and seismic noise. These cavities have recently been replaced by new cavities which are designed to have significantly reduced sensitivity to vibrations.[14] In the new design, material is cut out from the cylindrical spacer in the lower quadrants (Fig. 2) and the cavity is supported at four points which are symmetrically placed on either side of the cavity symmetry axes. For certain cavity support positions and cutout depth, the distance between the centres of the two high reflectivity mirrors where the laser beam enters and leaves the cavity remains invariant when a force is applied. The cutouts compensate for vertical forces whilst the symmetrical mounting provides insensitivity to horizontal acceleration. For the new 674 nm cavities, the sensitivity to vibrations has been measured to be less than 40 kHz/(ms^{-2}) for three perpendicular directions, a factor of ten improvement compared to the original cylindrical cavities. Further improvements may be possible through optimization of the support positions.

With the effects of vibration suppressed and with good temperature control at the zero thermal expansion point of the ULE spacer, the frequency stability of the probe laser will become limited by thermal noise in the mirror substrates and their coatings.[15–17] However it may be possible to reduce this limit by means of a careful choice of cavity geometry and materials.

4. Frequency stability

The stability and reproducibility of the $^{88}Sr^+$ optical clock is being assessed by comparisons between two similar standards. To enable stability data to be obtained out to long averaging times, significant effort has been put into automating the operation of the standards. In particular, software has been implemented for automatic monitoring and relocking of the 422 nm cooling laser and the 674 nm probe laser by monitoring the transmission signals through the reference cavities which are used to stabilize these laser systems. In addition, software control of the 1092 nm repumper laser drift

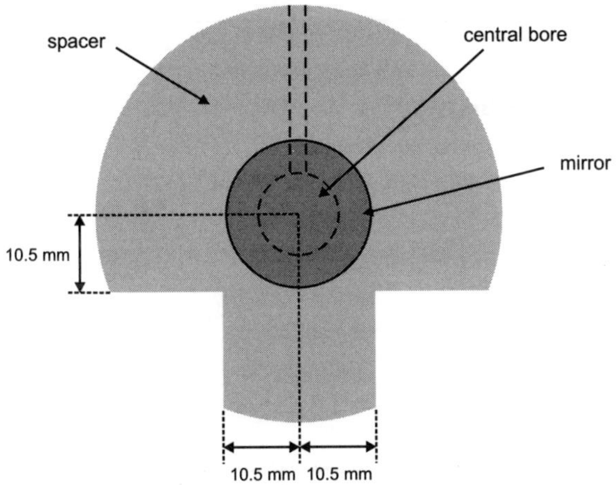

Fig. 2. Geometry of the new cutout cavities used for stabilization of the 674 nm probe laser. The ULE spacer is 99.8 mm long with a diameter of 60 mm and the cavity finesse is about 200 000.

has been demonstrated, based on maximizing the fluorescence signal from the trapped ion. These improvements have enabled the $^{88}Sr^+$ standards to be operated continously for periods exceeding 14 hours.

Most of the two-trap comparison results to date have been obtained using a probe pulse length of 10 ms, for which a relative frequency instability of approximately 10^{-14} has been observed for an averaging time of 100 s, scaling as $\tau^{-1/2}$ for longer timescales (Fig. 3). This agrees well with the predictions of a model which was developed to optimize the parameters of the probe laser lock to the clock transition. This model is based on a NAG routine that generates random numbers with Poisson statistics to mimic the quantum jump data from the clock transition. Improvements in the stability of the $^{88}Sr^+$ optical clock are expected to result from increased probe times; for example for 200 ms probe pulses the model predicts a relative frequency stability of 7×10^{-17} for an averaging time of 20 000 s.

5. Absolute frequency measurements

The largest source of systematic frequency shift for the $^{88}Sr^+$ optical clock arises from the electric quadrupole shift of the reference transition,[6] which is due to the interaction between the electric quadrupole moment of the atomic states and any residual electric field gradient present at the position of the trapped ion. The residual dc electric field gradient can be determined

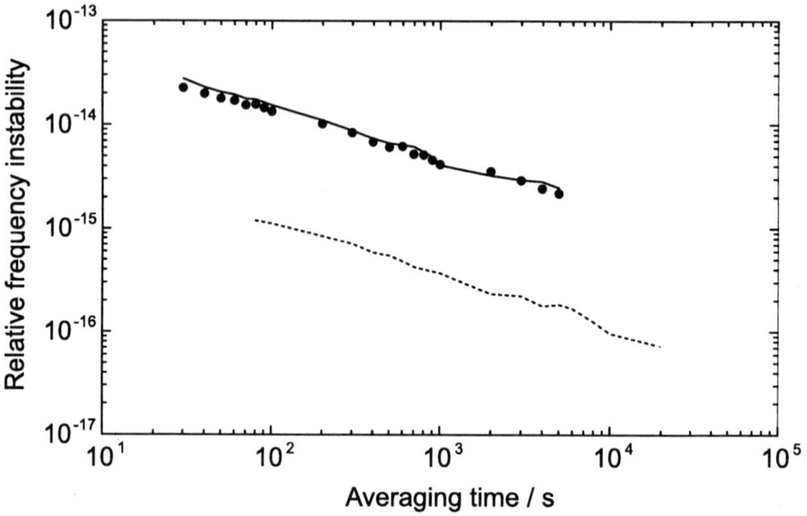

Fig. 3. Relative frequency instability observed in the comparison between two similar $^{88}Sr^{+}$ optical clocks. The circles show the experimental results for a probe pulse length of 10 ms, while the solid line shows the corresponding predictions of the servo model. The dotted line shows the model prediction for a probe pulse length of 200 ms.

from measurements of the secular frequencies of the ion motion within the trap and minimized by adjusting the dc voltages applied to the outer endcap electrodes, which typically reduces the quadrupole shift to the level of a few hertz.

However the quadrupole shift can be nulled by averaging frequency measurements made either for three mutually orthogonal orientations of the applied dc magnetic field[18] or for three different Zeeman component pairs.[5] Using these two techniques, an absolute frequency measurement of the 674 nm clock transition frequency was carried out in 2004, using the NPL caesium fountain primary frequency standard[19,20] as the reference. The result of 444 779 044 095 484.2(1.7) Hz,[4,21] with a fractional frequency uncertainty of 3.8×10^{-15}, is in good agreement with earlier, less accurate measurements[7,22] and a more recent measurement from NRC[5] (Fig. 4).

The dominant sources of uncertainty in this frequency measurement were technical in nature and have since been reduced by refinements to the experimental arrangement. As a first stage (2005) significant improvements were made to the extinction of the 422 nm cooling laser during the probe laser interrogation periods, reducing the ac Stark shift arising from this radiation to a negligible level. Improvements to the probe laser linewidth,

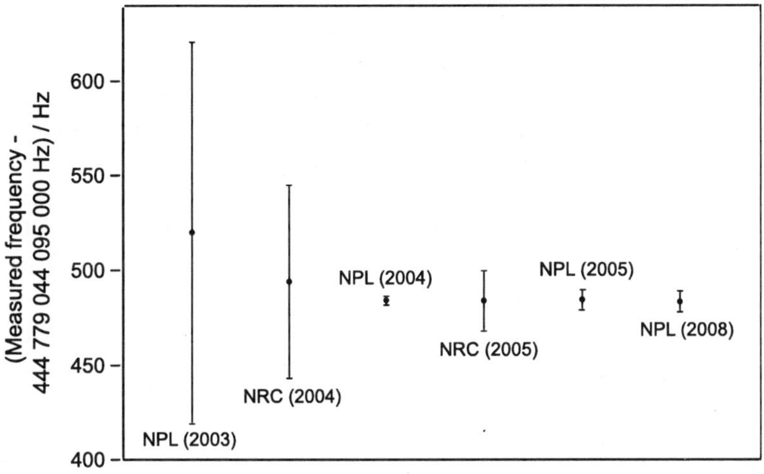

Fig. 4. Absolute frequency measurements of the 674 nm optical clock transition in $^{88}Sr^+$, showing the excellent agreement between results from NPL and NRC. [NPL (2003) and NRC (2004): first femtosecond comb measurements at NPL[22] and NRC;[7] NPL (2004) and NRC (2005): measurements in which the quadrupole shift was nulled;[4,5] NPL (2005): measurement with reduced ac Stark shift; NPL (2008): preliminary result derived from simultaneous frequency measurements of two $^{88}Sr^+$ standards using two femtosecond combs.]

discussed above, and reductions in the nonlinear component of the ULE cavity drift rate should also lead to lower uncertainty in the measured clock transition frequency by reducing servo errors in the lock to the centre of the Zeeman structure.

A new series of absolute frequency measurements of the 674 nm clock transition is now underway. Two independent strontium ion standards are being operated simultaneously, and for the data obtained so far, the two agree at the 2 Hz level, limited by statistics. Simultaneous frequency measurements are also being carried out using two independent femtosecond optical frequency combs, one based on a Ti:sapphire laser[23] and the other based on a fibre laser.[24] For the measurements carried out so far, the two combs agree at the sub-Hz level, again limited by statistics. This preliminary data gives a value for the absolute frequency of the clock transition which is in good agreement with our 2004 result (Fig. 4), but has an uncertainty approximately three times larger because the frequency reference used was a hydrogen maser linked to UTC rather than a caesium fountain.

6. Summary and outlook

At the current stage of development, a relative frequency instability of about 2 parts in 10^{15} has been observed for an averaging time of 5000 s in the comparison between two similar $^{88}Sr^+$ standards. These results were obtained with a probe pulse length of 10 ms, with significant improvements expected to result from increased probe pulse durations. Improved stability will enable detailed studies of systematic frequency shifts of the clock transition frequency to be carried out, targeting relative frequency stabilities and reproducibilities at the few parts in 10^{-17} level. With these improvements, a new measurement of the clock transition frequency will be limited by the uncertainty of the caesium fountain primary frequency standard used as the reference. A second NPL fountain, which will implement a recently demonstrated scheme for cancellation of the collisional frequency shift,[25] is nearing completion and should enable a frequency measurement to be carried out with a relative frequency uncertainty below 1 part in 10^{15}. When combined with the 2004 measurement and frequency measurements of other optical standards being developed around the world, this will lead to tighter constraints on the time variation of fundamental physical constants, in particular the electron to proton mass ratio.[26]

Acknowledgments

This work is supported by the UK Department of Innovation, Universities and Skills as part of the NMS Pathfinder Metrology Programme.

References

1. T. Rosenband, D. B. Hume, P. O. Schmidt, C. W. Chou, A. Brusch, L. Lorini, W. H. Oskay, R. E. Drullinger, T. M. Fortier, J. E. Stalnaker, S. A. Diddams, W. C. Swann, N. R. Newbury, W. M. Itano, D. J. Wineland and J. C. Bergquist, *Science* **319**, 1808 (2008).
2. C. Tamm, B. Lipphardt, H. Schnatz, R. Wynands, S. Weyers, T. Schneider and E. Peik, *IEEE Trans. Instrum. Meas.* **56**, 601 (2007).
3. K. Hosaka, S. A. Webster, P. J. Blythe, A. Stannard, D. Beaton, H. S. Margolis, S. N. Lea and P. Gill, *IEEE Trans. Instrum. Meas.* **54**, 759 (2005).
4. H. S. Margolis, G. P. Barwood, G. Huang, H. A. Klein, S. N. Lea, K. Szymaniec and P. Gill, *Science* **306**, 1355 (2004).
5. P. Dubé, A. A. Madej, J. E. Bernard, L. Marmet, J.-S. Boulanger and S. Cundy, *Phys. Rev. Lett.* **95**, 033001 (2005).
6. G. P. Barwood, H. S. Margolis, G. Huang, P. Gill and H. A. Klein, *Phys. Rev. Lett.* **93**, 133001 (2004).
7. A. A. Madej, J. E. Bernard, P. Dubé, L. Marmet and R. S. Windeler, *Phys. Rev. A* **70**, 012507 (2004).

8. P. Gill and F. Riehle, p. 282 in *Proceedings of the 20th European Frequency and Time Forum* (Braunschweig, Germany, 2006).
9. C. A. Schrama, E. Peik, W. W. Smith and H. Walther, *Opt. Commun.* **101**, 32 (1993).
10. D. J. Berkeland, J. D. Miller, J. C. Bergquist, W. M. Itano and D. J. Wineland, *J. Appl. Phys.* **83**, 5025 (1998).
11. J. E. Bernard, L. Marmet and A. A. Madej, *Opt. Commun.* **150**, 170 (1998).
12. E. Peik, T. Schneider and C. Tamm, *J. Phys. B* **39**, 145 (2006).
13. G. P. Barwood, P. Gill, G. Huang and H. A. Klein, *IEEE Trans. Instrum. Meas.* **56**, 226 (2007).
14. S. A. Webster, M. Oxborrow and P. Gill, *Phys. Rev. A* **75**, 011801(R) (2007).
15. K. Numata, A. Kemery and J. Camp, *Phys. Rev. Lett.* **93**, 250602 (2004).
16. M. Notcutt, L.-S. Ma, A. D. Ludlow, S. M. Foreman, J. Ye and J. L. Hall, *Phys. Rev. A* **73**, 031804(R) (2006).
17. S. A. Webster, M. Oxborrow, S. Pugla, J. Millo and P. Gill, *Phys. Rev. A* **77**, 033847 (2008).
18. W. M. Itano, *J. Res. Natl. Inst. Stand. Technol.* **7105**, 829 (2000).
19. K. Szymaniec, W. Chałupczak, P. B. Whibberley, S. N. Lea and D. Henderson, *Metrologia* **42**, 49 (2005).
20. K. Szymaniec, W. Chałupczak, P. B. Whibberley, S. N. Lea and D. Henderson, *Metrologia* **43**, L18 (2006).
21. H. S. Margolis, G. P. Barwood, K. Hosaka, G. Huang, H. A. Klein, S. N. Lea, A. Stannard, B. R. Walton, S. A. Webster and P. Gill, p. 92 in *Atomic Physics 20, Proceedings of the XX International Conference on Atomic Physics* (AIP, New York, 2006).
22. H. S. Margolis, G. Huang, G. P. Barwood, S. N. Lea, H. A. Klein, W. R. C. Rowley, P. Gill and R. S. Windeler, *Phys. Rev. A* **67**, 032501 (2003).
23. V. Tsatourian, H. S. Margolis, S. N. Lea, B. R. Walton, G. Marra, D. T. Reid and P. Gill, these proceedings.
24. B. R. Walton, H. S. Margolis, V. Tsatourian and P. Gill, *IET Optoelectron.* **2**, 182 (2008).
25. K. Szymaniec, W. Chałupczak, E. Tiesinga, C. J. Williams, S. Weyers and R. Wynands, *Phys. Rev. Lett.* **98**, 153002 (2007).
26. S. N. Lea, *Eur. Phys. J. Special Topics* **163**, 37 (2008).

A TRAPPED $^{171}Yb^+$ ION OPTICAL FREQUENCY STANDARD BASED ON THE $S_{1/2}$ - $F_{7/2}$ TRANSITION

P. GILL[1,*], S. A. WEBSTER[1], G. HUANG[1], K. HOSAKA[3], A. STANNARD[2], S. N. LEA[1], R. M. GODUN[1], S. A. KING[1,2], B. R. WALTON[1] and H. S. MARGOLIS[1]

[1]National Physical Laboratory, Hampton Road, Teddington, Middlesex, TW11 0LW, UK
[2]Clarendon Laboratory, University of Oxford, Parks Road, Oxford OX1 3PU, UK
[3]Present address: NMIJ, Tsukuba Central 3, Umezono 1-1-1, Tsukuba 305-8563, Japan
* *Email:* patrick.gill@npl.co.uk

The single cold $^{171}Yb^+$ ion has two optical clock transitions available. One of these is the $^2S_{1/2}$ - $^2D_{3/2}$ quadrupole (E2) transition at 436 nm. The other is the $^2S_{1/2}$ - $^2F_{7/2}$ octupole (E3) transition at 467 nm, which has an extremely narrow theoretical linewidth on account of the 6-year lifetime of its $F_{7/2}$ upper level. This paper reports progress on both these clock transitions at NPL. In particular, the absolute frequency of the observed octupole transition has recently been measured with an uncertainty of 12 Hz (1σ) for an octupole linewidth of < 40 Hz. Contributions from major shifts such as the AC Stark shift and second order Zeeman shift have been measured, and quadrupole and blackbody shifts estimated. The frequency ratio between the $^{171}Yb^+$ quadrupole and octupole transitions demonstrates one of the largest sensitivities for monitoring possible time variation of the fine structure constant. Ultra-stable probe laser light at 436 nm has recently been generated to interrogate the quadrupole clock transition, in preparation for measuring both clock transitions simultaneously.

1. Introduction

Single cold trapped ions held within rf quadrupole traps represent a class of optical frequency standards that are now demonstrating frequency instabilities and uncertainties that compete with and, in some cases, out-perform the corresponding capabilities of primary Cs fountain microwave standards [1-5]. The recent progress of these systems, together with the rapid progress now being achieved with cold atom lattice clocks, suggest optical clock capabilities which should reach uncertainty levels in the region of 10^{-17} – 10^{-18} within a few years. With demonstrated performance at these levels, the potential for a future redefinition of the second becomes increasingly likely.

One particular single ion candidate for an optical clock is the $^{171}Yb^+$ ion. This has two possible optical clock transitions. One is based on the $^2S_{1/2}$ - $^2D_{3/2}$ quadrupole (E2) transition at 688 THz (436 nm), with an upper state lifetime of 52 ms, and corresponding natural linewidth of 3 Hz. This has been studied at the Physikalisch Technische Bundesanstalt (PTB) in Germany [4]. The second is the ultra-forbidden $^2S_{1/2}$ - $^2F_{7/2}$ octupole (E3) transition at 642 THz (467 nm), which has an extremely narrow nanohertz theoretical linewidth on account of the 6-year lifetime of its $F_{7/2}$ upper level [6,7]. With such a narrow theoretical lineshape, any interrogation of the octupole transition will be limited purely by the linewidth of the probe laser or the Fourier transform of the probe pulse interrogation time.

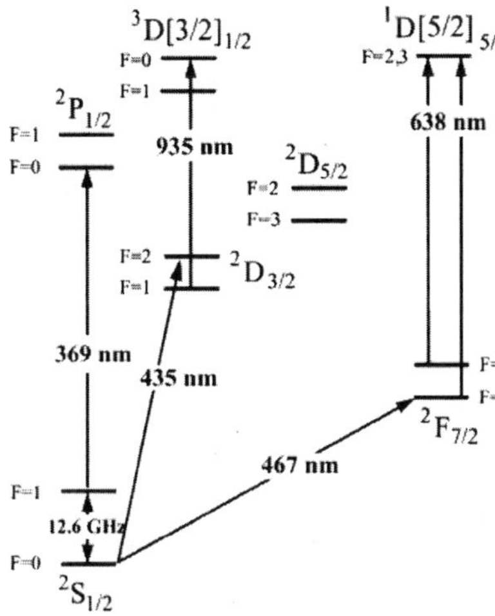

Fig. 1. Partial term scheme of the $^{171}Yb^+$ ion, showing the optical clock transitions (436 nm and 467 nm) and laser cooling, repumper and clear-out transitions.

The $^{171}Yb^+$ odd isotope hyperfine structure requires additional wavelengths, compared to the $^{172}Yb^+$ even isotope case, in order to laser cool and prepare the ion in the $^2S_{1/2}$ $F = 0$ ground state, and to recover the ion from the $^2F_{7/2}$ $F = 3$ state after driving the octupole clock transition. However, there is no linear Zeeman effect on the $^2S_{1/2}$ $F = 0$ - $^2F_{7/2}$ $F = 3$, $m_F = 0$ clock transition, and it has a very small quadratic Zeeman shift [8]. Additionally, the electric quadrupole shift is calculated to be small [9], some 15 times smaller than that for the $^{88}Sr^+$ ion, for example. The major difficulty is that the octupole transition is extremely weak, such that it requires a significant intensity to drive it, which gives rise to a dominant AC Stark shift under current conditions. However, by reducing the probe laser linewidth, the spectral intensity can be maintained whilst reducing the total intensity and hence the shift. For laser linewidths reduced well below 1 Hz, the AC Stark shift becomes less dominant, and its uncertainty can be reduced below the 10^{-17} level [9].

2. Experimental arrangement

Single ions of $^{171}Yb^+$ are produced by photo-ionisation of a low flux Yb atomic beam [8] and confined within an rf quadupole trap of the end-cap design [10]. The separation between inner end-cap electrodes is ~ 0.6 mm, and with a drive frequency of 14 MHz and 300 V amplitude, this gives rise to a time-averaged pseudo-potential with secular frequencies in the 1-2 MHz range.

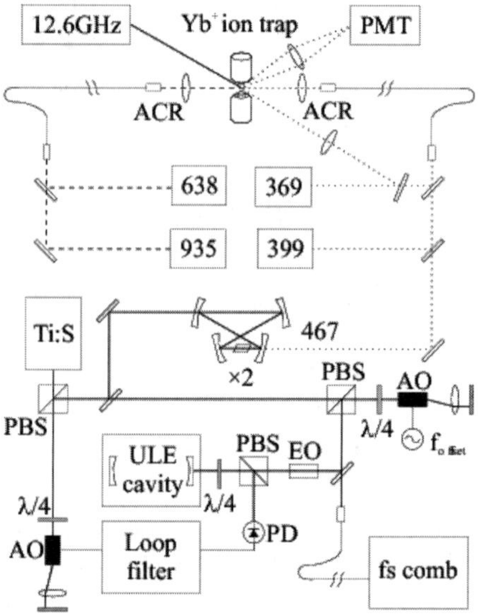

Fig 2. Experimental layout for driving the octupole transition; solid line, 934 nm probe laser fundamental wavelength; dotted lines, 467 nm probe light, 369 nm cooling light and 399 nm photoioization light; dashed lines, 935 nm and 638 nm repumper light; ACR, achromatic doublet; PMT, photomultiplier; AO , acousto-optic modulator; PBS, polarising beam splitter; EO, electro-optic modulator; PD, photodetector; Ti:S, titanium:sapphire laser.

The ion is Doppler cooled using the $^2S_{1/2}$ $(F = 1) - {}^2P_{1/2}$ $(F = 0)$ electric dipole transition at 369 nm. The $^2P_{1/2}$ level can decay to the $^2D_{3/2}$ $(F = 1)$ level with a branching ratio of 7×10^{-3}. Light at 935 nm is used to depopulate the $^2D_{3/2}$ $(F = 1)$ level in order to maintain the cooling cycle. Additionally, the ion can be driven to the $^2S_{1/2}$ $(F = 1)$ ground state by off-resonant pumping to the $^2P_{1/2}$ $(F = 1)$ state, and microwaves at 12.6 GHz are used to mix these ground hyperfine states. At the end of the cooling cycle, the microwaves are turned off for 30 ms to prepare the cold ion in the F=0 ground state. The $^2S_{1/2}$ $(F = 0)$ - $^2F_{7/2}$ $(F = 3)$ octupole and $^2S_{1/2}$ $(F = 0)$ - $^2D_{3/2}$ $(F = 2)$ quadrupole clock transitions are driven by light at 467 nm and 436 nm respectively. Quantum jumps in the 369 nm fluorescence provide the means for state detection of the ion following clock transition probing. Finally, once the clock transition is driven, the clock state upper level is quenched by light at 638 nm for the octupole, and 935 nm for the quadrupole transition, to recover the ion to the ground state and repeat the cooling/preparation/probe cycle. Figure 2 shows the experimental layout for the octupole transition.

3. $^{171}Yb^+$ cold ion octupole lineshape

Octupole probe light at 467 nm is generated by frequency doubling of 934 nm light from a Ti:sapphire laser. A doubled power of several mW is produced with a $KNbO_3$ non-linear crystal in a resonant enhancement cavity. The fundamental light is frequency stabilised to an ultra-low-expansion (ULE) glass reference cavity of 1.5 GHz free spectral range and 190,000 finesse by means of the Pound-Drever-Hall technique (PDH) [11]. The reference cavity is located within an evacuated chamber with two layers of temperature control, mounted on a vibration isolation platform and housed within an acoustically damped enclosure. The probe laser stability was monitored by heterodyning 934 nm beams stabilised to two separate ULE cavities. The results show a beat linewidth of ~ 5 Hz over 1 s acquisition times which broadens to 20 Hz for 100 s averaging after drift compensation. The frequency drift rate for the probe light locked to the ULE cavity was measured to be in the region of 50 – 500 $mHz.s^{-1}$, where the isothermal creep was ~ 50 $mHz.s^{-1}$.

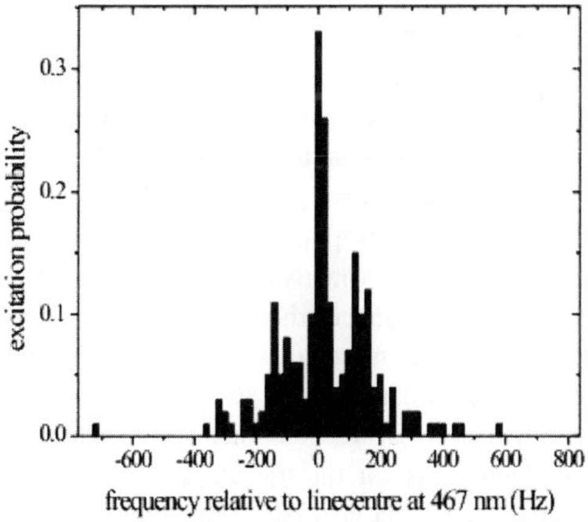

Fig. 3. Cold ion absorption spectrum for the $^2S_{1/2}(F = 0)$ - $^2F_{7/2}(F = 3, m_F = 0)$ octupole clock transition

The $^{171}Yb^+$ cold ion octupole lineshape recorded by scanning the frequency-doubled ULE cavity-stabilised probe light at 467 nm across the transition is shown in figure 3. This data comprises ten separate scans of 40 frequency steps of 20 Hz, where each step consists of 10 interrogations of 50 ms each. The cavity drift is monitored and subtracted from each scan before the scans are combined. The data shows a central carrier linewidth of less than 40 Hz, which corresponds well with the doubled linewidth of the 934 nm cavity-stabilised probe laser.

4. Absolute frequency measurement

In order to measure the absolute frequency of the octupole clock transition, the ULE cavity-stabilised and frequency-doubled light is scanned over the transition by monitoring the acousto-optic modulator (AOM) offset frequency from the ULE cavity mode whilst simultaneously measuring the frequency of the cavity mode using a femtosecond comb via synchronised timing between the two measurement data acquisition systems. Both the AOM cavity offset frequency and the femtosecond comb are referenced to the 10 MHz ouput from one of the NPL hydrogen masers which provide UTC (NPL). The cavity frequency was measured for extended periods of up to several hours with 10-s counter gate times and fitted with a 3rd order polynomial. The linear drift was measured to be 0.5 Hz s^{-1}. On combining the ULE cavity data with the corresponding $^{171}Yb^+$ frequency AOM offset data, the absolute frequency of each individual quantum jump is derived. For these frequency measurements, each scan comprised 100 x 20 Hz steps, with 10 x 50 ms interrogations at each step, resulting in ~ 100 quantum jumps per scan. In order to take account of the AC Stark shift, the scans were carried out at a number of probe laser powers and a linear fit to all quantum jump data made to provide the variation of clock transition frequency with laser power. This measurement procedure was repeated for four different focussing arrangements of the octupole light into the trap, and the resultant data fits extrapolated to zero power (intensity) to give a set of four values of the octupole transition frequency unperturbed by the AC Stark shift. Each value is separately corrected for NPL maser offset with an uncertainty of 6.4 Hz, as determined from Circular T data relevant to the particular measurement period. The four measurement values are combined as a rectangular distribution to provide an unweighted mean for the transition frequency unperturbed by AC Stark shift and corrected for maser offsets. The total 1σ uncertainty derived is 12 Hz, dominated by the statistical, AC Stark and maser uncertainties. The measurements were carried out with a magnetic field of $24 \pm 8\ \mu$T, and whilst there is no first order Zeeman effect for this $m_F = 0 - m_F = 0$ transition, a correction for the second order Zeeman shift needs to be applied. The 2nd order coefficient is -1.72 ± 0.03 mHz $(\mu\mathrm{T})^{-2}$ [8] giving a correction of +1 Hz for the zero field value. The final value for the $^2S_{1/2}\ F = 0$ - $^2F_{7/2}\ F = 3$, $m_F = 0$ octupole transition is 642 121 496 772 657 $\pm$ 12 Hz [12].

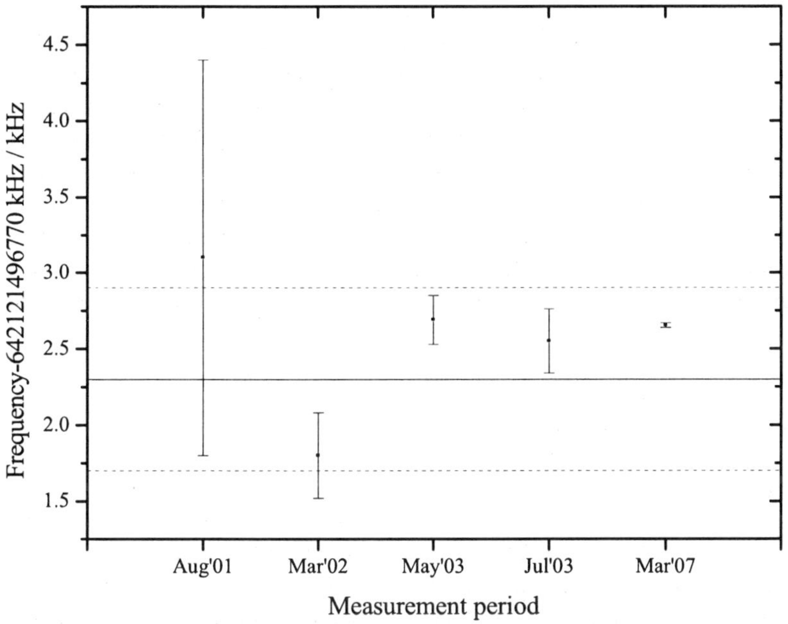

Fig. 4. Comparison of octupole transition frequency measurements over recent years. The solid line is the mean of measurements during 2002-3; the dashed lines show corresponding $\pm 1\sigma$ values.

Figure 4 shows the results of comb-based frequency measurements of the octupole transition over recent years. The recent 2007 value shows excellent agreement with the 2003 data based on 200 Hz wide octupole linewidths and good agreement with the 2002-3 global mean. The recent measurement has an accuracy 50 times better than the previously published value [8] based on the global mean, and corresponds to a relative uncertainty of 2×10^{-14}.

Whilst the AC Stark shift is extrapolated to zero in these scan measurements, extrapolations over extended periods may not be ideal when the interrogating light is servoed to the ion. In this latter case, as discussed earlier, the interrogating linewidth will need to be reduced below its current experimental value in order to reduce the AC Stark shift to a non-dominant value. By reference to the AC Stark shift coefficient [9], the shift appropriate to an interrogation linewidth of ~ 0.1 Hz is in the mHz region. Other systematic frequency shifts such as the electric quadrupole shift, blackbody Stark shift, 2nd order Doppler shift and low frequency Stark shift, do not contribute significantly to the overall 10 Hz measurement uncertainty, but will need to be evaluated as the uncertainty of future frequency measurements is reduced. The electric quadrupole shift arises from the interaction of the electric quadrupole moment (EQM) of the $^2F_{7/2}$ state with residual electric field gradients. The EQM has been calculated as $-0.2ea_0^2$ [9] which is 3-10 times smaller than EQM values in other quadrupole clock transitions, and with quadrupole shift nulling [13-15], should give rise to quadrupole shift uncertainties at the mHz level. The blackbody Stark shift (relative to 0 K) is ~ 150 mHz at room temperature, but with an uncertainty at the tens of mHz level due to lack of knowledge of the coefficient [16], and a few mHz for a 1 K temperature uncertainty of the environment. Doppler shifts and low-frequency Stark shifts due to residual thermal motion and micromotion can be reduced below 1 mHz with efficient Doppler cooling and well-minimised micromotion.

5. Quadrupole laser development

The availability of two separate optical clock transitions in $^{171}Yb^+$ offers the potential to make frequency ratio measurements of these transitions over extended periods of time. This is particularly interesting from a fundamental constants perspective in respect of the search for possible time variation in the fine structure constant α. The octupole transition in particular provides a theoretical sensitivity factor to monitor time variation in α of –5.95, which is the largest of existing microwave and optical frequency standards [17, 18]. When combined with the quadrupole transition sensitivity of +0.88 [19] , the frequency ratio between the two has a $d\alpha/dt$ sensitivity factor of 6.8. Not only is the sensitivity factor large, frequency ratio measurements between two clock transitions in the same species offers scope for common mode rejection of some frequency shifts, especially if these ratio measurements were made in the same cold ion. Assuming frequency ratio measurement of the octupole and quadrupole transitions at the 10^{-17} level of uncertainty, this would allow any time variation of α to be investigated at the level of 2×10^{-18} per year. Currently the most sensitive measurement of this type achieves a sensitivity of $-1.6 \pm 2.3 \times 10^{-17}$ per annum [1].

In preparation for $^{171}Yb^+$ octupole – quadrupole frequency ratio measurements, we are developing a 436 nm quadrupole frequency standard. Currently we have two $^{171}Yb^+$ end-cap traps where both transitions can be interrogated. Recently we have built a 436 nm source to probe the quadrupole clock transition. This is an extended cavity diode laser operating at 871 nm and frequency doubled to 436 nm via single pass frequency doubling in a temperature-tuned potassium niobate non-linear crystal. With 30 mW at the fundamental wavelength, $\sim 10\ \mu$W of doubled light at 436 nm is produced, sufficient to drive the quadrupole transition. The fundamental light at 871 nm has been locked to one of the new vibration-insensitive ULE cavities designed for use with the 934 nm probe laser. The very high reflectivities and associated high finesse of the 934 nm mirror coatings are also achieved for the 871 nm quadrupole case since the mirror reflectivity is centred on 900 nm. Independent monitoring of the 871 nm in-lock PDH amplitude noise suggests a cavity-stabilised fundamental laser linewidth ~ 1 Hz. Work is underway to verify this by heterodyning separate ULE cavity-stabilised 871 nm beams and make preliminary frequency scans across the 436 nm quadrupole clock transition.

6. Summary and outlook

The $^2S_{1/2}\ F = 0$ - $^2F_{7/2}\ F = 3,\ m_F = 0$ octupole clock transition at 642 THz has been driven with a frequency-doubled 934 nm probe laser which demonstrates a short term (1 s) linewidth of 5 Hz broadening to 20 Hz over 100 s. This correlates well with the observed $^{171}Yb^+$ cold ion lineshape of < 40 Hz at the doubled frequency. With this arrangement, the absolute frequency of the octupole transition has been measured via a maser-referenced femtosecond comb. The resulting octupole frequency value, unperturbed by AC Stark shift and corrected for the 2nd order Zeeman shift is 642 121 496 772 657 $\pm$ 12 Hz, where the uncertainty is dominated by statistics, AC Stark shift extrapolation and NPL maser offsets appropriate to the measurement periods.

Future comb-based absolute frequency measurements of both the octupole and quadrupole clock transitions will be made by reference to one of the NPL Cs fountains. We will investigate octupole frequency stability performance by heterodyning two independent 436 nm ULE-reference-cavity-stabilised sources servoed to single ions in separate traps. Additionally, these two traps will be used to evaluate the systematic frequency shifts relevant to the octupole transition. These measurement campaigns and systematics evaluations will be similarly carried out for the quadrupole clock transition. Frequency ratio measurements between the $^{171}Yb^{+}$ 467 nm octupole and 436 nm quadrupole transitions will contribute to improved sensitivity in the search for time variation of the fine structure constant. Whilst these frequency ratio measurements can be carried out on the octupole and quadrupole transitions in separate ions in the two co-located traps, an optimal measurement scenario would be simultaneous frequency measurements on a single $^{171}Yb^{+}$ ion in the same end-cap trap. With a sufficient period between consecutive frequency ratio measurements that achieve projected uncertainties, it is anticipated that sensitivities to monitor fine structure time variation would reach parts in 10^{18} per year.

Acknowledgements

This work is supported by the UK Department for Innovation, Universities & Skills as part of the NMS Pathfinder Metrology Programme.

References

1. T. Rosenband *et al., Science* **319,** 1808 (2008)
2. W. Oskay, W. M. Itano and J. C. Bergquist, *Phys. Rev. Lett.* **97,** 020801 (2006)
3. E. Peik, T. Schneider and C. Tamm, *J. Phys. B* **39,** 145 (2006)
4. T. Schneider, E. Peik and C. Tamm, *Phys. Rev. Lett.* **94,** 230801 (2005)
5. H. S. Margolis *et. al., Science* **306,** 1355 (2004)
6. M. Roberts *et al., Phys. Rev. Lett.* **78,** 1876 (1997)
7. M. Roberts *et al., Phys. Rev. A* **62,** 020501(R) (2000)
8. K. Hosaka *et al., IEEE Trans. Instrum. Meas.* **54,** 759 (2005)
9. P. J. Blythe, S.A. Webster, K. Hosaka and P. Gill, *J Phys. B* **36,** 981 (2003)
10. C. A. Schrama, E. Peik, W. W. Smith and H. Walther *Opt. Commun.* **101,** 32 (1993)
11. R. W. P. Drever *et al., Appl. Phys. B* **31**, 97 (1983).
12. K. Hosaka, S. A. Webster, A. Stannard, B. R. Walton, H. S. Margolis and P. Gill, *Phys. Rev. A* (submitted)
13. W. M. Itano, *J. Res. Natl. Inst. Stand. Technol.* **105**, 829 (2000).
14. G. P. Barwood, H. S. Margolis, G. Huang, P. Gill and H. A. Klein, *Phys. Rev. Lett.* **93,** 133001 (2004)
15. P. Dubé *et al., Phys. Rev. Lett.* **95,** 033001 (2005)
16. S. N. Lea, S. A. Webster and G. P. Barwood, *Proc. 20th European Frequency and Time Forum,* (Braunschweig, Germany, 2006).
17. V. A. Dzuba, and V. V. Flambaum, *Phys. Rev. A* **77,** 012515 (2008)
18. S. N. Lea, *Rep. Prog. Phys.* **70,** 1473 (2007)
19. V. A. Dzuba, V. V. Flambaum and M. V. Marchenko, *Phys. Rev A* **68,** 022506 (2003)

OVERVIEW OF HIGHLY ACCURATE RF AND OPTICAL FREQUENCY STANDARDS AT THE NATIONAL RESEARCH COUNCIL OF CANADA

A. A. MADEJ*, J. E. BERNARD, P. DUBÉ, AND L. MARMET

Frequency and Time Group, Institute for National Measurement Standards, National Research Council of Canada, Ottawa, ON, K1A 0R6, Canada

**Email: alan.madej@nrc.ca*

We describe progress towards ultra-accurate and stable standards of frequency and time at the National Research Council of Canada (NRC). Emphasis is placed on the research associated with creating a reliable and workable optical frequency standard and secondary realization of the second based on a single trapped and laser cooled ion of strontium at 445 THz. Work is described on recent results of probe resolution, stability, and the ability to connect this reference frequency with existing and developed RF standards at NRC. An overview is also presented of the group's progress in the development of a Cs fountain based standard for the realization of the second and upcoming determinations of the reference single ion frequency.

1. Introduction

The landscape of frequency standards at high accuracy and stability has undergone a profound transformation over the last few years. RF primary standards based on laser cooled Cs atomic fountains are now being used on a regular basis and have improved the accuracy of the SI second at a level of 10^{-15} to 10^{-16}. In addition, the decades old dream of a near ideal quantum based frequency standard using ultra-stable laser radiation referenced on transitions in laser cooled, trapped ions or atoms is coming to practical realization.[1,2] The impact of this work has been the further improvement in accuracy (10^{-17} – 10^{-19} level) with high stabilities such that very high precisions can be realized over short averaging times. Results from this work will provide continuing refinement of fundamental physical measurements such as the time variation of physical constants, precision measurements of relativistic effects such as the gravitational red shift, and in the future, space borne tests of new physics and phenomena.[3]

Within the myriad of new candidate atoms for potential realizations of quantum based time and frequency, we have been studying[4] since 1989 a standard based on a single trapped and laser cooled ion of $^{88}Sr^{+}$. The relative simplicity of the energy level structure and of the necessary laser wavelengths make it very suitable for a reliable operating standard and a promising system to be widely reproduced in labs elsewhere. This system was the first visible ion reference transition to be directly related to the SI realization of the second.[5] Based on these results and later comb measurements[6-8] by the NPL (U.K.) and our group, the transition was selected as an internationally accepted standard for frequency and vacuum wavelength and as a secondary realization of the second.

In this summary, we will focus on our recent results in advancing the NRC $^{88}Sr^+$ laser cooled single ion based optical frequency standard at 445 THz (674 nm) so that it can achieve its full potential in terms of accuracy, stability, and long term operation. Current work has focused in approaching the lifetime limited linewidth of the transition of 0.4 Hz (line $Q = 1 \times 10^{15}$) and increasing the control and reliability of the single ion standard. We will also provide an overview on the progress of the NRC Cs fountain frequency standard which will provide the operational realization of the SI second for Canada in the years to come and will serve as the reference for future accurate measurements of the strontium ion clock transition frequency and other cold atom standards.

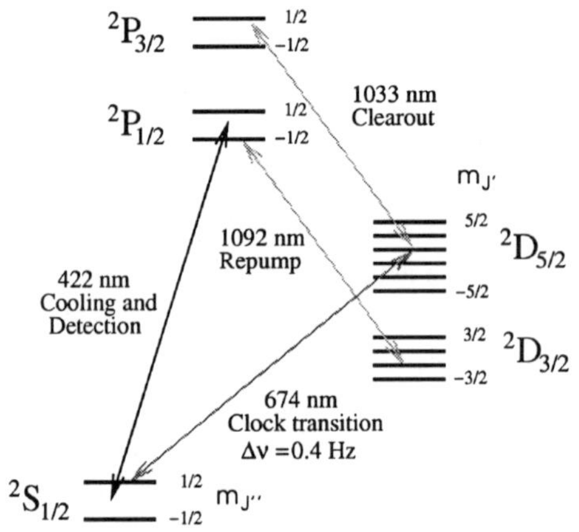

Figure 1: Energy level diagram for $^{88}Sr^+$ showing states and transitions used for the operation of the 445 THz single ion based optical frequency standard.

1.1. *Strontium Single Ion Frequency Standard*

The advantages of the strontium single ion system have been described in a number of previous works.[4-8] A schematic diagram of the lowest relevant energy levels of the system is given in Fig. 1. Briefly, the laser cooling (422 nm), state repumping (1092 nm) and probe (674 nm) transitions in the $^{88}Sr^+$ system are such that it is a very good match to current available laser technology with the reference 445-THz transition having a potentially high ultimate accuracy. All these sources can be obtained via commercially available diode and fiber laser technology either directly or by frequency doubling. The system shows great promise for the wide reproduction of such standards in many other laboratories. Results from the NPL (U.K.) have shown similar remarkable performance.[8]

2. NRC Strontium Ion Experiment

The present work utilizes a single ion of $^{88}Sr^+$ held within an electrodynamic quadrupole Paul trap with characteristic endcap to trap center distance of $z_o = 0.5$ mm and center to ring distance of $r_o = 0.71$ mm. The trap electrodes are made of tantalum and have a spherical form for the endcaps and a semi circular cross section for the ring electrode. All materials in the vicinity of the trap are made of non-magnetic materials with the chamber and window ports made with low permeability stainless steel. A single layer magnetic shield external to the vacuum enclosure is used to isolate the ion from external perturbations. The trap is operated at a frequency of 12 MHz and the ion moves within an effective potential well having a transverse secular frequency of 965 kHz and an axial secular frequency of 1930 kHz. A 12 μT bias field is applied to the ion to remove the degeneracy of the Zeeman energy levels of the ground and excited states of the clock transition. The transition frequency center is determined by measuring symmetric pairs of Zeeman components which eliminates the linear Zeeman shift.[7] By averaging the center frequency value of three pairs of components which connect to all sublevels in the $^2D_{5/2}$ excited state, the electric-quadrupole shift and the tensor terms of the quadratic Stark shift can also be effectively measured and removed.[6]

2.1. *422 nm Laser Cooling Source*

Recent work has been directed toward the development of a reliable and frequency stable laser cooling source for the strontium ion standard such that the ion can be maintained at low kinetic temperatures and probed for long periods of time (days, weeks) without adjustment. In our recent work, a diode laser system emitting at 422 nm has been used as the source for laser cooling and fluorescence excitation.[9] An external cavity diode laser which employs a GaN laser diode and a Littrow mounted diffraction grating is used to produce the source radiation. The diode laser is frequency narrowed from its free running linewidth of ~10 MHz to 2 MHz (FWHM) via stabilization to a reference Fabry-Perot cavity. Radiation is then sent to the ion trap experiment via optical fiber. The 422-nm source is also frequency stabilized to the 5s $^2S_{1/2}$ (F'' = 2)-6p $^2P_{1/2}$ (F'=3) ^{85}Rb transition which is only 400 MHz lower in frequency than the strontium ion S-P transition. A frequency shifted beam from an acousto-optic modulator is directed into a saturated absorption spectrometer arrangement containing a Rb cell having a vapor pressure of 14 mPa. The new system[9] has been shown to be very versatile and robust such that lock periods of several days are obtained with frequency drifts of the cooling source maintained below 100 kHz. Using such a source, stable and continuous cooling of the strontium ion sample has been achieved at ion kinetic temperatures of T = 10 mK for periods exceeding 24 hrs.

2.2. *674 nm Probe Laser System and Observed Resolution of Strontium S-D Transition*

The current probe laser system employs a 674 nm commercial diode laser that is frequency narrowed and stabilized using a cascaded arrangement of two successively higher finesse Fabry-Perot cavities and Pound-Drever-Hall locking. The second stage cavity with finesse of 160000 and linewidth of 3.7 kHz has its ULE spacer stabilized at the point where the material's coefficient of linear thermal expansion is zero and is thus exceptionally immune to thermal fluctuations. The residual laser linewidth fluctuations relative to the high finesse resonator have been determined to be below 1 mHz. The present system is very reliable and can remain stabilized for over 1 week. Current drift rates of the cavity relative to the ion resonance are at the level of 14 mHz/s when the table tilt is suppressed. A more detailed description of the probe laser is given within this volume.[10]

The stabilized laser radiation is sent to the ion trap via optical fiber with active fiber-noise cancellation. The observed linewidth of the ion S-D transition has been measured to be at the 4.3 Hz Fourier transform limit for a 200 ms probe time when the optical table supporting the cavities is suspended on a standard commercial pneumatic vibration damping system.

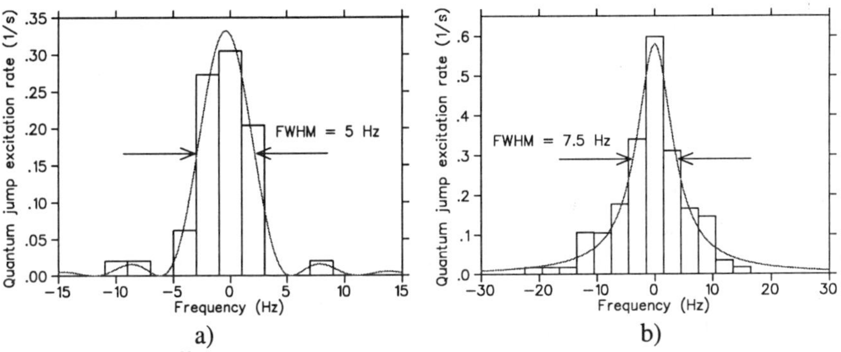

Figure 2: Observed $^{88}Sr^+$ single ion S-D spectral resolution using a two-stage probe laser system with a) Table with cavity on pneumatic suspension. Fitted curve is the Fourier transform of the applied temporal probe pulse, b) Table resting on laboratory floor. Fitted curve is a Lorentzian fit to the data.

The floating table based system has introduced tilt based sensitivity of the ULE reference cavity drift rate and more recent measurements have been performed using the table rigidly resting on the laboratory floor. Under these conditions, the observed linewidth increases to approximately 8 Hz width (FWHM) (see Fig. 2 b) but the laser exhibits a very linear isothermal drift. By measuring the frequency offset between the probe laser and the reference transition linecentre of the $^{88}Sr^+$ S-D transition, it was possible to carefully study the drift at high accuracy. Probe to ion lock periods of up to 9 hrs in duration have been obtained to date as shown in Fig. 3. When a piecewise linear drift is subtracted, the relative frequency fluctuations averaged down at a rate of $2.6 \times 10^{-14}\ \tau^{-1/2}$ for $\tau > 200$ s to a stability of 5×10^{-16} at 3000 s. The results show great promise in using such an ultra-stable probe together with our single ion standard to reach stabilities below the 10^{-16} level in a few hours and to achieve the ultimate accuracy of the ion. Further improvements in state preparation and laser linewidth reduction are expected to bring the stability to the quantum projection noise limited optimal stability of $3 \times 10^{-15}\ \tau^{-1/2}$.

3. Fiber Laser Frequency Comb Connection to the 455-THz Reference Transition

A significant element necessary to realize the new single ion standard will be the maintenance of a reliable long-term link with the RF (from our recently developed Cs fountain clock) or with other optical frequencies of interest. The development of optical frequency comb technology has revolutionized our ability to connect such diverse regions of the electromagnetic spectrum and has allowed the use of optical frequency standards in time and frequency metrology. Although Ti:Sapphire based frequency comb systems have been very successful in demonstrating measurement capability, ultra-accurate transfer, and performance, challenges arise when one considers applying such a technology for reliable, low-maintenance, and long-term relation of optical and RF frequencies extending to months. Our research has investigated the use of a fiber laser based frequency comb system developed in collaboration the University of British Columbia (UBC).[11,12] The femtosecond oscillator is based on a stretched pulse erbium doped fiber oscillator employing non-linear polarization rotation as the mode-locking mechanism. After leaving the oscillator, the output pulses are equally distributed into two parallel branches each with a fiber based chirped amplifier. One branch is used to obtain an octave spanning spectrum from 1000 to 2100 nm for detection and control of the carrier envelope phase of the comb.[11] The second measurement branch generates a narrower continuum with stronger output near 1348 nm. This radiation is frequency doubled in a periodically poled $MgO:LiNbO_3$ (PPLN) crystal to generate the desired 674 nm light to heterodyne against the light used in probing the ion. Measurements have obtained S/N=30 dB (100 kHz BW) for the heterodyne beat between the probe and the comb. With the use of tracking oscillators, continuous locking periods of the comb of over 3 days have been obtained recently. The instability of the measured frequency for 1 and 10 s gate times were at approximately the reference maser instability.[12] Allan deviation fits of recent ion frequency measurements have shown that the measured frequency exhibited a stability with averaging time of $\sigma(\tau) = 3.2 \times 10^{-13}\ \tau^{-1/2}$. Moreover, when the results of the Oct. 2008 data are compared with absolute frequency measurements of the ion frequency taken from February to May, 2004, a difference of $\Delta = -\ 0.4 \pm 4.5$ Hz (1σ) is observed confirming the high level of reproducibility of the single ion standard and the equivalence of the new frequency comb apparatus in the determination of the absolute frequency. The uncertainty in the above determined difference in frequency is dominated by the earlier 2004 measurements. The current measurements yield a 1σ uncertainty of 1.4 Hz dominated by the current knowledge of the reference maser frequency. No evaluation of the current systematic offsets has been performed at this time.

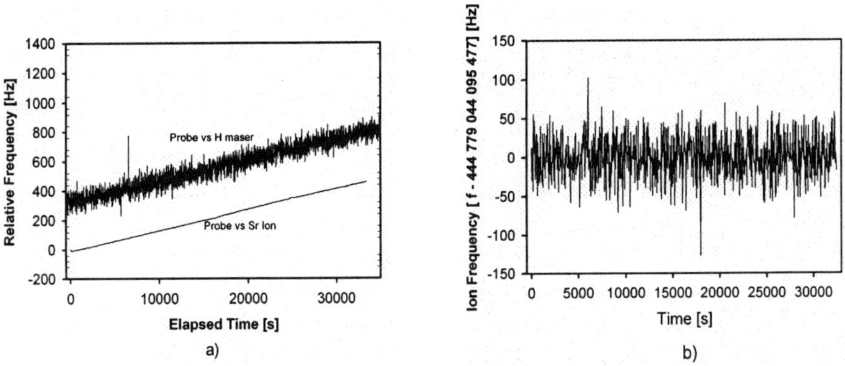

Figure 3: a) Upper trace - absolute frequency of the 445 THz probe measured with a fiber-comb system. Lower trace - measured offset of the probe laser vs. the ion centre frequency during the same period. b) By combining the comb measurements with the ion offset, a determination of the ion frequency is obtained.

4. Next Generation Ion Trap System for the NRC Ion Frequency Standard

A new single ion trap system is currently being developed which will further enable the reduction of systematic shifts and allow the full evaluation of the ion's motion. The new trap is of the end-cap design and is fabricated with non-magnetic and low permeability materials to ensure a stable B field environment for the ion. Probing and cooling will be obtained over three orthogonal directions. Monitoring and cancellation of micromotion will be accomplished via a number of techniques including photon correlation with the trap drive frequency and imaging the ion motion at low trap depth using a photon counting camera. The large fluorescence collection efficiency, well controlled magnetic environment, and the ability to monitor and null micromotion induced effects will allow the new trap system to be employed as an effective standard whose accuracy will be limited by knowledge of the blackbody radiation induced frequency shift.[7] Recent precision calculations of the alkaline-earth ion S-D black body shifts[13] indicate uncertainties of a few percent, thus allowing uncertainties at the 10^{-17} level.

5. NRC FCs 1 Cesium Fountain Primary Standard

The first cesium fountain primary standard built at NRC is nearing completion. A recent summary of the construction of the device has been given elsewhere.[14] The standard has been built using a number of different features which will test new approaches to the fountain design and seek improved performance. Some important aspects of the device are: a) a transverse C-field configuration for the drift region for deceased sensitivity to external longitudinal magnetic fields, b) electrostatic shutters for the effective isolation of light shift, reduction of dead time, and the ability to operate with square wave phase modulation to reduce cavity pulling, c) a cylindrical thermal shield in the drift region that can be varied from cryogenic temperatures to above 100 °C so that the blackbody radiation shift can be precisely evaluated, d) a split rectangular Ramsey microwave cavity for the attenuation of modes that would make the clock sensitive to Doppler shifts in the Cs atoms. The Cesium atoms are captured in a 110-beam geometry using a magneto-optic trap (MOT) whose coils have been arranged to minimize field interference in the fountain. State preparation into the lower hyperfine level of the clock transition is employed and the Ramsey cavity is placed in a well-shielded region located 77 cm above the MOT under three layers of magnetic shielding together with an active magnetic field stabilization system external to the shields. With the exception of the feed through connectors, the physical package is entirely built with non-magnetic materials. The expected short term stability of the standard is 7×10^{-14} at 1s averaging. The current estimated uncertainties for the fountain foresee a total uncertainty of below 1×10^{-15} when a careful evaluation of the standard is completed. The primary source of uncertainty at this stage appears to be due to the C-field inhomogeneity and spin exchange effects for Cs atom collisions. Current work is underway in the optimization of the trapped atom ensemble in the MOT and the improvement of the Ramsey fringe signal.

With the operation and full evaluation of the fountain, the system will then be able to effectively contribute to TAI together with the group's 3 hydrogen masers and ensemble of thermal beam Cs standards. An immediate goal will also be the absolute frequency measurement of the $^{88}Sr^{+}$ S-D reference transition. To date, the best value obtained has been by the group at the NPL (U.K.) with an assigned uncertainty of 1.5 Hz (1σ).[8] It is expected that the upcoming measurements with the NRC fountain together with the new NRC fiber based frequency comb system and upgraded trap system will allow improved measurement capability and further corroborate the extreme accuracy of the single ion standard at the current limits of the SI second.

Acknowledgements

The authors would like to thank S. Cundy and J.S. Boulanger for their contributions to the absolute frequency measurements, G. Humphrey and A.D. Shiner for essential contributions to the stabilized laser sources, J. Jiang and D.J. Jones for the development of the fiber laser based frequency comb, and R. Pelletier and B. Hoger for their excellent technical support in the development of the experimental apparatus. This work was partially supported by a Discovery research grant from the Natural Sciences and Engineering Research Council (NSERC).

References

1. G.K. Campbell et al., *Metrologia* **45**, 539 (2008) and references therein.
2. T. Rosenband et al. *Science* **319,** 1808 (2008) and references therein.
3. P. Wolf et al., *Experimental Astronomy* **DOI 10.1007/s10686-008-9118-5** (2008).
4. A.A. Madej. and J.D. Sankey, *Opt. Lett.* **15**, 634 (1990).
5. J.E. Bernard, A.A. Madej, L. Marmet, B.G. Whitford, K.J. Siemsen, and S. Cundy, *Phys. Rev. Lett* **82**, 3228 (1999).
6. P. Dubé, A.A. Madej, J.E. Bernard, L. Marmet, J.-S. Boulanger, and S. Cundy, *Phys. Rev. Lett.* **95**, 033001 (2005).
7. A.A. Madej, J.E. Bernard, P. Dubé, L. Marmet, and R.S. Windeler, *Phys. Rev. A.* **70**, 012507 (2004).
8. H.S. Margolis, G.P. Barwood, G. Huang, H.A. Klein, S.N. Lea, K. Szymaniec, and P. Gill, *Science* **306**, 1355 (2004).
9. A.D. Shiner, A.A. Madej, P. Dubé, and J.E. Bernard, *Appl. Phys. B* **89**, 595 (2007).
10. P. Dubé, A.A. Madej, and J.E. Bernard, "Ultra-stable Laser Source for the $^{88}Sr^+$ Single-Ion Optical Frequency Standard at NRC" in this volume.
11. J. Jiang, J.E. Bernard , A.A. Madej, A. Czajkowski, S. Drissler, and D. J. Jones., *J. Opt. Soc. Am. B* **24** , 2727 (2007).
12. J.E. Bernard, P. Dubé, A.A. Madej, S. Cundy, J.-S. Boulanger, J. Jiang, and D.J. Jones, "Development of an Optical Clockwork for the Single Trapped Strontium Ion Standard at 445 THz" in this volume.
13. M.S. Safronova, private communication (2008).
14. L. Marmet, B. Hoger, P. Dubé, A.A. Madej and J.E. Bernard, "Detailed Description of FCs1: NRC's Cesium Fountain Primary Standard" in *Proceedings of the 2008 Frequency Control Symposium* (IEEE Cat. No. CFP08FRE-PRT, ISBN 978-1-4244-1795-7), p.386 (2008).

Part VI

Optical Frequency Combs

EXTREME ULTRAVIOLET FREQUENCY COMBS FOR SPECTROSCOPY

A. Ozawa, M. Herrmann, A. Vernaleken, Ch. Gohle, B. Bernhardt, T. Wilken, W. Schneider, Ph. Vilar Welter, S. Knünz, V. Batteiger, R. Holzwarth, E. Peters, S. Reinhardt, T. W. Hänsch and Th. Udem

Max-Planck Institut für Quantenoptik,
85748 Garching, Germany
**E-mail: thu@mpq.mpg.de*
http://www.mpq.mpg.de/~ haensch/comb/index.html

Optical spectroscopy has matured to the most precise measurement tool in physics thanks to advances in single ion trapping and the possibility to directly measure the frequency of laser light. However, almost 50 years after the invention of the laser the spectral region that can be investigated in this way is still restricted to wavelengths below the near ultraviolet. The much larger spectral band of the extreme ultraviolet (XUV), where many fundamental transitions of say hydrogen like ions reside, is thus far unexplored by high precision laser spectroscopy. One possible route to narrow band radiation in this region could be the use of high order harmonics generated with short laser pulses of high repetition rate focused in a gas jet. Meanwhile μW power levels in the XUV at multi-MHz repetition rates have been demonstrated which are the main prerequisites for this method. The 1S-2S two photon transition at 60 nm in singly ionized helium is a rewarding candidate because it allows sensitive tests of quantum electrodynamics.

Keywords: Frequency Combs; Direct comb spectroscopy; high harmonic generation; trapped ions; quantum electrodynamics.

1. Introduction

The study of simple hydrogenic systems has played a central role in the development of physics. Theory is highly developed for such systems and can provide extremely accurate predictions to be compared with high-precision experimental data. While single mode lasers are currently employed for high resolution spectroscopy of transitions such as the hydrogen[1,2] 1S-2S, many transitions of fundamental interest occur at wavelengths too short for state of the art continuous wave lasers. The 1S-2S two-photon transition in singly ionized helium is a sharp resonance in a hydrogenic system that could extend the list of simple bound systems which have high potential to accurately test fundamental theories. Because (hydrogenic) energy levels scale with the nuclear charge Z^2 this transition requires 60.8 nm radiation when excited with two-photons (see figure 1).

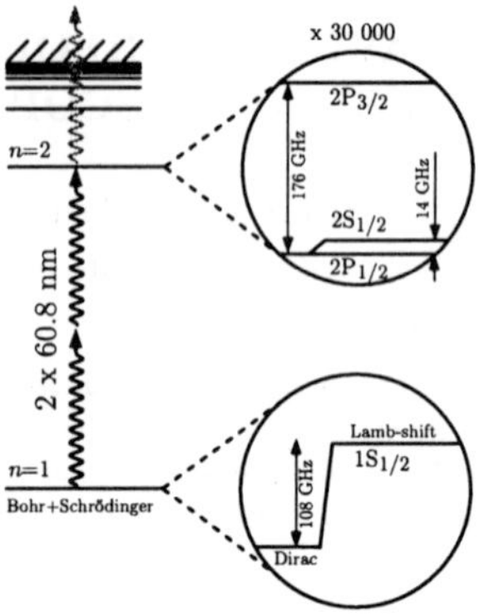

Fig. 1. Relevant Energy levels of $^4He^+$ including the fine structure and Lamb shift. After two-photon excitation from the ground state, the third, ionizing photon reaches 6.8 eV into the continuum.

So far, all experiments in high resolution laser spectroscopy were performed with spectrally narrow continuous wave lasers whose wavelengths reach down to the near ultraviolet. Lacking suitable continuous wave laser sources, the wavelength range beyond that is thus far unexplored by high resolution laser spectroscopy. One practical boundary is set by the transparency range of the existing non-linear crystals. The crystal material that is useful for the shortest wavelength is BBO (β-barium-borate), with a transparency wavelength cut-off at about 190 nm. Only with pulsed lasers it is possible to efficiently convert to much shorter wavelengths of the vacuum UV (200 – 10 nm), extreme UV (30 – 1 nm) and even to the soft X-ray regime ($<$ 10 nm). The established method for this is high harmonic generation (HHG) that takes place when intense infrared laser pulses are focused into a spatially localized gas jet.[3] For a long time pulsed lasers and high resolution spectroscopy seemed to exclude each other because of the large bandwidth associated with short pulses. However, coherent trains of pulses as compared to isolated or non-coherent pulses (say, from a Q-switched laser) have quite distinct spectra referred to as frequency comb.[4] If a coherent pulse train is used to generate high harmonics, the resulting spectrum is not simply a broad continuum but is believed to possess a regular array (frequency comb) of very sharp modes underneath it. Making these modes spectrally narrow as in any other continuous wave laser, they can be used for high resolution spectroscopy.

1.1. *Frequency Comb Spectroscopy*

The ideal source of a frequency comb (left side of Fig. 2) is the coherent pulse train generated by a mode locked laser. Its spectrum consists of narrow modes $\omega_n = n\omega_r + \omega_{CE}$ with the pulse repetition rate ω_r and the carrier-envelope frequency ω_{CE} that resides in the radio frequency domain. Thanks to the large integer n the optical frequencies of the comb ω_n could be determined since it became possible to measure ω_{CE} by a method subsequently dubbed self referencing.[5] The usual application is to exploit the

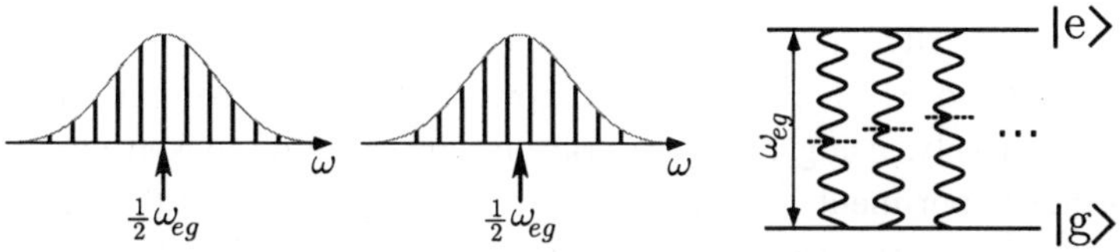

Fig. 2. The frequency comb consists of up to 10^6 modes that are equally spaced by the pulse repetition rate ω_r. The width of the spectral envelope is roughly given by the inverse pulse duration. For two-photon transitions the comb modes can add up pair wise to the proper transition energy if the pulses are Fourier limited. The condition is that either one of the central modes are resonant with the transition or that resonance occurs exactly half ways between two modes.

large integer n to convert from the optical frequencies ω_n to the radio frequency domain where comparisons can be conducted with a radio frequency reference. If the optical reference is to be taken as a frequency standard one could as well count the radio frequency signals in order to use the setup as an optical atomic clock.[6]

On the other hand, it is also feasible to use individual modes of the comb for direct frequency comb spectroscopy. One obvious problem with this approach is the limited power per mode in a broad frequency comb. Given an octave spanning comb, as required for simple self-referencing, this power can easily drop below 100 nW in practice. In addition the unused modes may cause an excess ac Stark shift that poses a problem in high accuracy measurements. Still these problems can be handled as demonstrated in reference.[8]

Yet an even better idea is to employ a two-photon transition as initially proposed by Ye. F. Baklanov and V. P. Chebotayev.[9] At first glance it seems that a two photon transition requires even more power. However, as sketched in Fig. 2, it is straight-forward to see that in this case the modes can sum up pairwise such that the full power of the frequency comb contributes to the transition rate: Suppose the frequency comb is tuned such that one particular mode $n\omega_r + \omega_{CE}$, say near the center, is resonant with the two-photon transition. This means that two-photons from this mode provide the necessary transition energy of $\omega_{eg} = n\omega_r + \omega_{CE}$. In this case the combination of modes with mode numbers $(n-1, n+1)$, $(n-2, n+2)$, $(n-3, n+3)\ldots$ are also resonant as they sum up to the same transition frequency. In fact for transform limited pulses all modes contribute to the transition rate in this way whereas a frequency chirp reduces the transition rate but does not shift the observed transition frequency. The same applies if the two-photon resonance occurs exactly halfway between two modes.

Experimentally, frequency comb spectroscopy has been pioneered by J. Eckstein and M. J. Snadden and coworkers[10,11] on sodium and rubidium, respectively, using direct excitation with a mode locked laser. While the former experiment was still a factor 2.5 short of the 1.6 MHz natural linewidth, the latter reached the natural linewidth of 300 kHz. The main difference between the two measurements is that Snadden and coworkers used laser-cooled atoms in a magneto-optical trap.

There may be a significant time-of-flight broadening because the Doppler-free signal only emerges from atoms within the pulse collision volume for counter-propagating pulse trains. Atoms that fly through this volume can only absorb a limited number of pulses that cause the line broadening effects. To obtain a narrow resonance it is therefore important to apply many pulses, ideally for a time that exceeds the inverse natural linewidth. When using fs pulses the collision volume may be smaller than 1 mm so that the atoms must be laser cooled and/or trapped as in reference[11] to reach the natural linewidth.

In the meantime, several other groups[12,13] including ours[14] have used the frequency comb directly from a mode locked laser, i.e. without any frequency up-conversion, for spectroscopy. However, realizing the above-mentioned disadvantages of frequency comb spectroscopy, it appears advisable to use a continuous wave laser whenever possible. On the other hand, if there is no laser of this kind, frequency comb spectroscopy can become a very powerful method. One of the first demonstrations with short wavelengths was done by S. Witte and co-workers who used UV pulses at 212 nm generated in conventional non-linear crystals to excite a two photon transition in krypton.[15] In this work only 3 pulses could be applied to the atoms so that the expected linewidth was about one third of the repetition rate or 23 MHz. Indeed that is about the observed linewidth which compares to the natural linewidth of 6.9 MHz. Even more exciting is the first experiment which uses a wavelength that cannot be generated by conventional crystals, i.e. continuous wave lasers. Such an experiment was performed by the same group on a single photon transition in xenon at 125 nm.[16] The required radiation was obtained as the third harmonic produced in a gas cell.

The experiments performed so far have been done with the laser fundamental or with thermal atoms that stay in the laser beam only for a few pulses. However, the actual advantage becomes obvious only when cold atoms or ions can be kept in the focus of the beam of harmonic radiation that can only be produced with pulsed lasers. Using HHG several octaves of laser radiation may be addressed without gaps. The large tunability of titanium:sapphire lasers allows in principle to shift between the harmonics seamlessly. When used on trapped ions the interaction time with the laser can be arbitrary long in order to not limit the observed linewidth.

2. XUV Sources

For high harmonic generation (HHG) one focuses an infrared femtosecond pulse train in a gas target employing its extreme non-linear response when subject to intensities exceeding $\sim 5 \times 10^{12}$ W/cm^2. As all known materials absorb in the XUV spectral region, crystals for frequency multiplication cannot be used and the lower density gas targets must be spatially limited to avoid re-absorption of the harmonic radiation. In most HHG experiments a nozzle emitting a jet of noble gas atoms is placed near the focus of an amplified femtosecond pulse train to produce up to the 1000th harmonic[17] in a nearly diffraction limited laser-like beam that travels collinearly with the generating pulse train.

In this process, a fraction of the wave function of the outer electron of the target gas escapes from the ion core due to the strong laser field and is accelerated by that field. Together with the remaining part of the wave function this generates a large dipole moment oscillating with the laser frequency but with significant deviations from the harmonic drive. Inversion symmetry of the gas target prevents even harmonics to emerge so that a HHG spectrum consists of odd harmonics that are distinct in three regions: the perturbative, the plateau and the cut-off harmonics with increasing harmonic order. In the perturbative regime, the individual harmonics fall off in intensity corresponding to their respective order until reaching the plateau region where the harmonics region obey a common scaling with intensity. As sketched in Fig. 3 we expect that each of these harmonics is

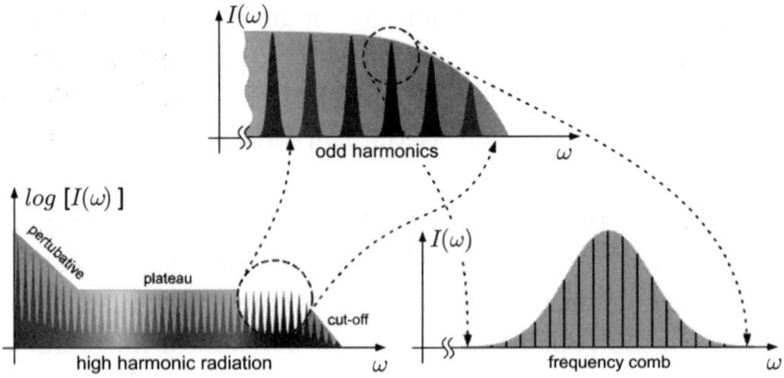

Fig. 3. The spectrum of high order harmonics is characterized by a plateau of constant intensity that reaches out to the cut-off energy. Interference between emission from each half cycle of one pulse leads to a peak structure with maxima at odd multiples of the driving frequency. Interference between the pulses leads to a frequency comb structure within each of the harmonics.

made up of a frequency comb so that two-photon spectroscopy as described in the previous section would become possible. So far, all data that have been recorded suggest that this process is extremely coherent. The resulting radiation is emitted in a collimated laser beam, clear interference fringes have been observed at the 21st harmonic,[7] and shot noise limited beat notes at the 3rd harmonic where conventional optics can still be used have been generated.[18,19]

As explained above, a high repetition rate is required so that the interrogated atoms or ions can resolve individual modes with their effective linewidth. Unfortunately, this is contradicting the requirement of reaching the necessary intensity for HHG. Up to recently, all available methods for HHG effectively concentrated the available average power in fewer pulses per second. For this the repetition rate is typically reduced to the kHz regime, i.e. mJ pulse energies. The modes of the resulting dense frequency comb would be very difficult to resolve. Even if the time-of-flight broadening could be reduced by using trapped particles that have transitions with narrow natural linewidths, the requirements on the laser system would be difficult to achieve. For this reason, a method allowing the production of high harmonic radiation with MHz repetition rates was sought after.

One solution to this problem was to use an enhancement resonator for the driving pulses with an intracavity gas jet for high harmonic generation. This method is similar to resonantly enhanced second harmonic generation that has been used for many years. However, there are several extra requirements that need to be fulfilled in order to resonantly enhance femtosecond pulses. In the frequency domain, the conditions are: The resonant modes of the cavity must be (i) equidistant, (ii) separated by the pulse repetition rate, and (iii) positioned correctly, i.e. matched to the CEO frequency of the driving laser. While the mode spacing and the position of the comb can be adjusted during the experiment, the irregularities of the cavity mode spacing are precompensated with chirped dielectric mirrors.[23] To achieve an accurate dispersion management, we first design a set of chirped test mirrors based on the best guess and then measure the residual dispersion by analyzing the spectral cavity enhancement.[24] Using this information, we generate a second set of mirrors with reduced dispersion.[20]

Output coupling the HHG radiation that is collinear with the driving laser field is achieved by total external reflection from an intra-cavity sapphire plate. Placed at Brewster's angle, it possesses very low loss for the infrared, while providing significant reflectivity in the XUV. Since there is no transparent medium in the XUV region the Brewster method output coupling scheme was central to our first intracavity HHG.[18] By reducing the repetition rate down to 10 MHz and by improvements of the enhancement process we are now able to generate μWs of HHG power.[20] This would produce Hz-level excitation rates provided this power can be focused to a μm-size spot size on the He ion. As this would be the proper excitation rate for a good single ion clock transition, it should be noted that available reflective optics for the XUV region and the necessary suppression of residual fundamental laser light superimposed on the XUV beam will significantly reduce the power that can be delivered to the ion.

3. He^+ Spectroscopy

The 1S-2S two-photon transition in He^+ has a linewidth of 84 Hz and extends the list of simple bound systems which have high potential to accurately test fundamental theories. Especially in comparison to the very successful QED tests in hydrogen does the 1S-2S in He^+ reveal its strengths: QED corrections of higher order, which are currently of greatest interest, scale in hydrogenic systems with large powers of the nuclear charge, Z^6, whereas the transition energy scales as Z^2. Hence, the sensitivity to these corrections is up to an order of magnitude larger. Second, since the ^{4}He nucleus carries no nuclear angular momentum, the centroid frequency is measured directly. But probably most important, He^+ is a charged particle and therefore may be trapped and (sympathetically) cooled. To analyze the

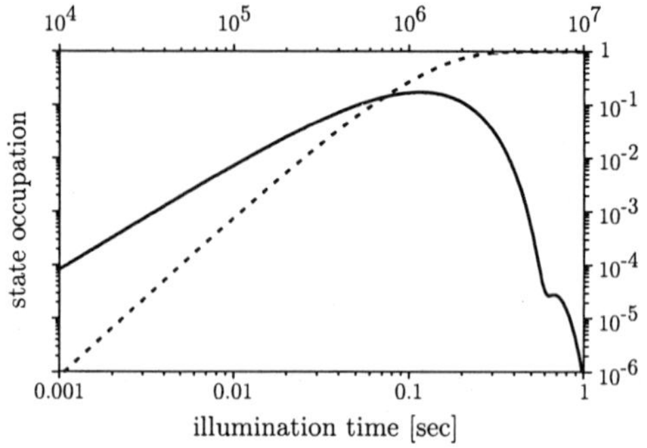

Fig. 4. Excitation (full line) and ionization (dashed) probabilities for a helium ion illuminated on resonance with a 60.8 nm laser of 1 μW power focussed down to a spot size of $w_0 = 1\ \mu m^2$ as a function of time by integrating the corresponding optical Bloch equations. The driving laser is assumed to operate at 10 MHz with 57 fs pulses (upper scale) as in reference.[20]

excitation dynamics we numerically integrate the optical Bloch equations for pulsed excitation and 3 levels: the ground, the excited, and the ionized state. A unique property of all hydrogen-like systems is that the photons that drive the 1S-2S transition have sufficient energy to ionize the 2S state. Therefore, ionization is almost inevitable unless one quenches the 2S state quickly enough with a pulsed electric field, for example. From the calculated occupation dynamics shown in Fig. 4, it becomes clear that only a very strongly damped Rabi oscillation takes place between the 1S and the 2S state for the intensity under consideration. Instead, the final state is always the doubly ionized state. In addition this plot shows that for short interaction times the ionization state occupancy grows with the cube of the laser illumination time while the 2S state occupation grows only quadratically.

The best strategy in terms of high event rates seems to be the use of the doubly ionized state as a signature for the 1S-2S transition.[22] The trap could be made stable for the He^{2+} ions which may be detected via a frequency change in the vibrational modes of the remaining ions that are used for sympathetic cooling. In this case, the signal could be accumulated for an arbitrary long illumination time with virtually zero background which could be essential for low excitation rate. A measurement cycle would then consist of several excitation trials with subsequent detection of the presence of He^{2+} for example via its specific secular frequency or relocalizations of the cooling ions. This is a well-established technique to detect impurities in an ion cloud or crystal. After reloading the ion trap the same procedure would be repeated with a different laser frequency setting.

As important as estimating the transition rate for a high resolution experiment is the determination of the expected systematic uncertainties. Since the properties of He^{+} can be calculated from first principles, one can derive coefficients that describe systematic shifts accurately. To estimate various effects we assume some reasonable conditions that should be achievable. Among the expected systematic shifts are the second order Doppler shift due to the finite temperature at the 10^{-17} level. The ac and dc Stark effect might be the dominating source of uncertainty at an estimated 10^{-16} level. Note that there is no quadrupole shift for this transition. More details of these estimations will become available in an upcoming publication.[22]

References

1. M. Fischer *et al.*, Phys. Rev. Lett. **92** 230802 (2004).
2. B. de Beauvoir *et al.*, Eur. Phys. J. D **12**, 61 (2000).
3. J. G. Eden, Prog. Quant. Electr., **28**, 197 (2004).
4. Th. Udem, R. Holzwarth, and T.W. Hänsch, Nature **416**, 233 (2002).
5. Th. Udem *et al.*, Proceedings of the Joint Meeting EFTF & FCS, Besançon France April 13th-16th April 1999, pp. 620.
6. T. Rosenband *et al.* Science **319**, 1808 (2008).
7. R. Zerne *et al.* Phys. Rev. Lett. **79**, 1006 (1997).
8. V. Gerginov *et al.*, Opt. Lett. **30**, 1734 (2005).
9. Ye.F. Baklanov, and V.P. Chebotayev, Appl. Phys. Lett. **12**, 97 (1977).
10. J. Eckstein, A.I. Ferguson, and T.W. Hänsch, Phys. Rev. Lett. **40**, 847 (1978).
11. M.J. Snadden *et al.*, Opt. Commun. **125**, 70 (1996).
12. A. Marian *et al.* Science **306**, 2063 (2004).
13. T.M. Fortier *et al.* Phys. Rev. Lett. **97**, 163905 (2006).
14. P. Fendel *et al.*, Opt. Lett. **32** 791 (2007).
15. S. Witte *et al.* Science **307**, 400 (2005).
16. R.Th. Zinkstok *et al.* Phys. Rev. **A 73**, 061801(R) (2006).
17. J. Seres *et al.*, Nature **433** 596 (2005).
18. Ch. Gohle *et al.*, Nature **436**, 234 (2005).
19. R. J. Jones *et al.*, Phys. Rev. Lett. **94**, 193201 (2005).
20. A. Ozawa *et al.*, Phys. Rev. Lett. **100**, 253901 (2008).
21. D.C. Yost, T. R. Schibli, and J. Ye, Opt. Lett. **33**, 1099 (2008).
22. M. Herrmann *et al.* submitted to Phys. Rev. A.
23. R. Szipöcs *et al.*, Opt. Lett. **19**, 201 (1994).
24. A. Schliesser *et al.*, Opt. Express. **14**, 5975 (2006).

DEVELOPMENT OF AN OPTICAL CLOCKWORK FOR THE SINGLE TRAPPED STRONTIUM ION STANDARD AT 445 THz

John E. Bernard*, Pierre Dubé, Alan A. Madej, Stan Cundy, and Jean-Simon Boulanger

Frequency and Time Group, Institute for National Measurement Standards
National Research Council of Canada, Ottawa, ON K1A 0R6, Canada
**E-mail: john.bernard@nrc.ca*

Jie Jiang and David J. Jones

Department of Physics and Astronomy, University of British Columbia
Vancouver, BC V6T 1Z1, Canada

An optical frequency comb, based on a stretched-pulse, mode-locked erbium fibre laser, has been developed to measure the clock transition frequency at 445 THz of a Sr^+ single-ion optical frequency standard. The comb employs separate amplifier branches for the control of the carrier-envelope offset frequency and for the measurement of the clock frequency of the optical standard. Continous locking periods for the comb of over three days have been observed.

Keywords: Optical frequency comb; Optical frequency standard; Single ion standard

1. Introduction

The optical frequency comb has made it possible to measure the frequency of optical frequency standards against state-of-the-art RF devices and to build optical clocks that are capable of generating highly-stable RF and 1-pulse-per-second outputs.[1] While combs based on modelocked Ti:sapphire lasers continue to be used for these applications, fi bre-based combs have advantages in terms of long-term operation, ease of use, and cost.[2,3]

We have developed a fi bre-based optical frequency comb that will serve as part of a stable link between the NRC Sr^+ single-ion optical frequency standard at 445 THz (674 nm)[4] and the NRC atomic clocks. It will be used both in extended measurements of the clock frequency of the single-ion standard with respect to the NRC cesium fountain clock, and as an optical clockwork that will permit the high stability and reproducibility of the ion standard to be used in the measurement and evaluation of other optical and RF standards.

2. Fiber Frequency Comb

The fiber frequency comb is shown in Fig. 1. It is similar to the comb described in Ref. 5 and is based on an erbium fiber oscillator which operates in the stretched pulse mode and employs nonlinear polarization rotation as the mode-locking mechanism.[6] Polarization control is through waveplates and an optical isolator in a free-space section. The oscillator is self-starting and produces chirped 1-ps pulses with a spectral width of 60 nm at a repetition rate of 102 MHz and average output power of 10 mW.

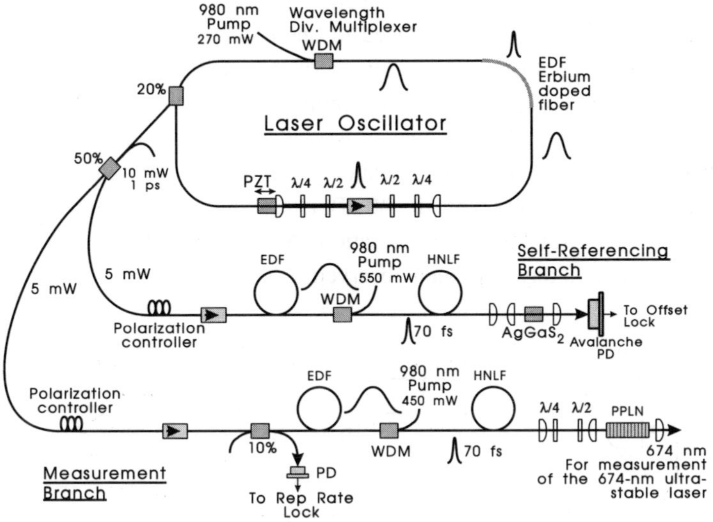

Fig. 1. Diagram of the fiber-based optical frequency comb used to measure the 674-nm ultra-stable laser.

The output from the oscillator is split by a 50:50 fibre coupler and sent to the self-referencing and measurement branches. Two branches are used to allow for optimization of the measurement branch for the desired application.[7] Each branch generates a comb of optical frequencies $f_n = nf_{rep} + f_o$, where f_{rep} is the laser oscillator pulse repetition rate and f_o is the offset of the comb elements from integer multiples of f_{rep}, also referred to as the carrier-envelope offset frequency.

In the self-referencing branch, the pulses from the oscillator pass through a SMF-28 fiber and an erbium fiber amplifier and are amplified and broadened to ~80 nm. They are then recompressed to < 100 fs in SMF-28 fibre before they are broadened to over an octave from 1000 to 2100 nm (total power = 75 mW) in highly nonlinear fiber (HNLF). The offset beat signal is detected with a common-path f-to-2f interferometer, that employs a 5-mm-long angle-tuned $AgGaS_2$ frequency doubling crystal. A short length of SMF-28 fibre is spliced to the output of the HNLF for dispersion compensation. The linewidth of the free-running offset beat signal is less than 20 kHz for a measurement time of less than 1 s.

The measurement branch is similar to the self-referencing branch but has been designed to generate a narrower comb spectrum with strong output at 1348 nm which is frequency doubled in bulk, periodically-poled MgO:$LiNbO_3$ (PPLN) to 674 nm. The total power contained in the comb is 60 mW and the power in the doubled spectrum (FWHM$\sim$ 0.8 nm) is 230 μW.

The 9th harmonic of the comb repetition frequency, f_{rep}, is locked to a signal from a synthesizer that is referenced to one of the NRC H-masers. See Fig. 2. This lock servo controls both the length of the oscillator through slow (travel$\sim$ 10 μm) and fast PZTs, and the temperature of the oscillator baseplate, for long term operation. The offset frequency is locked to a second synthesizer through feedback to the current of the oscillator pump diode laser. Although the S/N of the offset beat signal was typically 30 dB, a tracking oscillator and a division ratio of 64 were used for long term operation. A continuous locking period without dropouts of over 3 days has been observed.

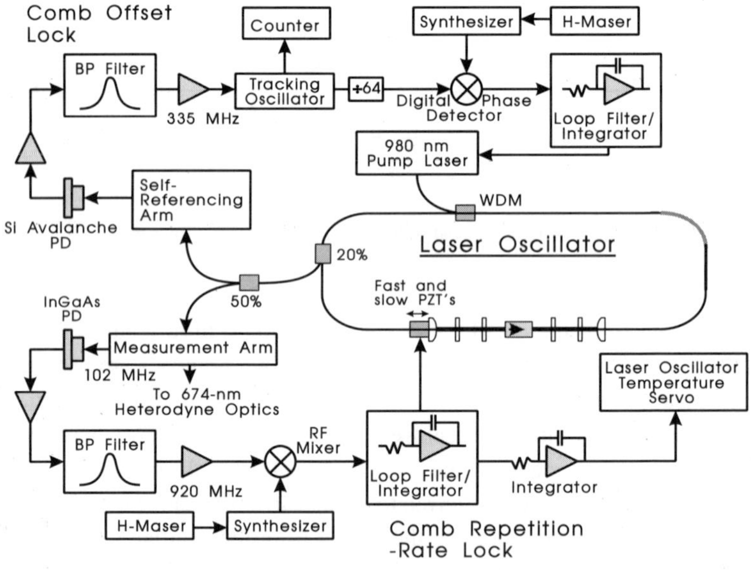

Fig. 2. Schematic of servos used to lock the comb repetition rate and offset frequency.

3. Optical Frequency Measurements at 674 nm

The outputs from the PPLN and the 674 nm laser are mixed on a Si avalanche photodiode to produce a heterodyne signal with a S/N of up to 30 dB (100 kHz BW). A tracking oscillator provides sufficient S/N for reliable counting. A measurement of the frequency of the free-running 674-nm laser for a period of over 30 hours with no drop-outs is shown in Fig. 3. The instabilities of the measured frequency for 1 s and 10 s gate times are approximately 1.5 times the maser instability. Tightening the repetition-rate lock resulted in maser-limited instabilities.

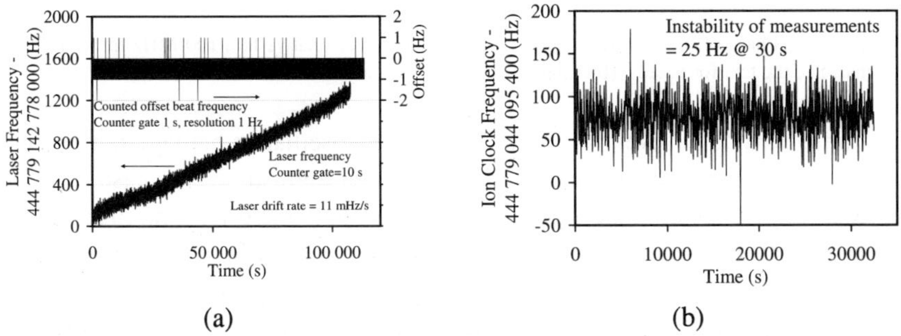

Fig. 3. (a) Measured frequency of the free-running probe laser system. The counted comb offset frequency is also shown. (b) Frequency of the clock transition of Sr^+ obtained by combining comb measurements of the probe laser frequency with simultaneous measurements of the offset frequency between the laser and the centre of the clock transition.

4. Stability of Transmitted RF Signals

The optical and RF frequency standards at NRC are in different parts of a building that only has modest environmental control. To study the stability of the RF link, a 10-MHz signal was sent through 155 m of coaxial cable (Heliax (35422-33/FSJ1, heat treated) from the maser to the reference input of a synthesizer (Agilent E4423B). The output from the synthesizer at 100 MHz was sent back to the maser laboratory through a similar cable. The results are shown in Fig. 4. Measurements with the cable alone show that the majority of the instability is due to the synthesizer.

5. Conclusions

A fibre-based optical frequency comb has been developed that permits measurements of the clock transition frequency of a single ion standard to be measured

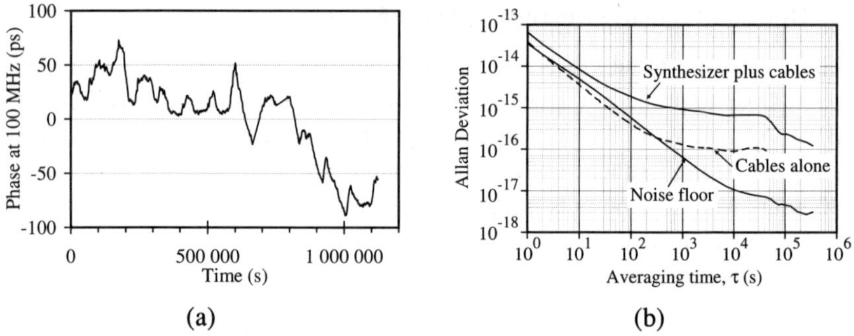

Fig. 4. (a) Phase of the 100-MHz output from a synthesizer relative to a 100-MHz maser reference signal. See the text for details. (b) Allan deviation of the data shown in (a).

for time periods of several days. While tracking oscillators help guarantee robust locking of the comb, they add to the complexity and can mask phase slips. Improvements to the S/N and signal stabilities for both the offset and measurement beat signals are possible that should eliminate the need for tracking oscillators. Better environmental control, particularly as it affects the repetition rate lock synthesizer, as well as co-location of the fountain and comb and active fibre noise cancellation, will be required to reach uncertainties of $< 1 \times 10^{-16}$ in the comb link.

6. Acknowledgements

The authors would like to thank R. Pelletier and B. Hoger for their technical assistance.

References

1. S. A. Diddams, T. Udem, J. C. Bergquist, E. A. Curtis, R. E. Drullinger, L. Hollberg, W. M. Itano, W. D. Lee, C. W. Oates, K. R. Vogel and D. J. Wineland, *Science* **293**, 825 (2001).
2. T. R. Schibli, K. Minoshima, F.-L. Hong, H. Inaba, A. Onae, H. Matsumoto, I. Hartl and M. E. Fermann, *Opt. Lett.* **29**, 2467 (2004).
3. N. R. Newbury and W. C. Swann, *J. Opt. Soc. Am. B* **24**, 1756 (2007).
4. A. A. Madej, J. E. Bernard, P. Dubé, L. Marmet and R. S. Windeler, *Phys. Rev. A* **70**, p. 012507 (2004).
5. J. Jiang, J. E. Bernard, A. A. Madej, A. Czajkowski, S. Drissler and D. J. Jones, *J. Opt. Soc. Am. B* **24**, 2727 (2007).
6. K. Tamura, E. P. Ippen, H. A. Haus and L. E. Nelson, *Opt. Lett.* **18**, 1080 (1993).
7. F. Adler, K. Moutzouris, A. Leitenstorfer, H. Schnatz, B. Lipphardt, G. Grosche and F. Tauser, *Opt. Express* **12**, 5872 (2004).

A PHASE-COHERENT LINK BETWEEN THE VISIBLE AND INFRARED SPECTRAL RANGES USING A COMBINATION OF CW OPO AND FEMTOSECOND LASER FREQUENCY COMB

E. V. KOVALCHUK and A. PETERS

Institut für Physik, Humboldt-Universität zu Berlin,
Hausvogteiplatz 5-7, 10117 Berlin, Germany
E-mail: evgeny.kovalchuk@physik.hu-berlin.de, achim.peters@physik.hu-berlin.de
http://www.physik.hu-berlin.de/qom

We report on the realization of a new technique for a phase-coherent link between the visible and infrared spectral ranges provided by a continuous-wave OPO in combination with a Ti:Sapphire femtosecond laser comb. We have developed a CH_4-based infrared molecular clock by phase locking the repetition rate frequency of a Ti:Sapphire femtosecond laser comb to the optical frequency of a He–Ne/CH_4 standard. We also performed a direct absolute frequency comparisons between an iodine stabilized laser at 532 nm and a methane stabilized laser at 3.39 μm.

1. Introduction

Femtosecond laser frequency combs have greatly facilitated precision optical frequency measurements and optical frequency synthesis.[1,2] Moreover, they have established favorable conditions for the implementation and direct frequency comparison of compact optical atomic and molecular clocks.

These developments have also opened up new perspectives for cw OPOs as multipurpose spectroscopic and metrological tools. Due to their relatively high output power, broadband tunability and narrow linewidth, OPOs have for a long time been considered for implementation as optical synthesizers.[3] Combining a cw OPO and a femtosecond laser frequency comb can increase stability and accuracy of such synthesizers by many orders of magnitude.

Cw OPOs developed in our laboratory for high-resolution Doppler-free molecular spectroscopy feature excellent spectral and tuning characteristics.[4] Together with the capability of phase locking to stable optical references this leads to a very high application potential of OPOs in the mid-infrared. Recently we have proposed and demonstrated how to provide a phase-coherent link between the visible and infrared spectral ranges using a pump enhanced singly resonant cw OPO.[5,6]

2. Method

Metrological applications of cw OPOs, combined with a femtosecond Ti:Sapphire laser comb (fs-laser comb), became possible by the following considerations: It is well known that femtosecond lasers produce combs of equidistant optical lines with frequencies $f_n = n\, f_{\rm rep} + f_0$, where $f_{\rm rep}$ is the repetition rate of the femtosecond laser (typical values for $f_{\rm rep}$ are 0.1–1 GHz) and $f_0 < f_{\rm rep}$ is the offset frequency due to pulse-to-pulse carrier-envelope phase shift, and $n \sim 10^5 - 10^6$ is an integer. It is less well known that under typical operating conditions cw OPOs also generate a comb-like spectrum of widely separated optical frequencies, as shown in Fig. 1.

Our OPO, pumped by a monolithic Nd:YAG laser (λ = 1064 nm) at the frequency P, emits not only the signal and idler waves with frequencies S and I respectively (P = S + I), but also a variety of linear combinations of these two frequencies in the visible spectral range also covered by the Ti:Sapphire femtosecond laser comb. These mixing products of the signal and pump waves, both resonated in the OPO cavity, are generally not phase-matched and therefore very week with power levels of the order of several microwatts or less. They are, however, strong enough to perform frequency beat note measurements with adjacent fs-laser comb lines.

The OPO output spectrum has two degrees of freedom, e. g., the pump and idler frequencies. Similarly, the fs-laser comb spectrum is determined by two microwave frequencies $f_{\rm rep}$ and f_0. If one of these frequencies (or a suitable combination of them) is well known or controlled, then the simultaneous measurements of three beat notes between OPO output lines and nearby comb lines provide the full information about all OPO and comb frequencies. Thus, the Ti:Sapphire femtosecond laser frequency comb can, for example, be referenced to an infrared optical frequency standard using the OPO and then used for absolute optical frequency measurements within its whole spectral range. Alternatively, the combination of cw OPO and fs-laser comb can also take advantage of the recent progress in the development of visible optical frequency standards based on atoms confined in optical lattices[7] or single trapped ion,[8] in order to synthesize highly stable infrared radiation everywhere within the extremely wide tuning ranges of the OPO signal (1.5–1.9 μm) and idler (2.4–3.7 μm) waves.

Providing such a phase-coherent link between widely separated opti-

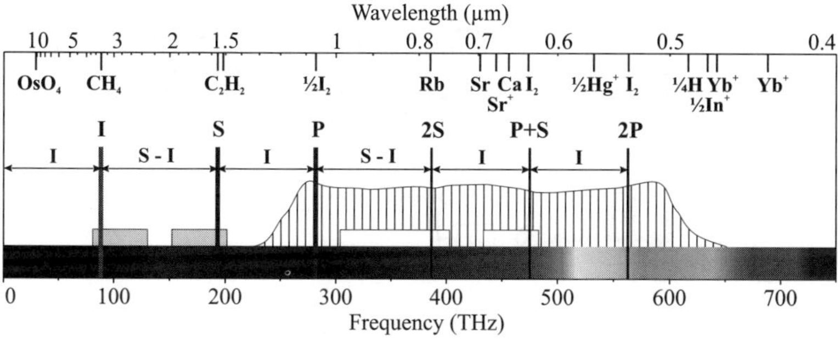

Fig. 1. Schematic showing the spectra of a Ti:Sapphire femtosecond laser frequency comb and OPO output lines alongside the molecular, ion and atomic reference lines employed in some modern optical frequency standards.

cal frequencies can simplify direct absolute comparisons between different molecular and atomic optical frequency standards and thus lead to new fundamental tests of physics.[8–14]

3. Experiment

The first step toward the realization of this concept was taken by a direct absolute optical frequency comparison between a methane stabilized He–Ne/CH_4 laser at 3.39 μm and an iodine stabilized frequency doubled Nd:YAG laser at 532 nm. This was reported in a previous paper,[6] which also gives further details on the experimental setup. Here we focus on the configuration representing an methane infrared optical clock, where the repetition rate frequency of a Ti:Sapphire femtosecond laser comb is phase locked to a He–Ne/CH_4 transportable optical frequency standard.[15]

Recently, several infrared molecular optical clockworks based on CO_2/OsO_4 or He–Ne/CH_4 systems have been demonstrated.[16–19] In these cases, the stability of infrared optical standards was transferred to the comb f_{rep} by making use of sum- or difference-frequency generation processes.

Our He–Ne/CH_4-based optical molecular clock has been realized using a cw OPO as a multi-frequency transfer oscillator from the mid-infrared to the visible optical range, as schematically presented in Fig. 2. First, we phase lock the OPO idler frequency to the He–Ne/CH_4 laser. Thereafter, we perform two heterodyne beat measurements between the 2S and P+S lines of the OPO output spectrum and the adjacent comb lines of a fs-laser comb. For this, the corresponding parts of the comb spectrum at 755 nm and 631 nm are split with a diffraction grating and optically filtered with

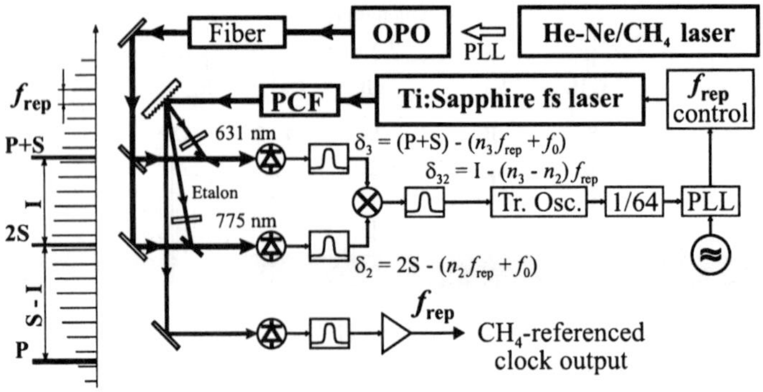

Fig. 2. Scheme of a He–Ne/CH_4-based infrared optical clock. PLL, phase lock loop; PCF, photonic crystal fiber; Tr. Osc., tracking oscillator.

Fabry-Perot etalons. Then, these two beams are overlapped with those of the 2S and P+S OPO spectral lines separated with a dichroic beam splitter and sent to avalanche photodiodes.

The generated beat notes at frequencies $\delta_2 = 2\mathrm{S} - (n_2 f_{\mathrm{rep}} + f_0)$ and $\delta_3 = (\mathrm{P} + \mathrm{S}) - (n_3 f_{\mathrm{rep}} + f_0)$ typically have signal-to-noise ratios at the order of (25 to 40) dB in a resolution bandwidth of 100 kHz. After appropriate amplification and filtering these signals are mixed in a double-balanced mixer in order to produce the frequency difference signal $\delta_{32} = \delta_3 - \delta_2 = \mathrm{I} - (n_3 - n_2) f_{\mathrm{rep}}$, fluctuations of which exactly mirror the fluctuations of f_{rep}. This difference signal is then strongly filtered using a bandpass filter and a tracking oscillator to further improve the signal-to-noise ratio. After prescaling it is fed to a phase lock loop (PLL) controlling f_{rep} by adjusting the length of the femtosecond Ti:Sapphire laser with a PZT.

Thus the repetition rate frequency of the femtosecond Ti:Sapphire laser comb becomes phase locked to the He–Ne/CH_4 optical frequency standard and can be directly expressed in terms of its frequency. It finally serves as the output of the CH_4-based infrared molecular clock. The stabilized microwave frequency f_{rep} possess the same stability as the methane standard and can be used as the reference for further frequency measurements. Future direct frequency comparisons with optical clocks of similar stability can reveal the actual performance of the implemented CH_4 clock.

4. Future Developments

Compact CH_4 clocks are really promising references for many precision applications, especially for navigation, atom interferometry and Very Long

Baseline Interferometry (see the contribution by S. Doeleman in these Proceedings). They exceed in short-term stability (up to 10–100 s) high performance microwave clocks (e. g., hydrogen masers) and exhibit extremely low phase noise.[18,19] A new generation of compact He–Ne/CH_4 standards based on a monolithic low-expansion Sitall glass cavity design should be available very soon[20] from the P. N. Lebedev Institute, Moscow, Russia.

To develop more reliable and compact CH_4 clock similar to that implemented by M. A. Gubin *et al.*,[19] we plan to combine a cw OPO and a femtosecond Er-fiber laser comb. In this case the comb covers the whole spectrum interval from the OPO signal to the pump frequencies. This will greatly simplify the clock scheme by phase locking $f_{\rm rep}$ directly to the frequency difference signal δ_{32} of the OPO pump and signal beat notes with the comb. Our method can also be easily modified to provide a phase-coherent link between the mid-infrared and telecom spectral ranges and lead to development of highly precise optical synthesizers.

References

1. D. J. Jones, S. A. Diddams, J. K. Ranka, A. Stentz, R. S. Windeler, J. L. Hall and S. T. Cundiff, *Science* **288**, 635 (2000).
2. T. Udem, R. Holzwarth and T. W. Hänsch, *Nature* **416**, 233 (2002).
3. K. Schneider, P. Kramper, S. Schiller and J. Mlynek, *Opt. Lett.* **22**, 1293 (1997).
4. E. V. Kovalchuk, D. Dekorsy, A. I. Lvovsky, C. Braxmaier, J. Mlynek, A. Peters and S. Schiller, *Opt. Lett.* **26**, 1430 (2001).
5. E. V. Kovalchuk, D. Dekorsy, A. I. Lvovsky, C. Braxmaier, J. Mlynek, S. Schiller and A. Peters, Doppler-free spectroscopy of CH_4 using a cw optical parametric oscillator, in *Proc. of the 6th Symposium on Frequency Standards and Metrology (St. Andrews, UK, 2001)*, ed. P. Gill (World Scientific, Singapore, 2002).
6. E. Kovalchuk, T. Schuldt and A. Peters, *Opt. Lett.* **30**, 3141 (2005).
7. A. D. Ludlow, T. Zelevinsky, G. K. Campbell, S. Blatt, M. M. Boyd, M. H. G. de Miranda, M. J. Martin, J. W. Thomsen, S. M. Foreman, J. Ye, T. M. Fortier, J. E. Stalnaker, S. A. Diddams, Y. Le Coq, Z. W. Barber, N. Poli, N. D. Lemke, K. M. Beck and C. W. Oates, *Science* **319**, 1805 (2008).
8. T. Rosenband, D. B. Hume, P. O. Schmidt, C. W. Chou, A. Brusch, L. Lorini, W. H. Oskay, R. E. Drullinger, T. M. Fortier, J. E. Stalnaker, S. A. Diddams, W. C. Swann, N. R. Newbury, W. M. Itano, D. J. Wineland and J. C. Bergquist, *Science* **319**, 1808 (2008).
9. J.-P. Uzan, *Rev. Mod. Phys.* **75**, 403 (2003).
10. S. Bize, S. A. Diddams, U. Tanaka, C. E. Tanner, W. H. Oskay, R. E. Drullinger, T. E. Parker, T. P. Heavner, S. R. Jefferts, L. Hollberg, W. M. Itano and J. C. Bergquist, *Phys. Rev. Lett.* **90**, p. 150802 (2003).

11. E. Peik, B. Lipphardt, H. Schnatz, T. Schneider, C. Tamm and S. G. Karshenboim, *Phys. Rev. Lett.* **93**, p. 170801 (2004).
12. E. R. Hudson, H. J. Lewandowski, B. C. Sawyer and J. Ye, *Phys. Rev. Lett.* **96**, p. 143004 (2006).
13. S. Herrmann, A. Senger, E. Kovalchuk, H. Müller and A. Peters, *Lect. Notes Phys.* **702**, 385 (2006).
14. H. Müller, P. L. Stanwix, M. E. Tobar, E. Ivanov, P. Wolf, S. Herrmann, A. Senger, E. Kovalchuk and A. Peters, *Phys. Rev. Lett.* **99**, p. 050401 (2007).
15. M. Gubin, A. Shelkovnikov, E. Kovalchuk, D. Krylova, E. Petrukhin and D. Tyurikov, Present performance of the transportable double-mode He-Ne/CH_4 frequency standards, in *Proc. 1999 Joint Meeting of The European Frequency and Time Forum and the IEEE International Frequency Control Symposium (Besançon, France, 1999)*, (IEEE, New York, 1999).
16. A. Amy-Klein, A. Goncharov, C. Daussy, C. Grain, O. Lopez, G. Santarelli and C. Chardonnet, *Appl. Phys. B* **78**, 25 (2004).
17. O. D. Mücke, O. Kuzucu, F. N. C. Wong, E. P. Ippen, F. X. Kärtner, S. M. Foreman, D. J. Jones, L.-S. Ma, J. L. Hall and J. Ye, *Opt. Lett.* **29**, 2806 (2004).
18. S. M. Foreman, A. Marian, J. Ye, E. A. Petrukhin, M. A. Gubin, O. D. Mücke, F. N. C. Wong, E. P. Ippen and F. X. Kärtner, *Opt. Lett.* **30**, 570 (2005).
19. M. A. Gubin, A. N. Kireev, A. V. Konyashchenko, P. G. Kryukov, A. V. Tausenev, D. A. Tyurikov and A. S. Shelkovnikov, *Quantum Electron.* **38**, 613 (2008).
20. M. A. Gubin, priv. comm. (2008).

IMPROVEMENTS TO THE ROBUSTNESS OF A TI:SAPPHIRE-BASED FEMTOSECOND COMB AT NPL

V. TSATOURIAN[1,2], H. S. MARGOLIS[1], S. N. LEA[1], B. R. WALTON[1], G. MARRA[1], D. T. REID[2] AND P. GILL[1]

[1]*National Physical Laboratory, Hampton Road, Teddington, Middlesex TW11 0LW, UK*
[2]*School of Engineering and Physical Sciences, Heriot-Watt University, Riccarton, Edinburgh EH14 4AS, UK*

This paper reports a number of improvements to a Ti:sapphire-based frequency comb. Changes to the spectral broadening set up, *f*:2*f* self-referencing arrangement and servo system are described, including a novel scheme for group-delay dispersion compensation using Wollaston prisms. In combination, these changes improved the signal-to-noise ratio of the carrier-envelope offset beat by 15 dB and increased its frequency stability by more than four orders of magnitude, as well as enabling it to be locked continuously for many hours without optical adjustment.

1. Introduction

Frequency combs provide a convenient and accurate way of phase-coherently linking optical and microwave frequencies, making them a key component of future optical clocks. The first comb constructed at NPL, based on a Kerr-mode-locked Ti:sapphire femtosecond laser with a repetition rate of approximately 90 MHz, has been used for many frequency measurements of optical frequency standards over the last seven years [1 – 4]. However, continuous measurement runs were limited in duration to around one hour by the need to make optical adjustments to maintain signal-to-noise in the beats. For continuous optical clock operation over many hours a robust comb arrangement is essential. To achieve this, several improvements to the comb design have been made.

2. Spectral broadening

The octave span required for detection of the carrier envelope offset frequency f_0 is obtained by spectral broadening in microstructure fibre. In the original system the most frequently required adjustment was the alignment of the laser light into the fibre, which has a core diameter of approximately 1.8 μm. The bare cleaved fibre has therefore been replaced by connectorized fibre made by Crystal Fibre A/S, consisting of 26 cm of NL-PM-750 fibre with collapsed holes at the output end and about 100 μm of single mode fibre (SMF) spliced to the input end. The 6 μm diameter core of the SMF offers improved tolerance to misalignment, enabling f_0 to be counted reliably for many hours without optical adjustment. It is also possible to clean both ends of the new fibre, eliminating the need for it to be periodically re-cleaved.

A further drawback of the original setup was the difficulty of obtaining good signal-to-noise ratio of both f_0 and the beat between the optical standard being measured and the nearest comb mode (f_b). This is because the octave-spanning comb generated in a single microstructure fibre must be optimised in three spectral regions simultaneously. To ease this constraint the light from the Ti:Sapphire laser is now split between two branches, each of which contains a microstructure fibre for spectral broadening. The first branch is used to create the octave spanning spectrum for f:2f self-referencing, while the second is used for the beat with the optical standard. This enables the signal-to-noise ratios of the two beat signals f_0 and f_b to be optimised independently.

3. *f*:2*f* self-referencing

In the original setup the signal-to-noise ratio of the f_0 beat was limited by the conversion efficiency of the critically phase-matched $KTiOPO_4$ (KTP) crystal used for frequency doubling. An alternative is periodically poled KTP (PPKTP), which has an effective nonlinear coefficient that is three times higher ($d_{eff} = 9.49$ pm/V), increasing the intensity of the frequency doubled light. A further advantage is the zero walk-off of PPKTP, which ensures that the beams at the doubled and fundamental frequencies are perfectly collinear after the PPKTP crystal [5]. The PPKTP crystal used is 5 mm long with 1×2 mm^2 cross-section and is phase-matched at 1064 nm (poling period 9 μm).

In order to observe the f_0 beat, the frequency-doubled and fundamental pulses at 532 nm must overlap in time at the detector. It is therefore essential to compensate for the group-delay difference between the visible and infrared spectral components of the pulse at the output of the microstructure fibre. Two different dispersion compensation schemes have been investigated, one based on a Michelson interferometer and the other using a pair of Wollaston prisms.

The optical design of the f:2f self-referencing scheme employing a Michelson interferometer for group-delay dispersion compensation is shown in figure 1. In this arrangement the infrared comb lines are separated from the visible comb lines by a short-wavelength pass dichroic beam splitter with an edge wavelength of approximately 950 nm. The difference in path length between the two arms of the interferometer is adjusted to achieve temporal overlap of the frequency-doubled and fundamental 532 nm pulses at the detector. The end mirrors of the interferometer are slightly tilted to enable the light passing back through the dichroic beam splitter to be picked off and directed into the PPKTP crystal for frequency doubling. The intensity of the frequency-doubled light is maximised using a 1064 nm half-wave plate. The f_0 beat frequency is detected using an avalanche photodiode after spectral filtering by an interference filter. A rotatable polarizer in front of the detector is used to optimize the beat signal.

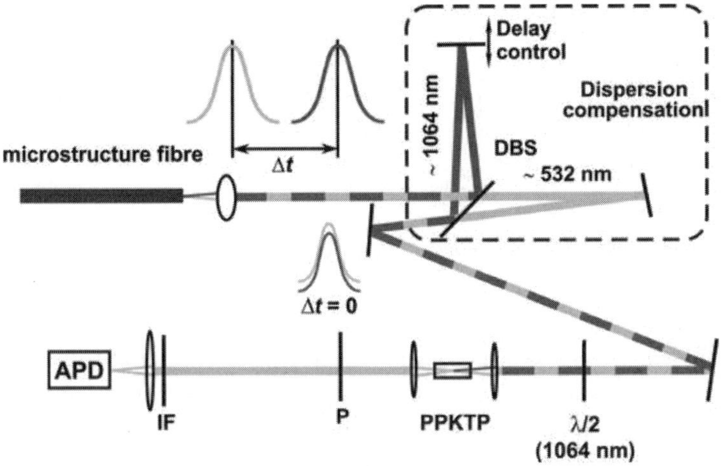

Fig. 1. Optical arrangement for *f*:2*f* self-referencing using a Michelson interferometer for dispersion compensation. DBS – dichroic beam splitter; λ/2 – half wave plate at 1064 nm; P – polarizer; IF – interference filter (λ= 532 nm; FWHM=3 nm); APD – avalanche photodiode.

An alternative *f*:2*f* self-referencing scheme in which Wollaston prisms are used to compensate for the group-delay difference between the visible and infrared spectral components of the pulse at the crystal is shown in figure 2. A Wollaston prism consists of two calcite prisms, cemented together to form two right angle prisms with perpendicular optic axes. It separates randomly polarized light into two orthogonally linearly polarised outgoing beams, which diverge from the prism with the angle of divergence determined by the wedge angle of the prism and the wavelength of the light. The relative delay between the two orthogonally polarised beams depends on the position of the separation point relative to the prism centre. Therefore, by adjusting the position of the separation point, the orthogonally polarised fundamental and frequency doubled 532 nm pulses separated by the first prism (WP1) and then recombined by the second prism (WP2) can be temporally overlapped at the detector. The PPKTP phase-matching geometry is such that the fundamental and frequency doubled light have the same polarization. Orthogonal polarisation of the 1064 nm and 532 nm pulses is obtained using a combination of a polarizer and 1064 nm half-wave plate at the input of the PPKTP crystal, with the group delay being compensated after the crystal.

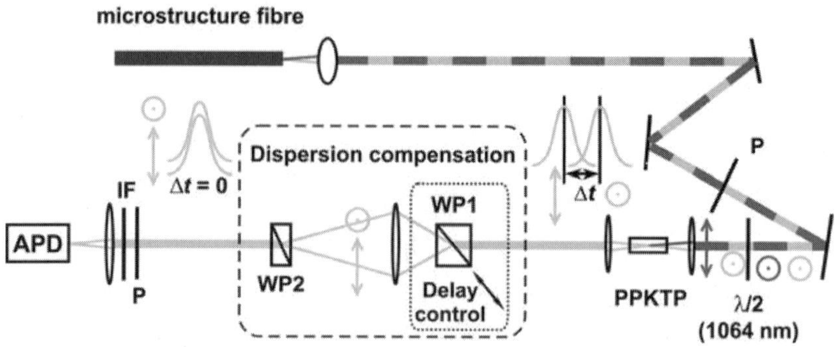

Fig. 2. Optical arrangement for f:2f self-referencing using Wollaston prisms (WP1, WP2) for dispersion compensation. P – polarizer; IF – interference filter (λ= 532 nm; FWHM=3 nm); APD – avalanche photodiode.

These new optical arrangements employing connectorized microstructure fibre with SMF spliced to the input end and the PPKTP crystal for frequency doubling improved the signal-to-noise ratio of the f_0 beat from typically 25 dB to 40 dB (in 250 kHz bandwidth) for both dispersion compensation schemes. However, the use of Wollaston prisms for dispersion compensation has the advantage over the more commonly employed Michelson or Mach-Zehnder interferometers in that a collinear beam path can be maintained. This means that the fundamental and second-harmonic pulses travel through the same optical components providing common-mode rejection of acoustic noise.

4. Offset frequency stabilisation

In the original setup the f_0 beat was stabilized by adjusting the tilt of the cavity end mirror in the Ti:Sapphire laser using a split tubular piezoelectric transducer (PZT). This has the advantage of being able to compensate for large changes in f_0. However, because of the low bandwidth of the actuator (a few kHz) the servo system is unable to remove high frequency noise. An improved servo system has therefore been implemented in which an acousto-optic modulator (AOM) is inserted into the pump laser beam to control its power [6]. The resulting modification to the non-linear refractive index of the crystal provides a means for fast control of f_0. The combination of the AOM and the PZT therefore enables both a tight lock and wide frequency range to be achieved. Figure 3 shows the resulting improvement of more than four orders of magnitude in the stability of the f_0 beat.

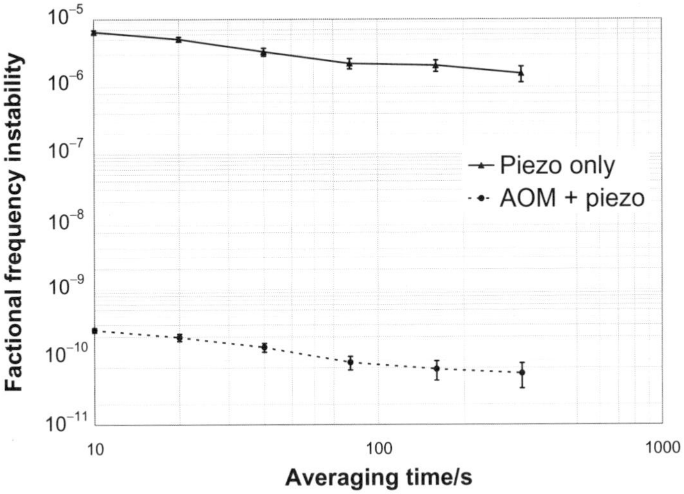

Fig. 3. Fractional frequency instability of the stabilized f_0 beat at 30 MHz for the original servo system using only a piezo actuator and for the new servo system, which also uses an AOM.

5. Conclusions

A number of improvements to the robustness of a Ti:Sapphire-based femtosecond comb have been demonstrated. Separate spectral broadening for f_0 and f_b provides independent control over the signal-to-noise ratios of these beats, while microstructure fibre with SMF spliced to the input end gives greatly improved tolerance to misalignment. The use of PPKTP for f:$2f$ self-referencing has yielded a 15 dB enhancement of the signal-to-noise ratio in the f_0 beat, and the addition of an AOM for fast frequency control of f_0 has improved the stability of the locked beat signal by more than four orders of magnitude. In combination, these improvements have increased by a factor of five the time for which the comb may be continuously locked without adjustments. Finally, a novel scheme for group-delay dispersion compensation using Wollaston prisms has been demonstrated.

References

1. S. N. Lea *et al.*, *Metrologia* **40**, 84 (2003).
2. H. S. Margolis *et al.*, *Science* **306**, 1355 (2004).
3. C. S. Edwards *et al.*, *Opt. Lett.* **29**, 566 (2004).
4. K. Hosaka *et al.*, *IEEE Trans. Instrum. Meas.* **54**, 759 (2005).
5. Y. Jiang *et al.*, *Metrologia* **42**, 304 (2005).
6. F. W. Helbing *et al.*, *Appl. Phys. B* **74** [Suppl.], S35 (2002).

Part VII

Atomic Microwave Standards

NIST F1 AND F2

T.P. HEAVNER, T.E. PARKER, J.H. SHIRLEY AND S.R. JEFFERTS
NIST – Time and Frequency Division 325 Broadway Boulder, CO 80305

The National Institute of Standards and Technology operates a cesium fountain primary frequency standard, NIST-F1, which has been contributing to International Atomic Time (TAI) since 1999. At the time of the last Symposium on Frequency Standards and Metrology in 2001, the uncertainty of the standard was $\delta f/f_0 \approx 1\times10^{-15}$, which was at that time the state-of-the art. During the intervening 7 years we have improved NIST-F1 so that the uncertainty is currently $\delta f/f_0 \approx 3\times10^{-16}$, dominated by uncertainty in the Blackbody radiation induced frequency shift. In order to circumvent the uncertainty associated with the blackbody shift we have built a new fountain, NIST-F2, in which the microwave interrogation region is cryogenic (80K) reducing the blackbody shift to negligible levels. We briefly describe here the series of improvements to NIST-F1 which have allowed its uncertainty to reach the low 10^{-16} level and present the first results from NIST-F2.

1. NIST-F1

Table 1 shows the error budget of NIST-F1 as of the summer of 2001. The type B frequency uncertainty of 1×10^{-15} at that time was the smallest achieved by fountain standards. Table 1 also shows the error budget of NIST-F1 as of July 2008. The type B frequency uncertainty of 3.2×10^{-16} defines the 2008 state of the art for frequency uncertainty in fountain frequency standards.

1.1. *Spin Exchange Frequency Bias*

It is apparent, from Table 1, that the frequency uncertainty in 2001 was dominated by the spin exchange shift from collisions between cold cesium atoms. In fact this shift was predicted to likely be the most "troublesome systematic effect of an atomic fountain" [1]. Since that time several new techniques have been brought to bear on the problem of estimating the spin-exchange shift in fountain frequency standards [2, 3]. The spin-exchange shift is no longer a dominant problem in the best fountain frequency standards in use today. In Table 1, we show the spin-exchange uncertainty as of 2008 reduced to $\delta f/f_0 = 7\times10^{-17}$, much smaller than the frequency uncertainty associated with the blackbody radiation shift and microwave effects. This trend is echoed in other cesium frequency standards in various laboratories.

In NIST-F1 we use a traditional extrapolation of the density to evaluate the spin exchange shift, along with a large optical molasses in order to make the density of the sample much smaller than that obtained with the use of a MOT. In addition we achieve temperatures of about 450 nK in the launched molasses. These low temperatures mean that approximately 80% of the atoms entering the Ramsey cavity for the initial microwave interaction eventually contribute to the signal. This allows significant reductions in the initial density (and hence spin-exchange shift) compared to returning atom fractions of 20% that are more typical with 1.5μK atoms. As a result we can achieve reasonable short term stability, $\sigma_y(\tau) \approx 2\times10^{-13}/\tau^{1/2}$, while keeping the uncertainty in the spin-exchange frequency shift well below $\delta f/f_0 = 10^{-16}$.

Table 1

The Type B Uncertainties of NIST-F1 in 2001 and 2008

Units are femtos of fractional frequency

Physical Effect	Magnitude	Uncertanity 2001	Magnitude	Uncertainty 2008
Second Order Zeeman	44.76	0.3	180.60	0.013
Spin Exchange	0.0	0.84	-0.41	0.07
Blackbody	-20.6	0.3	-22.98	0.28
Gravitation	180.54	0.1	179.95	0.03
Cavity Pulling	<0.1	<0.1	0.02	0.02
Rabi/Ramsey Pulling	<0.1	<0.1	10^{-4}	10^{-4}
Microwave amplitude	0	0.2	0.026	0.12
Cavity Phase	<0.1	<0.1	0.02	0.02
Light Shift	<0.2	0.2	10^{-5}	10^{-5}
Adjacent Transition	<0.1	<0.1	0.02	0.02
Microwave Spectrum	0	<0.1	0.003	0.003
Integrator Offset	0	<0.1	0	0.01
AM on microwaves	0	<0.1	0	10^{-4}
AC Zeeman (heaters)			0.05	0.05
Total Uncertainty		0.99		0.32

1.2. *Blackbody Frequency Bias*

As pointed out by Itano [4], the hyperfine splitting of the cesium atom is shifted by the ambient blackbody radiation field. This shift has recently been the subject of some controversy with several measurements being in disagreement [5, 6, 7]. The experimental measurements apparently stimulated a great deal of theoretical interest culminating in very high quality calculations of the blackbody frequency bias [8, 9]. At this point the situation seems to be that the shift is well characterized by

$$\delta\nu/\nu_0 = \beta(T/300K)^4\left(1+\varepsilon(T/300K)^2\right),\ \beta = -(1.710\pm0.006)\times10^{-14},\ \varepsilon{=}0.014,$$

where T is the blackbody temperature of the environment that the atoms in the fountain experience. NIST-F1 operates slightly above room temperature at about 46C and we estimate that the temperature uncertainty of the radiation field is, at most, 1K, leading to an uncertainty in the frequency bias from this source of $\delta f/f_0 = 2.8\times10^{-16}$. As shown in Table 1 this bias currently dominates the frequency uncertainty of NIST-F1. We note that the calculations referred to above depend on a measured value for the D.C. stark shift and that there have been no direct measurements of the blackbody bias with uncertainties close to the $\delta f/f_0 = 10^{-16}$ level.

1.3. *Microwave Induced Frequency Biases*

Ramsey interrogation using a TE_{011} cylindrical microwave cavity is universally used in cesium fountain primary frequency standards reporting to TAI. While this interrogation method is robust, it is still quite easy to introduce frequency biases as a result of various microwave effects. These include spurs in the microwave spectrum, microwave radiation leaking outside the microwave cavity and, position-dependent phase shifts within the microwave cavity itself. A glance at the current (in 2008) error budget in Table 1 shows that these microwave effects account for the second largest part of the uncertainty in the fountain. These effects are quite different from those in traditional thermal beam standards. Our group, as well as others, has investigated these effects in fountain style frequency standards extensively over the past several years [10, 11, 12, 13, 14]. We briefly review some of the conclusions here.

1.3.1. *Distributed Cavity Phase*

The phase of the microwave field within the Ramsey cavity of a fountain was first investigated by De Marchi et al. He developed a first order model of the phase field and explicitly linked the phase variations to losses and power flows within the cavity [15, 16]. Later many groups expanded on these results with various full 3 dimensional calculations of the phase gradients within the cavities and used these phase gradients to estimate the frequency bias. They assumed the microwave phase shift within the cavity caused a frequency bias given approximately by $\delta\nu/\nu \approx \delta\phi/\phi_{tot} = \delta\phi/2\pi\nu_0 T_R$, where $\delta\phi$ is the phase shift in the microwave field, $\nu_0 = 9.1926$ GHz is the frequency of the hyperfine splitting and T_R is the Ramsey time. This is however incorrect. What matters is not the phase variations of the microwave field within the cavity, but the effect on the cesium atom coherent superposition. As we first showed in [13] and was later reconfirmed in [17] the frequency bias shows a large dependence on microwave amplitude as shown in Fig 1 below. This dependence on amplitude allows one to search for frequency biases associated with distributed cavity phase in a relatively unambiguous way (subject to various assumptions discussed in [10, 12]).

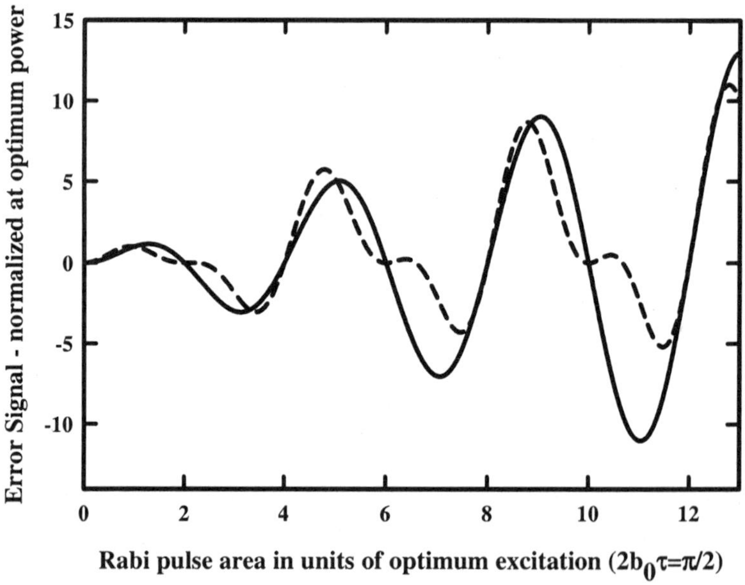

Figure 1 –Error signal caused by distributed cavity phase shift (dotted line) and that caused by microwave leakage after the second Ramsey interaction (solid line). These can be difficult to separate.

1.3.2. *Microwave Leakage*

Microwave fields interacting with the cesium atoms outside of the Ramsey interaction zone are a major source of frequency uncertainty in NIST-F1. These interactions can happen in two distinct places: first atoms can be subjected to a microwave interaction in the drift region above the Ramsey cavity and, second, the phase of the atomic superposition can be altered as well by interactions below the Ramsey cavity in the space between the Ramsey cavity and the detection zone. As detailed in [12] interactions above the Ramsey cavity in NIST-F1 are doubly forbidden by the physical structure of the drift region. The 2.5cm diameter drift tube is below cutoff for all microwave modes at 9.2 GHz except the dominant TE_{11} mode. The TE_{11} mode does not cause a frequency shift in first order because the azimuthal dependence of the mode averages to zero for a well centered atomic sample. Also, the drift tube is terminated on both ends and the length is chosen so that the resulting cavity is anti-resonant at 9.2 GHz.

It has been pointed out in [14] that if the two Ramsey pulses are, on average, different, then second order effects can be expected as well. This imbalance can be severe when a MOT is used in a fountain with a traditional "square" (diameter=height) microwave cavity. As a result of the tight MOT confinement and large thermal velocity of the sample, along with the ~20% variation of the microwave field amplitude over the aperture the atomic sample "sees" almost the maximum field in the microwave cavity on the way up with as much as a 10% average reduction on the way down. However, with the flattened cavity in NIST-F1, variation in the microwave field over the aperture is reduced by a factor of 2. Because the molasses is both cold ($V_{thermal}$~ 0.5 cm/s) and large (radius ~ 0.5 cm) , the two Ramsey excitations differ by less than 1% on average.

These considerations also apply to the use of the cancellation of the spin exchange shift as described in [3]. This cancellation is effected by carefully adjusting the amplitude of the first Ramsey interaction so that the frequency shift of the $|3,0\rangle$ and $|4,0\rangle$ components of the wavefunction (which have different signs at sufficiently low interaction energies) cancel. The very low interaction energies required necessitate the use of a MOT with tight confinement in order to introduce position-velocity correlations in the cloud very quickly. The tightly confined atomic sample has a large intrinsic spin-exchange shift. If the cancellation is not perfect a leakage field complicates the whole picture for two reasons. First, the two passages of the Ramsey cavity have unequal excitation for the reasons detailed above. Second, the leakage field changes the relative amplitude of the $|3,0\rangle$ and $|4,0\rangle$ components thereby affecting the cancellation of the spin-exchange frequency shift. In this case the frequency shift caused by a leakage field can be strongly "leveraged" by the spin-exchange shift with the result that the two effects are difficult to disentangle.

Microwave leakage after the second Ramsey interaction causes a frequency shift which maximizes in a fountain operated at optimum power. The signature of the effect is quite similar to that from distributed cavity phase (as shown in Fig 1). In NIST-F1 we combine these two effects in the error budget when we search for evidence of microwave effects by operating above optimum power. These combined effects (distributed cavity phase and leakage after the Ramsey interaction) dominate the uncertainty in the microwave amplitude shift at $\delta\nu/\nu_0 = 1.2\times10^{-16}$.

1.3.3. *Microwave Spectrum*

As a result of the pulsed operation of fountain standards, along with the possibility of operating well above optimum microwave excitation, spurs reveal rich and complicated features affecting frequency accuracy. We refer here to both incoherent and coherent spurs. Incoherent spurs are those spurs introduced onto the microwave spectrum by, for example, the 60 Hz line frequency: this type of spur generally has random and evolving phase with respect to the fountain cycle time. Coherent spurs are those introduced onto the microwave spectrum by the pulsed operation of the fountain itself. An example is a spur on the microwave spectrum caused by turning off the MOT coils just before launching the atom cloud. Careful study of the microwave spectrum using a spectrum analyzer can provide sufficient knowledge of incoherent spur amplitudes to hopefully eliminate spurs large enough to cause significant frequency errors. As discussed in [11] the magnitude of the frequency shift is difficult to estimate without detailed knowledge of the spur behavior at elevated microwave power. A full discussion of these effects is included in [11, 13] and the references contained therein.

Coherent spurs can cause frequency shifts far in excess of those predicted by the classical theory of spurs in [18]. We have developed in [13] a complete theory which agrees well with the experimental results presented there.

2. NIST-F2

We are developing a new fountain standard, named (imaginatively!) NIST-F2. NIST-F2 is designed to incorporate several unique features: cryogenic operation of the Ramsey interrogation region, Low-Velocity Intense-Source (LVIS) loading of cold atomic samples and multi-pulse operation. In its initial phase, NIST-F2 operates cryogenically. The cold atom loading and the multiple ball operation are not yet implemented.

NIST-F2 is shown schematically in Fig 2. The source is a pure optical molasses operated in a (1,1,1) geometry. Directly above the source region is a state selection cavity which is required for multiple ball operation. The detection region, also at room temperature, is between the source region and the cryogenic, magnetically-shielded Ramsey interrogation region. The magnetically shielded interrogation region is enclosed in a liquid nitrogen dewar

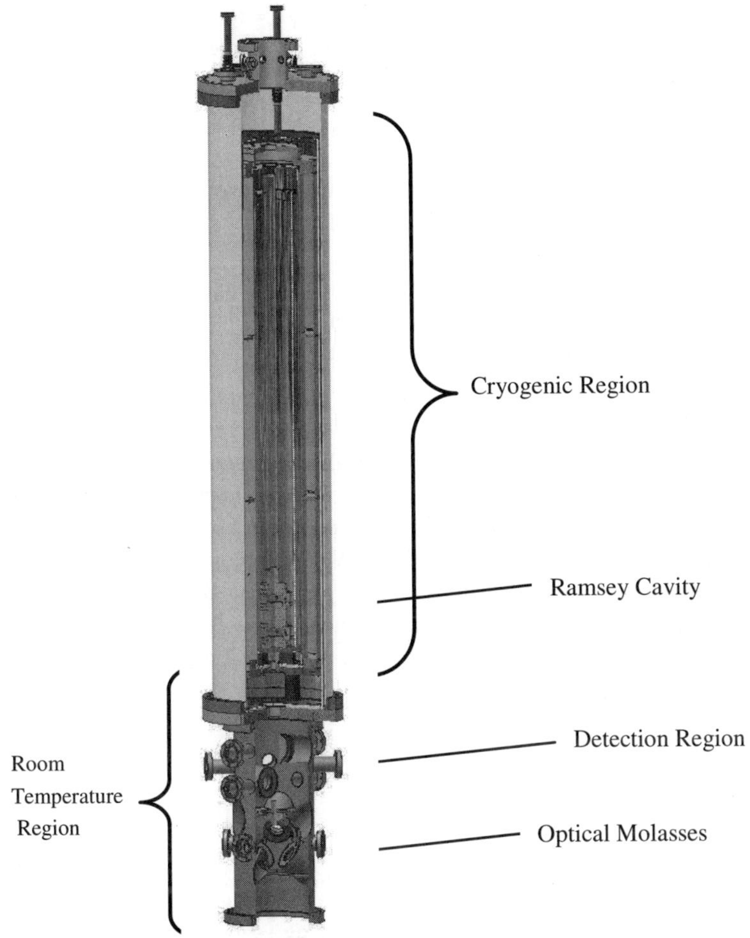

Figure 2 – This is a cutaway drawing of NIST-F2 showing the 1,1,1 molasses region, the detection region and the cryogenic microwave interrogation region. The overall height is about 2.5m.

and operates at about 80K. At these temperatures the blackbody shift, which is large in NIST-F1, is reduced in magnitude by a factor of about 250. The microwave interrogation and flight tube region is essentially identical to that of NIST-F1 except that the microwave cavities are tuned to be resonant at the cryogenic operating temperature and not at room temperature.

At the time of the symposium, NIST-F2 had operated for only a few hours. The frequency of the fountain, after correction for the Zeeman shift and the

Blackbody shift, was accurate at the $\delta\nu/\nu = 10^{-14}$ level supported by the statistical uncertainty of the limited data set. We expect to improve this over the next 12 months or so.

3. Conclusions

NIST-F1 is a mature standard and is unlikely to evolve much further. Its total systematic uncertainty of $\delta\nu/\nu = 3.2\times10^{-16}$ is strongly limited by the $\delta\nu/\nu = 2.8\times10^{-16}$ blackbody uncertainty.

NIST-F2 a cryogenic fountain has recently begun initial operation. This fountain is eventually expected to significantly improve on the current best results (as typified by NIST-F1) with a total uncertainty below $\delta\nu/\nu = 10^{-16}$

4. Acknowledgments

We have been fortunate to have many skillful collaborators over the period since the last symposium in 2001. We gratefully acknowledge long and fruitful collaborations with Elizabeth Donley and Filippo Levi who are responsible for much of the design of NIST-F2. We have enjoyed and benefited from collaborations with Bill Klipstein, John Dick, Eric Burt, Neil Ashby, Stefania Romisch, Dai-Hyuk Yu, Davide Calonico, Sasha Radnaev, Yaric Dudin and Paul Kunz. Don Sullivan, David Smith and Mike Lombardi provided valuable feedback on the manuscript.

References

1. K.Gibble and S. Chu, *Metrologia* **29** 201-212 (1992).
2. F. Pereira Dos Santos *et al*, *Phys Rev Lett.* **89** 233004 (2002).
3. K. Szymaniec *et al*, *Phys Rev Lett.* 98 1530023 (2007).
4. W. M. Itano, L.L. Lewis and D.J. Wineland, *Phys Rev Lett.* **25** 1233 (1982).
5. E. Simon, P. Laurent and A. Clairon, *Phys Rev A.* **57** 436-441 (1998).
6. F. Levi *et al,* Phys Rev A, **70** 033412 (2004).
7. A. Godone *et al,* Phys Rev A **71** 063401 (2005).
8. E.J. Angstmann, V.A. Dzuba and V.V. Flambaum, *Phys Rev Lett* **97** (2006).
9. K. Beloy, U.I. Safonova and A. Derevianko, *Phys Rev Lett* **97** 040801 (2006).
10. S.R. Jefferts, *et al*, *IEEE Trans. Ultrason.* **52** 2314-2321 (2005).
11. F. Levi *et al*, *IEEE Trans. Ultrason.* **53** 1584-1589 (2005).
12. J. H. Shirley *et al*, *IEEE Trans. Ultrason.* **53** 2376-2385 (2005).

13 J.H. Shirley *et al*, in press *IEEE Trans. Ultrason.*
14. S. Weyers, *et al Proc 20th Eur. Freq. Time Forum* 173-177(2006).
15. G Vecci and A. DeMarchi, *IEEE Trans. Instrum. Meas* **42** 434-438 (1993).
16. A Khursheed, G.Vecchi and A. DeMarchi, *IEEE Trans. Ultrason.* **43** 201-210
17 R. Li and K. Gibble *Proc IEEE Int. Freq. Cntrl. Symp* 99-104 (2005).
18. C. Audoin *et al*, *IEEE Trans Instrum Meas* **IM27** 325-329 (1978).

ATOMIC FOUNTAINS FOR THE USNO MASTER CLOCK

C. EKSTROM, S. PEIL, T. SWANSON, S. CRANE
Time Service Department, U.S. Naval Observatory
3450 Massachusetts Ave. NW
Washington D. C. 20392, USA

The U.S. Naval Observatory (USNO) has a program to field operational atomic fountain clocks for augmentation of the USNO timescale and master clock. We present the design and characterization of our engineering prototype and the initial results for the first two operational fountain clocks. In addition, we briefly discuss the new Master Clock Facility that houses the operational devices.

1. Overview

The U.S. Naval Observatory provides the timing reference for all operational systems within the U.S. Department of Defense. This common time reference, UTC(USNO) is currently generated with a large ensemble of commercial hydrogen masers and cesium beam clocks. We have developed rubidium fountain clocks to upgrade this ensemble.

Our program to develop rubidium fountains proceeded with a prototype NRF1 (Navy Rubidium Fountain 1). This device was used to test compact designs for the optics, electronics, and control systems, as well as the physics package. This testbed was also used to evaluate potential instabilities that would limit the long-term frequency stability of clocks based on this design.

After studies on NRF1, we proceeded with the construction of the first two operational fountains, NRF2 and NRF3. These devices were built and tested in our lab, then moved to our new operational clock building. This building will house the fountains at USNO along with roughly a third of the current USNO timing ensemble.

2. Rubidium Fountain Prototype – NRF1

Details of the construction [1] and initial systematic characterizations [2] of the engineering prototype have been described previously. The device was built to demonstrate the ability to house all the components in two full-sized equipment racks and one 85x85x210 cm^3 physics package. We will briefly review the layout, as it is largely in common with the operational devices.

The physics package has a long, narrow vacuum chamber built from titanium and copper. The lower regions for atom collection, state selection, and fluorescence detection are built out of titanium with welded-in sapphire optical viewports[3]. The upper region with the microwave spectroscopy cavity and the free precession drift region are built from copper. The system is pumped with a single 20 l/s ion pump at the top of the system, outside the magnetic shields.

Magnetic shielding is provided by three layers of shields that enclose the entire vacuum chamber with the exception of the pumping systems. A fourth inner shield encloses only the microwave spectroscopy cavity and free precession regions.

The optics systems that deal with optical frequency tuning and intensity control are housed in one of the equipment racks[4]. The optical systems are coupled to the physics package with polarization-maintaining single mode optical fibers that attach to monolithic beam shaping assemblies mounted to the vacuum chamber.

NRF1 produces a steered 5MHz output that is fed into our operational measurement systems, allowing us to measure its performance against any individual clock, or any calculated or physical timescale at USNO. The most stable of these timescales for internal comparison is often the maser mean. The short- and medium-term stability of NRF1 is typically characterized by white frequency noise, with an Allan deviation of 1.5×10^{-13} at one second.

Figure 1 shows the performance of NRF1 when measured against the maser mean at USNO. The data shows no behavior deviating from white frequency noise and shows an Allan deviation of 5×10^{-16} at an integration time of 1 day. This device has run, producing a continuous rf output, for times exceeding 30 days.

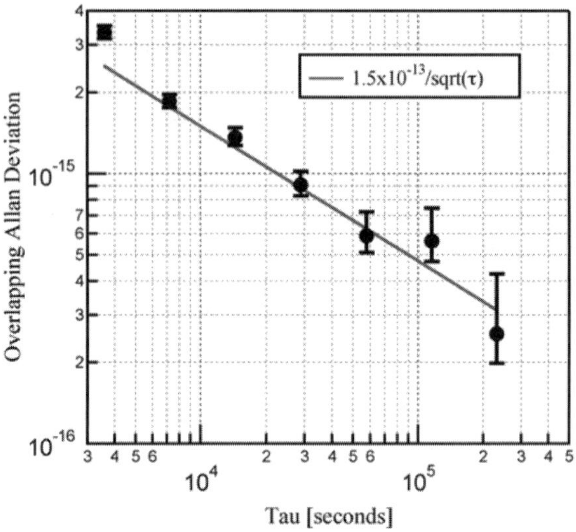

Figure 1. Stability of the engineering prototype NRF1 when measured against the USNO maser mean. The stability is consistent with $1.5\times10^{-13}/\tau^{1/2}$.

Since our last report[2], we have continued to investigate the issues and advisability of running our fountain clocks by loading from a magneto-optical trap (MOT). We have carried out non-leveraged tests of the impact of the MOT current on the fountain frequency by modulating the MOT current between different values, including zero (optical molasses loading). We find no detectible shift in the frequency for any of the different current values at the level of 1×10^{-15} for our full operating current, limited by the device's stability over the test times running with an optical molasses. In addition, we have run the same studies while operating the fountain on the magnetic field sensitive m_F=1 line. This provides a leveraged test for magnetic field effects, and shows a maximum shift on the clock transition below 1×10^{-17} for our full operating current. It is important to note that the field-sensitive test provide no leverage for synchronous modulation effects other than magnetic field that could be produced from cycling the MOT current and its associated power supply. Based on our tests and experience with NRF1, we plan on running with a MOT for all of the operational fountains.

3. Operational Fountains – NRF2 and NRF3

The first two operational rubidium fountains were built in parallel, keeping the construction as similar as possible. The devices have all components housed in two equipment racks and the physics package, and they incorporate several design changes to accommodate lessons learned in NRF1.

The vacuum chambers are entirely titanium inside the magnetic shields, including the microwave cavities. The lower Q of these cavities will require a reinvestigation[2] of distributed cavity shifts and vertical alignment sensitivity of the fountains. There is significantly more graphite in the detection regions to suppress resonant background scattering. In addition, we have added a graphite getter at the bottom of the collection region. This getter has a collection of small titanium tubes packed together vertically over the getter to limit the pumping speed to roughly 1 l/s. This structure is designed to suppress the vertical thermal beam coming from the collection region as seen in the detection region. This helps to suppress laser FM-to-AM conversion in the atom detection process.

The lasers are now housed in a 6U high rack-mounted enclosure. We are using two commercial diode lasers and a single semiconductor tapered amplifier to produce the trapping/launching/detection light and the repumper light. The laser light is coupled into two polarization-maintaining fibers and sent to the optical splitter/shifter table.

The splitting, shifting, intensity stabilization, and shuttering of the individual laser outputs is performed in another 6U high rack-mounted enclosure. Almost all of the power division and combining is carried out in polarization-maintaining evanescent-wave fiber splitters. The only free-space optical paths are three double-pass acousto-optical modulators (for trapping/launching and detection beams) and a single-pass acousto-optical modulator that has two independently-shuttered optical outputs (for collection and detection repumping). The heavy use of fiber splitters has greatly simplified the optical layout and reduced the number of optical alignments needed to attain and maintain operation.

The trapping master laser is locked with a modulation-transfer saturated absorption gas cell lock. The repumper laser is offset locked to the trapping laser with a phase-frequency comparator driven by the beatnote between the lasers. The phase-frequency lock provides a singe-valued recapture signal exceeding 3 GHz of tuning range.

Both NRF2 and NRF3 have demonstrated short-term stabilities of 1.5×10^{-13} at one second. Figure 2 shows preliminary results from both NRF2 and NRF3, obtained by comparing steered outputs from the fountains against the maser mean. At the time the data was taken, NRF2 had been running for longer than NRF3. The preliminary results are encouraging, but we are still optimizing laser, detection, and quartz crystal PLL parameters, which should lower the white frequency noise levels.

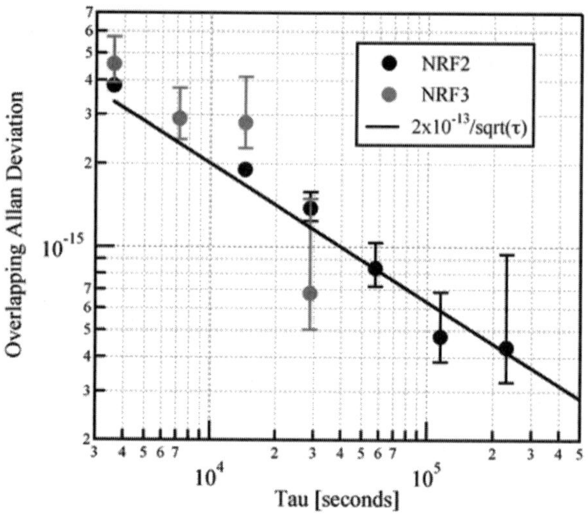

Figure 2. Preliminary medium-term stabilities of the first two operational fountains NRF2 and NRF3. The data points at Tau=7200 seconds lie on top of each other. This data was taken while the devices were still under construction.

4. New Master Clock Facility

The operational rubidium fountains are currently housed in a new building on the USNO grounds. This building was built for the sole purpose of housing and measuring atomic clocks for the USNO timing ensemble.

There are eight large (either 17'x20' or 15'x25') rooms, each of which has independent environmental controls. Six of these rooms are designed to house clocks, one contains measurement systems, and one houses computers.

The temperature stability spec for each of these rooms is +/- 0.1°C RMS from setpoint (nominally 23°C). The relative humidity spec is +/- 3% from setpoint (nominally 40%). The rooms have redundant terminal controls and air-supply ducts, and meet the temperature and humidity specs under failure of individual terminal control components like reheat coils and controllers.

Each clock room has redundant UPS power. The clocks are located on vibration isolated concrete piers, which have foundation footers that are independent of the building foundation.

5. Conclusion

We built and demonstrated three rubidium fountains, one prototype and two operational devices. The prototype has been used to characterize systematic shifts and sources of instability. Both operational fountains have been moved to our new clock facility at USNO, and we will be building 4 additional fountains for deployment at USNO and our Alternate Master Clock facility in Colorado.

Acknowledgments

The authors would like to acknowledge several organizations for their support of the work described in this paper. Initial work on NRF1 was supported by the Office of Naval Research. Characterization of NRF1 and all of the work on NRF2 and NRF3 were supported by USNO and SPAWAR.

References

1. S. Peil, S. Crane, T. Swanson, C. Ekstrom, *Proceedings of the 2005 IEEE Freq. Control Symp.,* Vancouver, Canada, p.304 (2005).
2. S. Peil, S. Crane, T. Swanson, C. Ekstrom, *Proceedings of the 2007 Joint meeting of the European Freq. and Time Forum and the IEEE Freq. Control Symp.,* Geneva, Switzerland, (2007)
3. S. Crane, C. Ekstrom, *Proceedings of the 2005 IEEE Freq. Control Symp.,* Vancouver, Canada, p.297 (2005).
4. S. Crane, S. Peil, C. Ekstrom, *Proceedings of the 2005 IEEE Freq. Control Symp.,* Vancouver, Canada, p.301 (2005).

THE TRANSPORTABLE CESIUM FOUNTAIN CLOCK NIM5: ITS CONSTRUCTION AND PERFORMANCE*

TIANCHU LI, PINGWEI LIN, MINGSHOU LI, PING WANG,
WEILIANG CHEN, NIANFENG LIU AND YIGE LIN

National Institute of Metrology (NIM), 18 Beisanhuandong Rd., Beijing 100013, China

The second laser cooling - Cesium fountain clock NIM5 at the National Institute of Metrology (NIM) China adopts the (1,1,1) direct OM (Optical Molasses) configuration. NIM5 has been running with a stability of 3×10^{-15} / d and an operation ratio of 99% since 2007. Preliminary evaluations of NIM5 in 2008 showed a typical combined uncertainty of 3×10^{-15}. The NIM5 clock is operating in parallel with NIM's first fountain clock NIM4. NIM4 and NIM5 are used to steer the frequency of the calculated NIM atomic time TA-c(NIM) and the first set of results are promising. We are now at the stage of comparing the frequency of NIM5 with UTC to support the independent frequency shift evaluations of NIM5 and contribute to the international atomic time in the near future.

1. Introduction

The first laser cooling - Cesium fountain frequency standard NIM4 at the National Institute of Metrology (NIM) China has been running since 2005 with an operation ratio (life time) of 98% (provides 98 days of valid data in 100 days) and a typical evaluation uncertainty of 5×10^{-15} [1,2].

From 2004 we started to build the second Cs fountain clock NIM5 based on experiences learnt from the NIM4 and references to the latest developments of fountain clocks worldwide.

NIM5 is designed and constructed to work accurately, reliably, sub-continuously and to be transportable. It can operate in either MOT (Magneto Optical Trap) or direct OM (Optical Molasses) mode to prepare the cold atoms. Unless otherwise specified, the experiments presented in this paper were performed in direct OM.

The paper describes the design, construction and performance of NIM5, especially those different from NIM4. This is because the design of NIM4 follows essentially the CsF1 clock of BNM-SYRTE France and the latter was presented in many papers, like [3,4].

* This work is supported by the Ministry of Sci. & Tech. China (Project No.: 2005CB724504).

2. Design and Construction of NIM5

The overall assembly of NIM5 consists of a 0.6×0.8×1.8 m^3 physical package, the laser-optics system on a 1.2×0.9 m^2 bench and electro-microwave system in two racks of 0.6×0.6×1.8 m^3.

2.1. *The Physical Package*

Figure 1 shows the schematic drawing of the physical package of NIM5 clock.

An OFC (oxygen-free copper) fountain tube and an OFC cylindrical TE011 Ramsey microwave cavity with Q = 9000 in the tube are surrounded with a four layer μ-metal magnetic shield system. Another OFC selective cavity (also TE011) is inserted into the top part of the MOT body just underneath the detection zone. The oscillation frequency of the Ramsey cavity was adjusted by controlling its temperature to around 50°C. Three thermometers are put on the outer surface of the fountain tube to measure the temperature with uncertainty of 0.4 K for the correction of black body radiation frequency shift.

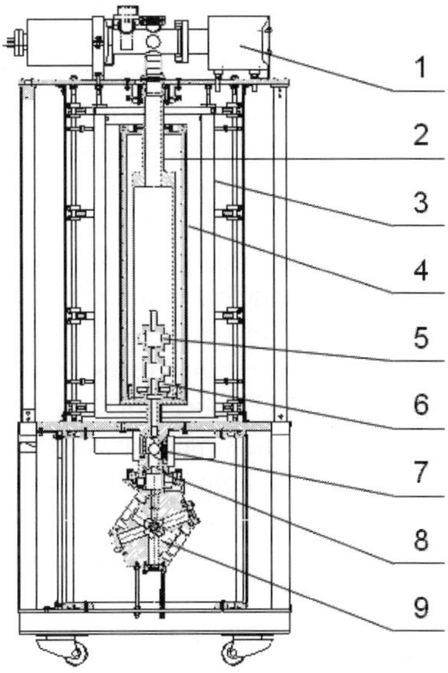

Figure 1. Schematic drawing of NIM5 physical package (1: top ion pump (bottom pomp not showed); 2: fountain tube; 3: magnetic shield; 4: C-field coil; 5 Ramsey cavity; 6: upper selection cavity; 7: detection part; 8 bottom selection cavity; 9: MOT/OM).

The detection assembly is located between the fountain tube on top and the MOT/OM body below.

The MOT/OM body was machined from an Aluminum alloy block. The angles of surfaces, on which the MOT/OM beam collimators (described later) are fixed, on the MOT/OM body were measured and corrected by means of manually polishing.

In NIM4 two ion pumps are connected to the MOT chamber. That leaves a dead space at the top part of the fountain tube, this situation is not preferred for good vacuum. While in NIN5 one pump is connected to the MOT/OM and another one to the top of the fountain tube to realize a better vacuum (better than 7×10^{-8} Pa) there so as to reduce the collision of the cold Cs atoms with background gases because the atoms spend more time in their ballistic movement on the upper part of the fountain tube.

2.2. *The Laser and Optics System*

An ECDL (external cavity diode laser) is frequency shifted and locked to the F=4 to F'=4 and F'=5 crossover transition of the Cs D2 line as the master laser [5]. One F-P slave diode laser working with 140 mW output, one single-pass AOM and one 1×3 SM (single mode) PM (polarization maintenance) fiber coupler constitute either channel of the up or down (1,1,1) MOT/OM optics. More than 11 mW optical power is available from each of the six outlets of the two 1×3 couplers. Each light is collimated to a beam with diameter of 26 mm as one of the six MOT/OM beams by an individual collimator assembly.

The fiber couplers were custom made. Their polarization extinction ratio was tested to be better than 1000 and the change of their optical splitting ratio with time less than 0.2% over 3 months. The fiber couplers are compact, robust and meet the requirements of the design.

The atom clouds are launched in an upwards moving-wave OM, which is realized by reversely frequency detuning of the three downwards and another three upwards OM lights. When a single-pass AOM performs the detuning it introduces an angle deviation on the outlet light. Fortunately, the angle deviation does not change the optical spot position on the fiber inlet face through the collimating - focusing optics (Figure 2). Experiment showed that the practical fiber coupling efficiency of the light into the fiber changes < 2% before and after detuning with the single pass AOM optics.

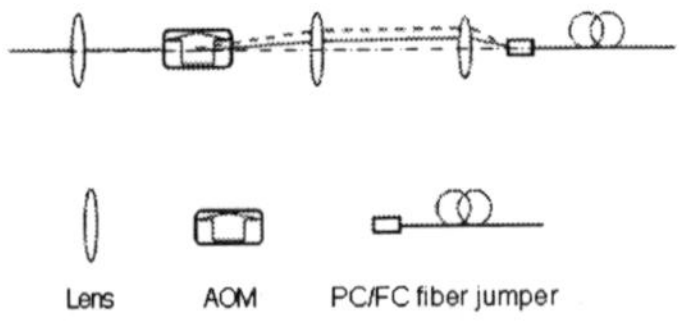

Figure 2. The AOM collimating and the fiber couple-focusing

A 5 mW light from the master laser is collimated and cleaved into two beams with rectangle section of 2×20 mm^2 as the detections.

The detection beam in horizontal is also used to push the |F=4, m≠0> atoms away while leaving the up-going |F=3, m=0> atoms unaffected after the selection of the atom states is performed just underneath the detection zone.

2.3. *The Electro – Microwave System*

Two dedicated Cs frequency synthesizers, synchronized to the master H-maser of the NIM time keeping clock assembly, generates the microwave for atom state selection and for atom interrogation respectively. The home made interrogation synthesizer, based on the scheme of reference [6], produces an extra stable in phase, super pure in spectrum and highly tunable in extremely fine step probing microwave. In order to eliminate the microwave power related frequency shifts it is preferable to switch off the microwave before the atom cloud enters and after it leaves the Ramsey cavity ($\pi/2$ microwave field) in a fountain circle. In another hand, the atoms in their up and down movement interact with two $\pi/2$ microwave fields in a Ramsey circle. In order to escape from the frequency shift induced by microwave phase difference, these two $\pi/2$ microwave have to be strictly phase continuous. An interference microwave switch [7] has been built on the output port of the interrogation synthesizer to produce the pulsed resonance microwave for the Ramsey interactions with the phase of the pulsed microwave remaining undisturbed. Another method based on the similar idea but realized through direct detuning the output frequency of the DDS (direct digital synthesizer) in the interrogation synthesizer is also under consideration. The key technique for the DDS method is also to realize the frequency flip flop while remaining the phase continuity among the pulsed interrogation microwaves.

The 9.19 GHz reference frequency is square-wave modulated with the DDS in the synthesizer under control of a PC to conduct the comparison between the reference frequency and the clock transition frequency. In NIM5 the comparison is performed with alternative order of the transition probabilities: R1-B1, R2-B1, R2-B2, R3-B2, R3-B3, ...(R*i* represents the red tuning transition probability and Bi the blue tuning; the transition probability $\eta = N_4 / (N_4+N_3)$ and N_4, N_3 is the number of atoms in F=4 and F=3 respectively). The servo-period is reduced to one fountain circle so the dead time in servo process is reduced in favor of improving the locking stability (reduces Dick effect) [8].

3. Experiments and the Preliminary Evaluations

NIM5 loads about 7×10^7 atoms in 0.6 s by a direct (1,1,1) OM, the atoms are further post-cooled to about 1.5 μK and launched up to 81 cm high from the center of the OM (30 cm above the center of the Ramsey cavity). Atoms loaded by direct MO on NIM5 is 6 to 8 times less than by MOT but fortunately with this reduced number of atoms a good enough stability of the frequency locking is still reached. OM fountain circle on NIM5 is 1.6 s, reduced by half compared to 3 s of MOT NIM4.

It is noted that the launching frequency detuning $\Delta\nu$ and $-\Delta\nu$ must be phase-locked to form the stable moving-wave OM but the average frequency of the two launching frequencies, $(\nu_L+\Delta\nu) + (\nu_L-\Delta\nu) = \nu_L$, needs not equal to the OM loading/cooling frequency ν_P. Instead it is better to have $\nu_P - \nu_L = 1$ MHz in NIM5.

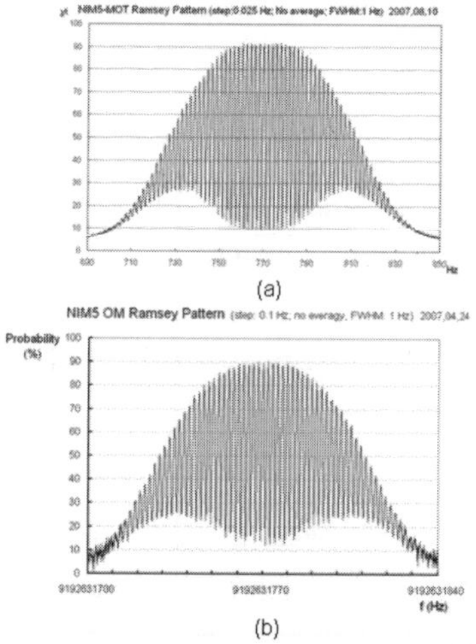

Figure 3. Typical MOT (a) and OM (b) Ramsey pattern with FWHM of around 1 Hz (frequency scanning step of 0.1 Hz without average).

The typical OM Ramsey pattern with frequency scanning step of 0.1 Hz without average is shown in Figure 3. A FWHM (full width half magnitude) linewidth of around 1 Hz with S/N (signal to noise ratio) of 400 is realized and more than 67 fringes can be clearly recognized.

The saturation optical power density for Cs cooling transition of F=4 to F'=5 is 1.1 mW/cm^2. Using the laser-optics mentioned above an OM with 6×11 mW optical power and 26 mm beam diameter is applied to capture more atoms for better S/N. In this operation with saturated optical power density the number of OM atoms is less sensitive to laser power fluctuations. Higher optical power density may heat the Doppler cooled atoms so the atom temperature after Doppler cooling can be above the 124 μK of Doppler cooling temperature limit. However the atoms are further cooled by the post polarization gradient cooling and experiment depicts a final temperature of 1.5 μK. In the end the number of atoms captured is increased and more stable at no sacrifice of atom cooling temperature.

The day to day sub-continuous frequency locking with operation ratio of 99% and a stability of 5×10^{-14} in 10 s (fountain circle of 1.6 s), 3×10^{-15} in 1 d (including the stability of H-maser) is approached in direct OM without using the extra low velocity atom source suggested by, say [9].

For evaluating the frequency shift introduced by atom collisions the atom density is changed by 3 to 4 times with direct OM by means of controlling the selection RF power. The low density OM is chosen as the normal operation mode and only for the collision evaluation the high density mode is used.

Table 1. NIM5 accuracy budget (2008,10).

	Physical Effect	Shift (10^{-15})	Uncert. (10^{-15})	
1	2nd Zeeman:	76.2	0.4	H_C=128.0 nT
2	Cold collision:	-1.1	0.8	
3	Microwave sractrum purity and leacage :	2.1	2.0	
4	Blackbody:	-23.2	0.9	Tempetature 51±3°C
5	Gravitation:	3.8	0.1	Above sea: 34.5m
6	Majorana:	0.0	0.1	
7	Light shift:	0.0	0.1	
8	Cavity pulling:	0.0	0.1	
9	Cavity phase difference:	0.0	0.1	Phase difference: 4 μradian
	Combined:	57.8	2.4	

The latest frequency shift evaluations on NIM5, following the commonly accepted evaluation process [10], is listed in Table 1 and we quote the final combined evaluation uncertainty for NIM5 as 3×10^{-15}.

The C-field was mapped by the magneto-sensitive

Rabi transition. The homogeneity of the C-field is less than 1 nT in the height range of 56 - 82 cm above the OM center.

The atomic density is switched alternatively and quickly between high density and low for atom collision shift evaluations. The zero density frequency is extrapolated according to the least-squire fit to the data obtained. The microwave power related shift is deduced in the similar way but with alternative interrogation microwave powers.

The dominate uncertainty source in Table 1 is the microwave spectrum purity and leakage frequency shift. Figure 4 shows the frequency shifts on NIM5 operating with interrogation microwave power 1, 9, 25, 49 and 81 times to the $\pi/2$ pulse and the curve fitted to these data. We expect that this shift could be eliminated and the related uncertainty reduced when the interference switch or DDS frequency tuning technique mentioned above functions properly before too long.

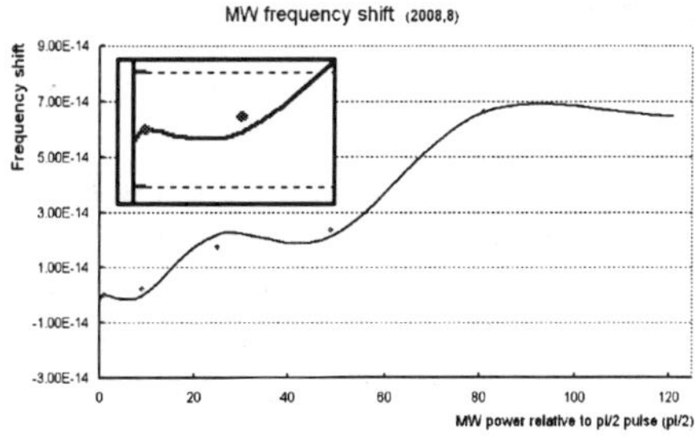

Figure 4. Evaluation of the microwave purity and leakage frequency shift.

4. Discussion and Future Work

The NIM5 is expected to resume operation with minor adjustments after its relocation to NIM's new Changping campus, which is 40 km away from the present Hepingli campus.

The NIM5 and NIM4 are going to run in parallel and compare with each other to form the primary frequency standard in China. Since the beginning of 2008 the NIM4 and NIM5 data are used to steer the frequency of H-maser through a phase micro stepper to generate the calculated atomic time TA-c(NIM) and the preliminary experiment results demonstrates significantly better stability than that obtained by the traditional Cs clock and H-maser assembly [11].

An experimental frequency transfer fiber system has been set up at NIM. A 1560 nm laser is phase modulated with 5 GHz and the 10 GHz beat frequency between the two modulation sidebands, traveling through the fiber, is taken as the transferred reference frequency. A stability of 3×10^{-15} / 1000 s is observed with active compensating of the fiber induced disturbances [12]. A 60 km fiber transfer linking the NIM Hepingli and Changping campus is under construction.

NIM has built a femto second optical comb based on Ti:Saphire laser [13] and acquired another fiber optical comb to establish the bridge between microwave frequency and optical frequency.

NIM started also the construction of an optical frequency standard based on laser cooling - Strontium lattice and the Sr blue MOT atom clouds were observed in July 2008.

Acknowledgments

The authors would like to specially thank Prof. Yiqiu Wang of Peking University China for his helps over years..

References

1. T.Li, M.Li, P.Lin, B.Huang *et al*, ACTA. Metrologica Sinica **25**, 193 (2004).
2. T.Li, M.Li, P.Lin *et al*, Chinese Phys. Lett. **24**, 1177 (2007).
3. A.Clairon, C.Salomon *et al* and W.Phillips, Europhys. Lett. **16**, 165 (1991).
4. A.Clairon, P.Laurent, G.Santarelli *et al*, IEEE Trans. **IM44**, 128 (1995).
5. Y.Lin, T.Li and W.Chen, CPEM (Torino) 592 (2006).
6. T.Heavner, S.Jefferts *et al* and T.Parker, IEEE IFCS (Vancouver) 308 (2005).
7. G.Santarelli1, G.Governatori *et al* and A.Clairon, EFTF (Braunschweig) 166 (2006).
8. J.Dick, Precise Time and Time Interval (Redondo Beach, CA) 133-147 (1987).
9. E.Donley, T.Heavner *et al* and S.Jefferts, IEEE IFCS (Montreal) 82 (2004).
10. S.Jefferts, J.Shirley, T.Parker *et al*, Metrologia **39**, 321 (2002).
11. Y.Gao, X.Gao, A.Zhang *et al*, V ITSA (San Fernando, Cádiz) (2008).
12. W.Chen, T.Li *et al* and Y.Lin, CPEM (Boulder), 614 (2008).
13. Zh.Fang, Q.Wang *et al* and T.Li, ACTA Physica Sinica **56**, 5684 (2007).

COMPENSATED MULTI-POLE MERCURY TRAPPED ION FREQUENCY STANDARD AND STABILITY EVALUATION OF SYSTEMATIC EFFECTS

E.A. BURT, S. TAGHAVI-LARIGANI, J.D. PRESTAGE, AND R.L. TJOELKER

Jet Propulsion Laboratory, California Institute of Technology,
4800 Oak Grove Drive, Pasadena, CA 91109

We have developed a compensated multi-pole Linear Ion Trap Standard (LITS) that eliminates nearly all frequency sensitivity to residual ion number variations. When operated with $^{199}Hg^+$, this trapped ion clock has recently demonstrated extremely good stability over a 9-month period. The short-term stability has been measured at $5x10^{-14}/\tau^{1/2}$ and an upper limit on long-term fractional frequency deviations of $<2.7x10^{-17}$/day was measured in comparison to the laser-cooled primary standards and to the post-processed ultra-stable version of TAI known as TTBIPM using GPS carrier phase time transfer. We have also made a first measurement of the Hg^+/Hg collision shift and place a limit of $+3.8(7.2)x10^{-8}$/Pa on the shift constant.*

1. Introduction

One of the several advantages of a spherical rf Paul trap is that it provides a nearly field-free region for a trapped single ion. Early on this was recognized as having great potential for atomic clocks [1]. Development of ion-trap-based atomic clocks subsequently followed two paths: laser-cooled single trapped ions with ultimate accuracy [2] and buffer-gas-cooled trapped ion clouds with long-term stability [3]. For the latter application it was soon realized that one could extend the advantages of a field-free region to larger numbers of ions trapped on a nodal line by using a linear ion trap [4-6], and ultimately to even larger numbers of ions trapped in a volume by using a multi-pole trap [7, 8]. Since the multi-pole trap approximates a square well potential, the ion-number-dependent second-order Doppler shift, which couples to ion micromotion was greatly reduced and largely accounts for the improved stability achieved using this type of trap. However this effect is still dominant in this clock.

In this paper we will describe a simple magnetic compensation technique that renders the number-dependent second-order Doppler shift almost insignificant in clocks employing large ion numbers in a multi-pole trap. We will also describe a long-term comparison between this compensated multi-pole ion trap clock and some of the world's most stable timescales using GPS carrier phase time transfer that demonstrates the improved stability achieved using the compensation scheme.

* This research was carried out at the Jet Propulsion Laboratory, California Institute of Technology, under a contract with the National Aeronautics and Space Administration.

Finally we will summarize a preliminary stability evaluation of this clock performed over the same long run that includes, among several new measurements not previously made, a first measurement of the Hg^+/Hg collision shift.

2. Multi-pole Ion Trap Clock Past Performance

In 2002 the first operational atomic clock based on ions in a multi-pole trap was installed at the US Naval Observatory (USNO). It began continuous operation there and was compared to the various time scales maintained by USNO. In 2003 the first results of these comparisons were published [9], which showed that over a 9-month period of uninterrupted operation the standard (known as LITS8) drifted by less than 2×10^{-16}/day relative to the USNO maser mean.

The most likely candidate for the small remaining drift was residual second-order Doppler shift. There are two contributions to this effect: 1) thermal motion and 2) driven motion resulting from the trapping fields. The latter is strongly related to the number of ions trapped because the size of the ion cloud determines the average trap field that it will sample. In 2002 we published data showing the sensitivity of the clock frequency to changes in ion number obtained by varying the well depth [10]. The derived sensitivity of approximately 4 mHz from an empty to a full trap was later verified by reanalyzing the USNO data during times when the mercury source oven temperature was changed [11].

For most of the long run at USNO, the ion number was changing far more gradually than in experiments designed to measure ion number sensitivity. Yet over a long period of time, the correlation between ion number and clock frequency is observed and explained by ion number sensitivity [12]. To the extent this sensitivity could be reduced, the overall stability of the clock would be improved.

3. Magnetic Compensation of the Number-Dependent Second-Order Doppler Shift

Since a number-dependent second-order Doppler shift in this clock is sensitive to the spatial volume occupied by the ions, number-dependent shifts may also result from the small magnetic inhomogeneity that exists in the trapping region. To test this, a small coil was wound at the interface between the quadrupole loading trap and the multi-pole trap. Finite element modeling suggested that a coil at this location with its axis parallel to the C-field axis, could significantly modify the small inhomogeneity that exists inside the C-field solenoid. Results of the modeling were confirmed by plotting the number-dependent frequency shift as a function of the current in this auxiliary coil [11]. A coil current was found for which the number-dependent shift is eliminated within the error estimate of the measurement over a wide range of trapped ion numbers. Figure 1 shows the frequency shift due to a 20% change in ion number with and without compensation, indicating more than an order of magnitude reduction in sensitivity.

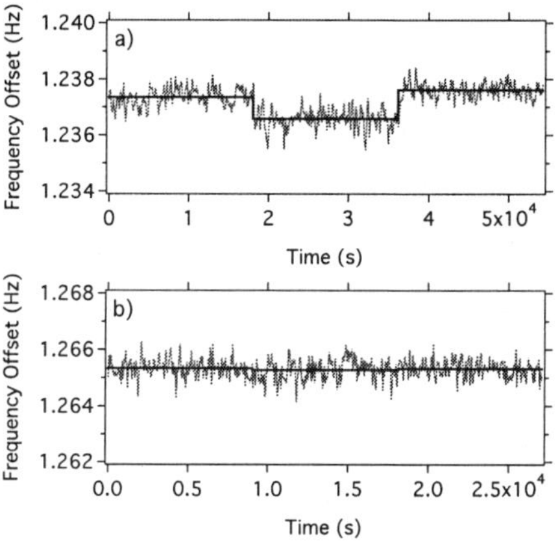

Figure 1. Clock frequency as a function of time with the ion number decreased by 20% for the middle third of run without magnetic compensation (a) and with (b).

This data indicated that the sensitivity to number-dependent effects could be eliminated in the short term with magnetic compensation. Figure 2 shows frequency offsets for the compensated multi-pole LITS9 from UTC over a 6-month period. The difference in sensitivity is clear, indicating that the compensation scheme is not transient, but stable over long periods.

4. Clock Stability and Long-Term Comparisons

A record room temperature microwave transition line Q of 5×10^{12} has been achieved with this clock [12]. With the current signal-to-noise in the multi-pole trap configuration, this corresponds to a short-term stability of $5 \times 10^{-14} \tau^{1/2}$.

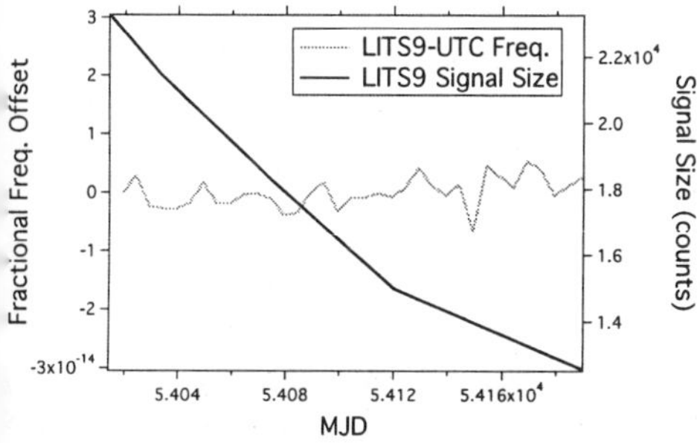

Figure 2. Frequency offsets for LITS9 vs. UTC (gray) and the LITS9 signal size (proportional to trapped ion number – black) over the same period.

At JPL the compensated multi-pole LITS9 is locally our most long-term stable clock. Therefore, to put limits on its long-term stability, we compare to time-scales or standards in other laboratories using GPS Carrier Phase Time Transfer (GPSCPTT) methods. The hydrogen maser that is currently used as the LITS9 LO is made the reference for a geodetic GPS timing receiver. In doing so we are able to compare LITS9 to any other standard used as a reference in the same way. In particular, we compare to data obtained from the receivers NISA, USN3, and PTBB, which are referenced to UTC(NIST), UTC(USNO) and UTC(PTB) respectively. Figure 3 shows one such comparison. For comparison on this graph we also show LITS9 vs. a local hydrogen maser, which demonstrates the LITS9 short-term stability of $5\text{x}10^{-14}/\tau^{1/2}$ (pair data); LITS9 vs. JPL's laser-cooled cesium fountain,which demonstrates an upper bound on the LITS9 stability floor of $2\text{x}10^{-16}$; and previous data comparing the uncompenstated LITS8 to the USNO maser mean. For the LITS8 data and the LITS9 vs. UTC(NIST) data, the short term noise is limited by the measurement system and GPS transfer noise respectively. However, in the long-term, the large reduction in drift for the compensated standard (LITS9) is apparent.

After processing the data using JPL's GIPSY software we can further double-difference the results with data published by the Bureau International des Poids et Mesures (BIPM) in the Circular-T [13], to get frequency differences between LITS9, UTC, and other timescales. The most stable of these is assumed to be TT(BIPM) [14], a post-processed timescale that very closely tracks the average of all primary standards reporting to the BIPM. Comparisons to TT(BIPM07) show that LITS9 deviates from this timescale by less than $2.7(0.4)\text{x}10^{-17}$/day [15]. Even if we assume that this is all due to LITS9 (and not, for instance, to time transfer noise), it represents an order of magnitude improvement over the previous multi-pole LITS8 performance.

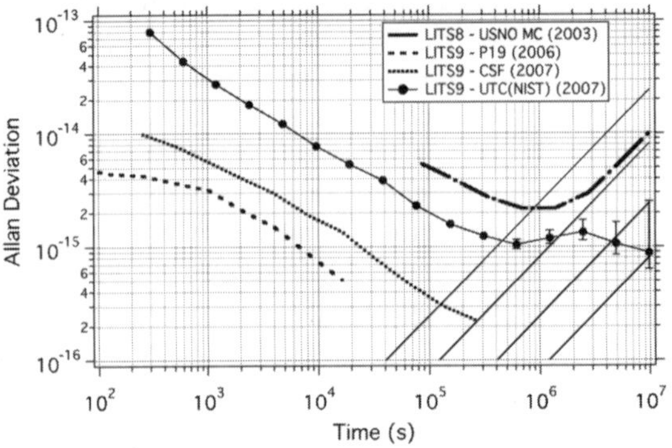

Figure 3. Long-term Allan deviation of frequency differences between LITS9 and a local hydrogen maser (dashed line), the JPL laser-cooled cesium fountain (dotted line), UTC(NIST) via GPSCPTT (line connecting solid circles) and for reference, the previous 2003 frequency differences between LITS8 and the USNO maser mean. The straight lines are drift rates ranging from $3\text{x}10^{-16}$/day to $1\text{x}10^{-17}$/day for reference.

Combining the measured short-term LITS9 stability of $5\text{x}10^{-14}/\tau^{1/2}$, the upper bound on the LITS9 stability floor of $2\text{x}10^{-16}$, and the upper bound on the LITS9 drift rate of $2.7\text{x}10^{-17}$/day, it is clear that the noise in the LITS9 vs. UTC(NIST) data shown in Figure 3 is limited by the time transfer method for averaging times less than $2\text{x}10^{6}$ seconds. Between $1\text{x}10^{5}$ and $2\text{x}10^{6}$ seconds (1-23 days), the Allan deviation reaches a stability floor of about $1\text{x}10^{-15}$ that is not limited by LITS9 and probably not limited by UTC(NIST) (the latter is a steered standard, with steers applied approximately monthly). Other researchers have reported time transfer stabilities in the mid to low 10^{-16} range [14], however we were not able to reliably do so over longer averaging intervals. While mid to low 10^{-16} performance may be possible using GPS time transfer methods, it is not routine and may require special attention to the multi-path environment and monument set up among other factors.

Finally, Figure 4 shows frequency offsets for LITS9 and several other standards differenced to UTC. In this graph short-term noise is dominated by GPS time transfer, however in the long term the graph shows that LITS9 can serve as an excellent stand-alone free-running single-clock representation of UTC.

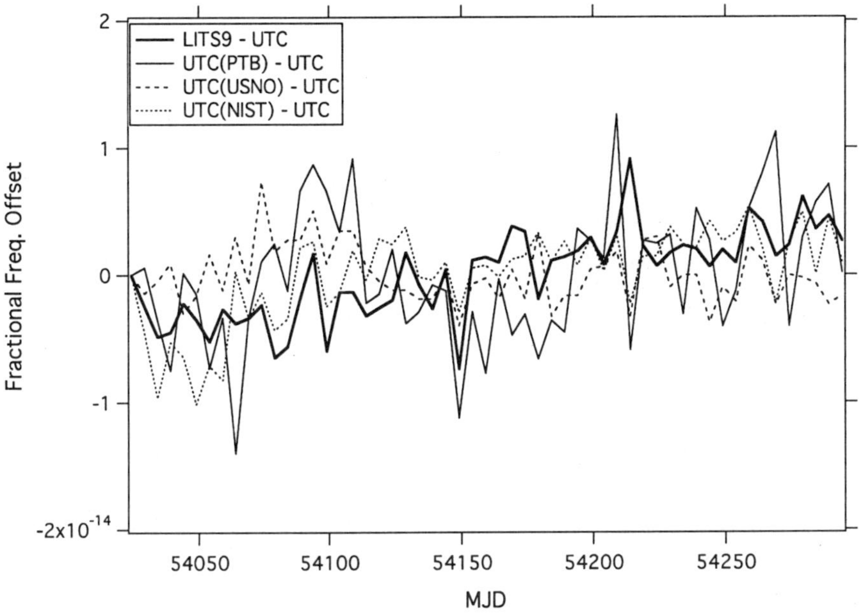

Figure 4. Fractional frequency offsets for LITS9 vs. UTC (thick black line). For comparison are also shown UTC(PTB) vs. UTC (thin black line), UTC(USNO) vs. UTC (long dashed line), and UTC(NIST) vs. UTC (short dashed line).

5. Stability Evaluation

To determine the possible causes of the small residual long-term instability in LITS9, we performed a stability evaluation for each of the systematic effects that could impact the overall stability of the clock. Some of the results are shown in Table 1. The effects are listed from largest to smallest. It is particularly notable that the top 3 effects are now all related to pressure instabilities (variations in the temperature-dependent second-order Doppler shift are due to variations in buffer gas pressure and to higher order, variations in ion number), and that the number-dependent second-order Doppler shift is now 4th on the list. This is the first time this effect has not been the leading cause of instability in this clock and is due to the compensation scheme described earlier. The following measurements were made during this stability evaluation for the first time: Hg/Hg^+ collision shift, AC Zeeman shift due to trap potential aging, microwave power, and the light shift (AC Stark shift due to the discharge lamp).

Table 1. Preliminary stability evaluation results for LITS9 during the MJD54023 to MJD54296 run.

Effect	Δf/f ($x10^{-17}$/day)	Uncertainty ($x10^{-17}$/day)
T-dependent second-order Doppler shift	-1.5	3.4
Collision shift (neon)	-1.1	0.6
Collision shift (CH_4)	--	0.94
N-dependent second-order Doppler shift	-0.84	0.23
Second-order Zeeman shift (filament V)	-0.35	0.14
Collision Shift (Hg/Hg^+)	-0.22	0.40
Second-order Zeeman shift (cur. source)	-0.21	0.13
Second-order Zeeman shift (ext. field)	--	0.14
Temperature Sensitivity	+0.11	+0.01 -0.09
AC Zeeman shift (trap potential)	< 0.00001	< 0.00001
Microwave power	--	0.05
AC Stark Shift (BBR)	-0.0001	0.00004
AC Stark Shift (light)	+0.0001	0.0001
AC Stark Shift (trap rf)	< -0.00002	0.00002

In previous versions of this clock, the largest effects were due to the second-order Doppler shift, magnetic shifts, pressure shifts, and the light shift. Magnetic shifts are attenuated simply by applying good magnetic shield design, the light shift is virtually eliminated by using a dual trap geometry together with ion shuttling, and the Doppler shift is now compensated. To the extent that we can improve background pressure stability in future versions of the clock, the overall clock stability will be even further improved. An effort in this direction is already under way. A high temperature bakeout (>450° C) physics package based on the LITS technology has been built to greatly reduce vacuum outgassing [16]. This system has already demonstrated stable signal-to-noise over several years of operation.

Also notable in this evaluation is the absence of any single large effect. All measured effects are very small indicating that the overall measured drift of the clock relative to the laser-cooled primary standards is not a fortuitous coincidence, but rather inherent in the technology. If the device was replicated, similar performance could be expected.

6. Pressure Effects

Since pressure effects are now the largest components of instability, we describe them in more detail. Of the background gases present in LITS9, neon is the most significant because it is intentionally introduced into the vacuum system as a buffer gas. We have determined that the total change in neon pressure (due to capillary leak depletion) was -3.6×10^{-7} torr over the run. Applying the pressure shift coefficient of 8.5×10^{-9}/torr [17], this corresponds to a frequency drift rate due to neon of -1.1×10^{-17}/day.

Many other background gases are present in the vacuum chamber. All are only in trace amounts, but the most important of these is methane (CH_4) due to its high shift coefficient in Hg^+: -3.6×10^{-5}/torr. Using calibrated residual gas analyzer (RGA) measurements, we were able to place a limit of $\pm1\times10^{-10}$ torr on the change in methane pressure during the run. This corresponds to a contribution to long-term clock stability of $\pm1\times10^{-17}$/day. Contributions due to other background gases will be much smaller than this.

Another important background gas in the LITS9 vacuum system is neutral mercury (ions are generated by electron bombardment of neutral mercury). The background neutral mercury pressure can be varied through the temperature of the mercury source oven. We search for a frequency shift as a function of changing neutral mercury pressure in this way. We verify that no other background gas pressure changes significantly as a result of oven temperature changes by monitoring the pressure with a RGA. Ion number was also held constant during changes in oven temperature by changing the load trap parameters (multi-pole trap parameters were unchanged). With a mercury pressure change of $+1.6\times10^{-8}$ Pa, a null frequency shift of $+5.9(11.1)\times10^{-16}$ was observed. This bounds the fractional frequency pressure shift constant in the $^{199}Hg^+$ hyperfine clock transition due to neutral ^{199}Hg at $+3.8(7.2)\times10^{-8}$/Pa. For the 9-month comparison, the change in neutral mercury pressure was -1.6×10^{-8} Pa, which would translate into a drift of $-2.2(4.1)\times10^{-18}$/day in the clock.

7. Conclusions

The multi-pole linear ion trap frequency standard has demonstrated high long-term stability for several years. We have shown here that a simple modification to the magnetic field homogeneity in the trap improves that long-term stability by about an order of magnitude. Comparisons to the laser-cooled primary standards via GPS and to the timescale TT(BIPM) show a deviation of less than $2.7(4)\times10^{-17}$/day, making this compensated multi-pole standard among the most stable continuously running clocks in existence. Preliminary results from a subsequent stability evaluation show that the second-order Doppler shift, once the largest systematic effect in this clock, is now 4th on the list. The largest effects are now related to background pressure. We have shown that variations in background neon and methane pressures are well controlled and have described a first measurement of the Hg^+/Hg collision shift resulting in a null shift coefficient of $+3.8(7.2)\times10^{-8}$/Pa.

References

1. H.G. Dehmelt, Bull. Am. Phys. Soc. **18**, 1521 (1973).
2. For instance, T. Rosenband, et al., Science **319**, 1808 (2008).
3. L.S. Cutler, et al., Proceedings of the 13th PTTI, p. 563 (1981).
4. J.D. Prestage, et al., J. Appl. Phys. **66**, 1013 (1989).
5. D.J. Wineland, et al., IEEE Trans. UFFC **37**, 515 (1990).
6. R.L. Tjoelker, et al., Proceedings of the 5th Symposium on Frequency Standards and Metrology, p. 33 (1995).
7. J.D. Prestage, et al., Proceedings of the 1999 Joint EFTF-IFCS, p. 121 (1999).
8. R.L. Tjoelker, et al., Proceedings of the 6th Symposium on Frequency Standards and Metrology, p. 609 (2001).
9. R.L. Tjoelker, et al., Proceedings of the 2003 Joint IFCS-EFTF, p. 1066 (2003).
10. E.A. Burt, et al., Proceedings of the 2002 IFCS, p. 463 (2002).
11. E.A. Burt and R.L. Tjoelker, Proceedings of the 2005 Joint IFCS-PTTI, p. 466 (2005).
12. E.A. Burt, et al., Proceedings of the 2007 Joint EFTF-IFCS, p. 1041 (2007).
13. See http://www.bipm.org/jsp/en/kcdb_data.jsp#nohref.
14. See G. Petit in these proceedings.
15. E.A. Burt, et al., to be published in IEEE Trans. UFFC, **55** (2008).
16. See J.D. Prestage, et al., in these proceedings.
17. S.K. Chung, et al., Proceedings of the 2004 IEEE IFCS, p. 130 (2004).

RESEARCH OF FREQUENCY STANDARDS IN SIOM---- ATOMIC FREQUENCY STANDARDS BASED ON COHERENT STORAGE *

B. YAN, H. D. CHENG, Y. S. MA, W. Z. ZHANG, L. LIU, Y. Z. WANG**,

Key Laboratory for Quantum Optics, Shanghai Institute of Optics and Fine Mechanics, Chinese Academy of Sciences, Shanghai, 201800, The People's Republic of China.

In this paper we proposed a frequency standard based on pulsed coherent storage (PCS), the basic idea of the PCS scheme is combining the information storage technique with the optical pumping method. It shows that the Ramsey pattern is printed on the coherence of the atom states, then is detected in the form of the retrieval light. The proposed PCS frequency standards are free of light shifts, low Q requirement, high signal contrast and low noise background. The estimated frequency stability is 2×10^{-14} in 1 second. Meanwhile we report the progress in laser cooling of atoms in an integrating sphere, some experimental results are given. This cooling method provides a robust source which is suitable to construct a compact PCS clock.

**yzwang@mail.shcnc.ac.cn

1. Introduction

Atomic frequency standards are very important in our human life and modern society, they are the basis of many key technical fields such as precise atomic time scale, time-keeping system, GPS system and especially in space sciences and fundamental physics. In order to improve the performance of the frequency standards, researchers search for new physical concepts to overcome the problems in the traditional atomic clocks. In recent years, the applications of new physics in atomic clocks strongly push the progress of this field. Many new physical concepts and technologies have been explored, such as electro-magnetic induced transparency (EIT) frequency standards[1], coherent population trapping (CPT) masers[2], pulsed optical pumped (POP) frequency standards[3]. Combining the quantum optical information storage technique with the optical pulsed method, we have proposed a pulsed coherent storage (PCS) frequency standards[4]. In this paper, first we will report some

* This work is supported by The Major State Basic Research Development Program of China under Grant No. 2005CB724507 and 2006CB921202， The Key Oriential Project of Chinese Academy of Science under Grant No. KGCX2-SW-110, and KGCX3-SYW-405.

experimental progress in our lab about laser cooling of atom gas in an integrating sphere. It is a useful cold atom source to construct the PCS frequency standards. Then we will sum up the basic ideas of the PCS method, and finally discuss the experimental realization of the PCS frequency standards.

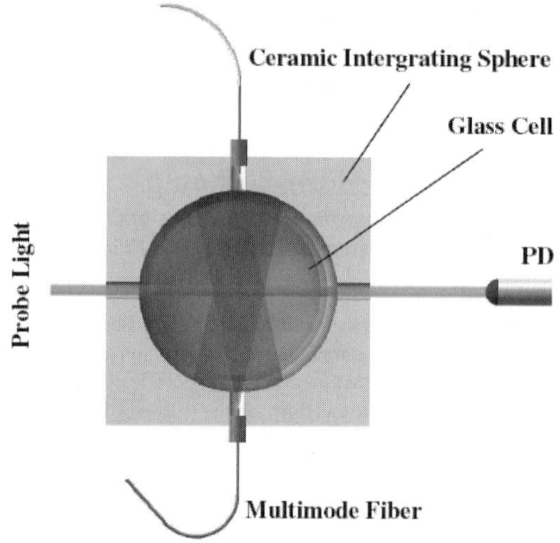

Fig. 1: the schematic of our experimental setup for diffuse light cooling.

2. Laser cooling in an integrating sphere

Laser slowing of an atomic beam by diffuse light in an integrating sphere was proposed at an early time[6]. Diffuse laser cooling was first realized by Ketterle *et al.* by slowing and cooling of an Na atomic beam [7]. Later Batelaan *et al.* used a similar setup to observe slowing of an ^{85}Rb atomic beam [8]. Chen *et al.* successfully slow and cool a sodium atomic beam [9, 10]. Guillot *et al.* carried out sub-Doppler cooling of cesium atoms to temperature as low as 3.5 μK with diffuse light [11]. Tremine *et al.* used this technique to develop a compact cold atom clock [12].

We demonstrate an experiment on laser cooling of ^{87}Rb atoms from vapor background in an integrating sphere. Fig. 1 shows a schematic diagram of the experimental setup. A spherical glass cell with an inner diameter of 43 mm is mounted on a vacuum pump and connected to a Rb reservoir. The glass cell is surrounded by an integrating sphere made from ceramic material whose diffuse reflection coefficient at the inner surface is 98% at 780 nm. The combined cooling and repumping light is split into two beams that are each coupled into multimode fibers.

To detect cold atoms in the glass cell, a probe beam, with power of 1 μW, passes through the center of the glass cell via two 2 mm diameter holes and is detected by a photodiode. An absorption spectrum is recorded by scanning the frequency, as shown in Fig. 2. Three hyperfine transitions F=2 to F'=1,2,3 are clearly shown in the spectrum (FWHM is about 10 MHz) that provides convincing evidence that atoms are cooled by the diffuse light in the integrating sphere. The temperature and number of cold atoms are estimated to be about 150 μK and 2×10^9, respectively. The temperature of about 26 μK is obtained by further sub-Doppler cooling.

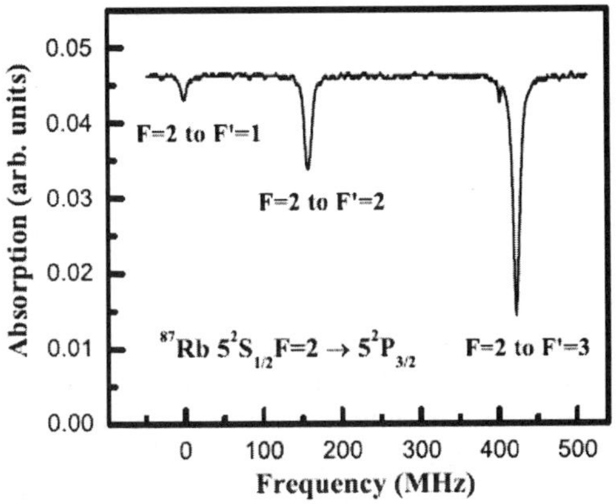

Fig. 2: The absorption spectrum when scanning the probe light frequency. This Doppler free spectrum is a direct evidence of diffuse cooling

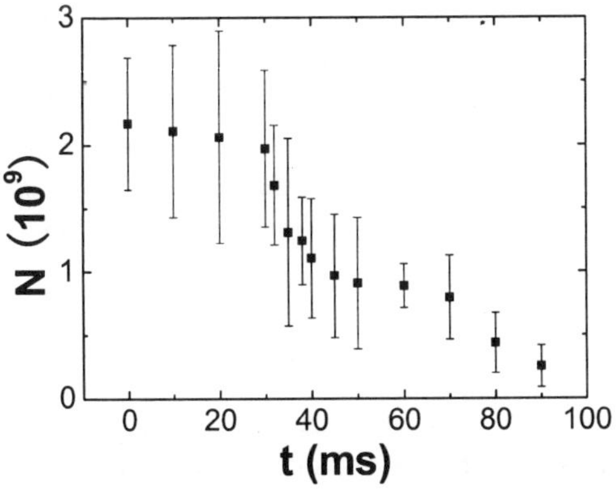

Fig. 3: The lifetime of intergrating sphere cooling atoms. It is about 80ms.

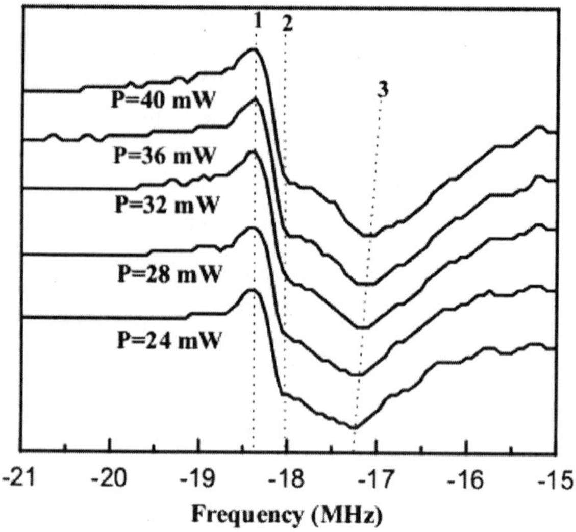

Fig. 4: The nonlinear spectroscopy of cold atoms in the diffuse light with the power changing of the injected cooling laser. The cooling light is locked and the probe light is scanning frequency.

In Fig. 3, we show the lifetime of these cold atoms after the diffuse light is turned off. It takes 80ms for the atom number to decay from 2×10^9 to 5×10^8. Our results show that the laser cooling of atoms by the diffuse light is robust and very useful as a cold atom source to construct the compact PCS atomic frequency standards.

We also demonstrate the recoil induced resonance of cold atoms in the diffuse light and investigate its tendency with the power changing of the injected cooling laser, as shown in Fig. 4. It is found that the positions of peak 1 and peak 2 are not changing with the injected cooling laser. This can be explained as the recoil induced resonance which corresponds to the velocity distribution of cold atoms. However, peak 3 becomes closer to peak 2 with the decreasing of the injected cooling laser because of the smaller light shift. Peak 3 may correspond to the electromagnetically induced absorption (EIA), but further theoretical investigating is needed.

3. The pulsed coherent storage frequency standard

Pulsed methods of frequency standards are made great progress recently. Among them, POP is very interesting and well established. The experiments of POP show that a frequency stability of $1.2\times10^{-12}\tau^{-1/2}$ can be reached, and a shot noise limit of $5\times10^{-14}\tau^{-1/2}$ has been proposed[13]. The basic idea of

POP can be understood as follows: First a laser pulse pumps atoms to a certain ground state. Then the population of this state is manipulated by double microwave pulses, which prints the Ramsey pattern on the population. Finally, the population of atoms is detected which gives the Ramsey fringe. The pulsed method which separates the state preparation and interrogation time eliminates the light shifts, because phase memory is destroyed by optical pumping[4].

We have proposed a pulsed frequency standard with a similar time sequence with POP, but with essential different physical meanings. We use the coherence not the population to carry the Ramsey pattern. The basic idea of the pulsed coherent storage frequency standards (PCS) is combination of the coherent information storage technique and the optical pumping method to completely eliminate light shift effects. It can be understood as follows: As shown in Fig. 5, first, two pulses of a signal laser and a coupling laser are applied, then shutting down the coupling laser which control the group velocity of coupled excitations (Dark-state Polaritons)[5], and the signal pulse is coherently converted into a spin wave, and stored into the medium. In this case there is no photon existed. We then manipulate the atom states by two microwave pulses which prints the Remsay pattern on the coherence. Finally, switching on the coupling laser, the stored signal light pulse is retrieved, and then detected to obtain the Remsay pattern. The central Ramsey fringe can be used as a frequency reference of atomic frequency standards.

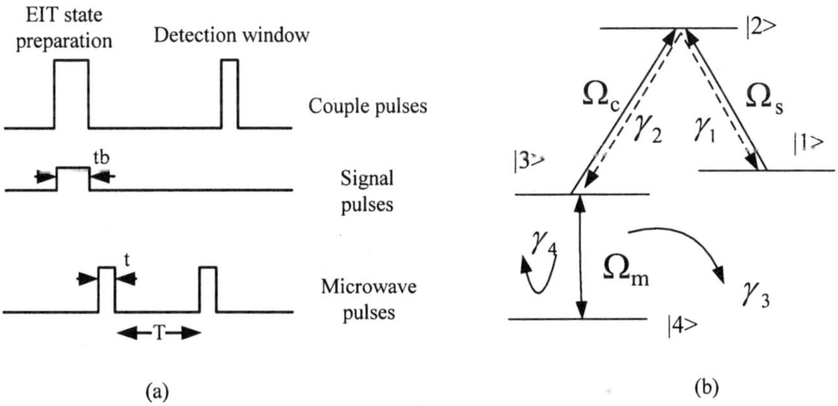

Fig. 5: (a) is the timing sequence for the PCS scheme. (b) is the four-level model of our PCS theory.

Here, we sum up the basic results, more details can refer to [5], the theory is based on a four-level model as shown in Fig. 5 (b).

In the first stage, a strong coupling laser pulse and a week signal laser pulse are applied, pump atoms to an EIT state, especially, the coherences between |1> level and |3>, |4> levels become

$$\rho_{31}^{0} = -\frac{\Omega_s}{\Omega_c}, \rho_{41}^{0} = 0. \tag{1}$$

In the second stage, double microwave pulses are applied, a Ramsey pattern is printed on the coherence, the coherence between |1> level and |3> level becomes

$$\rho_{31}(T+2t) = e^{i\frac{\Delta}{2}(T+2t)-\gamma_3 T}[e^{-i\frac{\Delta}{2}T}(\cos\xi t - i\frac{\Delta}{2\xi}\sin\xi t)^2 - e^{i\frac{\Delta}{2}T}(\frac{\Omega_m}{\xi}\sin\xi t)^2]\rho_{31}^{0}, \tag{2}$$

where $\xi = \sqrt{(\Delta/2)^2 + \Omega_m^2}$, t is the microwave pulse duration, T is the duration between two microwave pulses, and Δ is the detuning between the microwave frequency ω_m and the clock transition ω_{34},

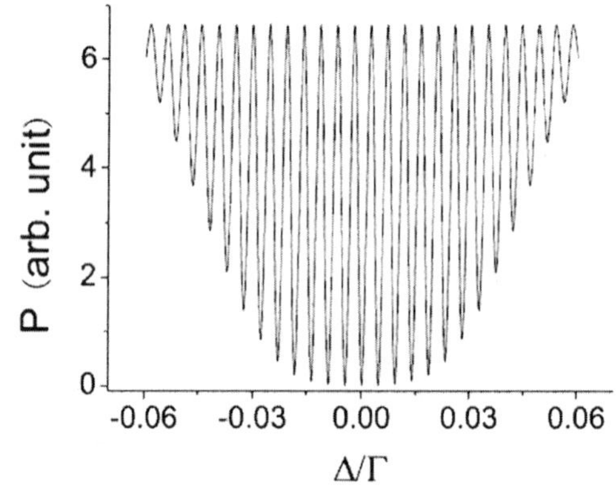

Fig. 6: The Ramsey fringe of the detected signal when $\Omega_m t = \pi/4$, $\gamma T = 400\pi$.

In the final signal detection stage, we apply a strong coupling laser pulse to get back the signal light pulse again, the detected signal is

$$P = P_0 e^{-2\gamma_3 T}[1 - 2\cos^2\xi t \sin^2\xi t(1+\cos\Delta T)]. \tag{3}$$

This exhibits a Ramsey pattern. The full-width at half maximum of the central Ramsey fringe is

$$\Delta\omega_{1/2} = \pi / T . \tag{4}$$

We plot a numerical result in Fig. 6 when $\Omega_m t = \pi/4$, $\gamma T = 400\pi$. When the microwave pulse is a $\pi/4$ pulse, i.e, $\Omega_m t = \pi/4$, the signal has the maximum signal to noise ratio. Since the central signal of this Ramsey fringe is zero, it is a homodyne detection. When $\Omega_m t = \pi/2$, the Ramsey pattern vanishes. The physical meaning is that, we detect the mode of ρ_{31}, if the pulses are $\pi/2$ pulses, there will be a Ramsey pattern printed on the phase of ρ_{31} which can not be shown in the intensity detection.

This PCS scheme has several advantages: First, light shift is eliminated due to the pulsed method; Second, it only needs a low Q cavity, the cavity pulling effect can be neglected. Comparing with POP, PCS scheme will be 10^{-3} smaller than the cavity pulling of POP. Third, the detection is homodyne detection, making the signal contrast quite high. And it is quite immune to laser noise.

The frequency stability of shot noise limit is

$$\sigma(\tau) = \frac{1}{\pi Q_a S / N} \sqrt{\frac{T_{total}}{\tau}} , \tag{5}$$

Using the data listed in[5], we get an estimated value of the shot noise limit,

$$\sigma(\tau) = 2.0 \times 10^{-14} \tau^{-1/2} . \tag{6}$$

4. Proposal of the experiment of PCS

The setup of our proposal is shown in Fig. 7. The cell, which contained Rb atoms and buffer gas, is placed in a microwave cavity. The cavity is tuned to the resonant frequency of the clock transition of ^{87}Rb atoms. In Fig. 7, we plot two possible energy levels for our proposal. The (b) (c) structures are both widely used in EIT and coherent light storage experiments.

The coherent time is a very important parameter. A long coherent time can lead to a long interrogation time, thus decreases the width of the central Ramsey fringe, and increases the signal intensity. So a high signal to noise ratio can be obtained. Usually, the coherent time in a cell is typically below 1ms. Using the coated cell, the light storage time can be much longer, 2ms and longer coherent time are reported[15, 16]. Another choice is using cold atoms as the atom source.

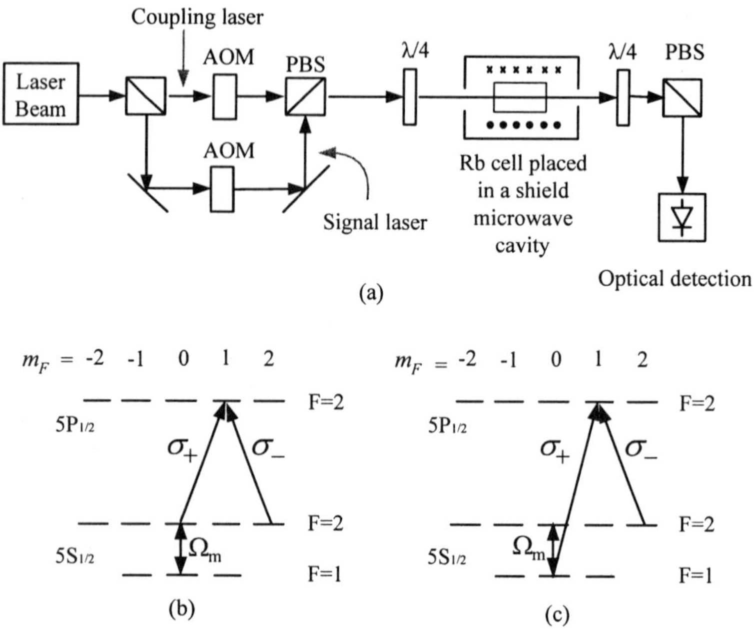

Fig. 7: The setup of our proposal (a), and two possible energy levels used in PCS scheme (b), (c).

Such as cold atoms in the integrating sphere. In this way, the light storage time can be much longer. But the atom density is lower, which requires a much sensitive detection to get this signal.

Conclusion

In conclusion, we report the researches about atomic frequency standards in SIOM. The cold atom gas in the intergrating sphere is a novel source which can be used for the space and compact atomic clocks. And we have proposed a pulsed frequency standard based on coherent light storage and retrieval which is light shift free, low Q required, homodyne detection, and immune to laser noise. This method can also be applied to other elements and cold atoms. It is very promising to build a compact atomic clock with high frequency stability by using the PCS method.

References

1. J. Vanier and C. Mandache, *Appl. Phys. B* **87**, 565 (2007).
2. T. Zanon, S. Guerandel, E. de Clercq, D. Holleville, N. Dimarcq, and A.

Clairon, *Phys. Rev. Lett.* **94**, 193002 (2005).
3. J. Vanier, *Appl. Phys. B* **81**, 421 (2005).
4. A. Godone, S. Micalizio, and F. Levi, *Phys. Rev. A* **70**, 023409 (2004).
5. B. Yan, Y. Ma, and Y. wang, *Arxiv:0810.0623* (2008).
6. Y. Z. Wang, in *National Symposium on Frequency Standards*, Chengdu, China, (1979).
7. W. Ketterle, A. Martin, M. A. Joffe, and D. E. Pritchard, *Phys. Rev. Lett.* **69**, 2483 (1992).
8. H. Batelaan, S. Padua, D. H. Yang, C. Xie, R. Gupta, and H. Metcalf, *Phys. Rev. A* **49**, 2780 (1994).
9. H. X. Chen, W. Q. Cai, W. Shu, F. S. Li, and Y. Z. Wang, *Chin. J. Lasers* **21** (1994).
10. Y. Z. Wang and L. Liu, *Aust. J. Phys* **48** (1995).
11. E. Guillot, P.-E. Pottie, and N. Dimarcq, *Opt. Lett.* **26**, 1639 (2001).
12. S. Tremine, S. Guerandel, D. Holleville, A. Clairon, and N. Dimarcq, in *Frequency Control Symposium and Exposition, 2004. Proceedings of the 2004 IEEE International*, 2004), p. 65.
13. A. Godone, F. Levi, S. Micalizio, E. K. Bertacco, and C. E. Calosso, *Ieee Transactions on Instrumentation and Measurement* **56**, 378 (2007).
14. M. Fleischhauer and M. D. Lukin, *Phys. Rev. Lett.* **84**, 5094 (2000).
15. I. Novikova, A. V. Gorshkov, D. F. Phillips, A. S. Sorensen, M. D. Lukin, and R. L. Walsworth, *Phys. Rev. Lett.* **98**, 243602 (2007).
16. M. Klein, I. Novikova, D. F. Phillips, and R. L. Walsworth, *J. Mod. Opt.* **53**, 2583 (2006).
17. H. Gao, M. Rosenberry, J. Wang, and H. Batelaan, *J. Phys. B* **38**, 1857 (2005).
18. I. Novikova, M. Klein, D. F. Phillips, and R. L. Walsworth, in *Advanced Optical and Quantum Memories and Computing II* (SPIE, San Jose, CA, USA, 2005), p. 87.

THE PTB FOUNTAIN CLOCK ENSEMBLE PRELIMINARY CHARACTERIZATION OF THE NEW FOUNTAIN CSF2

N. NEMITZ*, V. GERGINOV, R. SCHRÖDER, S. WEYERS and R. WYNANDS

Physikalisch-Technische Bundesanstalt,
Braunschweig, Germany
**E-mail: Nils.Nemitz@ptb.de*

We present a first uncertainty estimate for CSF2, the second Cs fountain clock at PTB. Initial measurements of the collisional shift and the quadratic Zeeman effect have recently been completed. Simultaneous measurements with CSF1 show a short term relative instability of the frequency difference of $2.6 \times 10^{-13}\,(\tau/\mathrm{s})^{-1/2}$ for averaging time τ and a relative stability of better than 10^{-15} after 7×10^4 s.

1. Introduction

PTB has operated the caesium fountain clock CSF1 since 2000, regularly contributing to the international time scale TAI.[1,2] A second fountain, CSF2, has recently become operational, thus providing a stable reference for internal comparisons. This will allow for an easier investigation of systematic effects than a comparison to the pair of hydrogen masers at PTB. It will also make it possible to work on implementing further improvements on one fountain while the other can still contribute to TAI.

2. Setup of CSF2

The construction of CSF2 is based on the design of CSF1, but incorporates some improvements.[3] Figure 1 shows a schematic.

The beam geometry was changed to a (111) geometry that allows for larger trapping beams of 2.2 cm $1/e^2$ diameter. These create optical molasses in a (lin $\perp$ lin) polarization configuration that is directly loaded from the background gas.

Additional field coils near the detection zone ensure a controlled quantization axis for optical pumping into the stretched Zeeman state. This reduces hyperfine pumping, resulting in an improved contrast of the Ram-

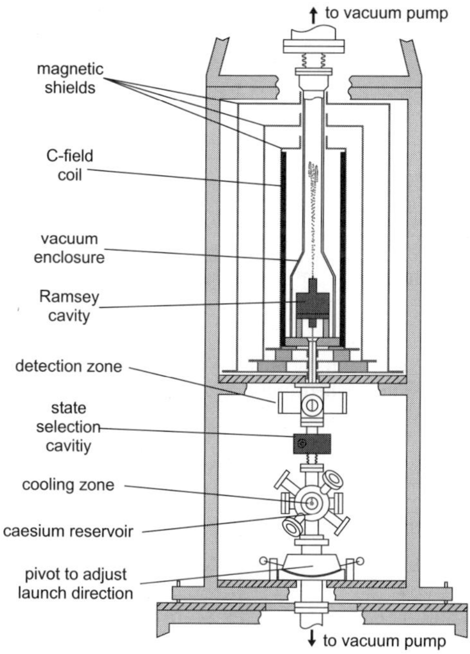

Fig. 1. Schematic of the CSF2 caesium fountain clock

sey fringe of up to 96%.

Two vacuum pumps, mounted at the top and the bottom of the vacuum enclosure, reduce the background gas pressure. At a cloud temperature of less than 1 μK, as calculated from the vertical expansion of the cloud, up to 30% of the state selected atoms return to the detection zone.

A pivot for the lower part of the vacuum system makes it possible to align the launching direction independently from the orientation of the drift tube. This has already been used to great success to reduce atom losses due to a non-vertical launch.

3. Stability and Preliminary Uncertainty Budget

The stability of CSF2 was tested against CSF1 by comparing the 5 MHz output signals from the quartz oscillators of the two clocks. Figure 2 shows the resulting Allan deviations. The short term instability of the difference is $\sigma_y^{\rm diff}(\tau) = 2.6 \times 10^{-13}(\tau/{\rm s})^{-\frac{1}{2}}$. The Allan deviation reaches 10^{-15} at approximately 7×10^4 s with no sign of leveling off. Taking into account CSF1's known instability value of $\sigma_y^{\rm CSF1}(\tau) = 1.4 \times 10^{-13}(\tau/{\rm s})^{-\frac{1}{2}}$ and as-

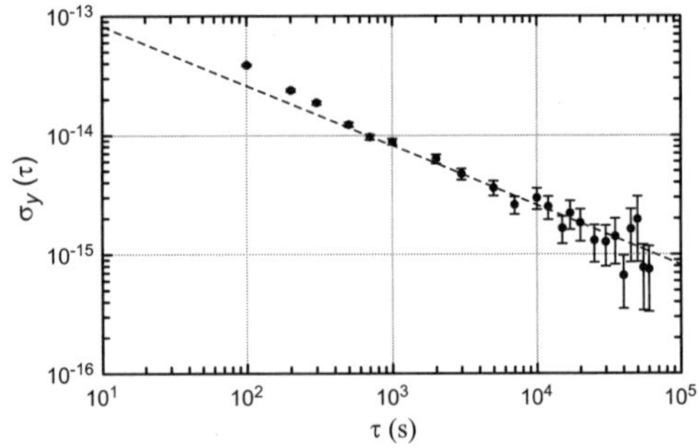

Fig. 2. Allan standard deviation $\sigma_y(\tau)$ for averaging time τ of relative frequency differences y(CSF2-CSF1). The dashed line indicates an Allan standard deviation $\sigma_y(\tau) = 2.6 \times 10^{-13}(\tau/\mathrm{s})^{-1/2}$.

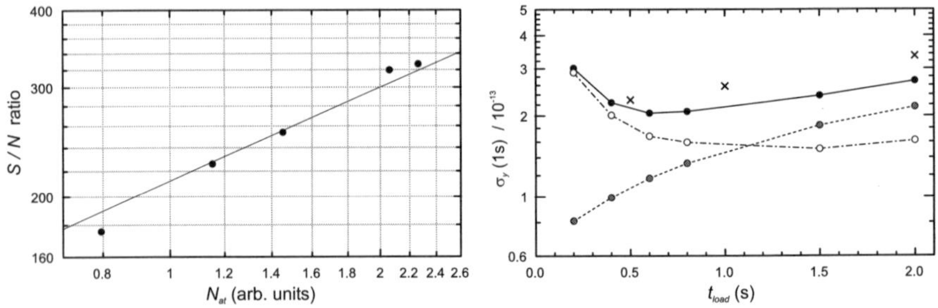

Fig. 3. **(left)** Measured signal-to-noise ratio S/N over detected atom number N_{at}, measured with resonant $\pi/4$ pulses. The solid line corresponds to the quantum projection noise limit for $45000 \times N_{\mathrm{at}}$ atoms.
(right) Contributions to the instability of CSF2 for different loading times. Open circles show contributions from the measured signal-to-noise ratio, gray circles show the calculated Dick effect contribution. Solid black points show the combined result and crosses indicate measured stability values.

suming uncorrelated noise, the instability of CSF2 alone was calculated to be $\sigma_y^{\mathrm{CSF2}}(\tau) = 2.2 \times 10^{-13}(\tau/\mathrm{s})^{-\frac{1}{2}}$.

This is consistent with expectations based on measurements of the signal-to-noise ratio for various loaded atom numbers and calculations of the Dick effect, as shown in Figure 3. The signal-to-noise ratio was measured with resonant pulses to reduce the influence of quartz oscillator noise. The pulse area of each interaction in the Ramsey cavity was reduced to $\frac{\pi}{4}$ such that the total transition probability was still approximately 50%. Its stability then appears to be mostly quantum-projection noise limited, indicating around 50000 detected atoms during normal operation.

Using standard methods for measuring the field in the drift tube, the collisional shifts and other contributions, a first uncertainty budget has been compiled for CSF2. Table 1 compares this to the latest known values of CSF1.

Table 1. Frequency biases and uncertainties for CSF1 and CSF2

physical effect	CSF1 bias (10^{-15})	CSF1 uncert. (10^{-15})	CSF2 bias (10^{-15})	CSF2 uncert. (10^{-15})
second order Zeeman shift	46.2	0.1	100.3	0.1
AC Stark shift	-16.6	0.1	-16.6	0.1
cold collisions	-1.8	0.3	-0.3	0.7[a]
gravitational red shift	8.6	0.1	8.6	0.1
cavity phase	—	0.5	—	0.5[a]
Majorana transitions	—	0.1	—	*0.1*[b]
Rabi and Ramsey pulling	—	0.1	—	*0.1*[b]
microwave leakage	—	0.1	—	*0.1*[b]
electronics	—	0.2	—	*0.2*[b]
light shift	—	0.1	—	0.1
background gas collisions	—	0.1	—	0.1
microwave power dependence	—	0.6	—	0.4
total type B uncertainty		0.9		1.0[a]

Note: [a] improvement expected from future measurements
[b] still under investigation, preliminary values taken from the evaluation of CSF1

4. Planned Measurements of the g-Factor Ratio in Caesium

Having two fountain clocks greatly simplifies precision measurements. An experiment to measure the g-factor ratio g_J/g_I in caesium is currently in preparation.

The relative frequencies of the $|F=3,m\rangle$ to $|F=4,m'\rangle$ transitions can be calculated using the Breit-Rabi formula

$$f(m',m) = \underbrace{\frac{f_{\rm hfs}}{2}\left(\sqrt{1+\frac{m'}{2}x+x^2}+\sqrt{1+\frac{m}{2}x+x^2}\right)}_{\rm (a)} + \underbrace{(m'-m)}_{\rm (b)}\frac{g_I\,\mu_B}{h}B$$

$$\text{with } x = \frac{(g_J-g_I)\,\mu_B}{h\,f_{\rm hfs}}B \quad . \tag{1}$$

Term (a) is left unchanged if m and m' are exchanged, while term (b) undergoes a change of sign. The difference between $f(0,-1)$ and $f(-1,0)$

is proportional to $g_I\, B$, while the sum eliminates term (b):

$$f(0,-1) + f(-1,0) \approx 2\, f_{\text{hfs}} - \frac{(g_J - g_I)\, \mu_B\, B}{4\, h} \quad . \tag{2}$$

This makes it possible to eliminate the influence of the magnetic field and calculate the g-factor ratio g_J/g_I. A large improvement in accuracy over the literature value[4] is expected. Figure 4 shows the required measurements.

The Ramsey cavity was designed to excite $\Delta m = 0$ transitions. However, the mode of the RF-field invariably has orthogonal components that allow the excitation of the $\Delta m = \pm 1$ transitions required for these experiments. First tests show Ramsey fringes with good contrast for these transitions if the RF-field amplitude is increased, as seen in Figure 4. The split envelope of the fringes is due to the distribution of the horizontal magnetic field components in the resonator.[5]

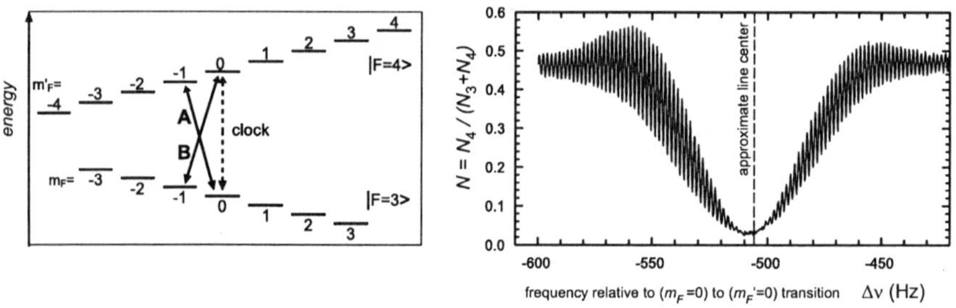

Fig. 4. **(left)** Transitions probed in the proposed g-factor measurements. **(right)** Ramsey fringe taken for transition A. The RF-field amplitude was increased by a factor of 19 for this measurement.

References

1. S. Weyers, U. Hübner, R. Schröder, C. Tamm and A. Bauch, *Metrologia* **38**, 343 (2001).
2. S. Weyers, A. Bauch, R. Schröder and C. Tamm, The atomic caesium fountain CSF1 of PTB, in *Proceedings of the 6th Symposium on Frequency Standards and Metrology*, (World Scientific, 2001).
3. R. Wynands, D. Griebsch, R. Schröder and S. Weyers, Current status of PTB's new caesium fountain clock CSF2, in *Proceedings of the 20th European Frequency and Time Forum (EFTF)*, 2006.
4. C. W. White, W. M. Hughes, G. S. Hayne and H. G. Robinson, *Phys. Rev. A* **7**, 1178 (1973).
5. A. Bauch, B. Fischer, T. Heindorff and R. Schröder, *Metrologia* **35**, 829 (1998).

THE PULSED OPTICALLY PUMPED CLOCK: MICROWAVE AND OPTICAL DETECTION*

SALVATORE MICALIZIO, ALDO GODONE, FILIPPO LEVI, CLAUDIO CALOSSO

Istituto Nazionale di Ricerca Metrologica, Str. delle Cacce 91
Torino, 10132, Italy

We present theoretical and experimental studies concerning the realization of a rubidium vapor cell frequency standard based on the pulsed optical pumping (POP) technique. This technique is extremely powerful since it avoids in principle the light shift effect and the noise conversions from the laser to the clock signal. The detection of the clock transition in the microwave domain (the so called POP maser) has been extensively investigated at INRIM and a frequency stability of $1.1\times10^{-12}\,\tau^{-1/2}$ has been measured until integration times of 10^5 s, reaching the level of 5×10^{-15} (drift removed). The optical detection is still under evaluation but this technique may lead in effect to some advantages with respect to the microwave detection, such as a lower or negligible cavity pulling, a lower operation temperature (which implies a lower spin exchange contribution) and a higher signal-to-noise ratio. All these features may lead to a better frequency stability performance than that previously reached.

1. Introduction

The pulsed operation mode of the vapor cell frequency standards (^{87}Rb, ^{133}Cs) optically pumped by a semiconductor laser has provided better performances with respect to the continuous mode of operation [1, 2]. In order to exploit the full capabilities of this approach we report here a theoretical multilevel description of the atomic medium that takes into account the full Zeeman manifold of the involved energy states, the laser absorption along the cell, the real microwave cavity mode and its feedback action on the atomic ensemble.

The theoretical predictions are compared with the experimental results obtained with a laboratory prototype based on ^{87}Rb atoms in buffer gas. A very satisfactory agreement is found, proving the validity of the theory here reported. Following the guidelines here reported it appears possible to approach the short-medium term frequency stability performances of the H-maser with an easier and more reliable technical implementation of basic importance for on-board space applications.

---

* This work has been supported by ASI (Italian Space Agency, contract I/057/06/0).

2. Theoretical background

We suppose to have a sample of ^{87}Rb atoms contained in a cell with a proper buffer gas. We refer to an 11-level system composed of the Zeeman manifold of the ground state (8 levels) and of the Zeeman manifold of the excited state (3 levels).

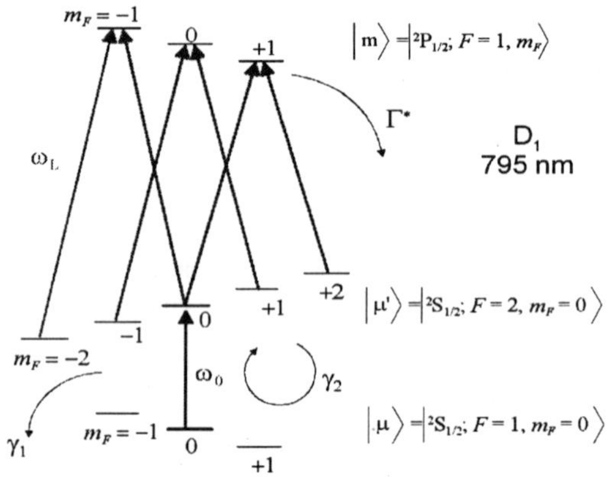

Figure 1. Atomic levels and interaction fields involved in the POP clock operation; ω_0 and ω_L are the angular microwave and laser frequencies; γ_1, γ_2 and Γ^* are the relaxation rates of the population inversion, of the coherence and of the excited state respectively

To fully describe the interaction depicted in the figure three general equations have been considered (with usual meaning of the symbols):

$$\frac{\partial \hat{\rho}}{\partial t} = D\nabla^2\hat{\rho} + \frac{1}{i\hbar}\left[\hat{H},\hat{\rho}\right] \tag{1}$$

$$\nabla\times\nabla\times\vec{H} + \frac{1}{c^2}\frac{\partial^2\vec{H}}{\partial t^2} = -\frac{1}{c^2}\frac{\partial^2\vec{M}}{\partial t^2} \tag{2}$$

$$\nabla\times\nabla\times\vec{E} + \frac{1}{c^2}\frac{\partial^2\vec{E}}{\partial t^2} = -\mu_0\frac{\partial^2\vec{P}}{\partial t^2} \tag{3}$$

Equation (1) is the Liouville equation for the density matrix. in this equation the Hamiltonian takes into account also of the collisions with buffer gas: the interaction atom-buffer gas is described through a random magnetic field and consequently selection rules among the sublevels take place [3].

Equations (2) and (3) are the Maxwell equations that account for microwave and optical transitions occurring in the operation of the clock.

3. Principle of operation and experimental set-up

At first the atoms experience a strong laser pulse to invert the population between the two ground state hyperfine levels which define the clock transition. Then the atoms are interrogated with two microwave pulses separated in time (Ramsey interaction scheme), and finally a detection window is enabled so that the atoms that have made the clock transition are detected.

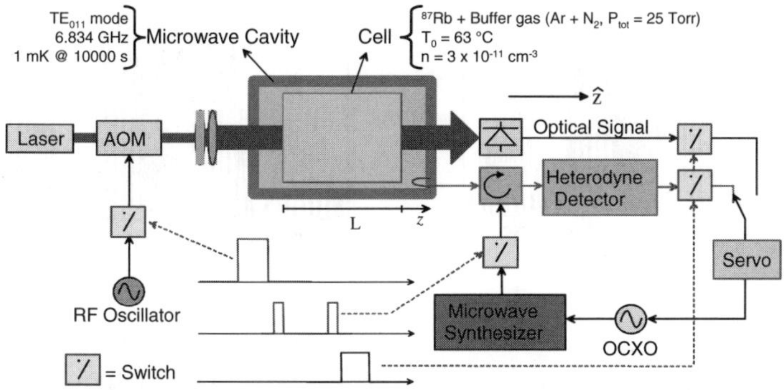

Figure 2. Experimental set-up of the POP clock.

The laser frequency is locked to the optical transition using an external cell containing ^{87}Rb only. An AOM works as optical switch and frequency tuning device. A quartz cell contains the ^{87}Rb vapor and the buffer gas mixture. The cell is placed in a temperature controlled microwave cavity. A double oven is used, allowing thermal fluctuations to stay below 100 μK. A solenoid provides the quantization magnetic field; three μ-metal shield are used in order to guarantee sufficient field uniformity and shielding. The interrogating microwave is obtained from (OCXO) by means of a direct synthesis chain. A coupling loop in the cavity allows the interaction of the atoms with the synthesized microwave as well as the detection of the emitted signal through a heterodyne receiver. The detected signal is acquired by a PLD that implements the frequency loop control. When all the apparatus operates as a clock, the 10 MHz OCXO is frequency locked to the atomic resonance via a digital servo loop.

4. Experimental results

When the clock transition is detected in the microwave domain (maser approach), the signal is proportional to the square absolute value of the microwave coherence $|\delta_{26}|^2$ generated between the two clock sublevels.

We report on the left the calculated $|\delta_{26}|^2$ versus the detuning of the (angular) frequency of the externally applied microwave with respect to the clock transition frequency, while the corresponding experimental curves are reported on the right. We consider the case of low temperature (40 °C, up) and high temperature (63 °C, down).

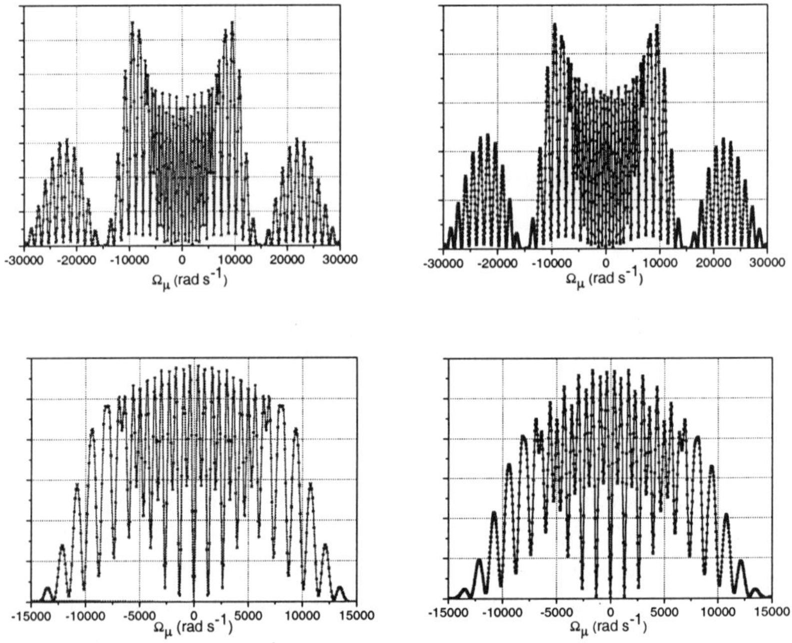

Figure 3. Computed (left) and experimentally observed (right) Ramsey fringes for two operation temperatures (vertical axis: arbitrary units).

The agreement between theory and experiment is very good; in more detail, the theory predicts a frequency stability for the POP maser of $1x10^{-12}\ \tau^{-1/2}$, in excellent agreement with the measurement reported in [4].

5. Optical detection

The optical detection can lead in principle to some advantages with respect to the maser approach. Among others, a high Q cavity is no more required, the clock transition being observed on the laser signal transmitted through the cell. The working temperature can be lower, reducing the spin exchange and the cavity pulling, two effects proportional to the atomic density.

Figure 4 shows the computed contrast C and the signal to noise ratio R versus the pumping rate of the laser during the detection phase for two temperatures corresponding to a transparent medium and to a dense medium. It turns out that provided $t_d << t_p$ high pumping rates can be used without affecting significantly the clock stability.

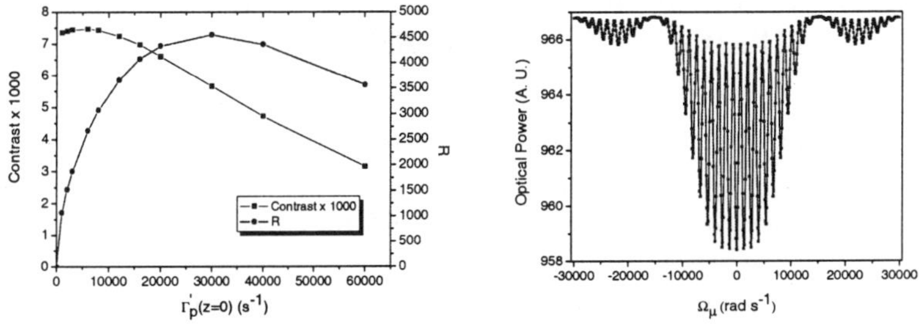

Figure 4. Computed contrast (squares) and signal-to-noise ratio (circles) of the Ramsey fringes shown on the right for an operating temperature of 40 °C. The central fringe has a linewidth of 100 Hz.

R is due to the shot noise related to the optical photons detection process. These photons are converted into a photocurrent by a photodiode. Assuming a quantum efficiency of 0.8 it turns out R=4000. Evaluating the Allan deviation as $\sigma_y(\tau) = \frac{\sqrt{T_C}}{\pi Q_a R}\tau^{-1/2}$ (being T_C the cycle time and Q_a the atomic quality factor), it turns out $\sigma_y(\tau) = 2\text{x}10^{-13}\,\tau^{-1/2}$.

6. Conclusions

In conclusion, we have realized and operated a prototype of a pulsed optically pumped maser obtaining very interesting stability performances both on the short and medium term. Excellent agreement with theoretical calculation was achieved. The theory suggests that several improvements can be achieved with pulsed operation and optical detection [4]. A new prototype is currently under development.

References

1. A. Godone *et al.*, *Phys. Rev.* **A74**, 043401 (2006).
2. S. Guerandel *et al.*, *IEEE Trans. Instrum. Meas.*. **56**, 383 (2007).
3. J. Vanier and C. Audoin, *The quantum physics of atomic frequency standards*, Adam-Hilger, Bristol (England, 1989).
4. S. Micalizio, A. Godone, F. Levi, C.E. Calosso, to appear on *Physical Review A*.

RESEARCH ON CHARACTERISTICS OF PULSED OPTICALLY PUMPED RUBIDIUM FREQUENCY STANDARD *

JIANLIAO DENG, ZHENGFENG HU, LIN LI, HUIJUAN HE

Key Laboratory for Quantum Optics, Shanghai Institute of Optics and Fine Mechanics, Chinese Academy of Sciences, Shanghai, 201800, The People's Republic of China.

In this paper we present some experimental results obtained with a laboratory setup of a pulsed optically pumped rubidium vapor cell with buffer gas frequency standard. A central Ramsey fringe with FWHM of $\Delta\nu$ =140±10Hz is observed in the transmitted optical signal, and a central Ramsey fringe with FWHM of $\Delta\nu$ = 60±10Hz is also observed in the free-induced decay signal at the end of the second microwave pulse. The effect of the microwave pulse power upon the Ramsey fringe is studied.

1. Introduction

The atomic clock is operated in a pulsed regime, all of the pumping light, the interrogating microwave signal, and the detection phase are occur at different times, but not simultaneous. The pulsed regime is the successful techniques. The original concept of pulsed optically pumped (POP) is conceived by Alley [1] in the early 1960s and has been applied to frequency standards for the first time by Arditi and Carver in 1964 [2]. The pulsed regime has been developed by Godone and his team group in recent years. They discussed both theoretically and experimentally a pulsed optically pumped (POP) frequency standard based on optical pumping of a diode laser, on a temporal Ramsey interaction for the microwave interrogation, and on the observation of the atomic free-induction decay to detect the clock signal [3] [4].

The operation of the POP frequency standard includes three steps. The first one: a strong diode laser pulse tuned to the optical D_1 transition of Rb^{87} pumps the atoms in the ground state; the second one: the atoms are then interrogated with two Ramsey microwave pulses with duration t_m and separated by a time T; the third one: a weak laser pulse detects the atoms that have made the clock transition in the optical domain or a free-induced decay signal is observed in the microwave domain.

---

* This work is supported by the Major State Basic Research Development Programme of China under Grant No 2005CB724507.

The full width at half maximum of the central Ramsey fringe is expressed as follows[3].

$\Delta\nu_{1/2} = 1/2T$ optical detection

$\Delta\nu_{1/2} = 1/4T$ microwave detection

These three steps of operation of the POP frequency standard are controlled by timing pattern as shown in Fig. 1.

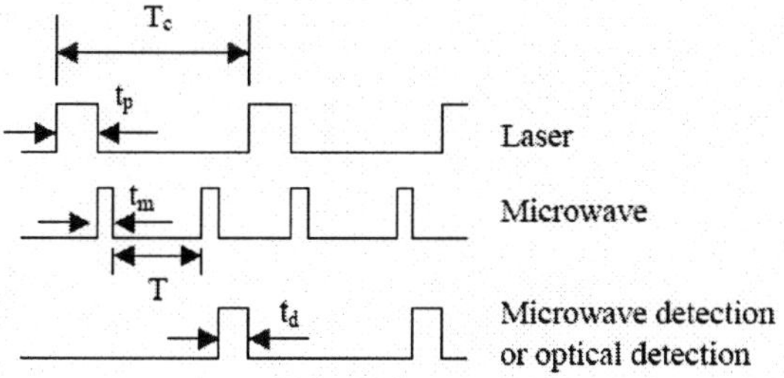

Figure 1. Timing pattern: T_c cycle period; t_p optical pumping time; t_m duration of the microwave pulse; T separation of two microwave pulses; t_d detection time. The cycle time is then $T_c = t_p + T + 2t_m + t_d$.

2. Experimental setup

The experimental setup is shown in Fig. 2. A tunable external cavity semiconductor diode laser (DL100, Toptica) at 795 nm (D_1 Rb transition) is used for the optical pumping. It is linearly polarized and its frequency is locked through a servo loop via saturated absorption spectrum to the $|{}^2S_{1/2}; F=2\rangle - |{}^2P_{1/2}; F=2\rangle$ transition. The main part of the laser beam then pass an acousto-optics modulator (AOM) operating in the double pass configuration. The AOM produces the laser pulses for optical pumping. The laser beam is expanded to a diameter 20mm by a beam expander before entering the cell. The aluminum cavity resonates at 6.8 GHz in TE_{011} mode. The Rb cell has a length of 30 mm with diameter of 30 mm. The Rb cell contains Rb^{87} vapor (98%) and a thermally compensated buffer gas mixture (15.5 Torr of Ar and 9.5 Torr of N_2). The cavity is temperature controlled and placed inside a three layered magnetic shields. A longitudinal static magnetic field of 50mG is applied to resolve the Zeeman levels. A photodiode is used to detect the transmitted optical signal. The circulator allows the interrogating Ramsey microwave pulses provided by the RF synthesizer coupling into cavity and detection of the free-induced decay signal through a spectrum analyzer. These all pulses are controlled through a data acquisition card PCI-6733 (National Instruments, USA). The detection signal is record by a computer data acquisition system containing USB6251 card and Labview software.

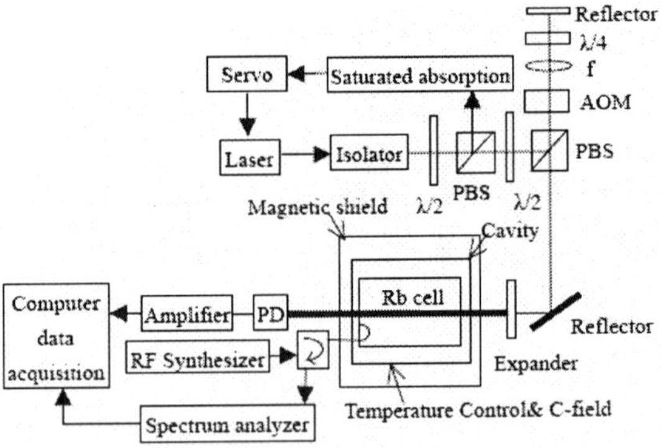

Figure 2. Experimental setup.

3. Experimental Results

Ramsey fringes can be observed in the transmitted optical signal by measuring the transmitted power of the weak laser probe pulse versus the microwave detuning. Ramsey fringes shown in Fig. 3 are obtained with the following parameters: T=3.5ms,t_m =0.4ms, t_p =4ms and t_d = 1.5ms. The laser intensity at the entrance of the cell is about 5mW/cm^2. The line width of the central Ramsey fringe is about $\Delta\nu$=140±10Hz, in agreement with the well-known relation. $\Delta\nu_{1/2}$ = 1/2T. A.Godone *et al.* introduced some theoretical consideration about the optical-detection scheme[5], but no experimental results were reported. In this paper we report the experimental results in optical domain.

We have studied the Ramsey fringes in the microwave domain by measuring the output power of the cavity at the end of the second Ramsey pulse versus the microwave detuning. The line width of the central Ramsey fringe as function of T is measured. The line width obviously becomes narrower when T is increased. Ramsey fringes shown in Fig. 4 are obtained with the following parameters: T=4.5ms, t_m =0.4ms, t_p =4ms and t_d = 2ms. The line width of the central Ramsey fringe is about $\Delta\nu$=60±10Hz.

Then the effect of the microwave pulse power upon the Ramsey fringe is studied. The Ramsey fringes with different microwave power are shown in Fig. 5. We find in Fig.5b that the peak of the Ramsey fringes begin to split into two peaks when the microwave power is increasing. If the microwave power continues to increase, then the peak of the Ramsey fringes absolutely split into two peaks, see Fig.5c. Therefore the line width of central Ramsey fringes is in agreement with the expression $\Delta\nu_{1/2}$ = 1/4T at higher microwave power, but the line width of central Ramsey is in agreement with the expression $\Delta\nu_{1/2}$ = 1/2T at lower microwave power.

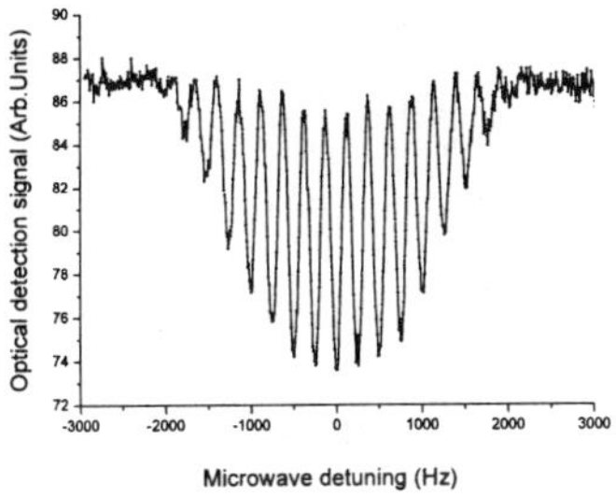

Figure 3. Ramsey fringes as observed in optical domain versus microwave detuning with the following parameters: T=3.5ms,t_m =0.4ms, t_p =4ms and t_d = 1.5ms.

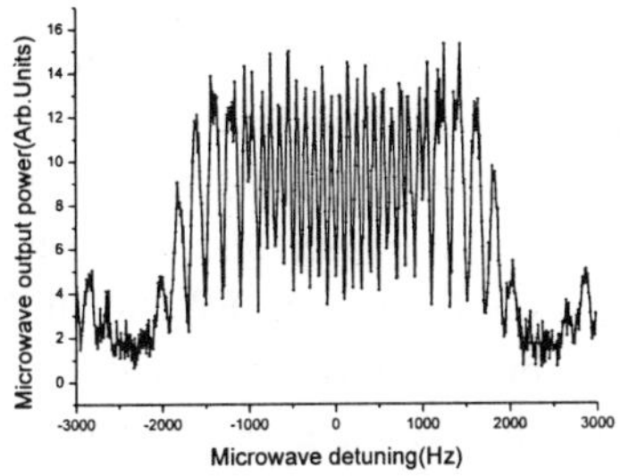

Figure 4. Ramsey fringes as observed in microwave domain versus microwave detuning with the following parameters: T=4.5ms,t_m =0.4ms, t_p =4ms and t_d = 2ms.

(a)

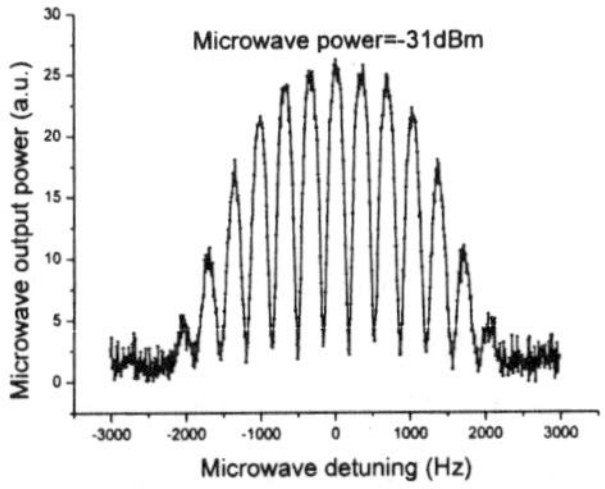

(b)

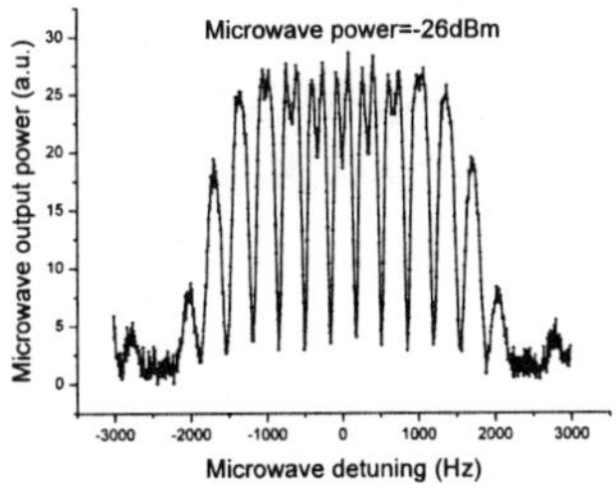

(c)

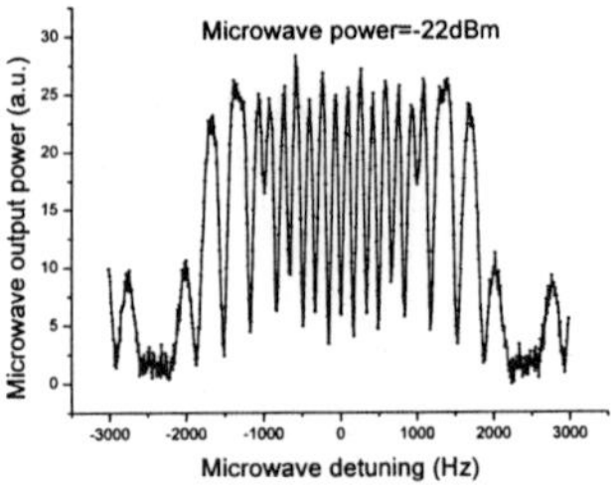

Figure 5. Ramsey fringes in microwave domain at different microwave power with the following parameters: T=2.5ms,t_m =0.4ms, t_p =4ms and t_d = 2ms.

4. Conclusions

In this paper some experimental results of POP rubidium vapor cell with buffer gas frequency standard are reported. The Ramsey fringes by optical detection mode and by microwave detection mode are studied respectively. A central Ramsey fringe with FWHM of $\Delta\nu$ =140±10Hz is observed in the transmitted optical signal. A central Ramsey fringe with FWHM of $\Delta\nu$ = 60±10Hz is also observed in the free-induced decay signal at the end of the second microwave pulse. The experimental results are in good agreement with the theoretical prediction reported in [4]. The experimental results in microwave domain allow to double the quality factor of the atomic transition with respect to the traditional observation and without laser background signal in the atomic response. The results show the POP frequency standard is extremely attractive for space applications.

References

1. C. O. Alley, in *Quantum Electronics*, edited by C.H. Townes (Columbia University Press, Now York, 1960)
2. M. Arditi, and T. R. Carver, *IEEE Trans. Instrum. Meas.***13**, 146 (1964).
3. A. Godone, S. Micalizio, and F. Levi, *Phys. Rev. A* **70**, 023409 (2004).
4. A. Godone, S. Micalizio, F. Levi, and C. Calosso, *Phys. Rev. A* **74,** 043401 (2006).
5. A. Godone, Filippo Levi, Salvatore Micalizio, Elio K.Bertacco, and Claudio Eligio Calosso, *IEEE Trans. Instrum. Meas.* **56**, 378 (2007).

STATUS OF THE CONTINUOUS COLD FOUNTAIN CLOCKS AT METAS-LTF

A. JOYET, G. DI DOMENICO, G. K. GULATI, P. THOMANN

Laboratoire Temps-Fréquence (LTF), University of Neuchâtel, Rue A.–L.–Breguet 1
2000 Neuchâtel, Switzerland

A. STEFANOV

Swiss Federal Office of Metrology (METAS), Lindenweg 50
3003 Bern-Wabern, Switzerland

Our alternative approach to primary fountain standards is based on a continuous beam of laser-cooled atoms. Besides circumventing stability limitations inherent to pulsed operation, a continuous beam is also interesting from the metrological point of view since it provides a different accuracy budget where density related frequency shifts are negligible. Two continuous fountains — FOCS–1 and FOCS–2 — have been designed and assembled. This contribution gives the present status of this work.

1. Introduction

All primary frequency standards working with laser-cooled atoms that currently contribute to TAI are based on a pulsed scheme of operation: atoms are sequentially laser-cooled, launched vertically upwards and interrogated during their ballistic flight before the cycle starts over again. This approach has spurred important advances in time and frequency metrology. Nowadays, state-of-the-art fountains are operated in National metrology institutes[1] at an accuracy level better than 10^{-15} in relative units.

Our alternative approach to primary fountain standards is based on a continuous beam of laser-cooled atoms. Besides making intermodulation effect negligible, a continuous beam is also interesting from the metrological point of view. Indeed, the relative importance of the contributions to the error budget is different for a continuous fountain than for a pulsed one, notably for density related effects (collisional shift and cavity pulling), which are an important issue if high stability and high accuracy are to be achieved simultaneously.

The proposal to build a frequency standard based on a continuous beam of laser-cooled atoms was made several years ago and two such standards — FOCS–1 and FOCS–2 — have since been designed and assembled.

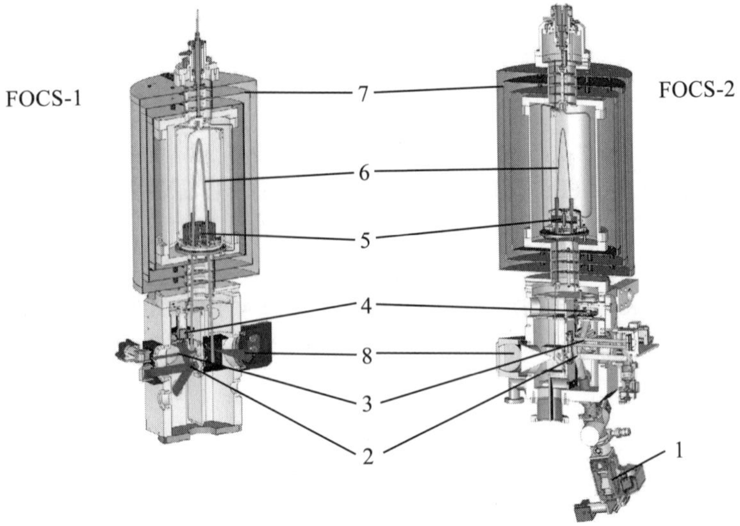

Figure 1. Schematic view of FOCS–1 and FOCS–2. 1: 2D–MOT pre-source; 2: optical molasses capture and launch region; 3: transverse cooling, two-stage collimation in FOCS–2; 4: light-trap; 5: rotatable microwave cavity; 6: atomic beam; 7: magnetic shielding; 8: detection.

2. The continuous fountains FOCS–1 and FOCS–2

The main structural difference between pulsed and continuous fountains lies in the parabolic — rather than linear vertical — trajectory of the atoms, which provides the required physical separation between atomic preparation and detection regions. Figure 1 shows a cut through the atomic resonators of FOCS–1 and FOCS–2[2]. The lower part of the vacuum vessels is separated in two compartments: source and detection. The atoms are prepared in the source region, launched upwards and detected after their parabolic flight in the detection region. The upper part of the resonator is surrounded by three layers of magnetic shields. A fourth layer (not shown in Fig. 1) encloses the whole setup, including the source and detection part. The overall height of the resonator is about 160 cm. The distance from the continuous source of atoms to the apogee of the parabola is 75 cm of which 30 cm are above the microwave cavity. The two microwave pulses of the Ramsey interrogation are thus separated by 0.5 s, yielding 1 Hz–wide Ramsey fringes. FOCS–2 is an improved version of FOCS–1, including a 2D–MOT pre-source[3] and a more efficient transverse cooling[4] both designed to increase the atomic flux by a factor of 40.

3. Frequency stability of FOCS–1

The frequency stability of the fountain has been measured and a typical result obtained at maximum flux against an H–maser is shown in Fig. 2 a). At 1s, the phase noise of the maser dominates over the free running VCXO, whereas for sampling times $\tau > 200$ s, the Allan deviation is $3 \cdot 10^{-13}\ \tau^{-1/2}$ and lies near the limit associated with the measured signal-to-noise ratio. As shown in Fig. 2 b), the absolute noise density is proportional to $\phi^{1/2}$, which is a clear signature of atomic shot noise. We conclude that in contrast to pulsed fountains whose stability is limited by the local oscillator, the stability of continuous fountains can be improved by increasing atomic flux. This is the goal of FOCS–2.

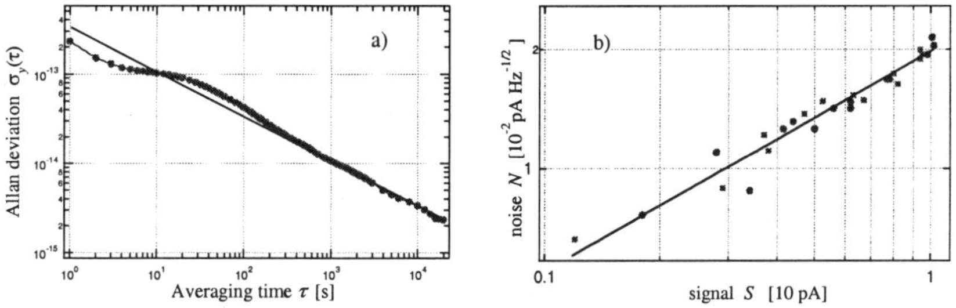

Figure 2. a) Allan deviation σ_y of the VCXO locked to FOCS–1 compared to the reference maser, as a function of the averaging time τ in seconds. The points represent the experimental data and the line indicates $3 \cdot 10^{-13}\ \tau^{-1/2}$. b) rms values of the total noise density N of the photodetector signal *vs.* the contribution S of the atomic fluorescence to the DC signal. The straight line is a fit by a power law with exponent 0.56 ± 0.02.

We have also evaluated to what extent aliasing or intermodulation will affect the frequency stability of our continuous fountains. For that purpose, we have developed a simple model which applies to a Ramsey type interrogation[5]. For a monokinetic atomic beam, the model predicts that no frequency instability due to intermodulation will be added to the locked local oscillator if the modulation frequency f_{mod} is equal to the resonator linewidth $\Delta\nu \approx 1$ Hz. To demonstrate experimentally the removal of the intermodulation effect, it was thus necessary to degrade deliberately the phase noise performance of the local oscillator[6]. We chose to inject white phase noise with an amplitude such that the intermodulation effect contribution is ~ 2 times the atomic shot noise for an experimentally sensible deviation of f_{mod} from 1 Hz (or its odd harmonics). Figure 3 shows the results as a function of the modulation frequency. There is a good agreement with the calculated intermodulation effect. The most important feature is the suppression of the effect when the modulation frequency equals an *odd* multiple of the Ramsey fringe linewidth.

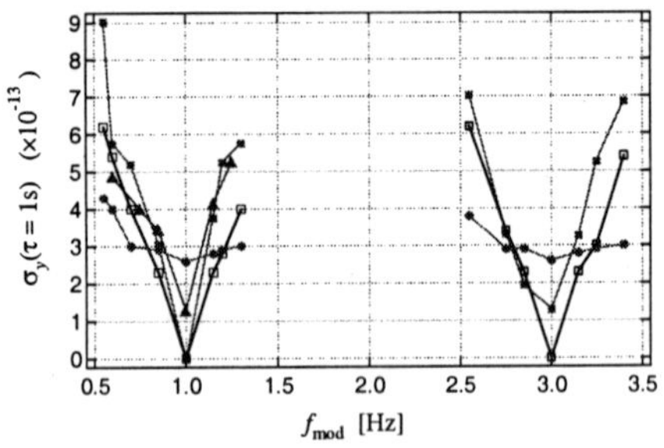

Figure 3. Allan deviation σ_y (1s) vs. interrogation modulation frequency f_{mod}: (filled circles) with square-wave phase modulation without injected phase noise, used as reference; (filled squares and triangles for square-wave phase and square-wave frequency modulations, respectively) with injected phase noise after quadratic subtraction of the reference; (open squares) predicted intermodulation effect for injected noise and square-wave modulation.

4. Accuracy issues of FOCS–1

In a continuous fountain, on the contrary to pulsed fountain, the lasers must remain on at all times and the associated fluorescence light shifts the clock frequency ($+ 1.2 \cdot 10^{-12}$ measured in FOCS–1). To solve this problem, we have implemented a continuously rotating light-trap driven by an electrostatic motor[2,7]. With a measured attenuation factor for scattered light larger than 10^4, it will reduce the total bias due to the light shift to below 10^{-16}.

The Zeeman shift correction is a topic that is dealt with differently in a continuous fountain than in a pulsed one. The problem is not so much to ensure that $\langle B^2 \rangle$ is equal to $\langle B \rangle^2$, which only requires an homogeneity of the C–field at the 4 nT level, but to ensure the proper identification of the central fringe in the $m_F \neq 0$ Ramsey patterns, in order to measure $\langle B \rangle$. Motivated by the experimental observation that Ramsey patterns for $m_F \neq 0$ lose contrast and become asymmetrical due to inhomogeneities, Ramsey fringes have been calculated for different configurations and criteria have been developed to minimize the deviations from a uniform magnetic field and thereby setting an upper limit to the uncertainty on the identification of the central fringe[8]. The present magnetic field situation (gradient < 2 nT / m near the apogee) allows us to state the corresponding uncertainty to be about 10^{-15}, and is designed to be significantly lower in FOCS–2.

5. Conclusion and outlook

With our alternative approach to primary frequency standards based on a continuous beam of laser-cooled atoms, intermodulation effects and limitations on the short term stability have been experimentally shown to play a negligible role in continuous Ramsey resonators. As a result, we have achieved, with our fountain FOCS–1, a short term stability of $3 \cdot 10^{-13}$ at 1s (Allan deviation) with a quartz local oscillator, limited by the available atomic flux only.

The second improved fountain FOCS–2 is nearing completion and should bring an expected increase of the atomic flux by a factor of 40.

The continuous beam techniques (high flux, collimated continuous sources, light shift suppression) developed for FOCS–1 and FOCS–2 have made substantial progress and show promise for other applications of cold atoms, such as for example atom interferometers used as inertial sensors.

References

1. R. Wynands, *et al.* Atomic fountain clocks. *Metrologia* **vol. 42**, S64–S79 (2005).
2. A. Joyet, *et al.* An Alternative Approach to Primary Cs Frequency Standards. *Proc. of the 22th Eur. Freq. and Time Forum*, 23–25 April 2008, Toulouse, France.
3. N. Castagna, *et al.* A novel simplified two-dimensional magneto-optical trap as an intense source of slow cesium atoms. *Eur. Phys. J. Appl. Phys.* **vol. 34**, p. 2130 (2006).
4. G. Di Domenico, *et al.* On the stability of optical lattices. *Laser Phys.* **vol. 15**, no. 12, pp. 1674–1679, 2005.
5. A. Joyet, *et al.* Theoretical study of the Dick effect in a continuously operated Ramsey resonator. *IEEE Trans. Instrum. Meas.* **50**(1), pp. 150–156, 2001.
6. J. Guéna, *et al.* An experimental study of intermodulation effects in an atomic fountain frequency standard. *Eur. Phys J. Appl. Phys.* **388**, 183–189 (2007).
7. F. Füzesi, *et al.* An electrostatic glass actuator for ultrahigh vaccum: A rotating light trap for continuous beams of laser-cooled atoms. *Rev. Sci. Instrum.* **78**, 103109 (2007).
8. A. Joyet, Aspects métrologiques d'une fontaine continue à atomes froids. *Ph.D. thesis. University of Neuchâtel* (2003).

EXPERIMENTS WITH A NEW $^{201}Hg^+$ ION CLOCK

E.A. BURT[1], S. TAGHAVI-LARIGANI[1], S.N. LEA[2], J.D. PRESTAGE[1], AND R.L. TJOELKER[1]

[1]*Jet Propulsion Laboratory, California Institute of Technology, 4800 Oak Grove Drive, Pasadena, CA*
[2]*National Physical Laboratory, Teddington, United Kingdom*

In this paper we describe a new clock based on $^{201}Hg^+$. All previous mercury ion clocks have been based on $^{199}Hg^+$. We have recently completed construction of the $^{201}Hg^+$ clock and will describe modifications to the design of our existing $^{199}Hg^+$ clocks to accommodate the new isotope. We will also describe initial spectroscopic measurements of the hyperfine manifold, and possible future experiments. One experiment could place a limit on variations in the strong interaction fundamental constant ratio m_q/Λ_{QCD}.*

1. Introduction

Mercury has two stable odd isotopes, $^{199}Hg^+$ and $^{201}Hg^+$, which have ground state hyperfine structure. Virtually all atomic clock work in mercury ions has focused on $^{199}Hg^+$ [1-8]. Building on our $^{199}Hg^+$ ion clock experience, we have begun a program to investigate the viability of $^{201}Hg^+$. In contrast to $^{199}Hg^+$, both ground state hyperfine levels in $^{201}Hg^+$ have non-zero total angular momentum (F) resulting in a more complex level structure (see Fig. 1). This leads to some challenges in state preparation and state readout not present in the other isotope. However, while systematic sensitivities can be estimated for $^{201}Hg^+$, they have not been measured and until that is done it is possible that some of these sensitivities could be smaller in $^{201}Hg^+$ than in $^{199}Hg^+$. Consequently a significant part of our work will be devoted to making these measurements. In addition, hyperfine structure measurements can also be a useful probe of nuclear structure. In particular, one experiment might exploit different sensitivities in $^{199}Hg^+$ and $^{201}Hg^+$ to variations in the strong interaction ratio, m_q/Λ_{QCD} in order to place a limit on temporal variation of this ratio using a long-term clock comparison [4]. In this paper we will describe a $^{201}Hg^+$ linear ion trap clock that we have built, initial spectroscopic data showing the entire ground state hyperfine manifold, and possible future experiments.

2. Experimental Apparatus

The new $^{201}Hg^+$ clock is an adaptation of an existing $^{199}Hg^+$ linear ion trap clock [3]. A quadrupole linear ion trap situated within a UHV vacuum enclosure is used to trap

* This research was carried out at the Jet Propulsion Laboratory, California Institute of Technology, under a contract with the National Aeronautics and Space Administration. Government sponsorship acknowledged.

approximately 10^7 ions. Helium buffer gas derived from a heated quartz leak is used to cool ions to near room temperature. An oven containing HgO enriched for the 201 isotope generates neutral mercury, which is then ionized by an electron beam. A rf discharge lamp containing ^{198}Hg is used to generate the 194 nm radiation used to drive the optical S to P transition used for optical pumping and state readout (The ^{198}Hg spectrum overlaps the $^{201}Hg^+$ $S_{1/2}$ F=2 to $P_{1/2}$ F=2 transition, but does not couple to the $S_{1/2}$ F=1 state, so that a ^{198}Hg lamp can be used with $^{201}Hg^+$ in exactly the same way that a ^{202}Hg lamp can be used with $^{199}Hg^+$). For our initial studies we use a tunable 30 GHz microwave synthesizer to directly interrogate the microwave transitions, however in the future we will multiply a fundamental frequency with a step recovery diode and mix in an offset frequency from a tunable synthesizer at the end of the frequency chain. We have recently completed construction of a $^{201}Hg^+$ clock to be used for initial systematic studies and have performed initial spectroscopy on the ground hyperfine manifold: $S_{1/2}$ F=1, m_F to $S_{1/2}$ F=2, m_F' (see Fig. 1). There are 9 possible hyperfine transitions, two pairs of which are degenerate, leaving the 7 lines visible in Fig. 1. The m_F=0 to m_F=0 clock transition is in the middle of the spectrum. All lines are highly power broadened to simplify the initial search. Also apparent in the spectrum are motional (secular) sidebands at approximately ±30 kHz offset from each atomic line.

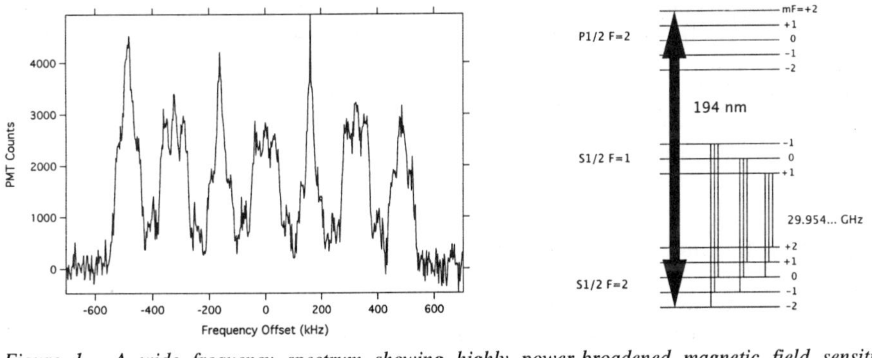

Figure 1. A wide frequency spectrum showing highly power-broadened magnetic field sensitive transitions. From lower to higher frequency offset these are F=2, m_F to F=1, m_F' with (m_F, m_F') equal to (+2,+1); (+1,+1); (0,+1); (0,0 – the clock transition); (0,-1); (-1,-1); and (-2,-1) respectively. Secular motion sidebands offset at ±30 kHz from each line are also visible. On the right is shown a level diagram, including the optical transition, and the 9 possible hyperfine transitions (not to scale). Of these, 2 pairs are degenerate leaving the 7 lines resolved in the spectrum on the left.

3. An Experiment to Place a Limit on Fundamental Constant Variation

The first atomic clock search for variations in fundamental constants [9] was based on measurements of the ratio of two hyperfine transition frequencies. The hyperfine structure depends on the fine structure constant α and the nuclear magnetic moment

μ_I, that is, on both the electromagnetic and strong interactions. The ambiguity in interpretation of the result is overcome by combining limits to the variation of the absolute frequency of several optical clock transitions, yielding limits on variation of α and m_e/m_p [10]. Direct comparisons between optical clock transitions yield more stringent limits on variation of α [3] but give no information on the strong interaction. In contrast, a comparison between the hyperfine frequencies of two isotopes of the same atom has no α-dependence since α depends only on the common electronic structure [11]. Flambaum and Tedesco have calculated the sensitivity of nuclear magnetic moments to variation in the dimensionless strong interaction parameter m_q/Λ_{QCD}, where m_q is the mean light quark mass and Λ_{QCD} is the QCD scale [12]. Comparisons between the hyperfine clock transitions in two isotopes of the same atom, in this case $^{199}Hg^+$ and $^{201}Hg^+$, will provide a stand-alone unambiguous limit on variations in m_q/Λ_{QCD} through the equation

$$\frac{\partial}{\partial t}\ln\frac{f_{201}}{f_{199}} = \frac{\partial}{\partial t}\ln\frac{\mu_{201}}{\mu_{199}} = [B_{201} - B_{199}]\frac{\partial}{\partial t}\ln\frac{m_q}{\Lambda_{QCD}}, \tag{1}$$

where the B_i are sensitivity factors. For the two mercury isotopes, the overall sensitivity factor is estimated to be +0.24 [12, 11]. The current best limit, obtained by combining a limit to variation of the Rb/Cs hyperfine frequency ratio (sensitivity factor –0.025 [12]) with limits on variation of optical transition frequencies, is $+13(9) \times 10^{-15}$/year [13]. Our experiment would compare the hyperfine clock transitions of $^{199}Hg^+$ and $^{201}Hg^+$ with systematic sensitivities in a frequency ratio measurement as low as the 10^{-16} level over long periods (the sensitivities are likely to be very similar for both isotopes) with an associated reduction in the limit on m_q/Λ_{QCD} variation by a factor of 10.

4. Systematic Uncertainties in a $^{201}Hg^+/^{199}Hg^+$ Difference Measurement

The three primary systematic effects in the mercury ion clock are: pressure shifts, the second-order Zeeman shift, and the second-order Doppler shift.

Using a theoretical model for the fractional frequency shift in the hyperfine transition of an alkali-like ion due to collisions with a background gas ("pressure shift") [14] we estimate that a difference measurement between $^{199}Hg^+$ and $^{201}Hg^+$ should vary by less than 1% due to pressure shifts. Frequency deviations due to pressure shifts in our $^{199}Hg^+$ clock have been constrained to < $1\text{x}10^{-17}$/day [15], so the sensitivity to pressure shifts in a difference measurement should be < $4\text{x}10^{-17}$/year.

The second-order Zeeman shift for $^{199}Hg^+$ is $(\Delta f/f_0)_{199} = 4.78\text{x}10^{-15}/\text{mG}^2\ B\Delta B$, while the difference in the nuclear magnetic moment for $^{201}Hg^+$ leads to the sensitivity $(\Delta f/f_0)_{201} = 8.76\text{x}10^{-15}/\text{mG}^2\ B\Delta B$ in that isotope. Thus the sensitivity to changes in magnetic fields in a *difference* measurement is $3.98\mathrm{x}10^{-15}/\mathrm{mG}^2$ BΔB. With good magnetic shielding, the largest source of magnetic fluctuations is the C-field current source with a stability due to aging over long periods of less than $-3.4\mathrm{x}10^{-8}$/day, limited by the stability of the meter. This gives an upper bound on clock variations of $-4.9\mathrm{x}10^{-16}$/year due to C-field variations. We can use field-sensitive Zeeman lines to easily measure the field at the 10 nG level. Therefore the total frequency difference between the $^{199}Hg^+$ and $^{201}Hg^+$ clock transitions due to magnetic field variations will always be known at the $5\mathrm{x}10^{-18}$ level and the sensitivity to magnetic field effects in the difference measurement should be $< 5\mathrm{x}10^{-18}$/year.

The ion temperature at equilibrium is determined by a combination of rf heating effects, buffer gas pressure, and vacuum envelope temperature. In a stable laboratory environment, differential fluctuations in ambient temperature and buffer gas pressure will be quite small so that changes in the temperature-dependent second-order Doppler shift, given by $(\Delta f/f)_T = -3k_BT/2mc^2$, are attenuated by the mass difference of 1%. Variations in the $^{199}Hg^+$ temperature-dependent second-order Doppler shift have been measured at $(\Delta f/f)_T^{199} < -3.4\mathrm{x}10^{-17}$/day [15], so in an isotope difference measurement we expect the sensitivity to be $< 1.2\mathrm{x}10^{-16}$/year.

Variations in the number of ions trapped can also lead to a second-order Doppler shift: Coulomb repulsion will cause different size ion clouds to sample different trap rf amplitudes and execute different average micromotion. Ion density within the trap is affected by the different isotope masses so this effect is slightly different for the two isotopes. Using numerical methods we estimate a difference in sensitivity of 0.0086 for the two isotopes. The sensitivity to the number-dependent second-order Doppler shift for $^{199}Hg^+$ using a compensated version of this 12-pole trap is $(\Delta f/f)_N^{199} = -0.84\mathrm{x}10^{-17}$/day [15], so the sensitivity in an isotope difference measurement is estimated to be about $-2.6\mathrm{x}10^{-17}$/year.

6. Expected Sensitivity to Fundamental Constant Variation.

In the worst case where all systematic effects add in the same direction, we estimate using the sensitivities derived above that a difference measurement between two isotopes will be able to discriminate a fundamental fractional frequency difference with a sensitivity of $<2.6\mathrm{x}10^{-16}$/yr. Using Flambaum's model and Eq. 1 we derive a sensitivity to variations in m_q/Λ_{QCD} of,

$$\frac{\partial}{\partial t}\ln\left(\frac{m_q}{\Lambda_{QCD}}\right) < 1.1\times10^{-15}\,/\,\mathrm{year}, \qquad (6)$$

which is about an order of magnitude improvement over the current best limit [13].

7. Conclusions

We have described a new $^{201}Hg^{+}$ mercury ion clock that could be very sensitive to time variations in fundamental nuclear constants. We have estimated each of the largest systematic uncertainties in a difference measurement between the hyperfine clock transitions of $^{201}Hg^{+}$ and $^{199}Hg^{+}$ and found them all to be small and that such a measurement may be able to resolve a frequency variation between the two isotopes at the 10^{-16}/year level. Using a theoretical model this resolution in frequency could constrain the ratio m_q/Λ_{QCD} at the 10^{-15}/year level, which is an order of magnitude better than the current best limit. We have also performed initial spectroscopy on the ground state hyperfine manifold of trapped $^{201}Hg^{+}$ ions and located the clock transition.

References

1. L.S. Cutler, et al., Proceedings of the 13th PTTI, p. 563 (1981).
2. D.J. Berkeland, et al., Phys. Rev. Lett. **80**, 2089 (1998).
3. T. Rosenband, et al., Science **319**, 1808 (2008).
4. J.D. Prestage, et al., IEEE Trans. Instrum. and Meas. **40**, 132 (1991).
5. J.D. Prestage, et al., Proceedings of the 1999 Joint EFTF – IEEE IFCS, p. 121 (1999).
6. R.L. Tjoelker, et al., Proceedings of the 2003 IEEE IFCS, p. 1066 (2003).
7. J.D. Prestage, et al., Proceedings of the 2007 Joint EFTF – IEEE IFCS, p. 1113 (2007).
8. E.A. Burt, et al., to be published in the IEEE Trans. UFFC (2008).
9. J.D. Prestage, et al., Phys. Rev. Lett. **74**, 3511 (1995).
10. S.N. Lea, Rep. Prog. Phys. **70**, 1473 (2007).
11. S.N. Lea, Eur. Phys. J. ST **163**, 37 (2008).
12. V.V. Flambaum and A.F. Tedesco, Phys. Rev. C **73**, 055501 (2006).
13. See S. Bize in these proceedings.
14. J. Vetter, et al., Z. Physick A **273**, 129 (1975).
15. E.A. Burt, et al., Proceedings of the 2008 IEEE IFCS, p. 371 (2008).

OPTIMISING A HIGH-STABILITY CW LASER-PUMPED RUBIDIUM GAS-CELL FREQUENCY STANDARD

C. AFFOLDERBACH, F. GRUET, D. MILETIC, G. MILETI

Laboratoire Temps-Fréquence, Université de Neuchâtel, Rue A.-L.-Breguet 1
2000 Neuchâtel, Switzerland

We report on our development of a compact and high-performance laser-pumped Rubidium atomic frequency standard. The clock design is based on optical-microwave double-resonance using cw optical pumping, and a physical realization as simple as possible. Main development goals are a short-term instability of $\leq 6x10^{-13}\ \tau^{-1/2}$ and a flicker floor of $\leq 1x10^{-14}$ up to one day. Here we discuss our approaches for controlling the clock's main physical parameters in view of optimized frequency stability.

1. Introduction

Secondary atomic frequency standards such as vapor-cell frequency standards ("Rb clocks") are of high interest in applications where excellent stability is required, but volume, mass, and power consumption of the instrument are of concern. Here we report on our developments on laser-pumped Rb clocks that aim for frequency instabilities of $\leq 6x10^{-12}$ at 1s and $\leq 1x10^{-14}$ over 1 day, from a compact device (physics package volume $\leq$ 1 liter). Application examples for such frequency standards are, e.g., in satellite telecommunication, satellite navigation systems, or space science missions.

2. Clock optimization concept

Our Rb clock scheme is based on continuous-wave laser-microwave double-resonance spectroscopy of the ground-state "clock" transition of 87Rb atoms held in a vapor cell [1]. The clock is conceived in a modular way, where the pump light from a single-mode laser diode at 780 nm (Rb D2-line) is provided from a compact laser module [2]. In order to reduce frequency fluctuations of the pump light, the laser module includes frequency-stabilization to saturated-absorption lines from an internal reference cell.

The pump light is directed to a resonator module that houses the Rb vapor cell (volume $\approx$ 2 cm^3) containing a buffer-gas for line narrowing. Microwave radiation (around 6.835 GHz) from a low-noise synthesizer is applied via a magnetron-type TE011 resonator cavity placed around the cell. The clock transition is detected in the light transmission through the cell.

One important aspect of the clock is the simplicity of its scheme, using a minimum of complex or special components, avoiding e.g. light modulators, optical switches, etc. In addition, cw operation does not put too stringent requirements on the microwave local oscillator (LO). These aspects are of particular importance for the goal of a small clock volume, as well as in view of a future space qualification of the instrument.

2.1. *Short-term stability*

The highly selective optical pumping by a single-mode laser avoids re-pumping by spurious spectral components and allows signal contrasts of several tenths of percent and a relatively narrow resonance line (0.5 to 1 kHz width).

The standard's short-term limit is determined by the detection noise:

$$\sigma_y^{\text{det.}}(\tau) = \frac{noisePSD}{\sqrt{2}D\nu_{\text{res}}} \cdot \tau^{-1/2}, \tag{1}$$

where $D \approx 4A/\Gamma$ is the discriminator slope derived from the clock signal of amplitude A and width Γ, $\nu_{\text{res}} \approx 6.835$ GHz here, and noise PSD includes noise contributions from the detector circuit, detection shot-noise, and laser AM noise. For typical operating parameters and from the measured laser noise we project a short-term stability limit around $2\text{x}10^{-13}\ \tau^{-1/2}$. This limit does however not take into account FM-AM conversion, which can be suppressed by noise subtraction using a second resonance cell [3].

2.2. *Medium-term stability (1 day)*

The use of laser optical pumping allows one to separately measure and optimize the main relevant systematic shift parameters that influence the clock stability:

Temperature coefficient (TC) of the clock resonance cell: Use of cells filled with buffer-gas mixtures cancels out the linear shifts of pure buffer gases (~ 10^{-9} /K) and results in a TC $\leq 10^{-11}$ /K in a small interval around a chosen temperature [4].

Intensity-related light-shift: We operate the laser frequency close to the center of the optical transition, at a laser detuning of $\delta_{LS} \leq 1$ MHz from the condition $\alpha=\partial\nu_{res}/\partial I_{laser}=0$, which allows to reach $\alpha \leq 2\text{x}10^{-12}$ /µW. In this way no intensity stabilization of the laser is needed (passive stability $\Delta I/I \approx 10^{-4}$ over 10^4 s). The detuning δ_{LS} is set by controlling the total buffer-gas pressure in the clock cell taking advantage of the collisional shift of the optical transitions with respect to the saturated-absorption lines from the un-buffered cell used for laser frequency stabilization. This requires the buffer-gas pressure to be controlled to better than 0.2 mbar during cell production.

Frequency-related light-shift: With the laser frequency set to the center of the optical transition and the buffer-gas pressure fixed by the requirement of α=0, the frequency-related light-shift $\beta=\partial\nu_{res}/\partial f_{laser}$ is typically 2×10^{-11} /MHz. The magnitude of β then determines the required laser frequency stability to avoid clock instabilities arising from this parameter, see section 2.3.

The cavity pulling gives a negligible contribution on the 10^{-15} level, due to the low Q-factor (Q < 200) of the cavity.

2.3. *The pump laser and its frequency stabilization*

Stable single-mode laser emission at the required atomic wavelength (780 nm here) is obtained from commercially available DFB laser diodes. For our DFB lasers, we measure intensity noise (relative intensity noise, RIN) of < 2×10^{-14} /Hz and a free-running frequency noise of < 3 kHz/Hz$^{1/2}$ (both at ≈ 300 Hz, the modulation frequency of the clock loop). These noise levels give instability contributions of < 2×10^{-13} $\tau^{-1/2}$ [5], sufficiently small as not to degrade the short-term stability of the clock.

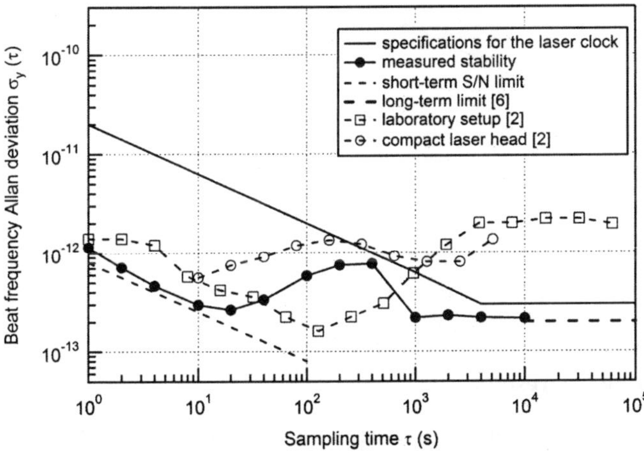

Figure 1. Measured laser frequency stability (in terms of Allan deviation of the beat frequency) for two independent diode lasers stabilized to simple saturated-absorption lines obtained from the new cell assembly. The bump observed around τ = 300 s is due to residual temperature variations.

The emission linewidth of the lasers is measured to be ≤ 2.5 MHz, from heterodyne beat-note signals between two identical DFB lasers. This linewidth is sufficiently narrow for resolving saturated-absorption resonance lines used as references for frequency stabilization. The reference lines are obtained from a compact (30mm x 30mm x 20mm) cell assembly, equipped with double-layer magnetic shields and a heater for improved temperature control, which allows to overcome previously reported limitations [6]. The cell assembly is integrated into our compact laser head [2]. Figure 1 shows the laser frequency stability measured by heterodyne beat-note measurement between two laser heads using the new cell assemblies. The short-term stability of $1x10^{-12}\ \tau^{-1/2}$ coincides well with the limit estimated from the signal-to-noise ratio. At 10^4 s a level of $2x10^{-13}$ is reached, fulfilling the requirements on laser frequency stability for our clock.

2.4. *Clock resonance cells*

In the presented clock approach, precise control of the clock cell's buffer-gas content is required to simultaneously achieve low intensity light-shift (via the total buffer-gas pressure) and low temperature coefficient (via the gases' mixing-ratio). We have produced Rb cells with a N_2/Ar buffer-gas mixture that fulfill both requirements, as shown by the results in Figure 2 for one of the recently produced cells. The intensity light-shift is $\alpha = 1x10^{-12}$ /μW, and the temperature coefficient TC $\leq 9x10^{-12}$ /K, when close to the operating temperature of 65°C.

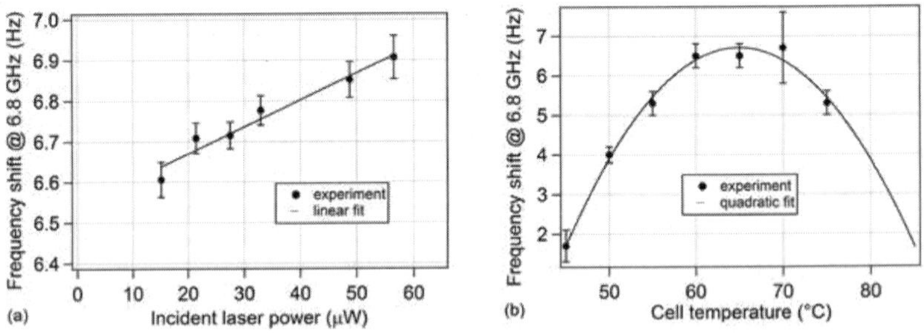

Figure 2. Evaluation example for one of the produced resonance cells. (a): Intensity-related light shift. (b): Temperature coefficient of the cell (corrected for light-shift contributions).

3. Clock stability results

Figure 3 gives the instabilities measured for the clock output using two different buffer-gas cells for the results labeled run 1 and run 2. In run 2, the short-term instability is limited by the local oscillator performance, only slightly above the signal-to-noise limit estimated for this measurement.

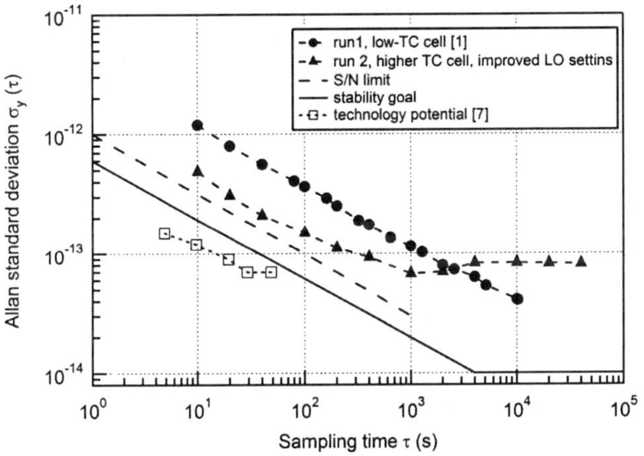

Figure 3: measured frequency stabilities of the Rb clock using two different buffer-gas cells (run 1 and run2, shown as circles and triangles, respectively).

At longer integration times ($\tau \geq 10^4$ s) the instability of run 2 is limited by a too high TC of the cell used (α_2=2x10^{-12}/μW, TC_2=2.5x10^{-10} /K), while no degradation is visible in run 1 but can be expected at the level of few 10^{-14} due to the not optimized α-coefficient for this cell (α_1=-9x10^{-12}/μW, TC_1=5x10^{-11} /K). With both coefficients sufficiently low for the new cells, we expect to improve the clock stability to the level of 1x10^{-14} at 10^4 s to 10^5 s when using these cells in the clock resonator. By improving essentially on the detection noise and the microwave LO, we expect also the short-term stability to reach the stability goal.

Acknowledgments

This work was supported by the European Space Agency, the Swiss National Science Foundation, the Canton and University of Neuchâtel. We also acknowledge contributions from Spectratime SA (Neuchâtel, Switzerland).

References

1. C. Affolderbach et al., *IEEE Trans. Instrum. Meas.* **55**, 429 (2005).
2. C. Affolderbach et al., *Rev. Sci. Instrum.* **76**, 073108 (2005).
3. G. Mileti et al., *Proc. IEEE FCS 1996*, p. 1066.
4. J. Vanier et al., *J. Appl. Phys.* **53**, 5387 (1982).
5. G. Mileti et al. *Proc. EFTF 1995*, p. 271.
6. C. Affolderbach et al., *Proc. EFTF 2004*, p. 84.
7. G. Mileti et al., *IEEE J. Quantum Electron.* **34**, 233 (1998).

RAMAN-RAMSEY Cs CELL ATOMIC CLOCK

R. BOUDOT, S. GUERANDEL, N. CASTAGNA*, O. KOZLOVA, E. DE CLERCQ, N. DIMARCQ and A. CLAIRON

SYRTE, Observatoire de Paris, 61 av. de l'Observatoire, 75014 PARIS, France.
** now in Fribourg University, Physics Department, Fribourg, Switzerland*
E-mail: stephane.guerandel@obspm.fr

A Raman-Ramsey Cs cell atomic clock based on Coherent Population Trapping is studied. High contrasts and narrow width Ramsey fringes are achieved by combining an original double-lambda scheme and a pulsed interrogation technique. The pulsed method allows a strong reduction of the light shift effect and avoids the power broadening. Removing a drift attributed to the cell, the clock frequency stability has been measured to be $7 \times 10^{-13}\ \tau^{-1/2}$.

1. Introduction

Coherent Population Trapping (CPT) is a promising way towards miniature size atomic clocks.[1] It also allows superior contrasts, narrower linewidths and smaller shifts[2] than in classical vapor cell standards. This motivated the present study whose recent results are reported in this article.

2. Experimental Set-Up

Figure 1 shows a schematic view of the Raman-Ramsey Cs cell atomic clock. The heart of the clock is based on a 2-cm diameter and 5-cm long pyrex cell containing Cs vapor and a 20 Torrs N_2 + Ar buffer gas mixture. The cell temperature is controlled at the 10 μK and 400 μK level for 1 s and 1000 s integration time respectively. A static magnetic field B parallel to the laser beam is applied with a solenoid to raise the degeneracy of the Zeeman sublevels.

The optical set-up is composed of two extended-cavity diode lasers resonant with the Cs D_1 line. Orthogonal linearly polarized 894 nm radiation fields are generated to create a so-called double Λ-scheme.[3] Compared to conventional CPT clocks where a single circularly polarized modulated laser is used to pump the atoms in the trap state, this original excitation scheme

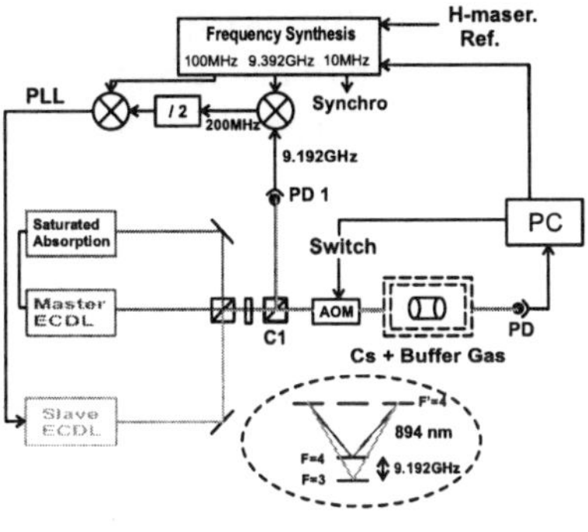

Fig. 1. The experimental set-up. AOM: acousto-optic modulator, PD: atomic signal photodetector, PD1: fast silicon photodiode, ECDL: External Cavity Diode Laser. The insert shows the atomic energy levels involved in the double Λ-scheme.

allows to increase dramatically the CPT resonance contrast.[4] The master laser is frequency locked to a hyperfine component of the Cs D_1 line by saturated absorption in an auxiliary vapor cell. An optical phase-locked loop controls the optical phase of the slave laser relative to that of the master laser with a difference frequency of 9.192 GHz by comparing the optically carried microwave beat-note to a low noise synthesized microwave signal.[5] Residual phase noise performances of the 9.192 GHz optically carried signal have been measured to be less than - 80 dBrad^2/Hz at 1 Hz of the carrier. In the time domain, the Allan standard deviation of the 9.192 GHz optical signal has been measured to be 2×10^{-14} at 1 s and better than 4×10^{-17} at 1 day integration.
Another key point of this experiment is to realize a Ramsey interrogation technique with a temporal pulsed interaction.[3] In the pulse train sequence, each pulse is used both for the atomic signal detection and for the coherence generation.

3. Influence of experimental parameters on the CPT resonance

The impact of the cell temperature on the amplitude of the central fringe is shown on Fig. 2(a). The signal is maximized for an optimal temperature of 29°C. This optimum comes from a trade-off between atomic density, optical thickness and spin-exchange collisions.[6]
Figure 2(b) plots the Ramsey fringes for different free evolution times T. Increasing T reduces the fringe linewidth but also the fringe amplitude

according to the coherence decay. In this experiment, the coherence lifetime T_2 has been measured to be about 10 ms. A Signal-to-Noise ratio as high as 100 is measured on a 33 Hz-wide fringe ($T = 15$ ms) with a 10 μs detection time.

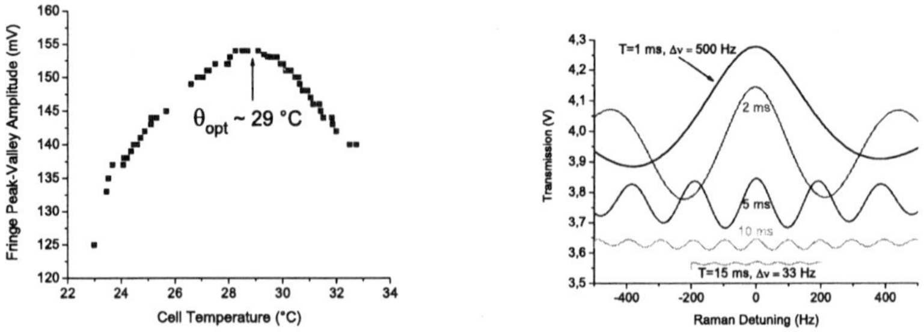

(a) (b)

Fig. 2. (a): Clock fringe peak-valley signal as a function of temperature. (b): Ramsey fringes obtained for different evolution times T (1 ms, 2 ms, 5 ms, 10 ms and 15 ms).

Conventional continuous CPT clocks work with low laser intensities ($< 100\ \mu W/cm^2$) to avoid the power broadening of the resonance.[7] On the contrary, using the pulsed interaction, the fringe width is almost insensitive to the laser intensity (see Fig. 3(a)) and only depends on the free evolution time T. Moreover, as shown on Fig. 3(b), the central fringe amplitude depends linearly on the laser intensity. It is then suggested that typical laser intensities optimizing the clock stability performances should be higher than in the continuous case.

It is known that several lambda-transitions are potentially involved by the double-lambda scheme[8] (see Fig. 4(a)). When the magnetic field is increased, the $\Delta m = 2$ transitions next to the clock transition are shifted in opposite directions. As a result and as shown on Fig. 4(b), the clock signal is maximized for periodic values of the magnetic field where the fringes of the different possible lambda-transitions add up constructively. The oscillation period of the fringe signal versus the static magnetic field directly scales as $1/(2T)$.

4. Frequency Shifts and Frequency Stability

The clock frequency stability is measured against an active hydrogen maser on Fig. 5(a). Removing an unexpected drift of 8×10^{-16}/s attributed to

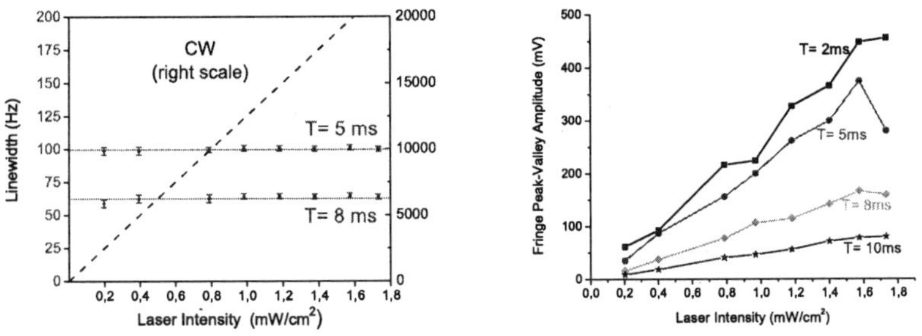

(a) (b)

Fig. 3. (a): Linewidth versus the total laser intensity. Dots: Ramsey technique for two evolution times T (T = 5 ms and 8 ms), experimental points (left scale). Dashed line: theoretical calculations in the continuous interrogation case (right scale). (b): Central fringe amplitude versus the total laser intensity for different values of T (2 ms, 5 ms, 8 ms, and 10 ms).

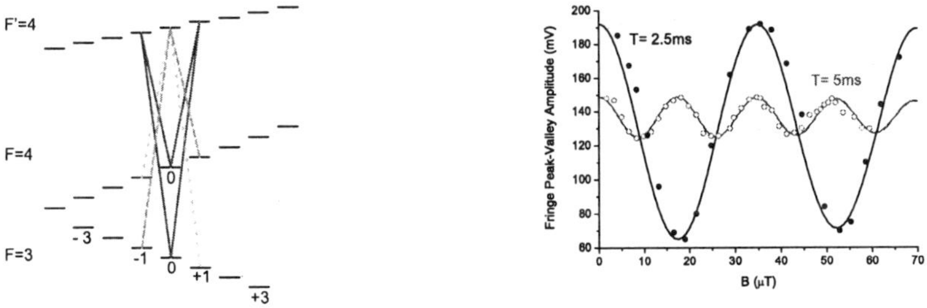

(a) (b)

Fig. 4. (a): Lambda-transitions quasi-resonant with the clock transition. Only the clock transition and $\Delta m = 2$ transitions are represented. (b): Central fringe amplitude versus the static magnetic induction for two free evolution times T (T = 2.5 ms, T= 5 ms). Circles: Experimental points, Full lines: calculations.

a supposed defective cell, the frequency stability is measured to be about $7\times10^{-13}\tau^{-1/2}$. This result is in agreement with a clock resonance Signal-To-Noise ratio of 1600 in a 1 Hz bandwidth and an atomic Q-factor of 9.2×10^7. The main contributions to the noise budget for the short averaging times are the laser intensity noise and the photodetector noise. In the mid-term, the frequency stability is strongly limited by the cell drift. The contribution of the light shift is estimated to be less than 1×10^{-14}. It is greatly reduced by the pulsed interaction technique[4] as shown on Fig. 5(b) with a different

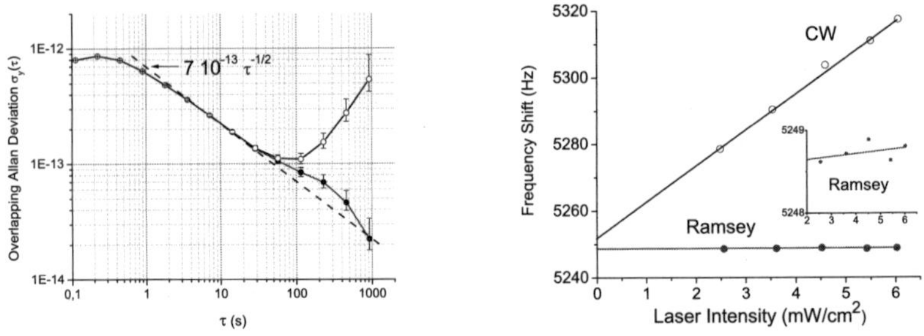

(a) (b)

Fig. 5. (a): Allan standard deviation $\sigma_y(\tau)$. Open circles: No drift is removed. Full circles: a drift of 8×10^{-16}/s is removed. (b): Resonance frequency shift versus the laser intensity. Open circles: Continuous interrogation. Full circles: Pulsed Interrogation.

N_2+Ar mixture. These results are really promising even if some points still have to be optimized. A critical point is to optimize the buffer gas mixture. In our case, a linear buffer gas pressure shift of 3.28 Hz K^{-1} has been measured, yielding a frequency stability limit of 3.5×10^{-13} for a 1 mK variation.

5. Conclusions

A pulsed CPT Cs cell atomic frequency standard has been presented. Assuming a removed drift, the clock frequency stability has been measured to be $7 \times 10^{-13}\ \tau^{-1/2}$. This technology reveals itself to be a promising way to develop compact high-stability frequency references.

References

1. J. Kitching et al., Proc. 6th Symp. Freq. Stand. and Metrology, P. Gill ed., World Scientific, pp 167-174, 2001.
2. J. Vanier et al., IEEE Trans. Instr. Meas., 52, 822 (2003).
3. T. Zanon et al., Phys. Rev. Lett., vol 94, p 193002 (2005).
4. N. Castagna et al., *Investigations on Continuous and Pulsed Interrogation for A CPT Atomic Clock*, to be published in IEEE Trans. Ultrason. Ferroelec. Freq. Contr.
5. R. Boudot et al. *Simple Design Low Noise NLTL-Based Frequency Synthesizers for a CPT Cs Clock*, to be published in IEEE Trans. Instr. Meas.
6. S. Knappe et al., Appl. Phys. B, vol 74, pp 217-222 (2002).
7. S. Brandt et al., Phys. Rev. A, 56, R1063 (1997).
8. S. Knappe et al., Phys. Rev. A, vol 61, 012508 (2000).

Part VIII

Microwave Resonators & Oscillators

SOLUTIONS AND ULTIMATE LIMITS IN TEMPERATURE COMPENSATION OF METALLIC CYLINDRICAL MICROWAVE RESONATORS

ANDREA DE MARCHI

Dipartimento di Elettronica, Politecnico di Torino,
C.so Duca degli Abruzzi 24, 10129 Torino, Italy

In trying to implement temperature compensation of high-Q metallic microwave cavities with the two-metal technique, crippling problems arise from the exceedingly high uncertainty with which the Coefficients of Thermal Expansion (CTE) of the two metals are known, and from their non linearity with temperature. A novel approach is here proposed, following which both problems can be separately addressed, opening the way for improvement of the resonance frequency temp-co into the 10^-10/K and possibly the 10^-11/K region.

1. *Introduction*

The key problem for the use of metal cavities as the reference in Ultra Stable Oscillators (USO), for which they are otherwise attractive because of the high power level they can handle, is the high temperature sensitivity of the metals themselves. In fact many metals, especially when suitably treated, can be immune to long term creep and relaxation effects and therefore otherwise extremely stable dimensionally.

Various temperature compensation techniques have been proposed in the past for different resonators, but the only one that appears promising for a cylindrical high-Q TE_{0nm} resonator is the simple two-metal technique proposed in [1], which produces a parabolic minimum in the frequency to temperature curve and thus, in principle, the possibility of realizing a vanishing temperature coefficient at the operating point.

However, problems arise for this solution from the high uncertainty with which the Coefficients of Thermal Expansion (CTE) of metals are known. Furthermore, their non linearity with temperature turns out to be an even bigger effect and is even less well documented in the literature. The first problem makes it impossible to even design an improvement of much more than an order of magnitude in resonator temp-co without resorting to accurate measurements of the CTEs or profusion of time and efforts in trial and error procedures. The second problem, if not considered (as in the analysis reported in [1]), can produce a second order term for the temp-co which is much bigger than expected, and impose unrealistic machining accuracies in the realization of metal parts. As a result, a temp-co much below 10 Hz/K at 10 GHz (or 10^{-9}/K) appears very hard to obtain for the resonance frequency with this approach.

2. *Short term stability*

A quite useful figure of merit for the short-term stability of oscillators is given by the resonator's Q factor squared multiplied by the circulating power. Although exactly how this figure plays in the short-term depends on the loop scheme that is used, it still helps in a rough estimate of the instability level that may be achievable with a given resonator.

A quick analysis of metal microwave resonators in this perspective returns a promising picture for the short-term stability of oscillators based on such devices, even when used at room temperature. In fact, the power that they can stand is very high, and this can greatly offset the somewhat limited help from their Q value, as compared for example to cryogenic whispering gallery (WGR) resonators. In order to carry out this comparison the power limitation to which each resonator is subject must be taken into account, and we will assume for simplicity that in both cases it is given by the dielectric breakdown of air, though in WGRs the breakdown may occur at lower fields at the sapphire surface.

Given the fact that the Q factor of any resonator is determined by the stored energy (W) to dissipated power (P) ratio ($Q=\omega W/P$), by substituting in the traditional above mentioned PQ^2 figure of merit one can easily find that a more useful figure of merit for comparison may be the product QV of quality factor and effective volume occupied by the radiation.

Since such volume can easily be a good fraction of a litre (e.g. 0.3 litre) for an air filled metal microwave resonator, while it is limited to the order of 1 cm^3 in WGRs, a factor of 300 can be recovered to offset the smaller in the Q-factor. It makes therefore a lot of sense to work on temperature sensitivity of metal microwave cavities in the perspective of developing high short and medium-term stability oscillators which could be competitive with the best quartz devices. Improvement is highly needed of course, if short term stability promises are to be made deliverable, because uncompensated cavities show a temperature to frequency conversion (tempco) equal to the linear expansion coefficient of their constituent metal, which places it around 10^{-5}/K so that frequency stability can hardly be better than 10^{-8} even with the best temperature stabilization scheme.

At least five orders of magnitude better tempco is needed in order to make metal microwave cavities a viable reference for high quality Local Oscillators. This is a problem because the expansion coefficients of metals are known with an uncertainty not even close to the score, which makes compensation hard.

3. *Previous art*

The thermal compensation technique proposed by A. Sen Gupta and F. L. Walls for a high-Q cylindrical TE_{023} cavity [1] calls for end caps made with a different metal from the cylindrical body, in such a way as to recover with their greater expansion coefficient the effects on resonance frequency produced by the temperature expansion of the latter. Their configuration is shown schematically in Fig.1, where symbols are also defined in a self explanatory way and the specific metals chosen are indicated. Condition for compensation in this case is

$$\ell_c/\ell=(\lambda_g/\lambda_0)^2\alpha_D/(\alpha_c-\alpha_D) \quad (1)$$

where the αs are the thermal expansion coefficients and the λs are wavelengths, of the guided field and of free propagation respectively.

The residual parabolic term reported in [1] for the resonance frequency variation around optimum temperature was calculated by considering quadratic terms in Taylor expansion of the resonance frequency formula when the given compensation condition is satisfied, and is given by

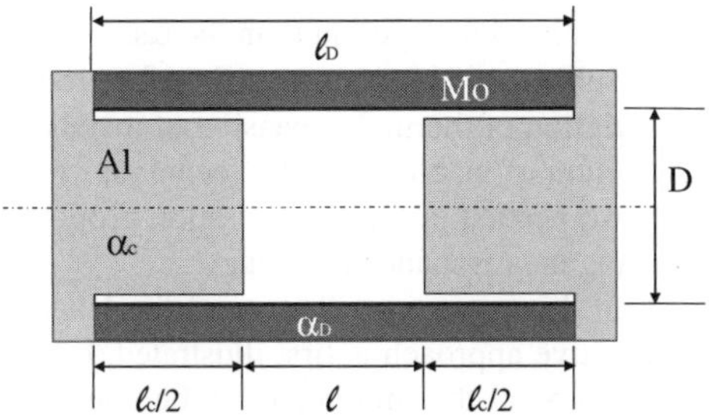

$$\Delta\nu/\nu = 3/2\,(\lambda_g/\lambda_c)^2\,\alpha_D^2\,(\Delta T)^2 \quad (2)$$

However, as it turns out, a much greater parabolic term derives in such a system from the linear Taylor series term of second order thermal expansion coefficients (β), which were not considered in that analysis. The complete parabolic term is

$$\Delta\nu/\nu = \{3/2\,(\lambda_g/\lambda_c)^2\,\alpha_D^2 + (\beta_c-\beta_D)[\alpha_D/(\alpha_c-\alpha_D)-\beta_D/(\beta_c-\beta_D)]\}\,(\Delta T)^2 \quad (3)$$

where the second order coefficients are those that appear in the series expansion

$$\Delta L/L = \alpha\,(\Delta T) + \beta\,(\Delta T)^2 + \gamma\,(\Delta T)^3 + \ldots (4)$$

of some linear dimension of materials around the optimum temperature, and turns out to be about 30 times greater than (2) for the configuration of Fig.1. The resulting resonance frequency offset is compared in Fig.2 with the theoretical prediction of (2) for the mentioned 10 GHz TE_{023} cavity. This

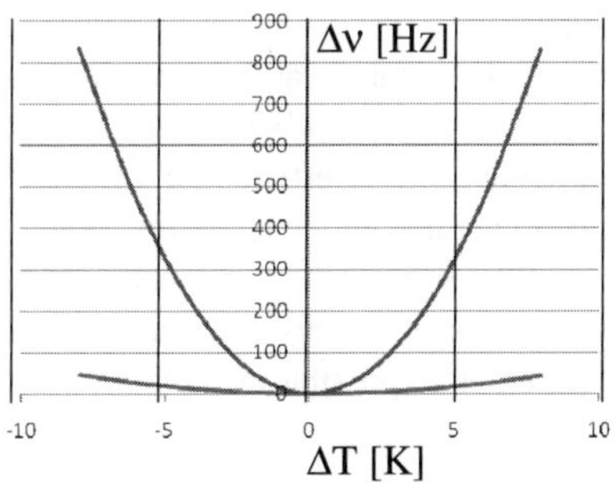

may well be the reason why obtained results disagreed from predictions in the work reported in [1].

In fact, design and realization of a real cavity, aimed at positioning both the optimum compensation temperature and the resonance frequency at the desired values, have limited success capability due to machining tolerances and uncertainty in the knowledge of metal thermal properties of. Since the obtainable uncertainty in position of the parabolic minimum is of the order of the temperature change needed to recover dimensional errors, a machining uncertainty of 10μm (or 10^{-4} over 100 mm) results in a shift of the minimum of the order of the tens of K, which in turn means that it's hard to take temperature sensitivity below 10^{-7}/K (or 1 kHz/K at 10 GHz). The task is made even harder by the relatively great uncertainty with which temperature expansion coefficients are known. The error in positioning of the parabolic minimum can easily exceed 50 K with 0.1% uncertainty in the αs.

Successful compensation therefore requires either a more helpful theoretical description of thermal expansion of metals and better machining accuracy, or the definition of viable trimming techniques requiring a single adjustment instead of the two needed in the configuration proposed in [1] to trim the parabola without moving the resonance frequency.

In this paper inroads toward the latter solution are presented. However, an alternative approach is first illustrated, based on the idea that the frequency shift curve around the minimum can be engineered to be much more stable, in this way reducing the necessity for trimming altogether.

But let's first analyze what theory can offer towards the task.

4. *Thermal expansion of metals*

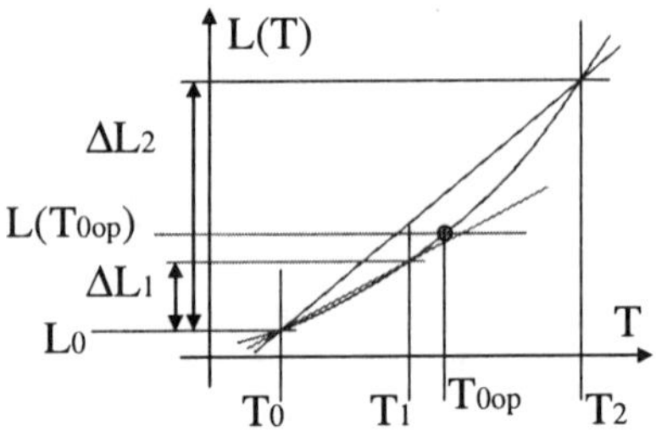

While in designing for temperature compensation one needs to know the actual values of local expansion coefficients for the metals involved, as defined in the Taylor series (4), at the operating temperature, what is available in the technical literature for each metal is at best a value, uncertain not much better than 1% , of the average Coefficient of Temperature Expansion (CTE=ΔL/(LΔT)) between two rather distant temperatures, as shown in Fig.3.

When more than one value of the CTE is available, relative to different temperature intervals with one extreme in common (like T_1-T_0 and T_2-T_0 in the figure), it is possible to obtain an estimate of the Taylor series around T_0 by solving for α and β a two equation system, or also for γ if three values of the CTE exist. Coefficients obtained in this way, however, are relative to T_0 and must be extrapolated in order to yield the coefficients at the operational temperature. Furthermore, the matrix to be inverted is extremely unbalanced, and their uncertainty is therefore high if the thermal behavior of the considered metal is not very smooth. In fact this method is based on the unverified assumption that the L(T) curve may be adequately described by a 2nd (or 3rd) order polynomial.

Unfortunately, theory is not very helpful because existing models don't quite match experimental data for most metals, particularly around room temperature, where one would want to operate cavities. In fact, the Mie-Grüneisen theory [2,3] predicts proportionality of $\alpha(T)$ to specific heat capacity c_T through the Grüneisen coefficient γ_G as given by

$$\alpha = \gamma_G \, c_T / (3E), \tag{5}$$

where E is the elasticity modulus (Young) of the metal. Reasonable agreement with experimental data is usually found below room temperature, particularly in the cryogenic region, but different degrees of disagreement are found as the Debye temperature is approached, which strongly limits the validity of this input at room temperature [4]. Since β and γ basically depend on the first and second derivative of $\alpha(T)$ with respect to T, the shape of the $\alpha(T)$ curve around room temperature is very important for gaining insight on the residual parabolic term after compensation and its possible minimization by balancing the two terms in equation (3). As it turns out, theory is not good enough to avoid trimming, but can be used to select materials useful for this latter task, as shown below.

5. *Second order compensation*

The residual parabolic term of Equation (3) can be forced to vanish if cavity end caps are realized in such a way as to provide as a whole, for α_c and β_c, a pair of "synthetic" values that extinguish it. It can be seen by inspection that for such a pair it must be $\beta_c/\alpha_c < \beta_D/\alpha_D$, and a judgment on metals usable to this aim can be made by considering the $\alpha(T)$ curve.

Experimental values for α and β for Tungsten and Molybdenum, which are the most likely metals for the tube section of the cavity, can be found in [5] and [6] respectively, and the choice of metals for the caps can start from there, with the mentioned constraint.

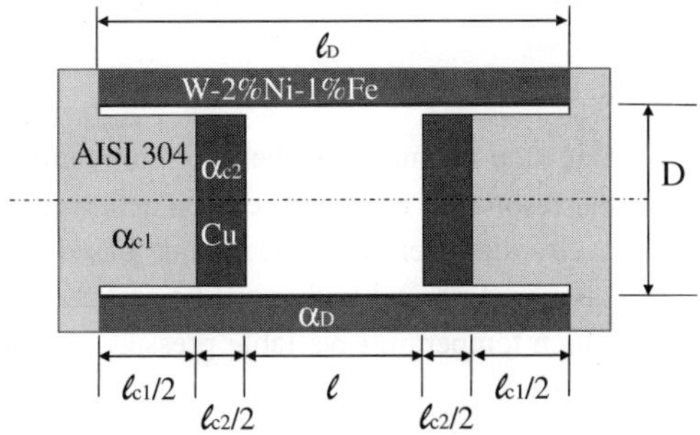

Synthetic values for caps coefficients can be obtained by compounding two metals in series, either in each cap, as in Fig.4, or by using two different metals for the two caps. Once the two cap metals are chosen, their length ratio ℓ_{c2}/ℓ_{c1} can be designed by imposing extinction of both linear and quadratic terms in (ΔT).

In Fig.5 the residual cubic shift is reported for the TE_{023} cylindrical cavity of Fig.4 (W body and Al/Cu caps in proper proportions). The third order thermal expansion coefficient γ of all metals is taken into account in this calculation. The parabolic curve to which this cubic is compared is the original incomplete theory of [1], just because the actual parabolic curve from the complete theory would be too sharp to help assessment by inspection. Therefore Fig.2 must be used also, in connection with this, for a realistic judgment of the improvement achievable with this second order compensation technique.

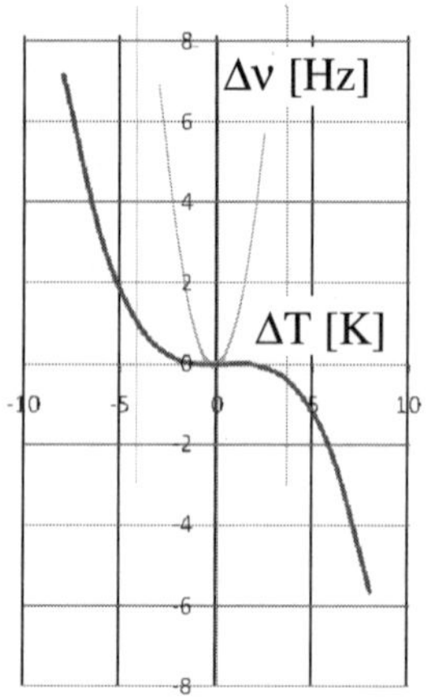

It can be seen that successful compensation is very effective for thermal sensitivity reduction, as it can offer better than 10^{-9}/K up to 10 K distance from the optimum compensation temperature, which is an improvement of 100 over linear-only compensation. The improvement is even greater if the optimum temperature is approached to better than 5 K, as the higher power of (ΔT) helps in the residual curve. In fact in that case sensitivity is below 10^{-10}/K, with a factor of 300 improvement. Succeeding in further moving that limit down to 1 K could in principle secure sensitivity well below 10^{-11}/K.

The question remains open as to how difficult is may be to trim dimensions in order to reach optimum compensation, so that this technique can be exploited at best, but as it turns out this task is less critical than in first order compensation because of the higher flatness of the residual curve.

Before we delve into that, the further comment may be in order that, given an adequate knowledge of the α(T) curves, third order compensation could in principle be attempted by using three metals in series for the caps, leaving only a residual fourth order term for the temperature sensitivity of resonance.

6. *Below 10^{-10}/K sensitivity*

One-step trimming of the optimum compensation temperature without affecting the resonance frequency can be achieved by regulating the thermal sensitivity of cavity diameter with a technique based on elastic deformation. To this aim, the cavity cylindrical body must be realized with a suitably thin metal sheet tube, on which temperature variable pressure can be applied inwards from the outside as in Fig.6. In this way, the diameter's tempco can be set at any value either side of zero just by regulating the pressure, and the task of compensating both first and second order terms is made that much easier because this single operational step can be performed without taking apart the cavity.

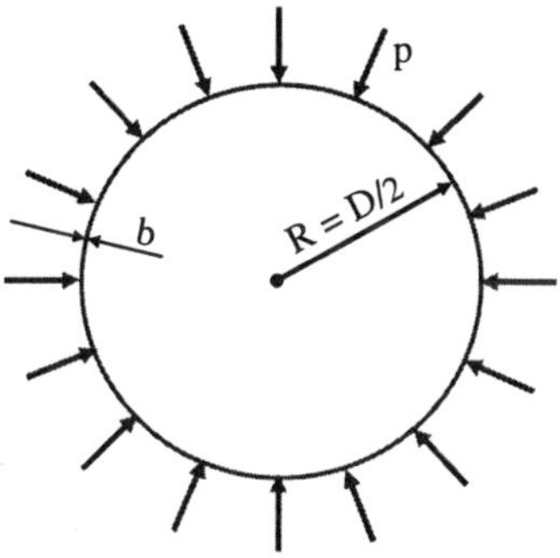

The distance between caps would still be held by a stable low expansion material in this case, in the form of an outer coaxial tube, and Molybdenum or Tungsten are best for the use.

Because the diameter's tempco is then essentially zero, the cap length needed for compensation becomes much shorter, as it only needs to compensate for length change. This is nice, as it reduces the overall length of the assembly, and makes it easier to obtain the necessary temperature uniformity.

The pressure needed to contrast thermal expansion of b thick tube diameter D is given by $\alpha E(b/R)\Delta T$, and it amounts to about 27 kPa/K for a ½ mm thick 80 mm diameter Beryllium-Copper tube.

Applying such variable pressure by means of some kind of fluid is out of the question due to the risk of Eulerian instability of the thin tube. A much better solution, suggested here, is the use of metal pushers leaning on the outer shell stiff cylinder that fixes the distance between the caps, and pushing radially inwards against the thin tube. Pressure regulation can be performed in this case by turning down slightly the outer shell to force it to take a greater share of the thermal load elastic deformation.

Assembly of the device is a critical point. Shrink-fit achieved with the use of liquid nitrogen is suggested. This imposes some 100 K equivalent thermal preload, and makes the choice of the thin tube material critical. In fact, the circumferential "hoop" stress that develops within such tube R/b greater than the applied pressure, and for the tube in the given example it would amount to more than 200 MPa with the 100 K preload. Furthermore, press-contact at the pushers' interface would prevent the shrinking of the tube in the clamped sections, leaving it free to give only in the interstitial sections between pushers. Therefore the load would be concentrate in those sections, calling for the use of a material with adequate pressure yield strength. Hardened Beryllium Copper alloy 25 offers pressure yield strength in excess of 1.2 GPa, and seems to be the best material for the task. A spatial duty cycle of 1/5 can be used in this case, as it would lift the stress at the gaps not far beyond 1GPa.

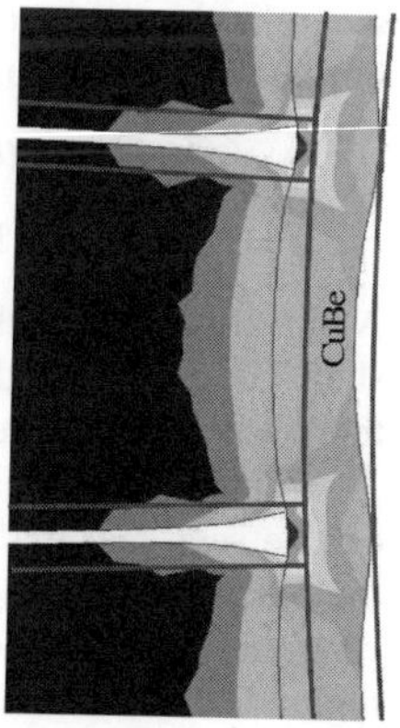

A finite element simulation of the stress distribution around the critical point of the assembly is shown in Fig.7, where the pushers are supposed to be made of AISI 304 stainless and the outer shell is in Nickel Iron loaded sintered Tungsten. The use of CuBe shoes between the pushers and the thin tube would be beneficial here as it would prevent the formation of the azimuthal waves shown in Fig.7 and the concentration of stress in the outer side of the gap sections.

An overall view of the whole assembly section is shown in Fig.8b, with an enlarged detail in Fig.8a, in which symbols used below are also defined.

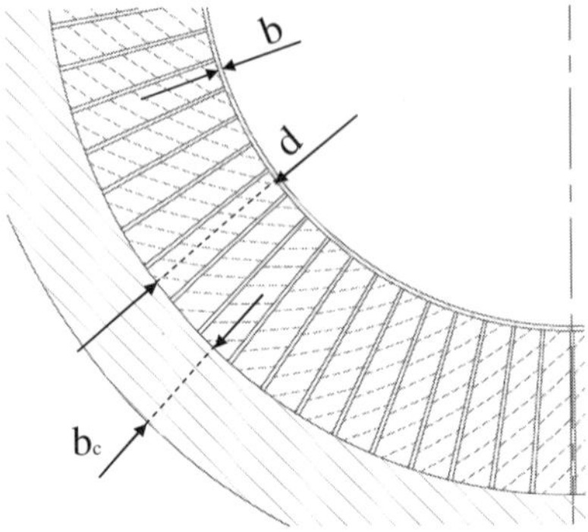

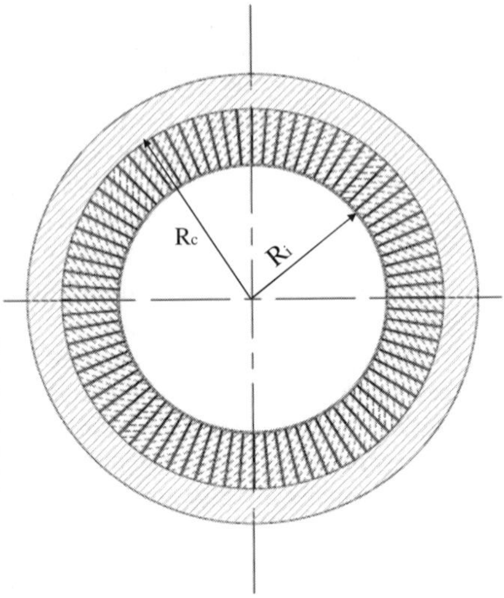

Figure 8a and 8b. Detail and overall view of the diameter compensation assembly

The complete theory of the compensation mechanism yields the following formula for the diameter temperature coefficient. Here metal 1 is the material of the inner thin tube, metal 2 is the pushers, and metal 3 is the outer shell, ν_1 is the Poisson ratio of material 1, and $x=d/R$; $y=b/R$.

$$\Delta D/(D\Delta T) = [\alpha_1(\xi-\nu_1 y)-\alpha_2 x(1+\nu_1 y)+\alpha_3(1+x)(1+\nu_1 y)]/[(1+\xi)(1-y)] \quad (6)$$

$$\xi = (E_1/E_2)\, y \ln(1+x) + (E_1/E_3)\,(b/b_c)\,(1+x) \quad (7)$$

Application of this theory to the case of the example assembly, with the outer shell Tungsten cylinder 13.2 mm thick, ± 1% in the three simulations, yields the three sensitivity curves shown in Fig.9 for the inner diameter.

In all cases the latter is smaller than $5\cdot10^{-9}$/K up to 5 K from the null, but it can be seen that the optimum compensation temperature is moved by trimming the outer shell. In Fig. 10 the residual cavity tempco is reported for the central case of Fig.9 and the possibility to push it below 10^{-10}/K into the low 10^{-11}/K region is clearly documented.

In Fig.10, a and b, the frequency-offset to temperature-offset curve for such optimally compensated cavity, and the corresponding temperature coefficient are shown respectively. It can be seen that with these provisions it shouldn't be exceedingly difficult to obtain a temperature coefficient in the low 10^{-11}s or even down into the 10^{-12}s per Kelvin.

One aspect of this new technique which still needs to be adequately addressed is the possible long term stability of metals subjected to a constant level of stress, as it happens in the described structure.

No useful information can in fact be found in literature on Stage-II creep behavior of metals at room temperature, which would be a mechanism of possible long term instability.

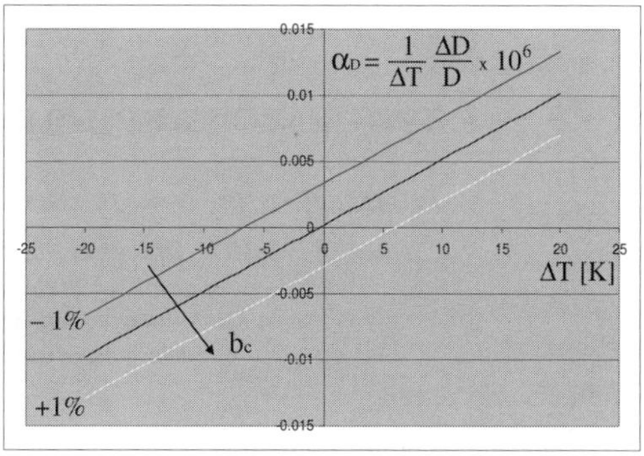

Figure 9. Diameter tempco as a function of b_c and ΔT

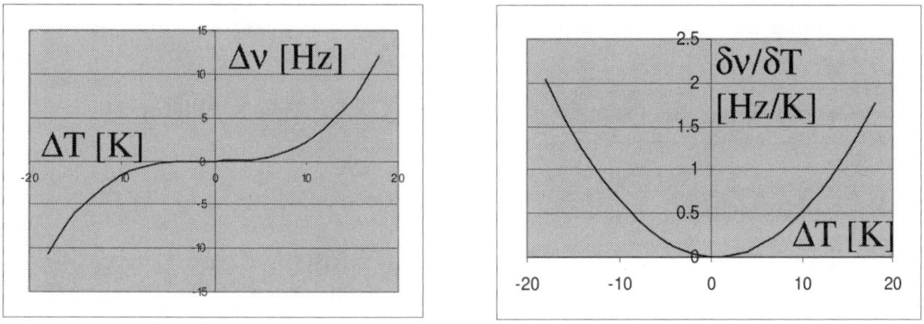

Figure 10, a and b. offset and tempco of optimally compensated cavity

References

1. M. Long, A. Sen Gupta, F. L. Walls, Proc. 2007 IEEE Intnl Frequency Control Symposium joint meeting with 21st EFTF, Geneva, June 2007
2. C. Kittel, *Introduction to Solid State Physics*, J.Wiley, N.Y.
3. V.L.Moruzzi, J.F.Janak, K.Schwarz, Phys. Rev. B 37, 790 - 799 (1988)
4. J.Garai, Computer Coupling of Phase Diagrams and Thermochemistry 30 (2006) 354-356
5. Worthing, Phys. Rev. 10, 638 (1917)
6. Gruneisen and Goens, Z.Phys. 29, (1924) 141

CRYOGENIC SAPPHIRE OSCILLATORS

J. G. HARTNETT*, E. N. IVANOV AND M. E. TOBAR
School of Physics, University of Western Australia,
Crawley, WA 6009, Australia
**E-mail: john@physics.uwa.edu.au*
www.uwa.edu.edu

Cryogenic sapphire oscillators, developed at the University of Western Australia, have now been in operation around the world continuously for many years. We describe and summarize their design, construction and performance. Fractional-frequency fluctuations below 6×10^{-16} at integration times between 10 and 200 s have been repeatably achieved at X-band.

Keywords: Cryogenic, sapphire, high stability, microwave, oscillator.

1. Introduction

Inspired by the early work of Braginsky at Moscow State University,[1] the cryogenic sapphire oscillator (CSO) project at the University of Western Australia (UWA) resulted in the creation of microwave signal sources with short-term frequency stability unmatched by any other type of electromagnetic oscillators[2] including laser-cooled atomic fountain clocks.

The first generation CSOs exhibited a fractional-frequency stability close to one part in 10^{14} over integration times of 10 to 300 seconds.[3] This was improved by more than an order of magnitude by the mid 1990's when the noise mechanisms affecting the CSOs became better understood.[4] The improvements in the CSO's frequency stability were assisted by the technological advances in growing ultra-high-purity single-crystal sapphire via the Heat Exchange Method HEMEX,[5] the availability of low-noise microwave components and the cryogenic operation of the vital components of the oscillator. This oscillator was the first to reach a fractional-frequency stability less than one part in 10^{15} over integration times of 10 s to 100 s.

In 2000, a UWA sapphire oscillator was used at Observatoire de Paris for the development of a laser-cooled cesium-fountain clock. These experiments enabled the first experimental observation of the quantum projection noise in atomic-fountain clocks.[6] In 2003, researchers at UWA and the Paris Observatory teamed up to undertake a search for the possible violations of Lorentz invariance by comparing the frequency of that CSO with that of a Hydrogen maser.[7] Also in 2003 another CSO was delivered to European Space Agency (CNES) to assist with the ground tests of the cold-atom space clock "PHARAO" intended to operate in a micro-gravity environment.[8] In 2005, at UWA a novel type of Michelson-Morley experiment involving two co-rotating CSOs was implemented to test the Lorentz invariance principle under the Standard Model Extension framework, and put the best limits on several Lorentz violating parameters.[9,10]

Between 2004 and 2006, four additional CSOs were developed by UWA. Two were built for the National Metrology Institute of Japan (NMIJ) at AIST, Tsukuba,[11,12] one for the National Institute for Information and Communication Technology (NICT) in Tokyo,[13] and the other to remain at UWA for the Frequency Standards and Metrology group. The fractional-frequency fluctuations of those oscillators are presently state-of-the-art approaching 5×10^{-16} at an integration time of 20 s and highly reproducible over long measurement time periods.[14]

2. The resonator

The heart of the CSO is a cylindrical sapphire-dielectric resonator machined with its rotational axis aligned with the crystal axis. See Fig. 1. From a broad continuum of eigenmodes only two families of quasi-TM and quasi-TE modes are of practical importance to the design of ultra-stable oscillators. These modes exhibit the highest electromagnetic field confinement to the dielectric and hence the highest electrical Q-factors. They are characterized by a small number of field variations along the axial and radial directions and a large number of field maxima around the azimuth, which is typical for modes characterized as 'Whispering Gallery' (WG), similar to acoustic modes observed by Lord Raleigh in St. Paul's Cathedral in London.[15]

A WG mode in a cylindrical dielectric resonator is a standing wave formed by a pair of traveling waves counter-propagating along the inside circumference of the dielectric cylinder and experiencing total internal reflection at the dielectric/air interface. Small geometrical imperfections of the sapphire cylinder, the presence of the coupling probes, as well as the misalignment of the crystal and rotation axes remove the degeneracy between the traveling wave "companions" and split the modes of resonance into two spectral components (doublets) having different coupling coefficients with frequencies separated by a few kHz. This effect is easily observed at cryogenic temperatures, where the frequency separation between the doublets can be a few hundred times larger than their individual resonance linewidths. A competition between doublets has never been observed in a loop oscillator due to the non-linearity of its gain stage. Oscillations always establish at the frequency of one of the doublets (with the lower insertion loss), which saturate the loop microwave amplifier.

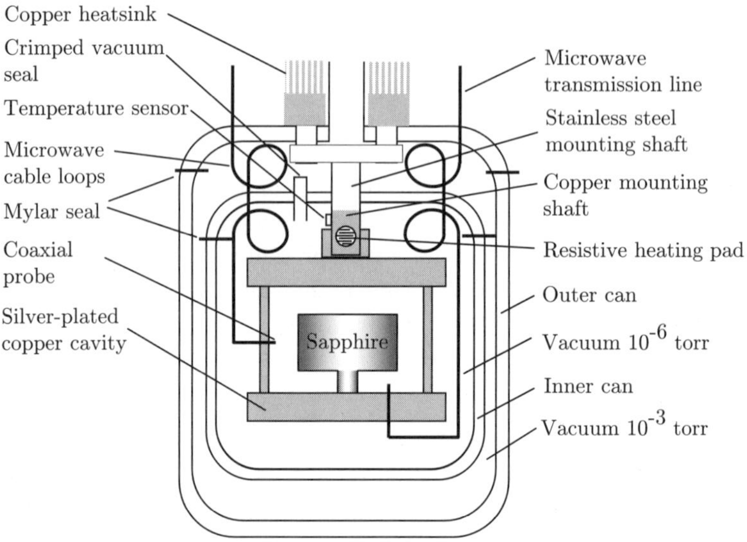

Fig. 1. Schematic diagram showing design features of the lower portion of the helium dewar insert comprising two vacuum cans and the sapphire loaded cavity.

The sapphire is housed in a cylindrical metal (usually silver-plated copper) cavity which protects it against the environmental perturbations and, most importantly for cryogenic applications, against surface contamination. In the absence of crystal defects and surface contamination, the Q-factors of the WG modes in SLC resonators are limited by the sapphire-dielectric loss tangent. Q-factors measured in the liquid-helium-cooled HEMEX-sapphire resonators at frequencies around 10 GHz are extremely high, of the order of 10^9 at $T \approx 4.2$ K, with the highest measured Q-factor of 1.8×10^{10} at 1.8 K.[16]

Even the highest quality HEMEX sapphires are not free from paramagnetic impurities such as Cr^{3+},[17] Fe^{3+},[18] Ti^{3+}, or Mo^{3+} ions.[4] In such resonators the paramagnetic ions, with only a small measurable effect on the Q-factor, strongly influence the frequency-temperature dependence of the WG modes by introducing a magnetic susceptibility to the sapphire crystal. The sign of the frequency shift resulting from the temperature induced change in the magnetic susceptibility is opposite to the frequency shift resulting from the temperature dependence of the sapphire dielectric permittivity. With the magnitudes of the competing frequency shifts becoming comparable below 15 K frequency-temperature compensation is observed in SLC resonators. Such phenomena exist in all WG modes in HEMEX resonators at frequencies below the electron spin resonance (ESR) frequency of the dominant ion.

The frequency-temperature turnover point depends on the concentration of paramagnetic impurities, their species type and the nature of the mode excited. Also, the flatness (or curvature $\kappa = (1/f_0)d^2f/dT^2$, where d^2f/dT^2 is the second derivative of the resonance frequency as a function of temperature and f_0 is the mean resonance frequency) at the frequency-temperature turnover point determines the required precision of the resonator temperature control. In HEMEX-crystal resonators the smallest values are around $\kappa \approx 10^{-9}\,K^{-2}$, which can be translated into resonator fractional-frequency fluctuations of order 10^{-16}, provided that the crystal temperature is kept within 1 mK of the turning point and the temperature fluctuations are controlled to be less than 0.1 mK. These conditions are not difficult to meet for liquid-helium-cooled resonators due to the availability of high-sensitivity, yet low-noise, carbon-glass temperature sensors.

Fig. 1 illustrates a typical approach to the temperature stabilization of a cryogenic SLC resonator. Here, the sapphire crystal is thermally insulated from the liquid helium bath by two evacuated stainless-steel vacuum cans. The temperature of the SLC resonator is stabilized to a precision of about 10 μK by a feedback control system based on a temperature bridge connected to a four-terminal carbon-glass thermometer. The latter is embedded into the copper mounting shaft, to which the SLC resonator is directly attached. The top end of the copper shaft is connected to a stainless-steel rod to ensure effective low-pass filtering of thermal fluctuations resulting from the boiling liquid helium. The outer can houses multiple microwave isolators which are needed to minimize the SLC resonator frequency pulling effect induced by the coaxial cable line resonances. Amplitude detectors, used in both the frequency and power control systems, are also located inside the outer can in the cryogenic environment to improve the oscillator's long-term frequency stability.

The SLC resonator in Fig. 1 is a single-spindle single-crystal sapphire cylinder clamped from below. This type of clamping was suggested in[16] to reduce mechanical stress on the active region of the sapphire resonator where most of the electromagnetic field is confined. Such stress was found to be responsible for the frequency drift of the first generation CSOs utilizing double-spindle sapphire cylinders squeezed between the lids of the metal cavity.

Three major noise mechanisms[19] limit the frequency stability of the CSO when it operates at the frequency-temperature turning point of the SLC resonator. The first noise mechanism is related to the intrinsic fluctuations of the Pound frequency discriminator.[20,21] The second noise mechanism is associated with AM-index fluctuations of the microwave interrogation signal incident on the resonator. The third noise mechanism is related to fluctuations of dissipated microwave power in the sapphire resonator due to radiation pressure induced expansion in a solid dielectric resonator.[22] The effect of the first two noise mechanisms on the spectral purity of the oscillator's signal can be greatly reduced by setting the SLC primary coupling close to critical. Such tuning, however, may require a few iterations (cool down cycles) due to the errors associated with the measurements of very small coupling coefficients and the strong dependence, near 4.2 K, of sapphire Q-factor on its surface cleanliness.

3. The oscillator

Fig. 2 shows a schematic diagram of a frequency stabilized cryogenic oscillator. It is based on a high-Q microwave SLC resonator which serves both as a band-pass filter of a self-sustaining loop oscillator and a dispersive element of a Pound frequency discriminator (FD). Operation of the Pound FD is based on rapid phase modulation of the incident microwave signal. If oscillator frequency is different from that of the high-Q resonator the signal reflected from the resonator acquires amplitude modulation (AM), the depth and the phase of which depend on the relative detuning between the oscillator and the resonator. By extracting the complex amplitude of the reflected signal (with an envelope detector), and comparing it to the original modulation signal (inside a lock-in amplifier as shown), an error voltage varying synchronously with the oscillator frequency is produced. The filtered error voltage is then applied to the voltage controlled phase shifter (VCP) in the loop oscillator steering its frequency to the chosen resonant mode of the high-Q resonator. The frequency discriminator, filter and VCP form a closed loop frequency control system which detects and cancels (within its bandwidth) fluctuations of the oscillator frequency from

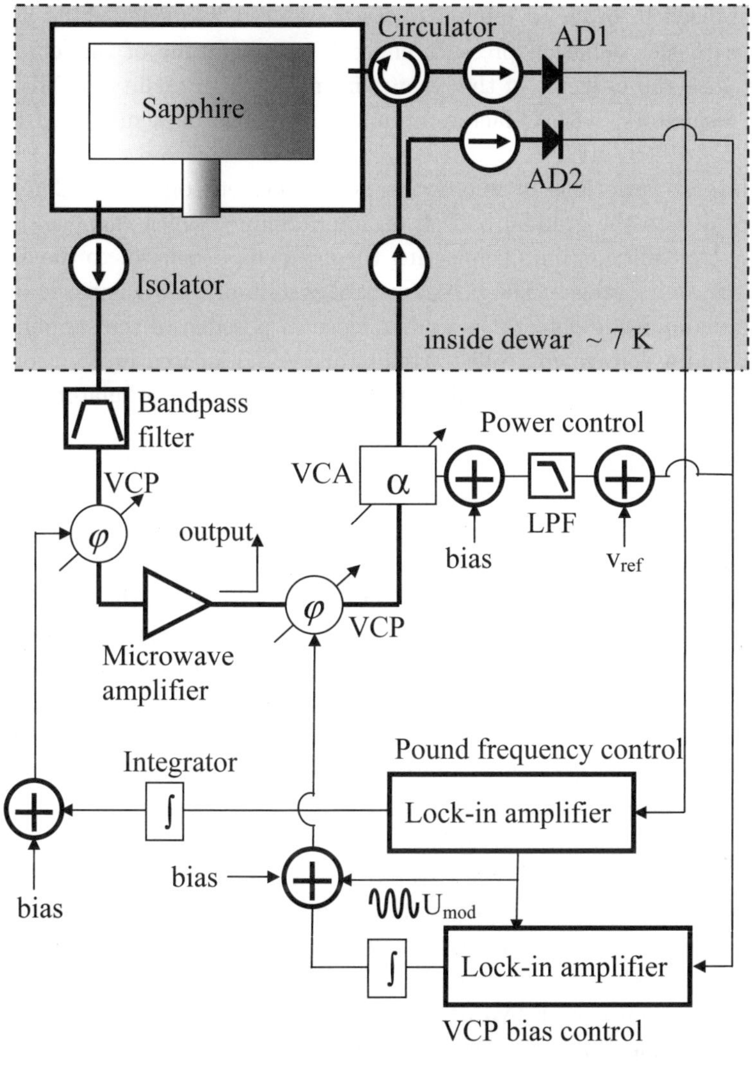

Fig. 2. Schematic diagram of a CSO showing three control systems stabilizing its frequency and power, and eliminating the spurious amplitude modulation of the interrogation signal. Thick lines indicate microwave coaxial cables. VCP = voltage controlled pahse shifter, VCA = voltage controlled attenuator, LPF = low pass filter.

the cavity resonance frequency. For a frequency control loop with the sufficiently high gain, the quality of oscillator frequency stabilization is entirely determined by the noise properties of the frequency discriminator.

The microwave power incident on the cryogenic sapphire resonator needs to be stabilized in order to minimize power induced frequency shifts associated with the radiation pressure effect. The magnitude of this effect increases with the Q-factor of the resonator and for high quality HEMEX sapphire resonators (with Q-factors around 10^9) it has been measured to be about 5×10^{-11}/mW.[22]

The largest variations of microwave power incident on the SLC resonator are due to the gradual boil-off of liquid helium, which changes the loss in the transmission lines connecting the cryogenic resonator to the oscillation sustaining stages. The power control system of a CSO is based on a cryogenic amplitude detector placed as close as possible to the sapphire resonator and a voltage controlled attenuator (VCA) located in the room temperature part of the loop oscillator next to the phase modulator (see Fig. 2).

Phase modulation of the microwave interrogating signal is often accompanied by spurious amplitude modulation due to the dependence of the VCP insertion loss on applied voltage. The AM sidebands in the spectrum of the interrogation signal are synchronously demodulated by the electronics of the Pound FD producing an error voltage proportional to the real part of the SLC reflection coefficient, as opposed to the demodulation product of the PM sidebands, which varies as the imaginary part of the reflection coefficient. As a result, the error signal of the Pound FD changes its sign at a frequency which is shifted from the SLC resonant mode frequency. This means that any noise mechanism affecting the spurious AM index will also affect the frequency stability of the CSO.

Two almost identical CSOs operating at 11.2 GHz were constructed to measure their short-term frequency stability. In each oscillator the temperature of the SLC resonator was maintained at its frequency-temperature turnover point (near 7 K). The beat frequency between the two oscillators (f_{beat} = 131.181 kHz) was measured with a frequency counter referenced to a 10 MHz signal from a hydrogen maser. The fractional-frequency uncertainty of the measurement system was limited by the counter triggering errors and is a diminishing function of the integration time: $\sigma_y^{counter} \approx 10^{-16}/\sqrt{\tau}$ (see curve 3 of Fig. 3).

Fig. 3 shows the Allan deviation of the fractional-frequency fluctuations of an individual CSO (σ_y) as a function of integration time. It behaves non-monotonically over the range of integration times $3 < \tau < 1000$ s reaching a minimum $\sigma_y^{min} \approx 5.6 \times 10^{-16}$ at $\tau \approx 20$ s and remaining below the level of 10^{-15}. This frequency stability is the state-of-the-art for modern-day microwave signal sources.

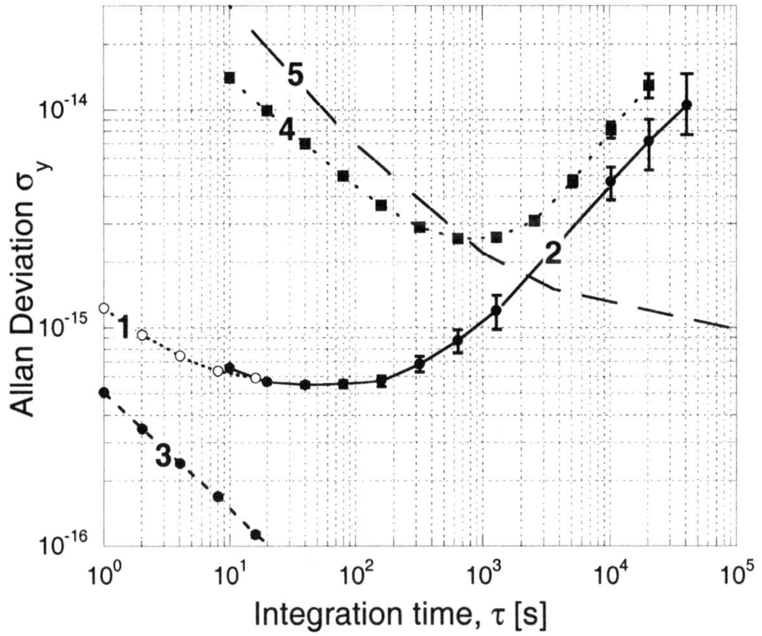

Fig. 3. Oscillator Allan deviation, σ_y, as a function of integration time, τ. Curve 1 is σ_y calculated from 15.9 hours of data (or 57317 samples) collected with a gate time of 1 s. Curve 2 is σ_y calculated from 16.7 hours of data (or 6022 samples) collected with a gate time of 10 s. Curve 3 is the frequency counter noise floor. Curve 4 is σ_y calculated from 9 days of data (or 76333 samples) also collected with a 10 s gate time. Curve 5 is the Kvarz hydrogen-maser stability specification, where the line represents the upper limit.

In addition, we also compared the "beat" between the CSO and a microwave frequency synthesizer phase-referenced to a hydrogen maser and operating near 11.2 GHz. See curve 4 in Fig. 3. As expected, the maser noise dominates the noise measurements at the relatively short integration times ($\tau < 2000$ s), while at larger integration times we only see the CSO. The data in Fig. 3 indicate that the long-term behavior of the CSO ($\sigma_y \propto \sqrt{\tau}$) is limited by the random walk of its frequency, which is, most likely, caused by the ambient temperature fluctuations influencing the oscillator's electronics. Data obtained early 2008 from the UWA CSO (which has been kept at or below 77 K for over 18 months) based at NICT[23] show an improvement in these results. The fractional frequency instability is 3×10^{-15} at 10^4 s rising to 10^{-14} at 10^5 s.

In contrast to Chang *et al*,[16] the frequency measurements reported here were performed over the relatively long periods of time (from a few hours to a few days) in order to avoid an over optimistic estimate of the oscillator frequency stability. The values of σ_y were calculated directly from the "raw" data without any post-processing such as a removal of the linear frequency drift typical for the earlier oscillators.[3,4,16] This was due to the extremely low drift rates exhibited by our oscillators (the lowest fractional-frequency drift rate was measured to be -2.2×10^{-15}/day). Such performance, we believe, was due to the optimal geometry of the shielded sapphire resonator characterized by a relatively sparse spectrum of spurious modes in the vicinity of the operational mode.

References

1. V.B. Braginsky, V.P. Mitrofanov and V.I. Panov, *Systems with small dissipation* (Univ. of Chicago Press, Chicago, 1985)
2. A.G. Mann, in *Frequency Measurement and Control: Advanced Techniques and Future Trends*, (Ed.) A.N. Luiten, (Springer-Verlag, Berlin, 2001), 37.
3. A.J. Giles *et al*, *Physica B* **165**(1): 145, 1990
4. A.N. Luiten, A.G. Mann, D.G. Blair, *IEEE Trans. Instr. Meas.* **44**: 132, 1995
5. See www.crystalsystems.com/hem.html
6. G. Santarelli, P. Lemonde, A. Clairon, *Phys. Rev. Lett.*, **82**(23): 4619, 1999
7. P. Wolf *et al*, *Phys. Rev. Lett.*, **90**(6): 060402, 2003
8. P. Lemonde *et al*, *IEEE Trans. Instr. and Meas.* **48**(2): 512, 1999
9. P.L. Stanwix, M.E. Tobar, P. Wolf, M. Susli, C.R. Locke, E.N. Ivanov, J. Winterflood, F. van Kann, *Phys. Rev. Lett.*, **95**(4): 040404, 2005
10. P.L. Stanwix, M.E. Tobar, P. Wolf, C.R. Locke, E.N. Ivanov, *Phys. Rev. D* **74**: 081101(R), 2006
11. K. Watabe, J.G. Hartnett, C.R. Locke, G. Santarelli, S. Yanagimachi, T. Shimazaki *et al*, *Jap. J. Appl. Phys.* **45**(12): 9234, 2006
12. K. Watabe, H. Inaba, K. Okumura, F. Hong, J.G. Hartnett, C.R. Locke, *et al*,*CPEM Special Issue IEEE Trans. on Instr. Meas.* **56**(2): 632, 2007
13. See www.nict.go.jp
14. J.G. Hartnett, C.R. Locke, E.N. Ivanov, M.E. Tobar, P.L. Stanwix, *Appl. Phys. Lett.*, **89**(20): 203513, 2006
15. L. Rayleigh, *Philos. Mag*, **20**: 1001, 1910
16. S. Chang, A. Mann and A. Luiten, *Elect. Lett.*, **36**(5): 480, 2000
17. M.E. Tobar and J.G. Hartnett, *Phys. Rev. D* **67**: 062001, 2003
18. P-Y. Bourgeois *et al*, *Appl. Phys. Lett.* **87**(22): 224104, 2005
19. C.R. Locke, E.N. Ivanov, J.G. Hartnett, P.L. Stanwix, M.E. Tobar, *Rev. Sci. Instrum.*, **79**(5): 051301, 2008
20. R.V. Pound, *Rev. Sci. Instrum.*, **17**: 490, 1946
21. R.W.P. Drever et al, *Appl. Phys. B.*, **31**: 97, 1983
22. S. Chang, A.G. Mann *et al*, *Phys. Rev. Lett.*, **79**(11): 2141, 1997
23. M. Kumagai, H. Ito, G. Santarelli, C.R. Locke, J.G. Hartnett, M. Hosokawa, to be published in Proceedings IFCS, 2008

ULTRA-STABLE OPTICAL CAVITY: DESIGN AND EXPERIMENTS

J. MILLO, S. BIZE, E. M. L. ENGLISH, P. LEMONDE, G. SANTARELLI
LNE-SYRTE, Observatoire de Paris, 61 Avenue de l'Observatoire
75014 Paris, France

We present the design geometry of an optical horizontal cavity. The Finite Elements Modelling analysis is used to predict and decrease the influence of vibrations on the cavity stabilized laser. Inspired by the results of this simulation, a horizontal cavity is constructed. The vibration sensitivity components measured for this cavity are equal to or better than $1.4 \times 10^{-11}/(m.s^{-2})$ for all spatial directions.

1. Introduction

Ultra-stable laser light is a key element for a variety of applications ranging from optical frequency standards, [1, 2], tests of relativity [3], generation of low phase noise microwave signals [4] and transfer of optical stable frequencies by fibre networks [5, 6], to gravitational wave detection (VIRGO, LIGO and LISA).

The laser beam wavelength is compared to that of a reference cavity, and locked using the Pound-Drever-Hall technique onto a narrow resonance (typically 1-10 kHz). The frequency noise spectrum of the laser is determined by the fluctuations of the optical length of the reference cavity. Classically cylindrical optical cavities lie on V-blocks and show a large sensitivity to vibration of $\sim 3 \times 10^{-10}/(m.s^{-2})$. Vibration isolation systems can minimize noise level, but commercial systems are not sufficient to reach a sub-Hz laser linewidth. One way to improve the spectral performances of stabilized lasers is to reduce vibration sensitivity by carefully designing the cavity geometry and its mounting. Several groups have proposed and successfully implemented low vibration sensitivity cavities [7-10]. We implement a horizontal design with new features, and use the Finite Elements Modelling to perform an extensive and refined study of the vibration response of the cavity.

2. Finite Elements Modelling of the Cavity

We restrict the analysis to the quasi-static response of the cavity. Mechanical resonances are in the kHz range, where acoustic and vibration isolation systems are efficient. In the finite elements model, spacer and mirror substrates are considered as a single rigid body. The cavity geometry is meshed with ~125000 prism elements, where each prism has 6 nodes. Finite element deformations are calculated within the elastic limit including those due to Poisson's effect.

When the cavity is accelerated, a length variation is induced by elastic deformations. Careful design allows the compensation of this variation using Poisson's effect. Deformations simulations have been done using the mechanical properties of Ultra Low Expansion glass (ULE™): density (2210 kg/m^3), Young's modulus (67.6 GPa) and Poisson's ratio (0.17). Vibration sensitivity of cavity is deduced by observing the deformations in the central region of the mirrors for a given acceleration value.

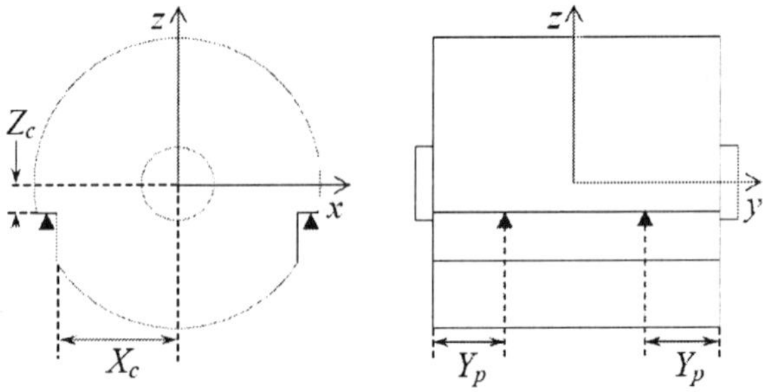

Figure 1. Front and side views of the cavity. The optical axis is aligned to the *y*-axis. The four support-points are represented with black triangles.

Two types of mirror deformations are important. The first type is the mirror translation along the cavity axis in order to analyze the distance between mirrors' centres. These length variations exist if the cavity has no symmetry plane orthogonal to the acceleration axis. The second type is the mirror tilt. For an ideal cavity where optical and mechanical axes coincide, tilt induced length variation is a second order effect. For a real cavity, the mechanical and optical axes are misaligned. In the worst case these two axes are parallel and separated by a distance *d*. Length fluctuations become proportional to both *d* and the tilt angle. We can write the cavity length fluctuations in the following way:
$\delta L/L = \boldsymbol{k}\cdot\boldsymbol{a}$ where $\boldsymbol{a}$ is the acceleration vector and $\boldsymbol{k}$ is the vector of the vibration sensitivity, where the components are expressed as:
$k_x = k_x^L + k_x^T(d)$, $k_y = k_y^L + k_y^T(d)$, $k_z = k_z^L + k_z^T(d)$, where k_i^L and $k_i^T(d)$ are respectively the sensitivity components to the mirrors' translation and tilt.

The geometry we consider is a 100 mm long cavity held horizontally. This configuration allows for fine adjustment of the positions of four support-points in order to experimentally minimize the vibration sensitivity (Fig. 1). Horizontal contact planes are obtained by machining two square "cutouts" along the cylindrical spacer (100 mm diameter). The four contact points are in a horizontal plane and placed symmetrically around the cavity. Spacer asymmetry is minimized by machining cutouts near the horizontal middle plane and the edge of the cylinder. The results of the simulations were used to finely tune the geometry of the spacer. The following results correspond to the implemented geometry ($X_c = 47$ mm and $Z_c = 3$ mm).

Due to the cutouts, the vertical acceleration vector is not orthogonal to a symmetry plane. Hence, the vertical vibration sensitivity component $k_z = k_z^L + k_z^T(d)$ is *a priori* non zero for both types of deformation. However, the component due to the mirrors' translation k_z^L can vanish for a specific position of the support-points. We emphasise that this component is strongly dependent on the contact model: the surface size and the constraints on the degrees of freedom. When the contact surfaces are constrained in all directions, a small surface size gives a very low k_z^L regardless of longitudinal position of contact points. For larger surface sizes, a solution with $k_z^L = 0$ does exist. When the contact with the cavity is only vertically constrained, the insensitive solution is independent of the support-point size (Fig. 2). The vertical acceleration also induces mirror tilt but the sensitivity component k_z^T is independent of the contact model. Assuming a rather large misalignment $d = 1$ mm, the dependence on the support-point position is low (~$1.3\text{x}10^{-12}$/m.s^{-2} per mm) with $k_z^T = 0$ for Y_p ~10 mm.

Beyond this evaluation of vertical axis vibration sensitivity, a considerable part of the work has been devoted to horizontal sensitivities. As a matter of fact, the residual horizontal vibration level can be the dominant source of noise in commercial anti-vibration platforms. Horizontal components k_x and k_y depend on the position of the support-points but are nearly independent of the contact point model. Due to symmetry, only tilting of the mirrors contributes to transverse horizontal sensitivity $k_x = k_x^T(d)$. This component can be zeroed for the support-point positions set at Y_p ~ 9 mm with a slope ~$2\text{x}10^{-12}$/m.s^{-2} per mm (Fig. 2). When an axial vibration is applied, the two mirrors are tilted in the same direction. The worst misalignment to consider is a rotation between the mechanical and optical axes. In this case, we consider a distance $z = d$ between axes on one mirror, and $z = -d$ for the other one. The axial vibration sensitivity component $k_y = k_y^T(d)$ also depends on d. This component can not be zeroed, however for $d = 1$ mm it is always below $3\text{x}10^{-12}$/(m.s^{-2}) and therefore negligible (Fig. 2).

The choice of geometry parameters $X_c = 47$ mm and $Z_c = 3$ mm were dictated by the consideration that regardless of the contact point model, it is always possible to find an optimum support spacing where k_z has a low value (~$5\text{x}10^{-12}$/(m.s^{-2})). Moreover this optimum spacing also corresponds to a low value of k_x (<$1\text{x}10^{-12}$/(m.s^{-2})).

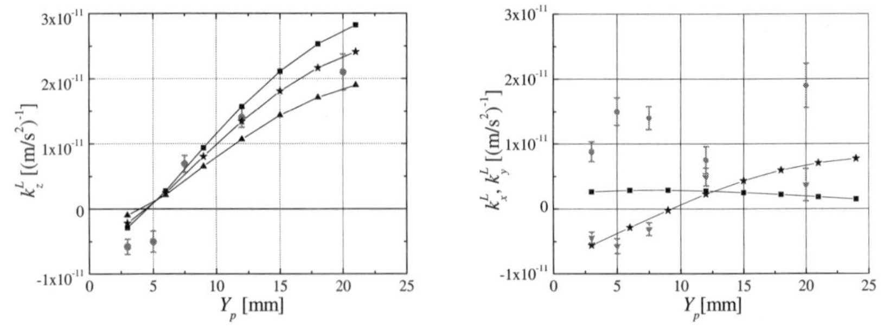

Figure 2. Left plot: comparison between measurements (circles) and simulation results for the vertical vibration sensitivity k_z as a function of Y_p. Contact point are vertically constrained with area 0.04 mm^2 (squares), 1 mm^2 (stars) and 4 mm^2 (triangles). Right plot: the horizontal sensitivity transversal measurements (down triangles) and simulations (stars) k_x, longitudinal direction measurements (circles) and simulations (squares) k_y. Both simulations are for $d = 0.3$ mm.

3. Experimental Set-up and Vibration Sensitivity Measurements

The spacers of two cavity configurations are machined from ULE glass rods. The mirrors have a high reflection coating operating at a wavelength of 1062.5 nm, and the fused silica substrates of the mirrors reduce the contribution of thermal noise floor [12]. The cavity has been optically contacted with a flat mirror and a concave mirror with a radius of curvature of 500 mm. The cavity shows a finesse of about 800000 and a fringe contrast better than 60%.

We measured the three vibration sensitivity components by shaking the cavity setup with sinusoidal signals in the frequency range 1-10 Hz. The cavity setup (cavity, vacuum chamber and optical table) is supported by an active vibration isolation platform. The cavity itself is supported under vacuum with four 2 mm^2 Viton pads 0.7 mm thick. The laser beam is stabilized to the cavity using the Pound-Drever-Hall technique. A reference cavity [13] is kept unperturbed while the cavity under test is shaken. The beat-note signal between the two lasers is demodulated by a frequency-to-voltage converter and analysed with a FFT analyzer. A low noise seismometer placed on the top of the vacuum chamber measures the accelerations of three spatial directions.

We excite sequentially two horizontal and the vertical directions measuring the amplitude of the induced frequency tone and the strength of the acceleration. These measurements are iterated for several support-point positions of the cavity. The active platform can apply accelerations in a given direction with amplitude of about 10^{-3}m.s^{-2} rms, however the coupling to other directions could be as much as 10%. The phase between the frequency response signal (filtered and amplified) and the vibration excitation signal is also measured and gives the relative sign of the cavity response. Measurement error bars have been estimated by considering the signal-to-noise ratio of the measurement, acceleration crosstalk effects and the calibration error on vibration measurement (total ~10%). Measurement results of the vertical response agree with the vertically constrained model and contact surfaces of about 1 mm^2. The lowest vibration sensitivity component observed is $5x10^{-12}/(m.s^{-2})$ and the dependence on the support-point position is very low: $1.6x10^{-12}/(m.s^{-2})$ per mm (Fig. 2). The difference between horizontal sensitivity components measured and calculated can be explained by the unequal restoring forces of the four support-points on the cavity. This asymmetry is not reproducible when support point positions are changed and can explain the dispersion of different measurements.

4. Conclusion

An enhanced version of a horizontal optical cavity has been studied to decrease influences of vibrations on the optical length. This cavity design has been realized and the vibration sensitivity measured. The vibration sensitivity components are about $10^{-11}/(m.s^{-2})$ or better in all directions. The vertical acceleration sensitivity component shows a very low dependence on the support-point positions of $1.6x10^{-12}/(m.s^{-2})$ per mm. Therefore, fine tuning of their positions is not necessary. A preliminary beat-note signal between two independent stabilized lasers shows a frequency stability of about $1x10^{-15}$@1s.

References

1. T. Rosenband *et al.*, *Science*, **319**, 1808 - 1812, (2008).
2. A. D. Ludlow *et al.*, *Science*, **319**, 1805 - 1808, (2008).
3. H. Müller *et al.*, *Phys. Rev. Lett*, **99** 050401, (2007).
4. A. Bartels *et al.*, *Opt. Lett.*, **30**, 667-669, (2005).
5. N. R. Newbury, *et al.*, *Opt. Lett.*, **32**, 3056-3058, (2007).
6. H. Jiang, *et al., J. Opt. Soc. Am. B*, **25** (2008).
7. T. Rosenband, J. Bergquist, private communication.
8. T. Nazarova, F. Riehle and U. Sterr, *Appl. Phys. B*, **83**, 531-536, (2006).
9. A. D. Ludlow *et al.*, *Opt. Lett.* **32,** 641-643, (2007).
10. S. A. Webster, M. Oxborrow, S. Pugla, J. Millo and P. Gill, *Phys. Rev. A* **77,** 033847, (2008).
12. K. Numata, A. Kemery and J. Camp, *Phys. Rev. Lett*, **93,** 250602, (2004).
13. J. Millo *et al.*, Proc. of the *IEEE Int. Frequency Symposium* (2008).

NEW RESULTS FOR WHISPERING GALLERY MODE CRYOGENIC SAPPHIRE MASER OSCILLATORS

K. BENMESSAI, P.Y. BOURGEOIS, N. BAZIN, Y. KERSALÉ* and V. GIORDANO

FEMTO-ST Institute, Time and Frequency Departement, Université de Franche-Comté,
Besançon, France
**E-mail: yann.kersale@femto-st.fr*

M. OXBORROW

National Physical Laboratory,
Hampton Road, Teddington, Middlesex, TW11 0LW, United Kingdom
E-mail: Mark.Oxborrow@npl.co.uk

D.L. CREEDON, J.G. HARTNETT, M.E. TOBAR*

University of Western of Australia, School of Physics,
6009 WA, Australia
**E-mail: mike@physics.uwa.edu.au*

We report here results obtained with the Cryogenic Whispering Gallery Mode Sapphire Maser Resonator-Oscillator. The first characterization against the UWA cryogenic sapphire oscillator exhibited frequency instability of orders 10^{-14} at 1 second limited by the readout system we used. We also report, for the first time, on the measurement of the fundamental limit (thermal noise) for a microwave resonator.

Keywords: Solid State Maser, Sapphire, Frequency Stability.

1. Introduction

The Cryogenic Whispering Gallery Mode Sapphire Maser Resonator-Oscillator is a very promising device for a microwave secondary frequency standard. Maser action results from a zero-field three level Electron Spin Resonance system created by residual Fe^{3+} ions within the crystal lattice[1,2], as oppossed to the intentionally doped crystal used in the fifties.[3–5] In our device the amplifier and the resonator mode are located in the same physical package which constitues a great advantage compared to the superconduct-

ing cavity maser oscillator developed by Dick et al.[6]
In this paper we present a review of the most important results obtained on our cryogenic sapphire maser oscillator.

2. The Whispering Gallery Mode Maser Oscillator.

2.1. *Principle*

Our maser exploits Bloembergen's classic three-level scheme,[7] but at zero applied DC magnetic field. The population inversion is created by applying a Pump signal at 31.35 GHz (over a 100 MHz span) with a power of order 17 dBm via a microwave synthesizer.
The presence of a high Q mode in the ESR lineshape[8] at 31.3 GHz enables us to saturate the transition $\mid 1/2 > \mapsto \mid 5/2 >$. Maser action results from the coincidence in frequency of the Electron Spin Resonance of Fe^{3+} ions with a very high Q ($\simeq 10^9$) Whispering Gallery Mode at 12.04 GHz as shown in Fig. 1.

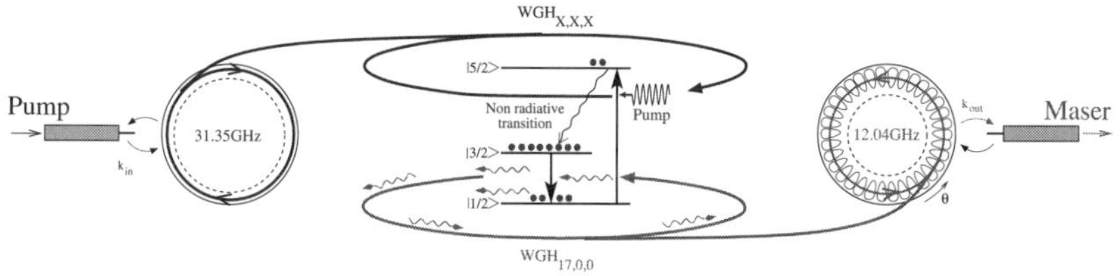

Fig. 1. Whispering gallery mode Maser principle.

2.2. *Frequency instability measurement*

The maser oscillating at 12.04 GHz (-56 dBm output power) is compared to the state-of-the-art Cryogenic Sapphire Resonator Oscillator[9] operating at 11.200 GHz.The beatnote between the CSRO and the maser produces a 838 MHz signal. Through a microwave synthesis chain[10] , the microwave beatnote is downconverted to some 100 kHz.
The maser frequency instability measurement is limited in the short term by the microwave synthesis chain at 10^{-14} up to 30 s, as shown in Fig. 2. For integration times higher than 30 s, the maser presents a random walk process (characterised by a $\tau^{1/2}$ slope.). This effect is probably due to the pump power fluctuations that need to be controlled with an appropriate design under development.

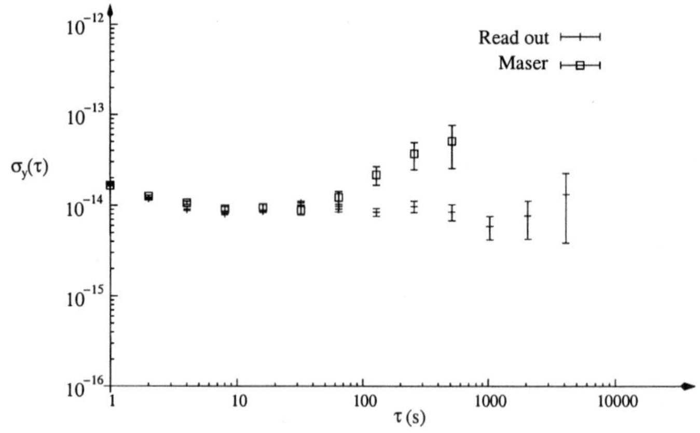

Fig. 2. Frequency instability measurement of readout system and maser.

For a microwave resonator at liquid helium temperature in the Nyquist thermal regime, the Schawlow-Townes limit on frequency instability is given in Eqn. (1).[11]

$$\sigma_y(\tau) = \frac{1}{Q_{mode}}\sqrt{\frac{k_B T}{2P_{mode}\tau}} \tag{1}$$

Here, Q_{mode} is the mode Q-factor, P_{mode} is the power in Watts, and τ is the measurement integration time in seconds.
For a -56 dBm output power this limit is of order $10^{-16}\tau^{-1/2}$. Recently, by annealing a sapphire in air (conversion of Fe^{2+} to Fe^{3+}), we obtained an output power of -37 dBm, leading to a potential frequency instability limit of the order $10^{-18}\tau^{-1/2}$.

3. Bimodal maser oscillations

High-Q Whispering Gallery Modes in sapphire resonators exist as doublets due to perturbations such as backscatter in the crystal due to Rayleigh scattering and probe perturbations.[12–15] The back scatter causes the two opposite traveling waves propagating around the circumference of the sapphire cylinder to exhibit strong coupling and split into two spatially orthogonal circumferential standing waves,[13] which in this case are separated by 10 kHz in frequency. Because the splitting is small (10 kHz) it is only observable in high-Q systems. We report here simultaneous maser action of the two modes as shown in Fig. 3 (bi-mode operation). In contrast, the bi-mode regime only occurs at low power levels between 10^{-15} to 10^{-12} dBm output power (at higher powers only one mode dominates).

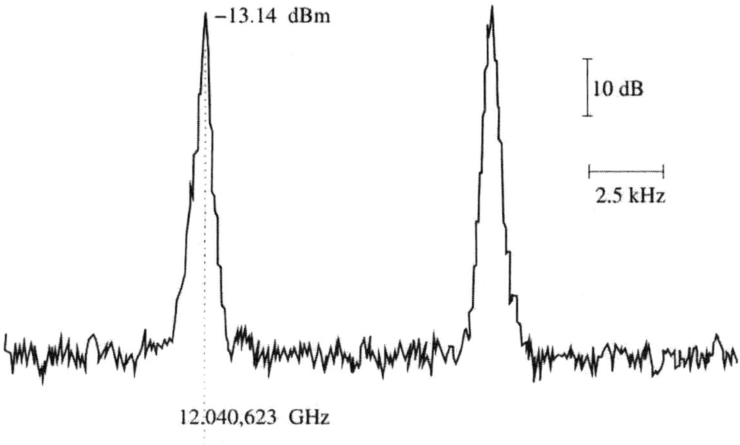

Fig. 3. Spectrum of the bimodal oscillation at 12.04 GHz using a readout with 80 dB gain.

The two signals, extracted from the same probe, are amplified by the same room temperature amplifier (two amplifiers with 40 dB gain each) and then sent to a quadratic detector (tunnel diode) to be mixed. A microwave band pass filter is placed between the two microwave amplifiers to reduce the thermal noise produced by the first amplifier. The 10 kHz beat note is band pass filtered and amplified by a low noise amplifier and then sent to a high-resolution counter, referenced to a hydrogen maser. The frequency instability in terms of Allan deviation is plotted in Fig. 4a for a maser output power of -93 dBm. The level of white noise is calculated from the coefficient of the $\tau^{-1/2}$ fit in the 1 to 10 second interval. Periodic effects cause clearly visible bumps in the Allan deviation curve, which can be ignored when calculating the white noise. This measurement was repeated for an output power of -98 and -96 dBm. The frequency instability at one second (white noise) is reported in Fig. 4b. The solid curve is the predicted stability assuming ideal amplification of the Maser. The calculation is done for a temperature of 5 K, Q-factor of 7×10^8 and averaging time of 1 second. From the comparison it is clear that we have measured the thermal noise limit[16] given by the Schawlow-Townes limit on frequency stability presented in Eqn. (1).

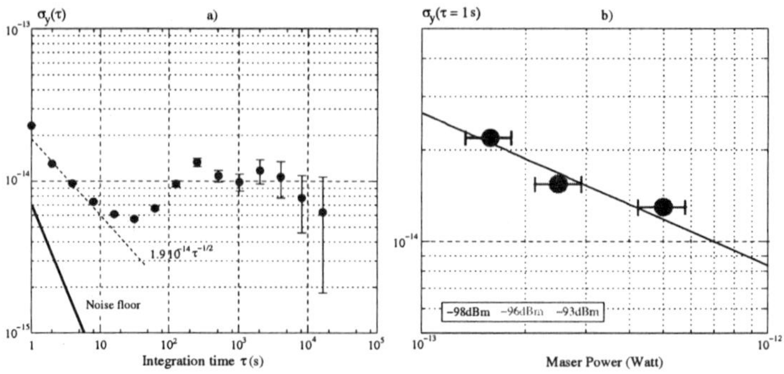

Fig. 4. a) : $\sigma_y(\tau)$ for a power level of P_{mode}= - 93 dBm. b) : Value of white noise component at 1 second for output powers of P_{mode}= -98, -96, - 93 dBm. The solid line is the value of the white noise calculated from eq.(1).

References

1. P. Y. Bourgeois, N. Bazin, Y. Kersalé, V. Giordano, M. E. Tobar and M. Oxborrow, *Appl. Phys. Lett.* **87**, 224104 (2005).
2. P. Y. Bourgeois, M. Oxborrow, M. E. Tobar, N. Bazin, Y. Kersalé and V. Giordano, *International Journal of Modern Physics B* **20**, 1606 (2006).
3. A. E. Siegman, *Microwave Solid State Maser* (McGraw-Hill, New York, 1964).
4. L. S. Kornienko and A. M. Prokhorov, *Sov. Phys. JETP* **36**, p. 649 (1959).
5. G. S. Bogle and H. F. Symmons, *Proc. Phys. Soc.* **73**, p. 531 (1959).
6. R. T. Wang and G. J. Dick, Improved performance of the superconducting cavity maser at short measuring times, in *Proc. IEEE Int. Freq. Cont. Symp. (IFCS 90)*, (Baltimore, USA, 1990).
7. N. Bloembergen, *Phys. Rev.* **104**, 324 (1956).
8. K. Benmessai, D. L. Creedon, P. Y. Bourgeois, N. Bazin, M. E. Tobar, Y. Kersalé and V. Giordano, Observation of a bimodal behavior in a whispering gallery mode maser oscillator, in *Proc. IEEE Int. Freq. Cont. Symp. (IFCS 08)*, (Hawai, USA, 2008).
9. J. G. Hartnett, C. R. Locke, E. N. Ivanov, M. E. Tobar and P. L. Stanwix, *Appl. Phys. Lett.* **89**, p. 203513 (2006).
10. K. Benmessai, P. Y. Bourgeois, Y. Kersalé, N. Bazin, M. E. Tobar, J. G. Hartnett, M. Oxborrow and V. Giordano, *Electron. Lett.* **43**, 1436 (2007).
11. A. L. Schawlow and C. H. Townes, *Phys. Rev.* **112**, p. 1940 (1958).
12. D. S. Weiss, V. Sandoghdar, J. Hare, V. Lefèvre-Seguin, J.-M. Raimond and S. Haroche, *Opt. Lett.* **20**, 1835 (1995).
13. T. J. Kippenberg, S. M. Spillane and K. J. Vahala, *Opt. Lett.* **27**, 1669 (2002).
14. M. L. Gorodetsky and A. D. Pryamikov, *J. Opt. Soc. Am. B* **17**, 1051 (2000).
15. P. Y. Bourgeois and V. Giordano, *IEEE Trans. Microwave Theory Tech.* **53**, 3185 (2005).
16. K. Benmessai, D. L. Creedon, M. E. Tobar, P. Y. Bourgeois, Y. Kersalé and V. Giordano, *Phys. Rev. Lett.* **100**, p. 233901 (2008).

Part IX

Advanced Techniques

FUNDAMENTAL NOISE-LIMITED OPTICAL PHASE LOCKING AT FEMTOWATT LIGHT LEVELS[1]

JOHN DICK, MEIRONG TU, KEVIN BIRNBAUM,
DMITRY STREKALOV and NAN YU

Jet Propulsion Laboratory, California Institute of Technology,
4800 Oak Grove Dr. Pasadena, CA 91109, USA

We describe an optical phase lock loop (PLL) designed to recover an optical carrier at powers below one picowatt in a Deep Space optical transponder. Previous low power optical phase lock has been reported with powers down to about 1 pW. We report the demonstration and characterization of the optical phase locking at femtowatt levels. We achieved a phase slip rate below one cycle–slip/second at powers down to 60 femtowatts. This phase slip rate corresponds to a frequency stability of 1×10^{-14} at 1 s, a value better than any frequency standard available today for measuring times equal to a typical two–way delay between Earth and Mars. The PLL shows very robust stability at these power levels. We developed simulation software to optimize parameters of the second order PLL loop in the presence of laser flicker frequency noise and white phase (photon) noise, and verified the software with a white phase noise model by Viterbi. We also demonstrated precise Doppler tracking at femtowatt levels.

1. Introduction

Deep Space navigation and Radio Science today use microwave transponders on spacecraft together with ultra-stable clocks on the ground presently for ranging. To improve the ranging accuracy over the past decades, microwave frequencies have steadily increased, leading to the X/Ka transponders used in the Cassini Mission for navigation, ring and atmospheric occultation experiments, and a search for gravitational waves. An obvious advantage of higher frequency is the reduction of diffractive losses on both earth-based and spacecraft-based antennas due to the shorter wavelength. Just as important, however, is a markedly

---

[*]This research was carried out at the Jet Propulsion Laboratory, California Institute of Technology, under a contract with the National Aeronautics and Space Administration.

increased immunity of higher frequency electromagnetic radiation to effects of the (ionized and unstable) solar wind. The culmination of this progression would be an "optical transponder" using light at optical wavelengths to provide a dramatic improvement in ranging performance compared to any microwave transponder [1].

We report here on the results of experiments, calculations and demonstrations designed to establish the feasibility of a Deep Space coherent optical transponder based on an optical phase locked loop (PLL) scheme. To meet the Deep Space application needs, such optical PLL will have to operate at extremely low light levels. As an example consider spacecraft for a Mars Mission, where the farthest distance from Earth is approximately 2.5 AU. For such a mission, and for base- and spacecraft-antennas of 1 meter and 10 cm diameters, respectively, a 5-Watt transmitter requires operation with received powers of 100 fW or below. To our knowledge, optical phase lock loop has not been demonstrated or characterized at femto-watt level. In this paper, we have modeled, developed and tested optical PLL's that operate at pico-watt and femto-Watt power levels. The excellent performance obtained; that is, low cycle-slip rates at very low optical power levels make the PLL-based optical transponder an attractive prospect for Deep Space Spacecraft. Implemented in a Mars Mission, an optical transponder could increase both ranging and Doppler accuracies by more than 10× compared to present technology, while reducing acquisition time by ≈ 100.

2. Laser Noise Characterization

In order to model and optimize the optical PLL at low powers, it is necessary to understand the noise characteristics of lasers used. We used 1550 nm fiber lasers in our experiments. The goal is to achieve near shot-noise limited PLL performance at femto-Watt light level.

In the presence of only shot noise, cycle-slips in a PLL can be reduced to any desired level by reducing the loop bandwidth [2]. However, with a free-running laser local oscillator (LO), optical PLL performance at the lowest possible optical power levels is always limited by the presence of low-frequency laser noise. This noise is typically white-frequency or flicker-frequency noise, which, with $1/f^2$ or $1/f^3$ spectra (respectively) at low frequency offsets, eventually dominates any shot noise value as the frequency is lowered. What this means for phase locking is that narrowing of the loop bandwidth fails at some point to improve the cycle-slip rate, and furthermore, that the rate will in fact increase if the bandwidth is further narrowed. Thus, a detailed knowledge of the laser phase noise is necessary in order to predict and optimize low power PLL performance with any real laser.

The test setup for phase noise measurements is shown in Figure 1. All measurements were obtained in a closed-loop setup for stability and reproducibility. The phase locked loop is a combination of RF and optical circuitry, after detection of a heterodyne of the weak signal with an added LO signal at about 100 MHz, the detected signal is phase locked to a synthesized signal by means of PLL circuitry that combines a second-order loop with variable parameters which is used in a fast loop that controls the frequency of an approximately 100 MHz VCO driving an acousto-optic modulator (AOM), and whose action is complemented by a very slow loop controlling the piezo frequency control incorporated into the laser.

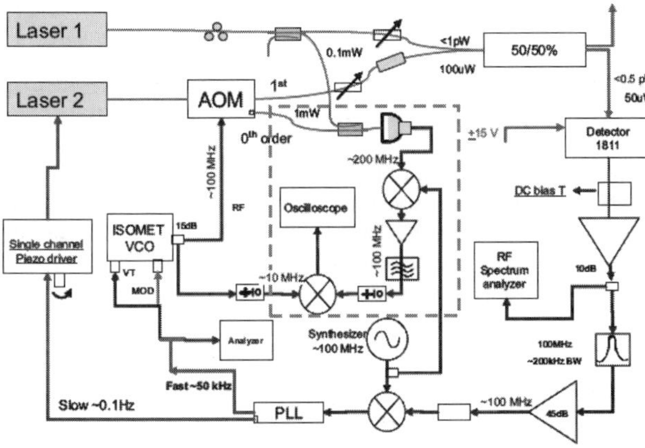

Figure 1. Block diagram of the general measurement test setup. Phase noise data is typically obtained from the "Analyzer" signal, while components in the dashed box are added in order to measure cycle–slips and their associated phase noise.

A summary of our laser phase noise characterization results is shown in Figure 2. The phase noise measurement at high laser power (green curve) shows a spectral density of phase fluctuations $S_\phi(f) \propto 1/f^3$ over a wide frequency range: The lower frequency results were measured as frequency fluctuations in a closed-loop setup with a relatively high loop bandwidth while the higher frequency curve were measured as phase fluctuations with a narrow loop bandwidth-effectively an open-loop setup. A straight-line value $S_\phi(f) = 60$ dB/Hz $\times$ $(1\text{Hz}/f)^3$ derived from this plot was used in our PLL simulations as the 2-laser phase noise. As will be shown later, simulations based this model gave good

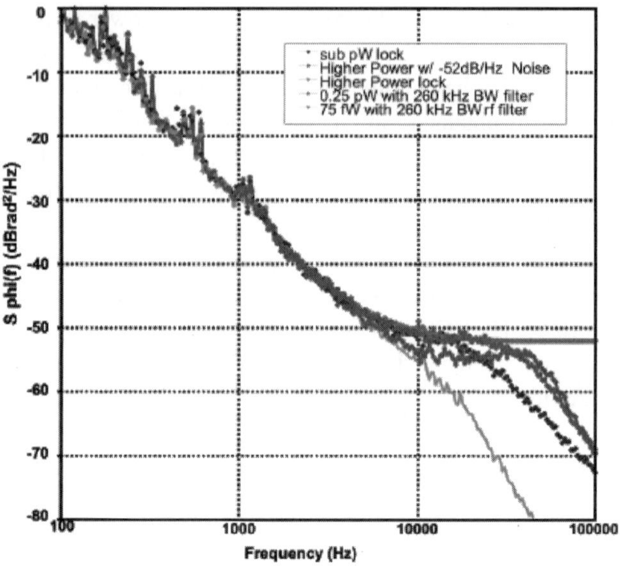

Figure 2. Measured phase noise for two fiber lasers. The data were combination of in- and out of loop measurements. The light green curve shows high power measurements when the laser phase noise dominating all measurement band. Measured phase noise at very low power showing the onset of shot noise. The -52 dB rad^2/Hz apparent asymptote for the 75 pW curve is just 6.5 dB above the theoretical value of -58.5 dB rad^2/Hz for that power level. Typical values obtained were 4 to 6 dB above the theoretical shot noise value.

agreement with experimental phase-slip measurements. By integrating the phase noise, we obtain a short-term laser linewidth of 2.8 kHz for a measuring time of 10 ms.

By reducing the power of Laser 1 into the detector, we can start seeing an addition of a constant value at the high frequency side to the curve in Figure 2. This constant noise floor is due to the shot noise of the laser. Calibrating the laser power and the noise floor observed at 75 fW, we conclude that the measured S/N is 6.5 dB poorer than the shot noise limit, This loss is primarily due to detector inefficiency and optical mismatch in the detector between the LO (heterodyne) signal and the weak 75~pW signal.

Our choice of an approximately 100 MHz heterodyne frequency for the optical PLL setups in Figure 1 is largely driven by amplitude noise (RIN) in our lasers noise. Figure 3 shows a plot of the measured RIN noise spectrum. The figure shows that the RIN noise is substantially sensitive to laser operation parameters, so that (e.g.) the 50 μW LO signal will have more RIN noise if it is

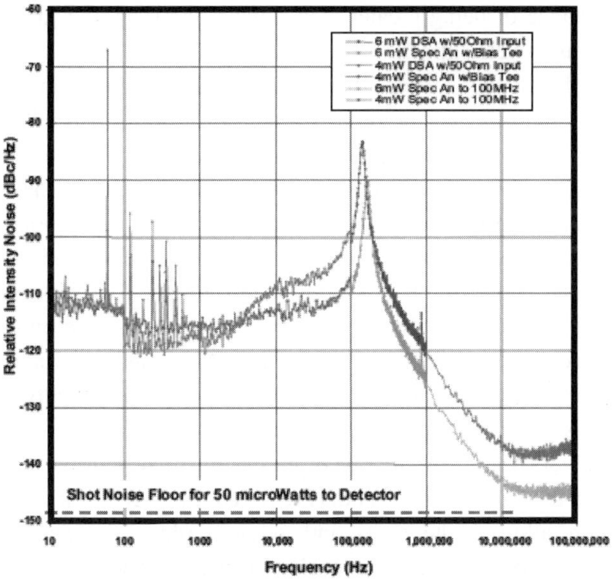

Figure 3. Relative Intensity Noise (RIN) for one laser. Notice that the peak near 100 kHz shifts with laser power. The approach to shot–noise limited performance between 20 MHz and 100 MHz points to this region for offset (heterodyne) and subcarrier frequencies. For this reason all phase–lock experiments were conducted with an injected signal that was offset by 50 MHz to 100 MHz.

derived from a 4~mW laser output than a similar signal derived from a higher laser output of 6 mW. We typically operate the lasers at an output level of 17 mW for best noise and stability performance.

While it is possible to reject RIN noise in the injected LO (heterodyne) signal to a substantial degree by the use of balanced and complementary phase detectors, achieving shot noise-limited performance is made easier by the use of a single high-quality optical detector. Amplitude noise is also nominally rejected by our RF mixer phase detector, but only by 20 dB or so. We have found that the RIN noise approaches the shot noise value for higher power outputs from the fiber laser at frequencies above about 50 MHz. Therefore, the 100 MHz heterodyne frequency is used.

3. Optical Phase Locking

In the low power phase lock loop, it is critical to optimize the loop performance. This requires not only the understanding of the laser noise behavior, but also the level of loop performance. One of the special issues of low SNR in a PLL is the cycle slip phenomenon. The phase noise fluctuation inside the loop bandwidth has a finite chance of being greater than π when the loop will jump lock to the next fringe, thus a cycle slip [2].

We have developed software that simulates PLL operation in the face of the various noise processes present in our system. PLL performance limitations due to photon shot noise has been previously studied by [2] for a first order phase-lock loop. Somewhat poorer results for a second order loop have also been reported. Because the fiber lasers used show flicker-frequency noise with a $1/f^3$ power spectrum over much of the frequency range, a second order loop is required, and the optimization of such a system for the lowest possible level of cycle-slips has not been previously reported. We verified the operation of our software by recovering the Viterbi result, and also found good agreement with LISA Project calculations for their NPRO lasers (which typically show "white frequency" noise with a $1/f^2$ power spectrum [3]).

A typical software run involved generating 1-4 million random phase variations with the various noise models required. For any given shot-noise level, the simulation allowed the first- and second-order loop parameters to be optimized so as to minimize the cycle-slip rate. Additionally, the highest possible shot-noise level for an optimized loop could be determined for any give cycle-slip rate. The results shown in Figure 4 have been optimized by adjustment of the loop parameters for a cycle-slip rate of 1.0 (per second), and indicate that this rate can be achieved with a shot noise level of -51 dB rad^2/Hz; which with a perfect (shot-noise limited) optical detector and an ideal RF system would correspond to a power level of 13 fW (at an optical wavelength of 1550 nm).

In addition to the "white" and "flicker frequency" noise processes shown in Figure 4, a small amount of "random walk frequency" noise with a $1/f^4$ spectrum was also added, in an amount too small to be seen on the graph; without such an added noise the simulated PLL would never actually "lose lock", behavior that is typically (and often) seen in the laboratory.

The block diagram of the experimental setup for the low power phase locking is shown in Figure 1. Phase noise data characterizing the cycle slips were measured at the "Oscilloscope" port in the figure; a comparison of a high-power phase reference signal at the right side of the dashed box with the phase-locked signal from the left. Both (approximately 100 MHz) signals are divided in frequency by a factor of 10, the output of the two dividers being square-wave signals at approximately 10 MHz. Combining these two signals in a double-balanced mixer gives a dc level that shows a triangle-wave dependence on the phase between the two input signals; with 5 cycles up and 5 cycles down as the differential phase varies by many cycles. We assume that cycle slips are random

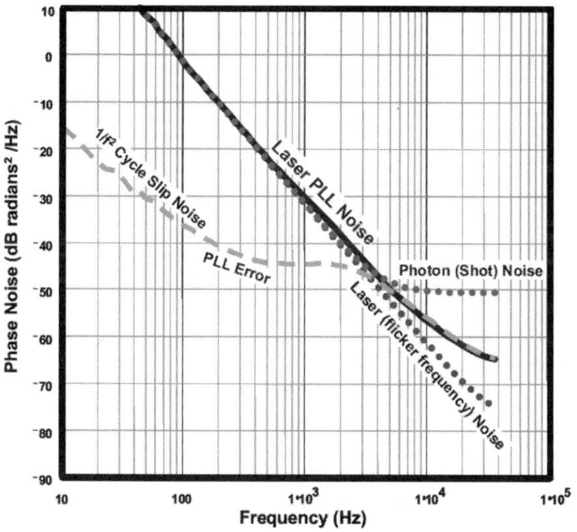

Figure 4. Simulation calculation of PLL performance optimized for the highest possible shot noise level (-51 dB/Hz) that allows a cycle slip rate of 1 cycle–slip per second as shown by a ~ $2/f^2$ (/Hz) noise contribution to the PLL error signal. The +60 dB /f3 flicker frequency noise imposed on the PLL simulation well matches the (two laser) measured noise values.

in direction, and so the sign inversion of the slope of the triangular mixer output every 5 cycles should not impact the measured noise values. However, there is a small (5%) correction that must be applied due to missed phase jumps at the turn-around points of this waveform.

In actual experiment, we have been able to achieve optical phase lock with actual optical power into the optical (heterodyne) detector of 40 fW, with no more than 1 cycle-slip per second. Previous work in support of the ASTROD [4,5] space mission study has reported optical PLL operation at powers down to ~2 pW. In Figure 5 we show a comparison between laboratory measurements at 300 fW optical power and the results of simulations as the loop gain is varied. While an increase in phase-slips with increasing gain may seem counterintuitive, these plots are qualitatively similar to the results of Viterbi-with the onset of phase slips occurring as the integrated phase noise within the loop bandwidth becomes greater than 1 radian^2. The simulations were chosen to match the observed unity gain frequency of 29 kHz at the lowest gain values (8 dB attenuation). A comparison of the data with the simulation calculation shows excellent agreement for the onset of phase slips, both occurring at the "4 dB" attenuation level, and with nearly identical values.

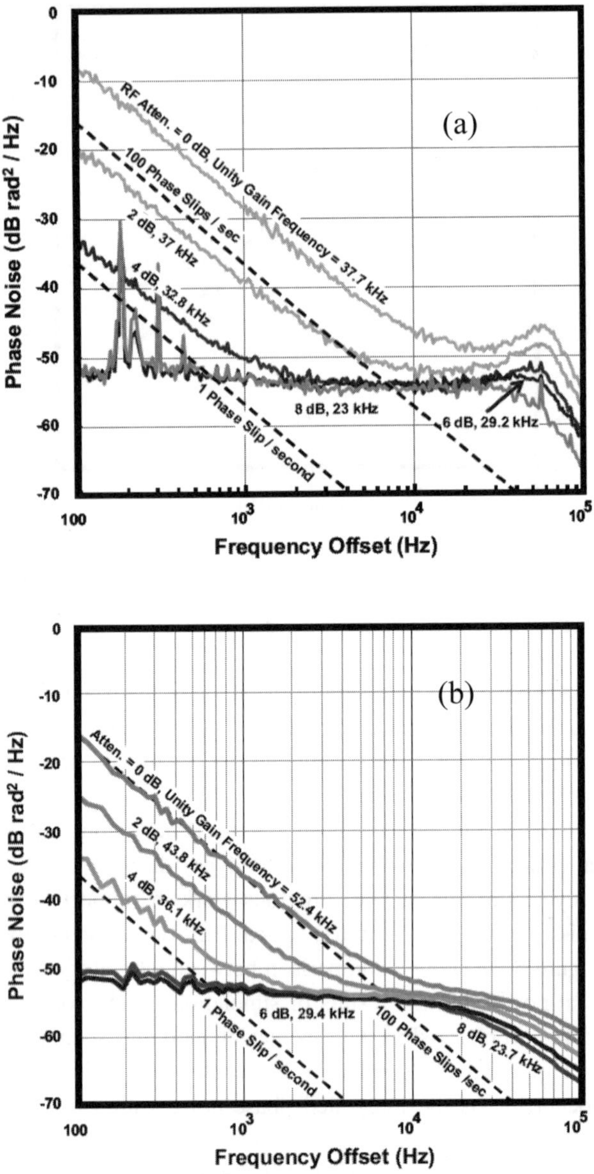

Figure 5. (a) Measured phase noise in the optical PLL, showing the onset of cycle–slips with varying loop gain for an optical power of 300 fW into the detector. (b) Simulation calculation of PLL cycle slip phase noise due to shot noise at -55.5 dB (rad^2/Hz) for loop gains (and unity gain frequencies) comparable to the experimental tests. Lines for several phase slip rates are indicated for each graph.

As the RF attenuation is reduced and the loop gain increased by this same factor, the observed unity gain frequency in the simulations scales roughly with this gain value, while for our measurements the unity gain frequency saturates at a value of about 38 kHz, with the data also showing increasing cycle-slip rates, compared to the simulation. We believe this discrepancy is due to the unmodeled RIN noise peak that can be seen at about 75 kHz for the higher gain values in Figure 5(a). These peaks showed strongly in the RF spectrum at higher gains, and gave rise to significant carrier reduction in the RF spectrum for the high gain values.

We expect that the utility of this technology will be at cycle slip rates of 1/second or lower where the impact of the RIN noise is small. This being so, the loop performance is practically equal to the theoretical prediction. Thus, while PLL performance is certainly limited by the $1/f^3$ laser phase noise, it is not practically impacted by the presence of laser RIN noise.

4. Doppler Ranging Demonstration

In addition to the PLL performance investigation, we have also performed several demonstration experiments that are designed to show that high performance can be obtained with the kind of time-varying range that a deep space transponder would face. Of course, we can't achieve space velocities in the laboratory, but we can achieve accelerations comparable to those experienced by spacecraft, and it is these higher order variations of phase that place significant burden on the optical PLL circuit.

Figure 6 shows the setup for a free space demonstration experiment that used an air track together with a corner-cube optical mirror to give excellent and stable optical performance together with large mechanical motion. The corner-cube reflector was found to be necessary in order to obtain good optical performance. Coiled springs allowed accelerations up to 0.3 g to be obtained. This setup succeeded in transferring 30% of the optical power from the source fiber to the output fiber, with a typical variation of 5% due to motion of the reflector.

In initial tests at 1 pW input power, the PLL transponder showed no evidence of cycle-slips when tested with free-space input from a movable corner-cube reflector mounted on an air track with springs for a period of ~1 Hz and path variation of ~1 m; conditions that simulate the acceleration (and associated Doppler rate) for typical orbits. These experiments were done without the use of predicts or feedforward that would be available in actual use. In these tests, the

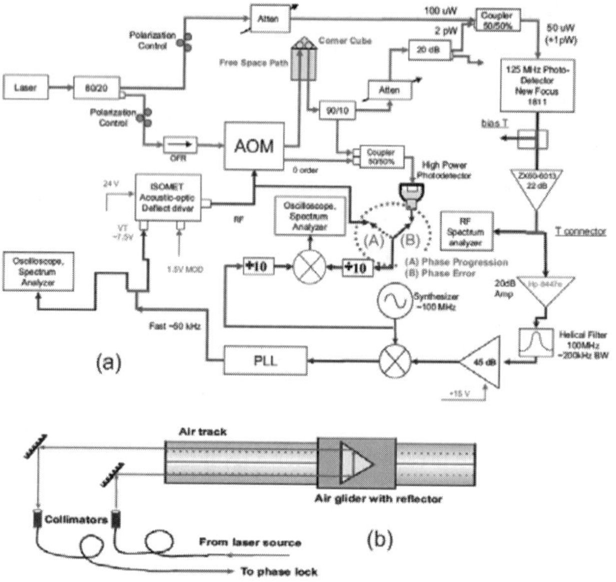

Figure 6. Details of the setup for free space measurements using a moveable corner–cube reflector on an air track. (a) Block diagram of the electro-optical setup; and (b) block diagram of the corner–cube setup.

PLL phase variation was less that 0.8 radians with mirror motion of 4 $\times 10^6$ radians.

A second demonstration experiment measured Doppler and Doppler errors at very low optical power (150 fW) using the moving mirror and air track, but without the springs. For this experiment the mirror moved at a relative smooth manner, bouncing from short springs at the ends.

High- and low-power Doppler measurements were obtained by simultaneously arming two counters. The raw data for both counters, showing the variation in Doppler frequency as the mirror traversed the air track for ≈7 out and back trips. A slight tilt to the track gave rise to the periodic velocity variation in the data. In addition, small variation can be seen between the two channels at the ends where the velocity reversed.

Figure 7 shows the difference between the high- and low-power Doppler signals, showing an RMS variation of 0.2 Hz for the one-second measurements. This corresponds to an optical frequency variation of about $\sigma_y(1s) \approx 1\times10^{-15}$, a value roughly comparable to the short-term frequency stability of the best optical frequency standards today.

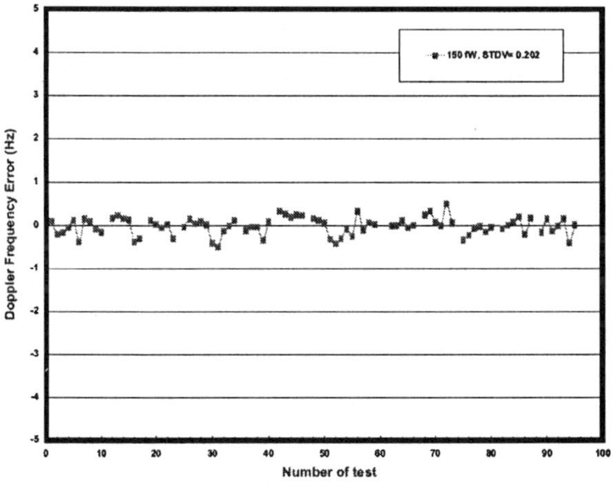

Figure 7. High-resolution data for the difference between the high- and low-power Doppler measurements in the counters. The RMS variation of the points is approximately 0.2 cycles, which for an optical frequency of 2×10^{14} Hz, corresponds to an optical frequency stability of approximately 1×10^{-15}.

5. Summary

Optical phase lock loops have been demonstrated for the first time at femto-Watt power levels. The optimized performance was enabled by development of a calculational methodology that simulated phase lock operation in the presence of realistic noise models for the laser, so that optimal loop parameters could be ascertained. Low cycle-slip rates have been demonstrated at power levels as low as 40 femto-Watts, and observed cycle-slip rates were found to be in good agreement with calculated values. Several demonstration experiments were performed using an air-track glider and a corner-cube optical reflector with approximately 1 meter of free movement. Low power phase lock was preserved with up to 0.3 g acceleration of the moving mirror, and Doppler accuracy of 0.2 cycles was demonstrated, corresponding to a stability for optical Doppler frequency measurements of 1×10^{-15}. The results suggest potential performance improvements in Deep Space ranging by the use of a coherent optical transponder.

Acknowledgments

The authors would like to thank Ertan Salik for helpful discussions.

References

1. P.W. Kinman, R.M. Gagliardi "Doppler and range determination for deep space vehicles using active optical transponders" Applied Optics, V. 27, 4487–4493 (1988).
2. Chapter "Phase–locked–loop behavior in the presence of noise," in Principles of Coherent Communication, by A.J. Viterbi (McGraw–Hill, 1966), pp. 77–120.
3. Robert Spero, JPL, Private Communication.
4. W.-T. Ni "ASTROD (Astrodynamical Space Test of Relativity using Optical Devices) and ASTROD I" Nucl. Phys. B (Proc. Suppl.) V. 166, 153–158 (2007).
5. A.-C. Liao, W.-T. Ni, J.-T. Shy, "Pico–Watt and Femto–Watt Weak–Light Phase Locking" Int. J. Mod. Phys. D, V. 11, 1075–1085 (2002).

MICROWAVE AND OPTICAL FREQUENCY TRANSFER VIA OPTICAL FIBRE

G. MARRA, S. N. LEA, H. S. MARGOLIS AND P. GILL
National Physical Laboratory, Hampton Road, TW11 0LW, Teddington, UK
giuseppe.marra@npl.co.uk

The radio frequency repetition rate of an infrared erbium-doped mode locked fibre laser is transferred over 50 km of spooled single mode fibre (SMF). The pulse broadening caused by the chromatic dispersion of the fibre is partially counteracted by incorporating a matched length of dispersion compensating fibre (DCF). The fibre-induced phase noise is cancelled by means of a fibre stretcher at the transmitter end. The transfer frequency stability delivered at the 'user' end is measured to be better than $5 \times 10^{-15}\ \tau^{-1/2}$ for integration times longer than 1 s.

1. Introduction

Transmission via optical fibre networks has recently been identified as the most viable alternative to more traditional satellite-based techniques for the transfer of high stability frequency references between remote locations. Two-way satellite time and frequency transfer techniques (TWSTFT) and GPS carrier phase (CP) can achieve a few parts in 10^{15} at one day [1]. These transfer techniques are already unable to preserve the stability delivered by state of the art caesium fountain primary frequency standards ($< 1 \times 10^{-15}$ at 1 day) [2, 3]. Moreover, recent advances of optical frequency standards have shown the potential for stability improvements of two or more orders of magnitude compared to the best microwave standards [4, 5], making even more urgent the need for better frequency transfer techniques. Satellite-based techniques are also unsuitable for most metrological and synchronization applications where high short-term stability is required.

Three methods of transferring frequency references via optical fibre for timing and metrological applications have been explored to date: microwave transfer by amplitude modulation of a continuous wave (cw) optical carrier [6, 7]; direct transfer of the optical carrier [6, 8, 9]; and simultaneous transfer of the optical and microwave references by transmission of an optical frequency comb [6, 10]. The best transfer stability is achievable when the optical carrier itself is transmitted, with a stability of $3 \times 10^{-16}\ \tau^{-3/2}$ over 251 km having been demonstrated recently [9]. However, if a microwave reference is to be transferred, one of the other two techniques has to be used. The amplitude modulated cw optical carrier technique has demonstrated stability of better than 1.5×10^{-15} at 1 s over 86 km [7]. The frequency transfer technique in which an optical comb is directly transmitted has shown a stability, measured after a round trip, of 9×10^{-15} at 1 s over 6.9 km, limited by the measurement noise floor [10]. In this work we report the transfer stability for the latter technique over a 50 km link of spooled single mode fibre (SMF), when different fibres are used for the forward and return optical path and the stability is measured at the 'remote' end. In an optical comb transfer experiment on a real long-distance fibre network the use of two fibres is imposed by the unidirectional optical amplifiers normally required to regenerate the optical signal.

2. Experimental setup

The set-up of the experiment is shown in figure 1. A 1.5 μm erbium-doped mode-locked fibre laser emits a train of sub-150 fs pulses at a repetition rate of 100 MHz. The optical spectrum of the comb extends over 100 nm. A 90:10 splitter is used to split the laser output beam into two optical paths, each constituting a measurement arm. The low power arm (lower arm) is the low noise reference against which the phase noise introduced by the long length of fibre in the upper arm is measured.

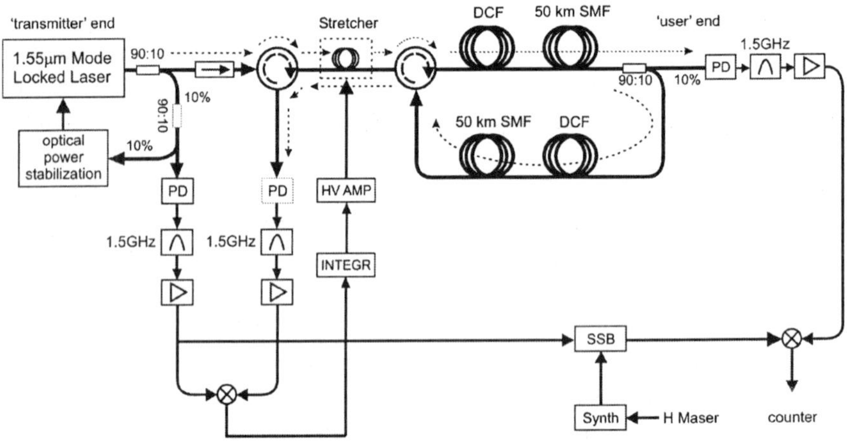

Fig. 1. Transfer frequency stability measurement set-up. PD: photodetector; SSB: single sideband modulator; DVM: digital voltmeter.

The higher power output of the splitter (upper arm) is connected to a circulator and a fibre stretcher. The fibre stretcher is used to compensate for the fast fibre-induced phase fluctuations of the detected repetition rate caused by environmental acoustic noise. It was constructed in-house by winding 250 turns of SMF-28 on a 5-cm diameter piezo tube and allows optical path length compensation of the order of a few millimetres. This range is sufficient to allow measurements of a few minutes duration to be carried out. A second circulator separates the forward and return optical paths. At the middle of the optical fibre link, which simulates the point where the transferred comb will be delivered to a 'user', a 90:10 optical splitter extracts the transferred repetition rate before sending the optical comb back through the return spool. After passing again through the fibre stretcher and the first circulator, the repetition rate is detected for comparison with the reference arm.

The chromatic dispersion of the single mode fibre causes the pulse to broaden to 300 ns duration over the full length of fibre. The overlap between consecutive pulses causes the light at the fibre output to become effectively cw and the repetition rate is no longer detectable. By incorporating a matched length of dispersion compensating fibre (DCF) in the optical link, we recover a pulse duration of 220 ps. Although maintaining a high temporal pulse separation is crucial to achieve sufficient signal to noise ratio of the detected repetition rate, full recovery of the initial sub-150 fs pulse duration is not a necessary condition to preserve the repetition rate stability over the link. Other noise processes, such as the excess phase noise introduced by the fibre or the noise introduced by the microwave amplifiers, may in practice be the dominant limitation to the transfer frequency stability achievable. The optical power launched into the upper arm is approximately 60 mW. Stimulated Brillouin scattering (SBS) does not occur as the optical power within the Brillouin gain bandwidth remains below the threshold for this process. The total optical loss of the SMF and the DCF is 30 dB, including connection losses. In the reference arm, the optical power is attenuated to avoid saturation of the photodetector. It has been observed that fluctuations in the laser power cause excess instability in the measured transferred frequency. For this reason a feed-forward type optical power stabilization scheme employing an acousto-optic modulator is used.

At the output of each photodetector, a comb of frequencies spaced by 100 MHz is detected which extends up to the upper limit of the photodetector bandwidth. As the detector thermal and shot noise is independent of the harmonic of the repetition rate detected, the highest possible harmonic should be chosen to minimize the detection noise contribution to the total phase noise. The 15th harmonic (1.5 GHz) is selected with a narrow bandwidth band pass filter and then amplified by low phase noise microwave amplifiers. The transfer stability of the optical link is determined by direct frequency measurement with a high-resolution frequency counter. The frequency detected in the reference arm is shifted by 50 kHz using a single sideband modulator (SSB) driven by a low phase noise direct digital synthesizer (DDS). The beat produced by comparison of the frequency-shifted reference to the frequency detected in the other arm is amplified, filtered and finally measured by a Λ-type high-resolution counter [11]. For averaging times of 1 to 10 s, the frequency stability was measured by changing the gate time for each individual measurement. For averaging times above 10 s, the frequency stability is calculated from the set of frequency readings taken at 10 s.

3. Results

The fractional frequency instability introduced during the transfer of the 15th harmonic of the repetition rate of the mode-locked laser over 50 km of spooled fibre is shown in figure 2. For integration times shorter than 1 s the fractional frequency instability scales approximately as $5 \times 10^{-15}\ \tau^{-3/2}$ and then scales approximately as $\tau^{-1/2}$ for longer time scales. The transfer stability measured when the phase noise compensation is turned off is also shown for comparison.

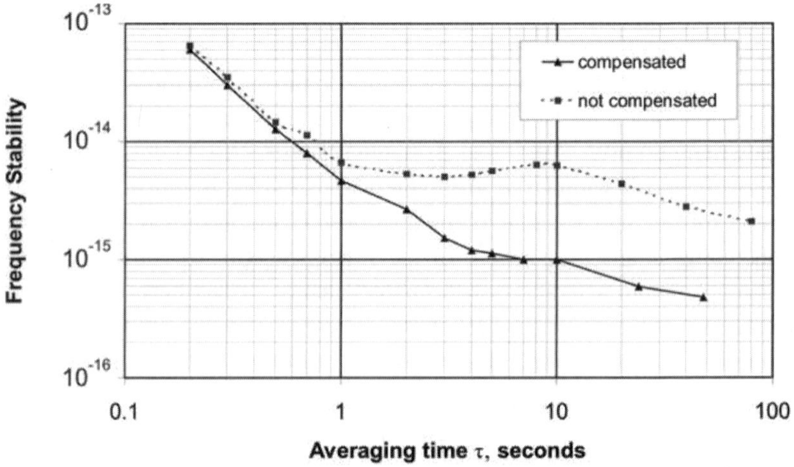

Fig. 2. Fractional frequency stability of the transferred repetition rate of the 1.5 μm erbium-doped fibre laser over 50 km of SMF and matched DCF, measured at the 15th harmonic. Squares: no phase noise compensation; triangles: fibre noise compensated

4. Conclusions

We have transferred the 100 MHz repetition rate of an erbium-doped fibre laser over 50 km of spooled fibre in the laboratory by transmitting the optical comb.

The pulse broadening caused by the dispersion of the fibre is partially compensated by incorporating a matched length of dispersion compensating fibre in the optical link, leading to recompression of the pulse duration by more than a factor of 1000. The excess phase noise introduced by the fibre is compensated using a fibre stretcher at the transmitter end. The transfer frequency instability of the 15th harmonic of the repetition rate is measured to be better than $5 \times 10^{-15}\ \tau^{-1/2}$. To our knowledge, this is the first time that the transfer stability has been measured at the 'user' end rather than after a round trip, for this technique. The measured transfer stability over 50 km is approximately a factor of two better than previously reported results for a 6.9 km round-trip installed network [10].

The duration of the measurement is currently limited by the small phase compensation range of the fibre stretcher. We plan to overcome this limitation by inserting a temperature-controlled short fibre spool in series with the fibre stretcher. Future work will also include testing the transfer technique on an installed fibre network.

Acknowledgements

This work is supported by the UK Department for Innovation, Universities & Skills as part of the NMS Pathfinder Metrology Programme.

References

1. A. Bauch *et al.*, *Metrologia* **43**, 109 (2006).
2. S. R. Jefferts *et al.*, *Proc. SPIE* **6673**, 667309 (2007).
3. F. Chapelet *et al.*, p. 467 in *Proceedings of the 2007 EFTF-FCS conference* (IEEE International, 2007).
4. T. Rosenband *et al.*, *Science* **319**, 1808 (2008).
5. A. D. Ludlow *et al.*, *Science* **319**, 1805 (2008).
6. S. M. Foreman *et al.*, *Rev. Sci. Instrum.* **78**, 021101 (2007).
7. O. Lopez *et al.*, *Ann. Phys. Fr.* **32**, 187 (2007).
8. G. Grosche *et al.*, paper CMKK1 in *Proc. Conference on Lasers and Electro-Optics (CLEO) 2007 Technical Digest* (OSA, 2007).
9. N. R. Newbury *et al.*, *Opt. Lett.* **32**, 3056–3058 (2007).
10. K. W. Holman *et al.*, *Opt. Lett.* **30**, 1225–1227 (2005).
11. E. Rubiola, *Rev. Sci. Instrum.* **76**, 054703 (2005).

ULTRA-STABLE LASER SOURCE FOR THE $^{88}Sr^+$ SINGLE-ION OPTICAL FREQUENCY STANDARD AT NRC

P. DUBÉ,* A. A. MADEJ, J. E. BERNARD

Frequency and Time Group, Institute for National Measurement Standards, National Research Council of Canada, Ottawa, ON, K1A 0R6, Canada
**E-mail: pierre.dube@nrc.ca*

We describe an ultra-stable laser system used to probe the clock transition of the $^{88}Sr^+$ single ion optical frequency standard at NRC. The laser performance is determined primarily by the reference Fabry-Perot cavity which is stabilized at the temperature where the thermal expansion coefficient of the spacer is zero. The frequency stability was studied by comparing the frequency of the laser locked to the reference cavity with that of the $^{88}Sr^+$ clock transition. After subtraction of a piecewise linear drift, the laser frequency reaches a stability of 5×10^{-16} at an averaging time of 3000 s. The laser linewidth was investigated by recording high resolution spectra of one of the Zeeman components of the clock transition. A Fourier-transform limited linewidth of ≈ 4.3 Hz, corresponding to a fractional frequency resolution of 1×10^{-14}, was observed for a 200 ms probe pulse length.

Keywords: Optical frequency standard; Ultra-stable laser; Fabry-Perot cavity; Single ion.

1. Introduction

An optical frequency standard can have very high stability compared to a microwave clock due to the high frequency and narrow linewidth of the reference transition. The quality factor, $Q = \nu_0/\Delta\nu$, of this optical clock transition is typically 10^{14} or higher for frequency standards based on either a single trapped ion or a lattice of trapped neutral atoms. For example, the $5s\,^2S_{1/2}$–$4d\,^2D_{5/2}$ clock transition of $^{88}Sr^+$ has a linewidth of $\Delta\nu = 0.4$ Hz at $\nu_0 = 445$ THz (674 nm), which provides a line Q of 10^{15}. The interrogation of such a narrow resonance requires a suitably stable and narrow linewidth probe laser source to approach the quantum projection noise (QPN) limit offered by the optical frequency standard.[1] In this paper, we give a brief overview of the design and performance of the laser system used to probe the clock transition of the $^{88}Sr^+$ single ion optical frequency standard developed at the National Research Council of Canada.

2. The $^{88}Sr^+$ Frequency Standard

A single ion of $^{88}Sr^+$ is confined in an rf Paul trap operated at a frequency of 12 MHz that has a characteristic endcap-to-center distance of $z_0 = 0.5$ mm.[2] The ion oscillates in the trap with a transverse secular frequency of 965 kHz and an axial secular frequency of 1930 kHz.[3] The ion is cooled using the strongly-allowed $5s\,^2S_{1/2} - 5p\,^2P_{1/2}$ dipole transition with a frequency-stabilized 422 nm laser source.[4] A fiber laser emitting radiation at 1092 nm prevents decay of the ion from the upper state of the cooling transition to the metastable $^2D_{3/2}$ state.

A 12 μT magnetic field is applied to the ion to split the $5s\,^2S_{1/2}{-}4d\,^2D_{5/2}$ clock transition into its ten Zeeman components. The clock frequency linecenter is obtained from the average frequency of three pairs of symmetric Zeeman components chosen to connect to all of the six sublevels of the metastable $^2D_{5/2}$ state. This technique removes the linear Zeeman shift, the electric-quadrupole shift, and the tensor terms of the quadratic Stark shifts.[3,5] The interaction between the probe laser light and a Zeeman component of the clock transition is monitored using the electron shelving technique where excitation into the $^2D_{5/2}$ state is detected with near unit efficiency by observation of quantum jumps with the strong 422 nm *S*–*P* fluorescence. A computer monitors the rate at which the quantum jumps occur to either lock the probe laser frequency to the $^{88}Sr^+$ clock transition linecenter or to record high-resolution spectra of the Zeeman components.

3. Probe Laser System

The probe laser system consists of a commercial diode laser at 674 nm that is frequency narrowed and stabilized using a cascade of two Fabry-Perot resonators as shown in the schematic diagram of Fig. 1. The performance of the frequency lock to the reference ULE cavity was investigated using the frequency noise density of the Pound-Drever-Hall error signal. The results indicate that the laser linewidth relative to a cavity mode is smaller than 1 mHz. This lock fidelity ensures that the laser acquires the stability of the reference cavity. Special care was taken to minimize drifts and to isolate the ULE cavity from external perturbations. The cavity was made exceptionally immune to thermal fluctuations by using two-stage temperature stabilization and by setting the vacuum chamber temperature at the value where the coefficient of thermal expansion (CTE) of the ULE cavity on its support structure crosses zero. The data for the determination of this temperature are shown in Fig. 2(a).

The short- and long-term isothermal frequency drift rates of the reference ULE cavity as a function of the time spent under vacuum are shown in Fig. 2(b). The so-called long-term rates were calculated from the change in the frequency of a

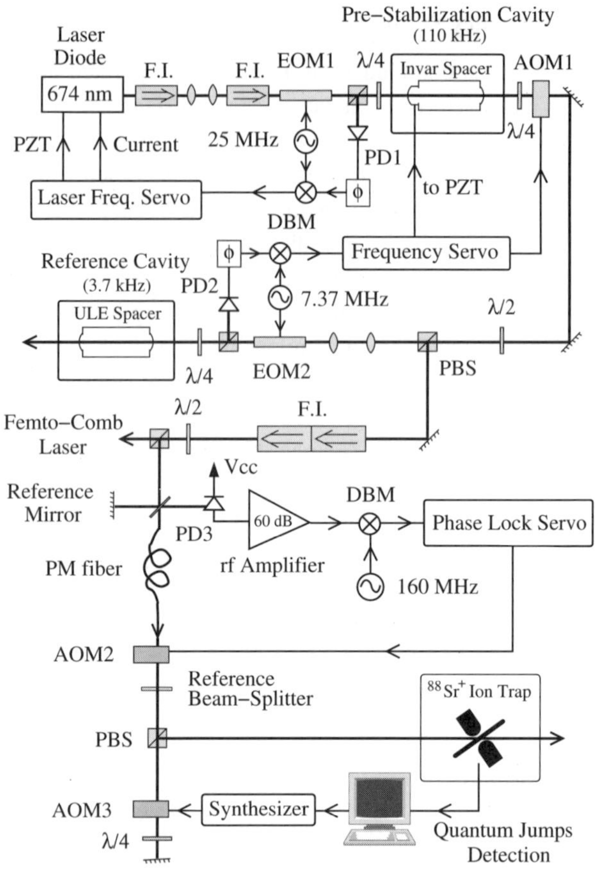

Fig. 1. Schematic diagram of the 674 nm probe laser system for the $^{88}Sr^+$ single ion optical frequency standard. EOM: electro-optic modulator; AOM: acousto-optic modulator; PZT: piezo-electric transducer; PD: photo-detector; $\lambda/4$: quarter-wave plate; $\lambda/2$: half-wave plate; PBS: polarizing beam-splitter cube; F.I.: Faraday isolator; DBM: double-balanced mixer; PM fiber: polarization maintaining fiber; ϕ: phase shifter.

cavity mode measured on different days divided by the time between these measurements. Only data with a minimum of two weeks between measurements were considered for the evaluation of the long-term rates. The short-term rates were obtained from data recorded during periods of a few hours. The frequency measurements were made predominantly with a Ti:Sapphire femto-second laser frequency comb referenced to a maser source. There is good agreement between the long- and short-term rates, indicating that the cavity frequency is not strongly perturbed by environmental changes or by laser power absorbed on the cavity mirrors. The

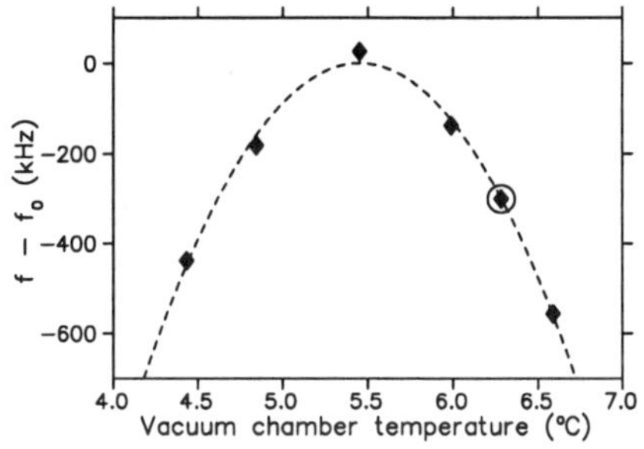

(a)

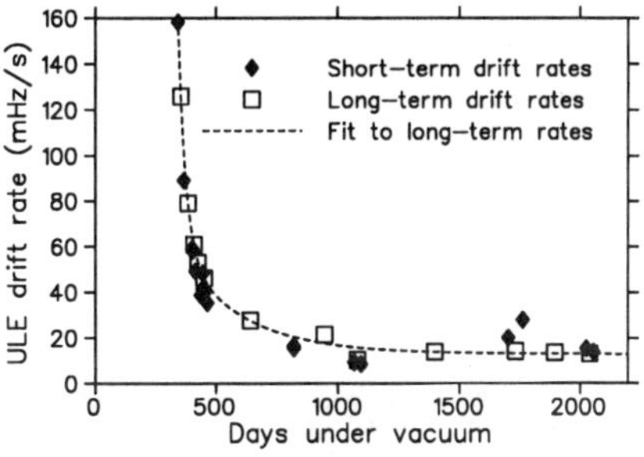

(b)

Fig. 2. (a) ULE cavity frequency as a function of the vacuum chamber wall temperature at thermal equilibrium conditions. The dashed line is a fit to the data. The coefficient of thermal expansion is zero at the extremum found at $T_0 = 5.45\,(1)°\mathrm{C}$. The circled data point was the first and last temperature point measured. The change in frequency between these two measurements was used to determine an average cavity drift rate of 22 mHz/s during these measurements. (b) Isothermal frequency drift rates of a cavity mode of the reference ULE cavity as a function of time spent under vacuum.

present long-term isothermal drift rate of the ULE cavity is 13.9(3) mHz/s based on the last four data points of Fig. 2(b). To our knowledge, this is the smallest drift rate reported in the literature for a ULE cavity.

The stability of the probe laser system was investigated by measuring the relative frequency between the probe laser system and the clock transition linecenter of $^{88}Sr^+$. This was achieved by monitoring the error signal of the probe laser locked to the ion linecenter.[5] The data are shown in Fig. 3(a). After removal of a piecewise linear drift (two linear sections) caused by the ULE cavity, the frequency fluctuations average down to 5×10^{-16} at 3000 s of averaging time as shown in Fig. 3(b). We attribute the high stability of the probe laser system at averaging times of a few thousand seconds to stabilization of the reference cavity at the temperature where the CTE crosses zero.

An upper limit to the laser linewidth was obtained by recording high resolution spectra of the $(m_{J''}, m_{J'}) = (1/2, 3/2)$ component. The statistical distribution of the linewidths observed with a 200 ms probe pulse are in agreement with a Fourier transform limited value of 4.3 Hz. This linewidth represents a fractional frequency resolution of 1×10^{-14} at the transition frequency of 445 THz.

4. Conclusions

This paper described briefly the ultra-stable laser system used at NRC to probe the clock transition of the $^{88}Sr^+$ single ion optical frequency standard. The stability of this laser was characterized using the Allan deviation of a 5 hr lock sequence to the

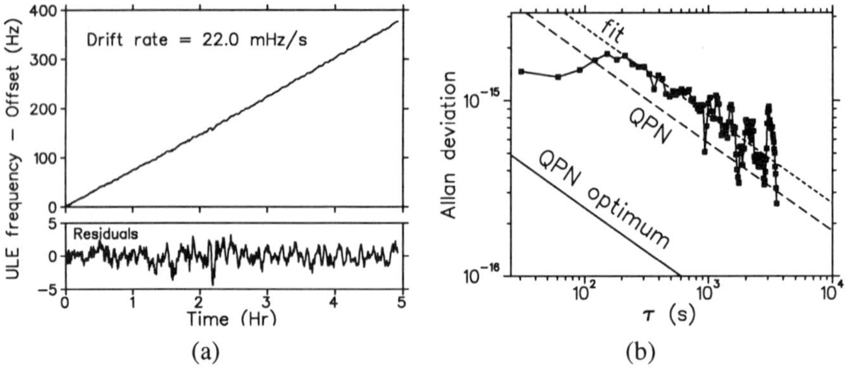

Fig. 3. (a) Upper graph: cavity drift measured by comparing the probe laser frequency with the ion clock transition. Lower graph: the frequency residuals after subtraction of a piecewise linear drift. (b) Allan deviation of the residuals. fit: $2.6 \times 10^{-14}/\sqrt{\tau}$; QPN: $1.9 \times 10^{-14}/\sqrt{\tau}$ (QPN-limited Allan deviation for the present experimental conditions); QPN optimum: $2.6 \times 10^{-15}/\sqrt{\tau}$.

ion linecenter. After subtraction of a piecewise linear drift, the stability of the cavity reached 5×10^{-16} at an averaging time of 3000 s. This level of in-stability also represents an upper limit estimate for that of the frequency lock to the ion linecenter. The laser linewidth was investigated by recording high-resolution spectra of a single Zeeman component of the $^{88}Sr^{+}$ ion clock transition. Fourier transform limited linewidths of ≈ 4.3 Hz were observed, corresponding to a fractional frequency resolution of 1×10^{-14}. The present performance of the laser system is thus adequate to measure the ion linecenter frequency at a level limited by the accuracy of a state-of-the-art Cs fountain clock.

Acknowledgments

The authors would like to acknowledge the important technical contributions of R. Pelletier and B. Hoger to the experiments reported in the present work.

References

1. W. M. Itano, J. C. Bergquist, J. J. Bollinger, J. M. Gilligan, D. J. Heinzen, F. L. Moore, M. G. Raizen and D. J. Wineland, *Phys. Rev. A* **47**, 3554 (1993).
2. A. A. Madej, J. E. Bernard, P. Dubé, L. Marmet and R. S. Windeler, *Phys. Rev. A* **70**, p. 012507 (2004).
3. P. Dubé, A. A. Madej, J. E. Bernard, L. Marmet, J.-S. Boulanger and S. Cundy, *Phys. Rev. Lett.* **95**, p. 033001 (2005).
4. A. D. Shiner, A. A. Madej, P. Dubé and J. E. Bernard, *Appl. Phys. B* **89**, 595 (2007).
5. J. E. Bernard, L. Marmet and A. A. Madej, *Opt. Commun.* **150**, 170 (1998).

CLOCK LASER SYSTEM FOR A STRONTIUM LATTICE CLOCK

T. Legero*, Ch. Lisdat, J.S.R. Vellore Winfred, H. Schnatz, G. Grosche, F. Riehle and U. Sterr

Physikalisch-Technische Bundesanstalt
Braunschweig
Germany
**E-mail: Thomas.Legero@ptb.de*

We describe the setup and the characterization of a 698 nm master-slave diode laser system to probe the $^1S_0-^3P_0$ clock transition of strontium atoms confined in a 1D optical lattice. The frequency noise and the linewidth of the laser system have been measured with respect to an ultrastable 657 nm diode laser with 1 Hz linewidth. The large frequency difference of more than 25 THz was bridged using a femtosecond fiber comb as transfer oscillator. In a second step the virtual beat was used to establish a phase lock between the narrow line 657 nm laser and the strontium clock laser. This technique allowed to transfer the stability from the 657 nm to the 698 nm laser.

Keywords: ultrastable lasers, lattice clock

1. Introduction

Optical lattice clocks with strontium show relative uncertainties below 10^{-15}.[1-3] With 10^6 atoms and a Fourier-limited linewidth of 1 Hz the instability due to the quantum projection noise would ultimately result in an Allan deviation of the relative frequency fluctuations of $\sigma_y = 10^{-18}/\tau^{1/2}$. However, to reach this instability an even better instability of the clock laser is needed. Therefore the improvement of optical sources which are phase stable during the required interrogation time of a few seconds is a key technology for optical frequency metrology. Until a few years ago, the investigation of the instability of a laser system was only possible by comparison with a laser system operating at the same wavelength. With the invention of the frequency comb technology it is now possible to measure the instability of an ultra narrow laser by comparison with a microwave frequency standard or a laser system operating at a different wavelength.

By using the femtosecond laser as a transfer oscillator[4] the stability of the narrow-linewidth laser can be transferred to the second laser that does not need a sophisticated frequency stabilization.

2. Setup of the clock laser

For probing the $^1S_0-^3P_0$ strontium clock transition we have set up a 698 nm master–slave diode laser system as shown in Fig. 1. The master laser is an extended cavity diode laser (ECDL) in Littman configuration operated at a diode temperature of 44°C and locked to a high finesse optical cavity by the Pound–Drever–Hall stabilization technique. The cavity is made of a 100 mm long ULE spacer with optically contacted mirrors made from the same material. The optical axis of the cavity is oriented horizontally. To minimize the sensitivity against vertical vibrations the cavity is supported at four points near to its horizontal symmetry plane.[5] The positions of the supporting points were determined by a finite elements simulation. The cavity shows a finesse of 330 000, corresponding to a linewidth of about 4.5 kHz. The resonator is mounted in a temperature stabilized vacuum chamber at a residual pressure of 10^{-7} mbar. The whole cavity setup is placed on a passive vibration isolation platform to further reduce the coupling to seismic vibrations.

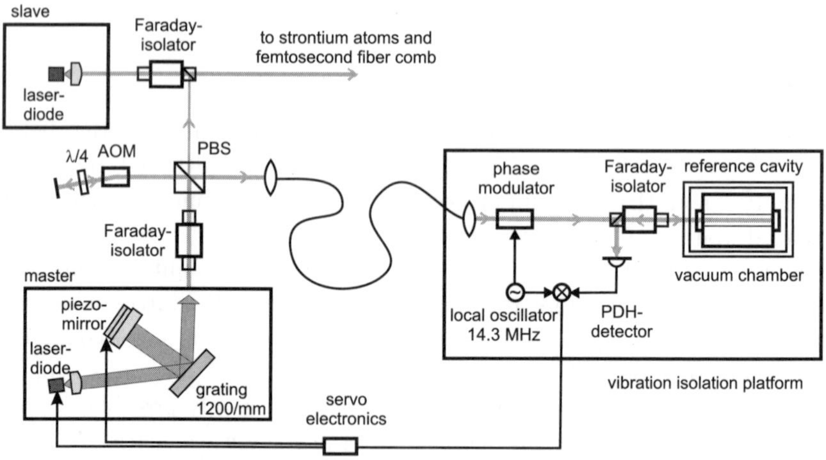

Fig. 1. Master–slave laser setup for a strontium lattice clock.

The output power of the master laser is not sufficient for the comb and to quickly excite ^{88}Sr atoms. Therefore a laser diode is injection-locked to the master laser. It delivers an output power of 23 mW which is sent to the strontium atoms and to the femtosecond fiber-laser comb by two optical fibers.

3. Characterization of the clock laser system

A commercial femtosecond fiber-laser comb was used to characterize the laser system. The laser is used as transfer oscillator[4,6] to compare the optical frequencies ν_{Ca} and ν_{Sr} of the 657 nm and 698 nm laser. For both laser frequencies the beat Δ_{Ca} and Δ_{Sr} with the nearest comb tooth of the frequency doubled femtosecond laser output is measured (Fig. 2, left). The frequencies of the frequency doubled comb teeth are given by the product of the integer mode number m and the repetition rate f_{rep} plus the twice the carrier envelope offset frequency ν_{CEO}. After dividing each signal by two, adding ν_{CEO} to each signal and multiplication of the lower branch by m_1/m_2 using a direct digital synthesizer (DDS) a signal is obtained that corresponds to a virtual beat of the two optical frequencies $\nu_{\mathrm{Sr}}/2$ and $\nu_{\mathrm{Ca}} \cdot m_1/2m_2$. This signal is independent from fluctuations of f_{rep} and ν_{CEO}.

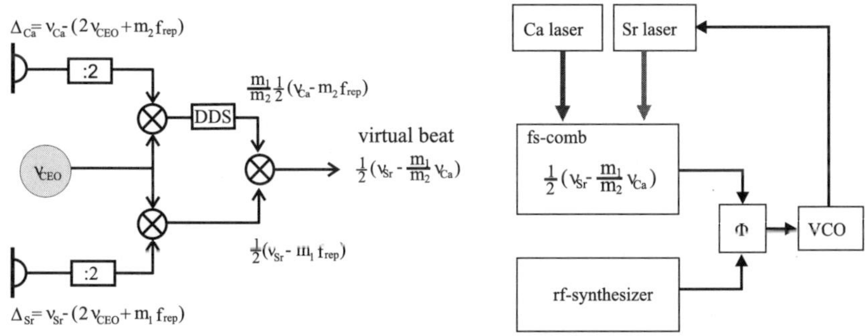

Fig. 2. Transfer-oscillator method (left) and setup for the phase lock of both lasers (right).

The virtual beat between the strontium clock laser and the 657 nm laser of the calcium clock[7] was used to characterize the strontium laser. First the vibration sensitivity of the cavity setup was investigated by manually shaking the platform and simultaneously measuring the acceleration and the corresponding frequency excursions. A sensitivity of about 140 kHz/ms^{-2}

to vertical vibrations was measured, which is far bigger than calculated from a finite element analysis. A difference of 1 mm from the optimal support point positions would result in a residual sensitivity of less than 8 kHz/ms^{-2}. Thus the observed sensitivity cannot be explained by tolerances of the support point positions. We therefore attribute the discrepancy to the spacer touching the surrounding heat shield. This will be corrected in the near future.

This high sensitivity to vibrations also leads to a linewidth of the virtual beat at half the strontium clock frequency of 30 Hz (Fig. 3, left) which is limiting the observed linewidth in the strontium clock.[8] To further reduce the laser linewidth, the virtual beat was phase locked to a stable radio frequency.[9] The clock laser frequency was adjusted by driving the offset AOM (Fig. 1) from a voltage controlled oscillator (VCO), which was controlled by the output of the phase detector (Fig. 2, right). This allows to transfer the linewidth and the spectral properties of the ultrastable 657 nm laser to the strontium clock laser within a bandwidth of 10 kHz, limited by the various tracking filters employed in the generation of the virtual beat. With the phase-lock-loop (PLL) closed, the spectrum of the virtual beat (Fig. 3, right) shows a narrow δ peak with a measured linewidth limited by the spectrum analyzer bandwidth of 1 Hz. This indicates that the strontium laser frequency precisely follows the one of the calcium laser.

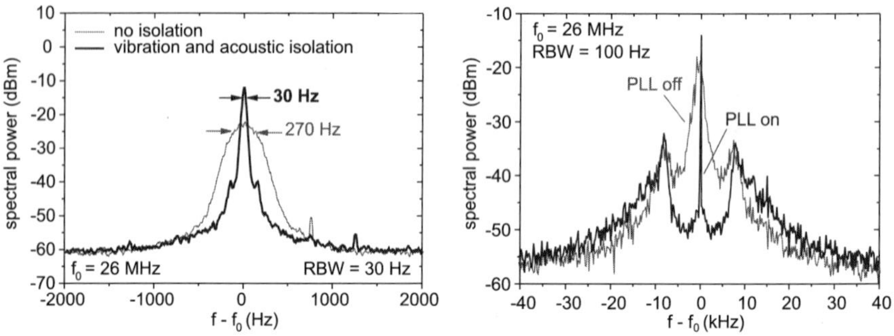

Fig. 3. Left-hand side: Virtual beat between the Sr and the Ca laser with and without vibration isolation of the reference cavity. Right-hand side: Virtual beat spectrum of the free-running laser (thin gray curve) and of the phase locked strontium laser (thick black curve). Without the phase lock, the spectral width is dominated by the frequency noise of the free running voltage controlled oscillator.

4. Conclusion and Outlook

We have set up a narrow-linewidth 698 nm laser system for a strontium lattice clock. Using a femtosecond comb as a transfer oscillator, the clock laser was characterized with respect to a 1-Hz linewidth 657 nm laser. The virtual beat that was generated with this setup also allowed to perform a phase lock of the strontium laser to the 657 nm laser. This method can be applied to avoid the construction of ultrastable cavities for every narrow linewidth laser. Instead a single, well optimized master laser can be sufficient to provide the stability to different lasers at nearly arbitrary optical frequencies.

Acknowledgments

The authors gratefully acknowledge the support by the Deutsche Forschungsgemeinschaft under SFB 407, by ESA, DLR, and the Centre for Quantum Engineering and Space-Time Research (QUEST).

References

1. X. Baillard, M. Fouché, R. L. Targat, P. G. Westergaard, A. Lecallier, F. Chapelet, M. Abgrall, G. D. Rovera, P. Laurent, P. Rosenbusch, S. Bize, G. Santarelli, A. Clairon, P. Lemonde, G. Grosche and B. Lipphardt and H. Schnatz, *Eur. Phys. J. D* **48**, 11 (2008).
2. M. Takamoto, F.-L. Hong, R. Higashi, Y. Fujii, M. Imae and H. Katori, *J. Phys. Soc. Jap.* **75**, 104302 (2006).
3. M. M. Boyd, A. D. Ludlow, S. Blatt, S. M. Foreman, T. Ido, T. Zelevinsky and J. Ye, *Phys. Rev. Lett.* **98**, 083002 (2007).
4. H. R. Telle, B. Lipphardt and J. Stenger, *Appl. Phys. B* **74**, 1 (2002).
5. T. Nazarova, F. Riehle and U. Sterr, *Appl. Phys. B* **83**, 531 (2006).
6. G. Grosche, B. Lipphardt and H. Schnatz, *Eur. Phys. J. D* **48**, 27 (2008).
7. H. Stoehr, F. Mensing, J. Helmcke and U. Sterr, *Opt. Lett.* **31**, 736 (2006).
8. J. Vellore Winfred, C. Lisdat, T. Legero, F. Riehle and U. Sterr, "Decoherence and losses by collisions in a ^{88}Sr lattice clock," in these proceedings.
9. T. Legero, C. Lisdat, J. S. R. Vellore Winfred, H. Schnatz, G. Grosche, F. Riehle and U. Sterr, "Interrogation laser for a strontium lattice clock," arXiv:0809.5140, to appear in IEEE Trans. Instrum. Meas. (2008).

MEASUREMENT NOISE FLOOR FOR A LONG-DISTANCE OPTICAL CARRIER TRANSMISSION VIA FIBER

G. GROSCHE[1*], O. TERRA[1], K. PREDEHL[1,2], T. HÄNSCH[2], R. HOLZWARTH[2], B. LIPPHARDT[1], F. VOGT[1], U. STERR[1], H. SCHNATZ[1]

[1]Physikalisch-Technische Bundesanstalt, Bundesallee 100, D-38116 Braunschweig, Germany

[2]Max-Planck-Insitut für Quantenoptik, Hans-Kopfermann-Str. 1 D-85748 Garching, Germany

** E-mail: Gesine.Grosche@ptb.de*

We investigated the measurement floor and link stability for the transfer of an ultra-stable optical frequency via an optical fiber link. We achieved a near-delay-limited instability of $3\text{x}10^{-15}/(\tau\cdot\text{Hz})$ for 147 km deployed fiber, and 10^{-20} ($\tau = 4000$ s) for the noise floor.

1. Introduction

Ultra-precise optical frequency standards (OFS) have reached a relative instability and accuracy below 10^{-16} [1], putting increased demands on methods for comparing standards, which are usually located in different laboratories. Long-distance transmission of either a modulated carrier [2] or an ultra-stable optical carrier frequency near 200 THz [3,4,5] over tens or hundreds of kilometers using standard telecommunication fiber is a promising, relatively new alternative to satellite-based methods. It allows remote comparison of OFS and may support precision experiments such as a search for variations of the fine-structure constant.

As part of a German research fiber network, we have set up a fiber link with two telecommunication fibers running in parallel which bridge almost 70 km distance between PTB and a laboratory of the Institute of Quantum Optics at the University of Hanover. At Hanover, a frequency standard based on neutral magnesium is located: a recent frequency measurement [6] was limited by the transportable microwave frequency reference (Cs clock) used. Thus, the aim of the fiber link is to deliver a precisely known and highly stable optical frequency to the remote optics lab, for a more precise frequency measurement of the intercombination transition of neutral magnesium and assessment of laser sources used in the spectroscopy set-up.

1.1. *Measurement noise floor and first measurements on 147 km link.*

We report two results [7]: to achieve optimal long-term stability of stabilized fiber links, we have investigated noise from fiber leads and components included in the interferometer set-up which is used stabilize the fiber link, and optimized the interferometer. Secondly, we report a stabilized fiber link of 147 km length, consisting of two underground 73 km-fibers put in series, where each fiber connects PTB with Hanover. We also outline future applications.

2. Method and experimental realization

2.1. *Laser stabilization and frequency reference*

As light source, we stabilize a 50 mW fiber laser at 1542 nm (Koheras Adjustik, free-running fast line-width ~ 2kHz) to a master laser, which is a cavity stabilized optical clock laser at 657 nm with a line-width of about 1 Hz. The clock laser is part of the Ca neutral atom frequency standard at PTB, and a fiber based fs-laser frequency comb acts as transfer oscillator to obtain a phase-lock between the two lasers, bridging > 200 THz [8]. From earlier independent measurements of the residual phase noise [8] we estimate a resulting coherence length of the stabilized fiber laser of more than 100 000 km; the contribution of laser noise to detected fiber link noise is therefore negligible.

2.2. *Fiber link and compensation interferometer set-up*

The fiber link and compensation interferometer are shown in fig. 1. Two parallel fiber links, F1 and F2 (in the same cable) connect our lab at PTB with the computer centre of the University of Hanover (RZ-H). Fibers F1 and F2 were connected with a short patch-cord at RZ-H, resulting in a fiber link starting and ending at PTB and going via RZ-H. The single-pass link length is approximately 147 km and incorporates one bi-directional erbium doped fiber amplifier. More recently, F1 and F2 continue from RZ-H over a distance of less than 1 km to the Mg optical frequency standard.

Perturbations leading to optical path length fluctuations appear as a Doppler shift of the light transmitted through the fiber link. These fluctuations are compensated with a scheme based on the well-established method described in [9]: detecting a double pass (return) signal and applying a correction signal derived from the return signal. As observed in our first experiments on a 86 km fiber link in Paris [3] and also reported e.g. by Williams [4], the long-term stability of a fiber link is limited by residual noise of the interferometer used to detect the fiber link noise. We have developed and tested different designs for the interferometer, using off-the-shelf fiber-optic components and minimizing out-of-loop fiber leads in a compact, portable set-up.

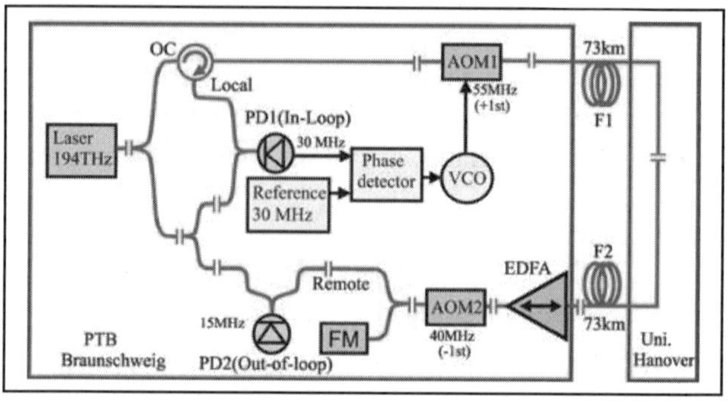

Fig. 1: Schematic set-up for active fiber noise compensation of 147 km fiber link. Laser: fiber laser stabilized to optical clock laser via fs-comb [8], OC: optical circulator, AOM: Acousto-optic modulator, VCO: voltage-controlled oscillator, PD1: in-loop photo-detector, PD2:out-of-loop photo-detector, EDFA: Erbium-doped fiber amplifier, FM: Faraday mirror.

3. Results for interferometer noise floor and 147 km fiber link

We detect beat signals between out-going laser light and returned light (In-Loop signal), and remote light (Out-of-loop signal). Using zero dead-time totalizing counters acting as Π-estimators [5,10], we calculate the Allan standard deviation (ADEV) as measure of frequency instability.

3.1. *Optimizing the compensation interferometer*

Unstabilized optical fiber in the lab with lengths ~ 1 m… 20 m showed phase noise from 10^{-3}… 10^{-1} $rad^2/Hz \times f^{-2}$, and an ADEV of up to 2×10^{-15} for integration times up to 100 s. Balancing the fiber length in the interferometer to minimize uncompensated fiber, we obtain a measurement noise floor around $10^{-7} rad^2/Hz \times f^{-2}$, or (with recording bandwidth > 10 kHz) 3×10^{-17} at 1s for the ADEV, reaching 10^{-20} after 1 hour, see green open triangles in fig.2.

3.2. *Realization of stabilized 147 km fiber link over deployed fiber*

Fig. 2 shows the relative instability of the remote signal when stabilizing the 147 km fiber link using the return signal (after 294 km), see full black squares. The accuracy of the frequency transfer was found to be within the ADEV value for the same integration time, reaching 10^{-18} after one hour. Phase-noise analysis (fig. 3) shows that our results are close to the fundamental delay-time limit [4].

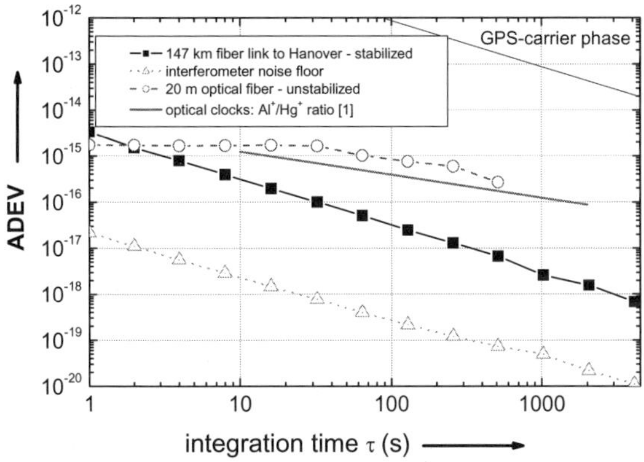

Fig. 2: Relative frequency instability given by Allan standard deviation (ADEV) using Π-type counters with a recording bandwidth > 10 kHz. Green triangles: interferometer noise floor (obtained by shorting the fiber link and detecting the stabilized remote signal); full black squares: remote signal of 147 km fiber link, stabilized. Open red circles: remote signal after 20 m of unstabilized fiber. For comparison: reported data for Al^+/Hg^+ clock comparison & for satellite based frequency transfer (GPS carrier phase)

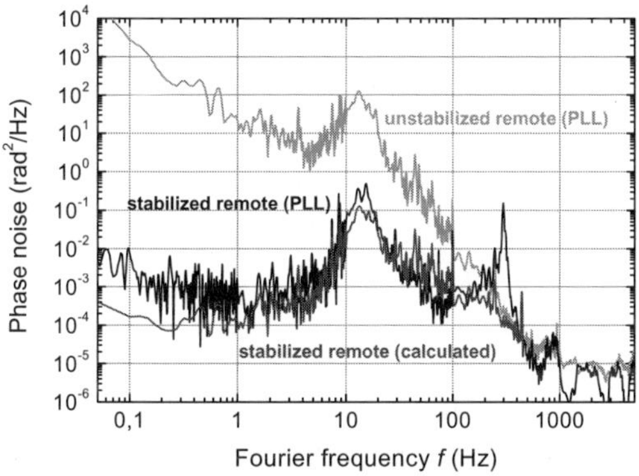

Fig. 3: Phase noise density (rad^2/Hz) for beat signal detected at remote end (after 147 km). Green line: without stabilization; black line with fiber stabilization; purple line: theoretical limit due to "delay-unsuppressed noise" [4]: comparison of black and purple line show close-to-optimal noise suppression.

4. Implication for remote frequency measurement, future extensions of fiber link and short distance frequency comparisons

We have characterized a fiber loop of 147 km length, consisting of two 73 km-fibers in series, where each fiber connects PTB with Hanover. We obtained a relative instability and uncertainty of approximately $3x10^{-15}/\tau$. This is significantly lower than the expected relative instability and uncertainty for the Mg-frequency standard; we are currently preparing a frequency measurement and laser characterization at university of Hanover via one 73 km fiber. We also aim to extend the current fiber network to a link spanning 900 km to the MPQ in Garching/Munich. Finally, our improved fiber interferometer noise floor will allow very low instability measurements (ADEV < $10^{-16}/\tau\cdot$Hz) up to a day over distances up to a few kilometers.

5. Future applications of optical carrier transmission via fiber

The transfer of a known, ultra-stable frequency with a relative instability in the 10^{-14} to 10^{-19} range over distances up to hundreds of kilometers between any "local" and "remote" laboratory connected by optical fiber makes precision frequency metrology available to a new "class" of end users. With commercially available fs-frequency combs, the delivered optical frequency may be mapped to any optical [8] or microwave [11,12] frequency in the remote lab.
Thus fiber-based frequency transmission can be used for the remote frequency measurement of optical [13] or rf-transitions, the remote generation of precision microwave signals, remote characterization of laser sources and remote spectroscopy and sensing.

Acknowledgments

This work was supported by SFB 407 / Deutsche Forschungsgemeinschaft.

References

1. T. Rosenband *et al.*, *Science* 28 March 2008, **319** (5871), p. 1808
2. O. Lopez, A. Amy-Klein, C. Daussy *et al.*, *Eur. Phys. J. D* **48**, 35 (2008)
3. G. Grosche, B. Lipphardt, H. Schnatz, G. Santarelli *et al.*, in: OSA Technical Digest, CLEO-USA, paper CMKK1 (2007)
4. N.R. Newbury, P.A. Williams, W.C. Swann, *Opt. Lett.* **32**, 3056 (2007); P.A. Williams, W.C. Swann and N.R. Newbury, *JOSA B* **25,** 1284 (2008)
5. H. Jiang, F. Kefelian *et al.*, arXiv:0807.1882; *JOSA B* **25,** 2029 (2008)
6. J. Friebe, A. Pape, M. Riedmann *et al.*, *Phys. Rev. A* **78**, 033830 (2008)
7. O. Terra, G. Grosche, K. Predehl *et al.*, in preparation
8. G. Grosche, B. Lipphardt and H. Schnatz, *Eur. Phys. J. D* **48**, 27 (2008)
9. L.S. Ma, P. Jungner, J. Ye and J L. Hall, *Opt. Lett.* **19**, 1777 (1994)
10. S.T. Dawkins, J.J. McFerran, A.N. Luiten, *IEEE Tr. UFFC* **54**, 918 (2007)
11. A. Bartels, S.A. Diddams, C.W. Oates *et al.*, *Opt. Lett.* **30**, 667 (2005)
12. B. Lipphardt *et al.,* arXiv:0809.2150 (2008) accepted *IEEE Tr. Inst. Meas.*
13. F.-L. Hong *et al.*, arXiv:0811.1816v1 (2008)

OPTICAL FREQUENCY TRANSFER OVER 172 KM OF INSTALLED FIBER

S. CRANE

U.S. Naval Observatory, 3450 Massachusetts Avenue NW
Washington, D.C. 20392, USA

H. JIANG, P. LEMONDE, G. SANTARELLI

LNE-SYRTE, Observatoire de Paris, CNRS, 61 avenue de l'Observatoire
75014 Paris, France

F. KEFELIAN, O. LOPEZ, A. AMY-KLEIN, CH. CHARDONNET

Laboratoire de Physique des Lasers (LPL), UMR 7538, CNRS and Université Paris 13,
99 av. J.-B. Clément, 93430 Villetaneuse, France

We transferred the frequency of an ultra-stable laser achieving a link distance of 172 km by implementing a recirculation loop on a continuous 86 km optical fiber fully installed in the urban environment of Paris. The link is fed with a 1542 nm cavity stabilized fiber laser having a sub-Hz linewidth. The fiber-induced phase noise is measured and cancelled with an all fiber-based interferometer using commercial off-the-shelf pigtailed optical telecommunications components. The compensated link shows an Allan deviation of a few 10^{-16} at one second and a few 10^{-19} at 10^4 seconds.

1. Long-Distance Coherent Optical Frequency Transfer over Fiber

Optical clocks using a trapped single ion or cold atoms confined in lattice are expected to reach a fractional frequency instability level of 10^{-17} [1,2] or better at one day and will consequently require very stable frequency transfer systems for making comparison between remotely located laboratories. However, the most commonly used method utilizing satellites, either directly by Two-Way Satellite Time and Frequency Transfer, or indirectly through the Global Positioning System carrier phase measurement are limited by instability of 10^{-15} after one day of averaging time [3].

To overcome satellite limitations, an alternate approach would be to take advantage of the globally pervasive telecommunications fiber network infrastructure for frequency standard dissemination. The transfer of microwave frequency over optical fibers by amplitude modulation of an optical carrier have demonstrated instability as low as 2×10^{-18} at one day over 86 km [4] by detecting and compensating for the fiber-induced phase noise that arises as a result of varying fiber optical length with temperature changes and acoustic disturbances.

When considering the ultimate link distance that can be achieved over an optical fiber, higher stabilities and lower phase noise will result from detection and compensation on the 1550 nm optical carrier itself [4]. Higher signal-to-noise-ratio (SNR) can be achieved through heterodyne detection using a frequency shifted return signal and will allow greater measurement resolution of the fiber-induced phase noise than can be achieved with amplitude modulation. Coherent optical frequency transfer also provides the versatility of comparing optical signals (e.g. from optical atomic clocks), microwave signals, or a mixture using optical phase-locking and microwave conversion with femtosecond frequency combs [5,6].

2. Experiment and Results

Our optical fiber link is composed of two cascaded 43 km fibers connecting two laboratories, LNE-SYRTE and LPL in the Paris area. Each fiber is composed of various sections of a commercial metropolitan fiber network spliced together and the attenuation in each fiber is about 12 dB. These fibers are then connected end to end to form a continuous section of optical fiber 86 km long. Previous long-distant link results have been demonstrated recently by adding spooled fiber to a section of installed fiber [7,8]. One would expect that spooled fiber has significantly different phase noise characteristics than installed fiber and especially for fiber installed in an urban environment such as Paris. Recently a 120 km urban link has been implemented in Tokyo which shows a rather high level of noise [9]. In an effort to determine the distance limitations of an optical link of this type while realistically representing an installed urban fiber link, we devised a simulated link of 172 km by implementing a recirculation loop to double the effective length of the urban fiber link.

Our full system for implementing optical frequency transfer is described in a recent publication [10]. The basic principle is rather simple and involves sending the ultra-stable laser light through an acousto-optic modulator (AOM) just before entering the fiber link where it experiences a phase change due to the one-way fiber transmission. After amplification with an Erbium Doped Fiber Amplifier (EDFA) part of this light is sent back through the fiber and AOM to be heterodyned with the original laser light forming a beat-note signal with approximately twice the phase shift experienced by the one-way light. The phase noise can be suppressed by applying a phase shift through the AOM that exactly opposes the measured two-way phase shift. Since the one way light only passes the AOM once, we can therefore expect the one-way to be nearly cancelled as well. The compensation system is entirely fibered and uses only commercial off-the-shelf pigtailed optical telecommunications components.

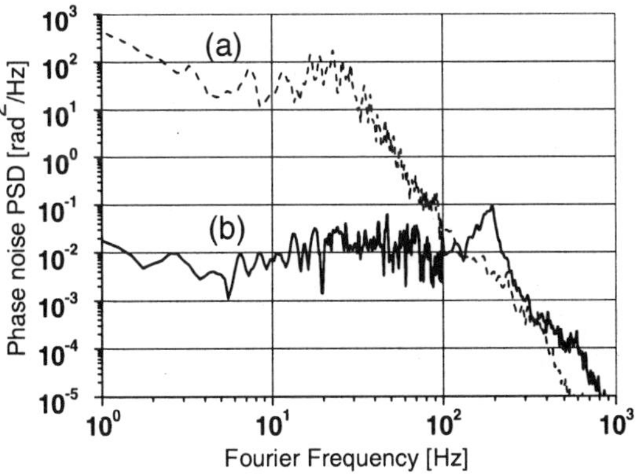

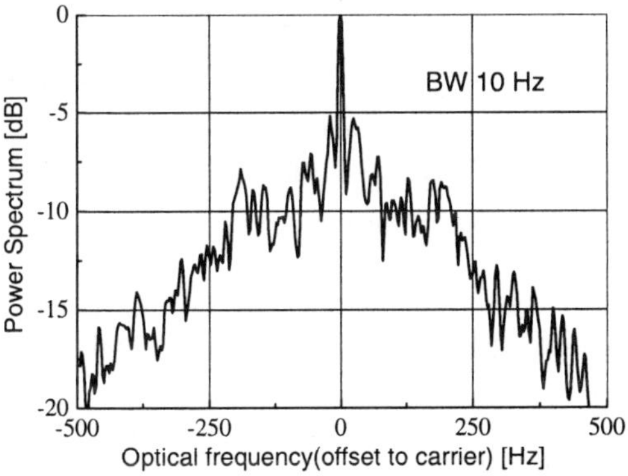

Figure 1. Upper : phase noise power spectral density of the (a) free running (dashed line), and (b) compensated 172 km link. Lower: Power spectrum of the optical transfer function for the stabilized link.

Figure 1 shows the 172 km link optical phase noise power spectral density, without and with compensation. It exhibits a roughly $1/f^2$ roll-off from 1 Hz to 100 Hz with a wide bump between 10 Hz and 100 Hz, and $1/f^4$ behavior above 100 Hz. The correction bandwidth is limited to $1/4t_{trip}$ (~ 280 Hz) where t_{trip} is the one-way trip delay in agreement with [8]. The integrated phase noise from 1 Hz to 1 kHz is equal to 53 rad and 2.5 rad (rms) for free running and compensated link respectively. In the same figure we report the spectrum of the optical transfer function of the stabilized link measured with the beatnote of optical signals at both ends, it has approximately 10% of the power in the central peak.

Figure 2 shows the fractional frequency instability of the 172 km link. The Allan deviation is about 4×10^{-16} at 1 second and in the range of 10^{-19} at 1 hour with a 10 Hz measurement bandwidth. The system floor is measured by replacing the urban fiber with an optical attenuator having the equivalent attenuation.

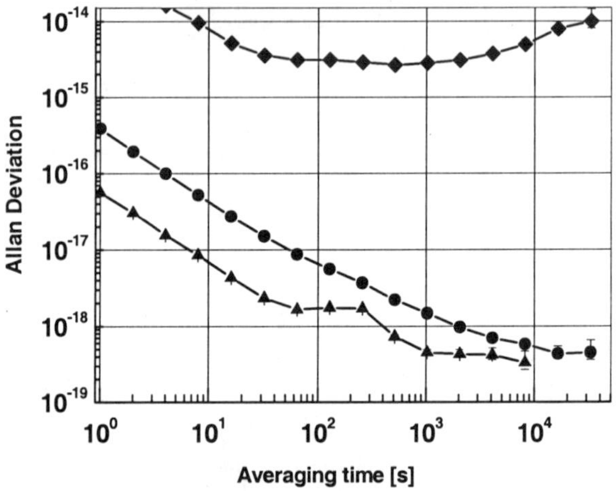

Figure 2. Fractional frequency instability of the 172 km free running link (diamonds), compensated link with 10-Hz measurement bandwidth (circles) and the experimental set-up floor (up triangles).

3. Link Length Limitations

It is conceivable to extend the length of the link to continental scales, 1000 km or more. In a single segment approach however, major difficulties will arise. Firstly, fiber attenuation will make mandatory in-line bidirectional amplifiers in order to provide sufficient power at the output and allow phase locking on the return signal. It should be noted that the bidirectional amplifier phase noise is cancelled within the control bandwidth. However, when cascading amplifiers, amplified spontaneous emission can be very large and will require the use of narrow band optical filters to overcome a significant degradation of the signal-to-noise ratio at both ends [8]. Secondly, the reduced bandwidth of the servo-loop and the reduced noise rejection level both due to the delay effect will degrade the performance of the transmission. We have empirically observed that the phase noise exhibits a $1/f^2$ dependence in the free running link. Assuming a linear dependence of the phase noise power on the length of the link L, the Allan deviation will scale as $L^{3/2}$ [8]. One could therefore expect an Allan deviation below 10^{-14} at one second for a 1000 km single segment link in a 10 Hz measurement bandwidth, averaging down as $1/\tau$ to the noise floor of the system. An alternative approach is to split the long distance link in shorter compensated segments using intermediate stations. Each one will achieve three functions, to send back part of the received signal to the previous station, to amplify (and filter if necessary) the received signal, to compensate the phase noise induced by the following segment. The maximum distance of each segment will depend on the fiber phase noise distribution, attenuation and on the total link length. This multiple segment approach allows for an increased correction bandwidth and enhances the resolution of the system. This would enable much better performance down to a level where the full coherence of the laser itself can be transferred.

4. Conclusion

We have demonstrated a long-distance link for optical frequency transfer using optical phase stabilization over a dedicated fiber that lies buried beneath the urban environment of Paris, France. Frequency transfer was demonstrated with instability of 4×10^{-16} at 1 s and a few 10^{-19} at 10^4 s over the 172 km urban link, which is better than the anticipated performance of optical clocks.

References

1. T. Rosenband, *et al*, *Science*, **319**, 1808-1812 (2008).
2. A. D. Ludlow *et al*, *Science*, **319**, 1805-1808 (2008).
3. Bauch, *et al*, Metrologia, **43**, 109-120 (2006).
4. O. Lopez *et al*, *Eur. Phys. J. D*, **48**, 35-41 (2008).
5. S. M. Foreman *et al*, *Rev. of Sci. Instrum.*, **78**, 021101 (2007)
6. S.M. Foreman, *et al Phys. Rev.Lett.* **99**, 153601 (2007).
7. G. Grosche, *et al*, in *CLEO*, paper CMKK1, *OSA*, (2007).
8. N. R. Newbury, P. A. Williams, W. C. Swann, ", *Opt. Lett.*, **32**, 3056-3058 (2007) and P.A. Williams, W.C Swann, and N.R. Newbury, *J. Opt. Soc. Am. B*, **25**, 1284-1293 (2008)
9. M. Musha, *et al*, *Optics Express*, **16**, 16459, (2008).
10. H. Jiang, *et al*, *J. Opt. Soc. Am. B*, **25**, 2029-2035, (2008).

Part X

Miniature Systems

CHIP-SCALE ATOMIC DEVICES: PRECISION ATOMIC INSTRUMENTS BASED ON MEMS

J. KITCHING, S. KNAPPE, V. GERGINOV[a], V. SHAH[b], P. D. D. SCHWINDT[c], B. LINDSETH[d], E. A. DONLEY, Y.-J. WANG, E. HODBY, M. EARDLEY[e], R. JIMENEZ[f], W. C. GRIFFITH, A. GERACI, J. PREUSSER, T. C. LIEBISCH, H. G. ROBINSON AND L. HOLLBERG

Time and Frequency Division, NIST,
325 Broadway, Boulder, CO USA

We describe recent work at NIST to develop compact, low-power instruments based on a combination of precision atomic spectroscopy, advanced diode lasers and microelectromechanical systems (MEMS). Designed to be fabricated in parallel in large numbers, these "chip-scale" atomic devices may eventually impact a wide range of applications, from the global positioning system to magnetic resonance imaging and inertial navigation. We focus here on recent work to develop compact, high-performance magnetometers.

1. Introduction

The miniaturization of atomic frequency references and related instruments continues in 2008 to support considerable activity within the frequency control community. Throughout the 1990s, compact atomic clocks were developed for synchronization of wireless communication systems[1]. This development effort led to the commercial availability of atomic frequency references with volumes near 100 cm^3, power requirements of about 10 W and a fractional frequency instability below 10^{-11} at one hour of integration. These frequency references were lamp-pumped instruments in which the atomic vapor cell was placed in a microwave cavity and the clock transition excited by directly applying a microwave field.

---

[a] Currently with Physikalisch-Technische Bundesanstalt, Arbeitsgruppe 4.41, Bundesallee 100, D-38116 Brauschweig, Germany

[b] Currently with the Department of Physics, Princeton University, Princeton, NJ 08544

[c] Currently with Sandia National Laboratory, Albuquequem NM 87185

[d] Currently with the National Center for Atmospheric Research, Boulder, CO

[e] Also with the Department of Physics and Astronomy, SUNY Stony Brook, Stony Brook, NY 11794

[f] Also with the Department of Physics, The University of Colorado, Boulder, CO 80309

At the time of the 6th Symposium on Frequency Standards and Metrology in 2001, new ideas related to miniaturized instruments were rapidly emerging and the last seven years has seen many of these ideas realized. In particular, the use of micromachining is now being explored extensively for application to atomic frequency references. In addition to a number of key laboratory experiments and proof-of-principle demonstrations, substantial progress has also been made in engineering easily manufactured systems, with the goal of bringing a new generation of miniaturized atomic clocks to the marketplace.

Micromachined vapor cell frequency references[2] are distinguished from their more conventional counterparts by the inclusion of key components fabricated by use of lithographically defined patterning and chemical etching of materials. The use of microelectromechanical systems (MEMS) allows not only for very small physics package size and correspondingly low power requirement but also makes possible the parallel fabrication of large numbers of units with a single process sequence. Since the first demonstration of a MEMS-based atomic clock physics package[3], significant advances in power dissipation[4], performance[5] and integration with electronics have been made[6-8]. At present, chip-scale atomic clocks are nearing commercial reality, and it is anticipated that large-scale manufacturing of units will begin soon.

At NIST, we are currently working to adapt the core technologies developed for chip-scale atomic clocks to use in other types of instruments, such as magnetometers, optical frequency references and gyroscopes. Micromachined alkali vapor cells[9], fabricated from etched silicon wafers, form the central component of all of these new types of instruments. We focus here on recent work to develop high-sensitivity magnetometers that share the size and power performance of their chip-scale frequency reference counterparts. Sensitivity levels to DC fields approaching those of magnetometers based on superconducting quantum interference devices (SQUIDs) have been achieved, opening the door to millimeter-scale uncooled sensors for a wide range of magnetic applications. These applications include magnetic anomaly detection, biomagnetic imaging, nuclear magnetic resonance and the measurement of magnetic fields in space.

2. Alkali Vapor Cells

The design and fabrication of MEMS-based alkali vapor cells at NIST has been discussed previously[9, 10] but is repeated here for completeness. Traditional methods for fabricating alkali vapor cells[11] are based on glass blowing techniques. This fabrication method has worked well for centimeter-scale physics packages for decades but suffers from some limitations with respect to highly miniaturized instruments, namely that cells must be fabricated one by one and that it is difficult to reduce the cell size below ~ 1 mm by use of a gas torch to heat the glass. While some alternative glass-blowing techniques have been developed in an attempt to overcome these problems[12, 13], the MEMS-based fabrication method described below still appears to be superior for most types of instruments.

The basic structure of the MEMS-based vapor cells is shown in Figure 1(a). A hole with a typical dimension of 1 mm is etched in a silicon wafer by lithographic patterning and chemical etching. Glass wafers are bonded onto the top and bottom of the silicon wafer, and the alkali atoms, along with a buffer gas, are confined in the enclosed volume.

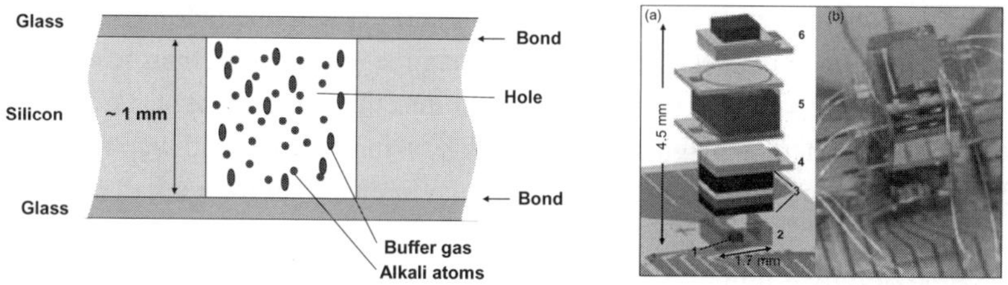

(a) (b)

Figure 1 (a) Basic structure of a micromachined alkali vapor cell. (b) Photograph of a chip-scale atomic magnetometer (after Ref. 14).

This MEMS-based cell structure offers several advantages when compared with glass blowing. The first is that small cells can be made quite easily, since the physical dimensions are defined by lithography. The size of the cells is also quite scalable: the fabrication of cells with dimensions smaller than 1 mm in principle requires changes only in the etch mask and the wafer thickness. At NIST, we have made cells with one transverse dimension as small as 100 μm. The second advantage is that the wafer-level fabrication potentially allows for large numbers of cells to be made with the same process sequence. While this parallel fabrication has yet to be demonstrated, at the wafer level, small arrays of cells have been demonstrated and the extension to full wafers is expected to be straightforward. Finally, the planar structure of the cell allows for easy integration with other optical components in a stacked assembly.

3. Chip-Scale Atomic Magnetometers

The cells shown in Figure 1(a) can be integrated with a laser, some optics, and a photodetector to form a number of different instrument physics packages. The first physics packages to be developed were for atomic frequency references[3, 4, 7, 8]. In these instruments, a coherent population trapping (CPT) resonance is excited by modulating the injection current of the diode laser and detected by monitoring the transmitted optical power with the photodiode.

The first chip-scale atomic magnetometer demonstrated at NIST[15] was in fact identical in structure to the NIST chip-scale atomic clock physics packages. The magnetic field was measured by exciting a magnetically sensitive hyperfine transition using CPT, and the sensitivity of this instrument was 40 pT/√Hz. A subsequent magnetometer physics package[14] based on the Mx geometry[16] showed an improvement in sensitivity to 6 pT/√Hz. In this magnetometer, shown in Figure 1(b), the spin precession about the magnetic field (Zeeman resonance) was excited with a pair of RF coils integrated into the component stack. Finally, we also implemented a magnetometer[17] using a microfabricated vapor cell in which spin-exchange collisions that broaden the linewidth at high alkali densities were suppressed[18, 19]. This magnetometer had a sensitivity of 70 fT/√Hz with a cell volume of only 6 mm^3.

Atomic magnetometers are widely used to detect magnetic anomalies in the context of geophysical surveying, unexploded ordinance detection and perimeter monitoring. While miniaturized atomic magnetometers will likely impact these existing application areas, we are optimistic that the considerable reduction in size and power offered by chip-scale magnetometers will significantly broaden their utility.

4. Advanced Chip-Scale Atomic Magnetometers

4.1. *Chip-scale atomic magnetometers with flux concentrators*

To improve the sensitivity of the magnetometers even further, we have begun integrating the micromachined alkali vapor cells with flux concentrators[20]. The effect of the concentrator is to capture flux from an area larger than the vapor cell cross-section and concentrate this flux through the cell. For a fixed sensitivity associated with the vapor cell, the concentrator therefore improves the overall system sensitivity by a factor approximately equal to the concentrator enhancement factor. While the size of the complete instrument increases with the addition of the concentrator, only the small vapor cell needs to be heated, implying reduced power consumption compared with a magnetometer with no concentrator but based on a large cell.

Two types of flux concentrators were used: a pair of thin triangular prisms machined from high-permeability metal (Figure 2(a)) and a pair of cylindrical rods made from ferrite (Figure 2(b)). It was expected that the additional noise introduced by the concentrators would be different for the two materials[21].

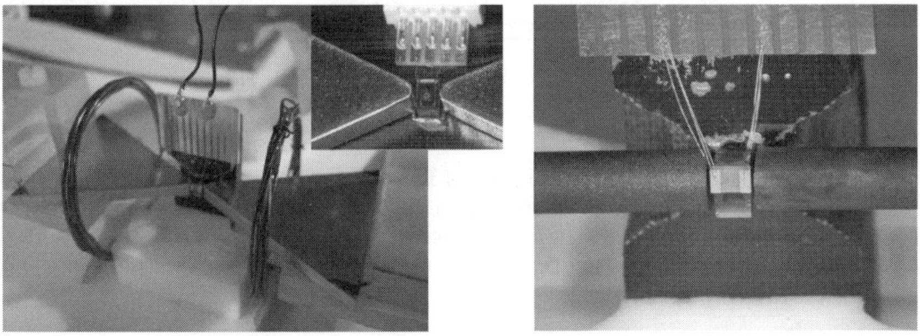

(a) (b)

Figure 2 Photographs of the flux concentrators used in the experiments. (a) Triangular prism concentrators machined from high-permeability metal. (b) Cylindrical concentrators made from ferrite.

The magnetometer was operated in the single-beam spin-exchange relaxation-free (SERF) geometry[17] at a cell temperature of 160 °C. The enhancement factor was measured as a function of the separation of the tips of the flux concentrators. The enhancement is expected to increase as the distance between the concentrators decreases, due to the increased conductivity of the magnetic path through the concentrators. The enhancement factors of the two concentrator systems as a function of tip spacing is shown in Figure 3(a). An enhancement of the magnetic field by a factor of about 20 is obtained for a tip spacing of 2 mm. The noise was determined by measuring both the dispersive zero-field resonance at the output of the lock-in detector and the noise at the same output with a FFT spectrum analyzer. The sensitivity is shown in Figure 3(b).

We find that the magnetometers incorporating the flux concentrators are limited by fundamental thermal noise associated with dissipation processes in the concentrator material. In the case of metallic concentrators, this noise is due to thermal currents in the conductive material, whereas for the ferrite concentrators, this noise is due to magnetic dissipation (hysteresis) in the material. Predictions from modeling following Ref. 21 are shown as dashed lines in Figure 3(b).

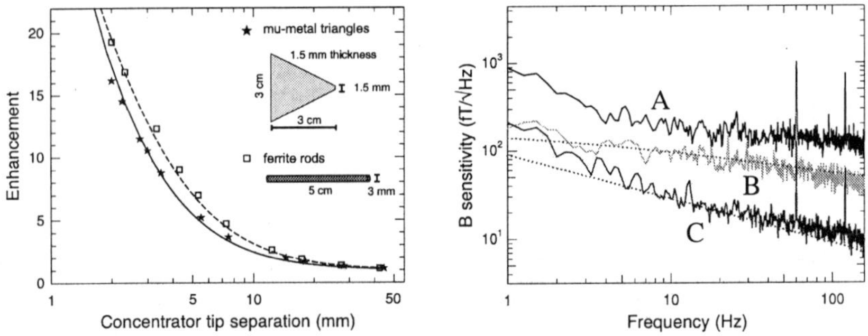

(a) (b)

Figure 3 (a) Enhancement factor for two types of flux concentrators. The lines are the results of simulations and agree well with the measured data. (b) Magnetometer sensitivity for a chip-scale magnetometer. Trace A is for the case with no flux concentrators, Trace B is for the triangular (high permeability metal) concentrators and Trace C is for the ferrite concentrators. The dashed lines show the expected sensitivity based on modeling described in Ref. 21.

4.2. *"Photonic" Magnetometer*

In many cases, atomic magnetometers themselves generate magnetic fields that can interfere with the sensor reading or the readings of surrounding sensors if implemented in an array. These spurious field sources include DC currents in the laser and cell heaters, thermal currents in conductive material near the cell, and even the drive field required to excite the spin resonance. For very low-field applications or use in arrays, it is therefore desirable to develop sensors that generate as little field of their own as possible.

We are therefore exploring a chip-scale atomic magnetometer design in which all inputs to and outputs from the sensor head are optical[22]. In this "photonic" magnetometer, the Larmor resonance is excited and detected with a modulated optical field[23]; no oscillating magnetic field is therefore needed and the light can be transmitted to and from the cell through optical fibers[24]. In addition, the cell is heated to its operating temperature by use of light from a laser transmitted through an optical fiber and absorbed by the silicon cell material. Laser-induced temperature control of a magnetic sensor based on a SQUID was demonstrated in Ref. 25.

The basic design of the photonic magnetometer sensor head is shown in Figure 4(a). The vapor cell is suspended in an evacuated enclosure with thin polyimide tethers similar to those described in Ref. 4. Light from a broad area diode laser at 915 nm is transmitted through multi-mode optical fiber into the cell enclosure and illuminates the cell, heating it to its operating temperature.

Light from a second semiconductor laser (either a vertical cavity surface emitting laser or distributed Bragg reflector laser) is coupled into single-mode polarization-maintaining fiber and reflected into the cell by a small cube beamsplitter. After passing through the cell, the light is collected with a large-core multi-mode fiber for transmittal back to a photodetector. Figure 4(b) is a photograph of the complete sensor head.

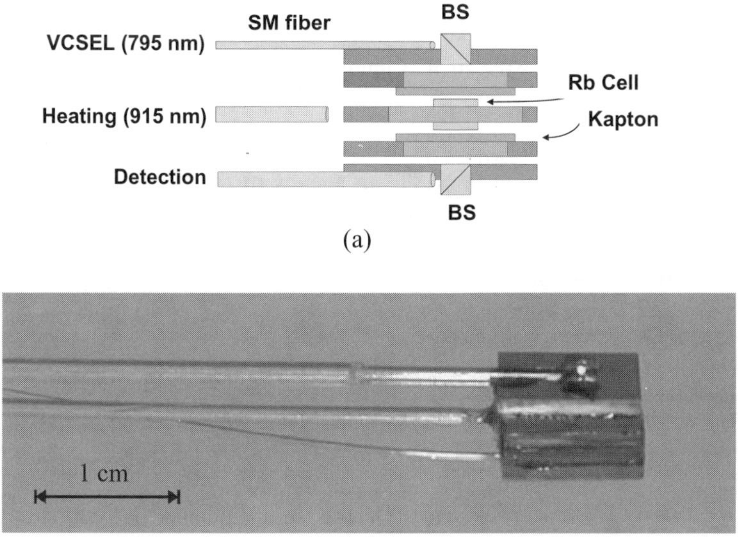

(a)

(b)

Figure 4 (a) Photonic magnetometer design. (b) Photograph of a completed instrument.

The sensitivity of the magnetometer measured in a magnetic field of 7 μT was approximately 2 pT/√Hz. This is comparable to the sensitivity of the chip-scale Mx magnetometer described in Ref. 14.

5. Conclusion and outlook

The technology developed for chip-scale atomic clocks is being adapted for a variety of other instruments. Among these, atomic magnetometers based on microfabricated alkali vapor cells are now achieving sensitivity levels comparable to those of SQUID-based magnetometers and offer the potential for high-performance sensing in a highly compact, noncryogenic instrument. Recent work to integrate flux concentrators with these magnetometers has resulted in potentially low-power instruments with a sensitivity of 10 fT/√Hz. In addition, novel designs such as the "photonic" magnetometer may allow microfabricated atomic magnetometers to be used in more unique situations, such as high-voltage environments or under water.

Beyond magnetometers, a variety of other instruments could be miniaturized by use of these same techniques. For example, chip-scale saturation spectrometers have already been demonstrated[26] and could prove useful in future general atomic physics experiments. Gyroscopes based on nuclear spins[27] also appear possible and would enhance inertial navigation, particularly on platforms with limited power and space such as unmanned aerial vehicles.

Acknowledgments

This work is a contribution of NIST, an agency of the US Government, and is not subject to copyright. This work was supported by NIST, the Defense Advanced Research Projects Agency (DARPA) and the Strategic Environmental Research end Development Agency (SERDP).

References

1. J. A. Kusters and C. A. Adams, *RF Design* **May**, 28 (1999).
2. J. Kitching, S. Knappe, and L. Hollberg, *Appl. Phys. Lett.* **81**, 553 (2002).
3. S. Knappe, V. Shah, P. D. D. Schwindt, L. Hollberg, J. Kitching, L. A. Liew, and J. Moreland, *Appl. Phys. Lett.* **85**, 1460 (2004).
4. R. Lutwak, J. Deng, W. Riley, M. Varghese, J. Leblanc, G. Tepolt, M.Mescher, D. K. Serkland, K. M. Geib, and G. M. Peake, The chip-scale atomic clock - low-power physics package, Proceedings of 36th Annual Precise Time and Time Interval (PTTI) Meeting, Washington, DC, 2004.
5. S. Knappe, P. D. D. Schwindt, V. Shah, L. Hollberg, J. Kitching, L. Liew, and J. Moreland, *Opt. Exp.* **13**, 1249 (2005).
6. R. Lutwak, P. Vlitas, M. Varghese, M. Mescher, D. K. Serkland, and G. M. Peake, The MAC - A Miniature Atomic Clock, Proceedings of Joint IEEE International Frequency Control Symposium and Precise Time and Time Interval (PTTI) Systems and Applications Meeting Vancouver, Canada, 2005.
7. D. W. Youngner, L. M. Lust, D. R. Carlson, S. T. Lu, L. J. Forner, H. M. Chanhvongsak, and T. D. Stark, A manufacturable chip-scale atomic clock, Proceedings of IEEE Transducers, Lyon, France, 2007.
8. J. F. DeNatale, R. L. Borwick, C. Tsai, P. A. Stupar, Y. Lin, R. A. Newgard, R. W. Berquist, and M. Zhu, Compact, low-power chip-scale atomic clock, Proceedings of IEEE Position Location and Navigation Symposium (PLANS), Monterey, CA, 2008.

9. L. A. Liew, S. Knappe, J. Moreland, H. Robinson, L. Hollberg, and J. Kitching, *Appl. Phys. Lett.* **84**, 2694 (2004).
10. S. Knappe, V. Gerginov, P. D. D. Schwindt, V. Shah, H. G. Robinson, L. Hollberg, and J. Kitching, *Opt. Lett.* **30**, 2351 (2005).
11. J. Vanier and C. Audoin, *The Quantum Physics of Atomic Frequency Standards* (Adam Hilger, Bristol, 1992).
12. S. Knappe, V. Velichansky, H. G. Robinson, J. Kitching, and L. Hollberg, *Rev. Sci. Instr.* **74**, 3142 (2003).
13. E. J. Eklund, A. M. Shkel, S. Knappe, E. Donley, and J. Kitching, *Sensors and Actuators A - Special Issue: MEMS 2007* **16**, 232 (2007).
14. P. D. D. Schwindt, B. Lindseth, S. Knappe, V. Shah, J. Kitching, and L.-A. Liew, *Appl. Phys. Lett.* **90**, 081102 (2007).
15. P. D. D. Schwindt, S. Knappe, V. Shah, L. Hollberg, J. Kitching, L. A. Liew, and J. Moreland, *Appl. Phys. Lett.* **85**, 6409 (2004).
16. A. Bloom, *Appl. Opt.* **1**, 61 (1962).
17. V. Shah, S. Knappe, P. D. D. Schwindt, and J. Kitching, *Nat. Phot.* **1**, 649 (2007).
18. W. Happer and H. Tang, *Phys. Rev. Lett.* **31**, 273 (1973).
19. J. C. Allred, R. N. Lyman, T. W. Kornack, and M. V. Romalis, *Phys. Rev. Lett.* **89**, 130801 (2002).
20. W. C. Griffth, R. Jimenez-Martinez, S. Knappe, J. Kitching, and V. Shah, *Submitted* (2008).
21. S. K. Lee and M. V. Romalis, *J. Appl. Phys.* **103** (2008).
22. J. Preusser, V. Gerginov, S. Knappe, and J. Kitching, A microfabricated photonic magnetometer, Proceedings of IEEE Sensors Conference, Lecce, Italy, 2008.
23. W. E. Bell and A. L. Bloom, *Phys. Rev. Lett.* **6**, 280 (1961).
24. J. Belfi, G. Bevilacqua, V. Biancalana, Y. Dancheva, and L. Moi, *Journal of the Optical Society of America B-Optical Physics* **24**, 1482 (2007).
25. C. E. Cunningham, T. R. DeYoung, and T. A. Bouma, *Physica B* **284**, 2111 (2000).
26. S. A. Knappe, H. G. Robinson, and L. Hollberg, *Opt. Exp.* **15**, 6293 (2007).
27. K. F. Woodman, P. W. Franks, and M. D. Richards, *The Journal of Navigation* **40**, 366 (1987).

CSAC – THE CHIP-SCALE ATOMIC CLOCK*

R. LUTWAK AND A. RASHED†
Symmetricom – Technology Realization Center, 34 Tozer Rd., Beverly, MA 01915, USA

M. VARGHESE‡, G. TEPOLT, J. LEBLANC, AND M. MESCHER
The Charles Stark Draper Laboratory, Group, Laboratory, 555 Technology Square, Cambridge, MA 02139, USA

D. K. SERKLAND, K. M. GEIB, AND G. M. PEAKE
Sandia National Laboratories§, P.O. Box 5800, Albuquerque, NM 87185, USA

The authors have developed a Chip-Scale Atomic Clock (CSAC) of sufficiently low power and small size to enable atomic timing in portable battery-powered devices. The collaboration of diverse research teams in clock technology, micro-electromechanical systems (MEMS), and optoelectronic devices has resulted in size and power reduction of atomic clock technology by more than two orders of magnitude. In 2007, we completed a pre-production build and evaluation of ten identical CSACs, with power <125 mW and short-term stability of $\sigma_y(\tau) < 2\times10^{-10}\ \tau^{-1/2}$ [1]. This paper reviews the physics and engineering considerations which underlie CSAC technology, describes in detail our CSAC implementation, and presents the results from the pre-production build.

1 Introduction

Atomic frequency standards play an essential role in the synchronization and accuracy of modern timing, communications, and navigation systems. To date, the relatively large size and power consumption of conventional technologies have prevented their deployment in portable battery-powered applications. Over the past decade, advances in micro-electromechanical systems (MEMS) and low-power semiconductor lasers and microwave electronics have enabled a new generation of atomic clocks, 100X smaller and lower power than previously possible. So-called "chip-scale atomic clocks" (CSACs) enable new capabilities in diverse applications, including secure communications and ad hoc networking, remote sensor networks, and high-integrity navigation.

---

* This work is supported by the Defense Advanced Research Projects Agency, Contract # NBCHC020050.

† Now at MKS Instruments, Andover, MA

‡ Now at MEMSIC Inc. Andover, MA

§ Sandia is a multiprogram laboratory operated by Sandia Corporation, a Lockheed Martin Company, for the United States Department of Energy's National Nuclear Security Administration under contract DE-AC04-94AL85000.

CSAC development was pioneered in the early 1990's by Chantry and Lieberman at Westinghouse Corporation[2]. Utilizing a laser-pumped cesium vapor cell contained in a dielectrically-loaded microwave cavity, they demonstrated a 250 mW 16 cm^3 physics package, with stability of $\sigma_y(\tau)<2x10^{-11} \tau^{-1/2}$ [3]. This effort never resulted in a fully functional CSAC due to the lack of availability, at the time, of suitably efficient laser diodes and microwave electronics, which have since become available, due to the demands and developments of the telecommunications industry. At the time of the 6th FSM, one research group reported on small low-power atomic oscillators[4]. Today, seven years later, and due in large part to significant investment by the U.S. Defense Advanced Research Projects Agency (DARPA), CSAC research and development is flourishing at academic and industrial laboratories worldwide.

The CSAC technology presented in this paper results from a highly successful collaboration, since 2001, between the clock development team at Symmetricom's Technology Realization Center, the Micro-electromechanical Systems (MEMS) group at the Charles Stark Draper Laboratory, and the Optoelectronic Components group at Sandia National Laboratories. Throughout the project, emphasis has been placed on autonomy, reliability, and robustness, in order to facilitate rapid system-level integration and capability demonstration. We demonstrated our first prototype in the Spring of 2005 with power consumption ≈ 150 mW and short-term stability of $\sigma_y(\tau) < 6x10^{-10} \tau^{-1/2}$ [5]. In 2007, we completed a pre-production build and evaluation of 10 identical CSACs, with power ≈ 125 mW, and short-term stability of $\sigma_y(\tau) < 2x10^{-10} \tau^{-1/2}$ [1].

2 CSAC Physics

2.1 Microwave Double-Resonance vs. Coherent Population Trapping

Prior to CSAC, the conventional rubidium oscillator (RBO) was the smallest and lowest power atomic clock. RBOs utilize a conventional double resonance interrogation technique in which the atomic ground state is optically pumped by an RF discharge lamp and repumped by microwaves at the ground state hyperfine frequency, ν_0. Typical RBOs consume 5-15 W, depending on ambient temperature. The majority of the power is required to heat the atomic resonance cell/microwave cavity assembly and to heat and excite the ^{85}Rb RF discharge lamp. It is straightforward, in principle, to replace the discharge lamp with a relatively low-power semiconductor laser source. Reducing the power consumption of the resonance cell assembly is more problematic. The cell must be heated in order to create sufficient vapor pressure of the metal in the atomic vapor phase. Moreover, the temperature must be stabilized in order to reduce temperature-dependent shifts of the clock frequency due to variability of the atomic collision rate. The power requirement is determined by the surface area of the cell as well as by the thermal resistance of the mounting system and insulation. In order to sustain stabilization, the operating temperature must be somewhat higher than the highest expected ambient temperature.

An alternative approach to atomic interrogation, commonly referred to as "Coherent Population Trapping," or CPT, was proposed in 1993 by Cyr and Tetu[6]. Rather than interrogating differences between ground state *populations*, Cyr proposed interrogating *coherences* between the states. The atomic sample is illuminated by a pair of coherent laser fields, with wavelength tuned to an optically-accessible dipole transition but with frequencies separated by ν_0, so as to simultaneously couple both ground states to a single excited state, thereby creating a coherence between the two ground states. The coherent state is sometimes called a "dark" state because the atom does not scatter photons. There is, thus, a resonant increase in the light transmitted through the vapor cell when the frequency difference between the two lasers ν_1-ν_2=ν_0. Implementing CPT interrogation in CSAC eliminates the microwave cavity, which is otherwise a significant barrier to size (and power) reduction.

Cyr and Tetu demonstrated that CPT resonances could be observed with two coherent laser beams generated from a single laser diode by modulating the laser bias current at a sub-harmonic of ν_0, thereby producing optical sidebands separated by ν_0. Today's high bandwidth semiconductor lasers, such as vertical cavity surface emitting lasers (VCSELs), exhibit modulation bandwidths in excess of 10 GHz and thus can be modulated directly at $\nu_0/2$ (=3.4 GHz or 4.6 GHz, for Rb or Cs, respectively) thereby producing a coherent pair of first-order sidebands separated by ν_0 (=6.8 GHz, or 9.2 GHz, respectively).

This geometry, CPT interrogation with a modulated VCSEL, lends itself to a simple and compact geometry for size and power reduction.

2.2 Choice of Atomic Species

The choice of atomic species for CSAC is dominated by the requirements for a simple two-level ground state structure, availability of suitable laser sources, and the need to produce adequate atomic vapor pressure at relatively low temperature. The logical choices for CSAC are either rubidium or cesium.

The principal advantage of rubidium is that the total (coupled) angular momentum of the nucleus is F=1, compared to F=3 for cesium, and thereby lower degeneracy of the ground state. For a given population of rubidium atoms, uniformly distributed amongst the three m_F magnetic sublevels, the number participating in the Δm_F=0 "clock" transition is 1/3, compared to 1/7 for cesium. This results in a signal level which is $\approx$7/3 higher for rubidium. Cesium, on the other hand, has higher transition Q, because the clock transition frequency is ν_0=9.2 GHz, compared to ν_0=6.8 GHz. The short-term stability of the CSAC is given by $\sigma_y(\tau = 1\text{ sec}) \approx [Q \times (S/N)_{1\text{ Hz}}]^{-1}$, where $(S/N)_{1\text{Hz}}$ is the signal/noise ratio of the recovered signal, measured in a 1 Hz bandwidth. In the case where $(S/N)_{1\text{Hz}}$ is determined by atomic shot noise, then $(S/N)_{1\text{ Hz}} = \sqrt{S}_{1\text{ Hz}}$ and there is no clear choice between cesium and rubidium because $\sqrt{7/3} \approx (9.2/6.8)$. However, in practice, the dominant noise source in a CSAC is laser frequency noise[7] and so, $(S/N)_{1\text{ Hz}} = S_{1\text{ Hz}}$ and there remains a small (1.7X) advantage of selecting rubidium over cesium.

2.3 D1 vs. D2 excitation

Early CSAC research employed the D2 transition of cesium, at 852 nm, for reasons of laser availability. In 2002, a clear advantage of utilizing the D1 transition was published independently by two groups, at NIST[8], and Agilent Laboratories[9]. The advantage of D1 over D2 excitation can be understood from the Cs energy level diagram of **Figure 1** (Rb is similar). The CPT coherence is excited, in either the D1 or D2 case, between the F=3,4 ground state levels and the F=3,4 excited state levels. In the case of D2 excitation, additional excited state levels, F=2,5 provide an alternate pathway for off-resonant excitation. This results in reduction of the coherence lifetime and reduction of steady-state signal levels. Following these publications, it was several years before VCSELs were developed at the Cs and Rb D1 wavelengths of 894 and 795 nm, respectively.

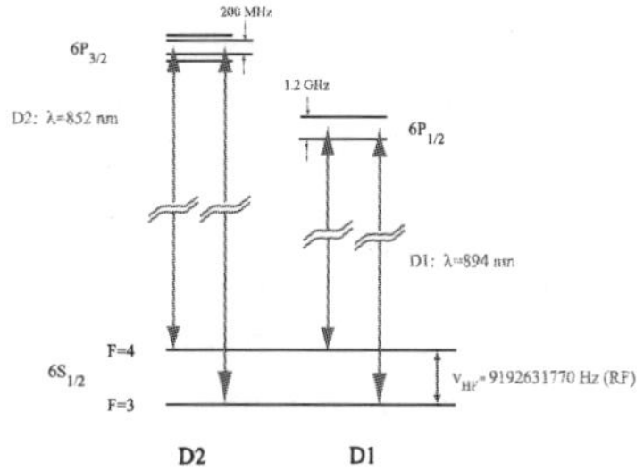

Figure 1: Cesium energy levels

The first VCSELs at the cesium D1 wavelength were developed at Sandia in 2003[10]. In 2003, our group published the results of CPT performance utilizing Cs D1 vs. D2[7]. At optimum laser power, D1 produces resonance features nearly 3X stronger than D2, resulting in improved clock performance.

3 CSAC Engineering

Despite the marginal advantage of rubidium, our CSAC employs the D1 transition of cesium for practical and historical reasons. Cesium has roughly 2X higher vapor pressure than rubidium at any given temperature, enabling reduction of cell size for a given operating temperature. More importantly, at the time we began the CSAC program, in 2001, VCSELs were available only at the cesium D2 wavelength, 852 nm, due to its coincidental overlap with a telecommunications wavelength standard. Now, however, as a result of CSAC,

VCSELs are available at all cesium and rubidium D1 and D2 wavelengths.

3.1 Physics Package

The basic components necessary to perform CPT interrogation are the laser source, an alkali vapor resonance cell, and a photodetector. Both the resonance cell and the laser must be temperature-controlled in order to stabilize the collision rates and wavelength, respectively. The power limitations of the CSAC prohibit the use of thermoelectric cooling and, as such, both must be stabilized at temperature higher than the operating ambient temperature of the CSAC. This presents an engineering challenge as VCSEL reliability decreases by roughly a factor of two for every 10°C increase in operating temperature. Thus it is desirable to run the VCSEL as cool as possible, but somewhat higher than the expected ambient in order to maintain thermal control. Since the resonance cell must also be stabilized above ambient, both the VCSEL and resonance cell may be operated at the same temperature. This leads to a significant reduction in the complexity of the physics package and control electronics, because the VCSEL and resonance cell, and, in fact the entire physics package, can be combined into a single temperature-stabilized assembly. This creates an additional, but inevitable, challenge for the VCSEL specifications, which must tune to the atomic resonance at a particular temperature.

The CSAC physics package is illustrated in **Figure 2**. The VCSEL is located at the top of the figure. The light passes through a λ/4 plate and the resonance cell before arriving at the photodetector at the bottom of the figure. The resonance cell is fabricated by sealing a small amount of cesium, along with a temperature-compensating buffer gas mixture, inside an anodically-bonded glass-silicon-glass "sandwich," using a technique similar to that pioneered by Liew, et al[11].

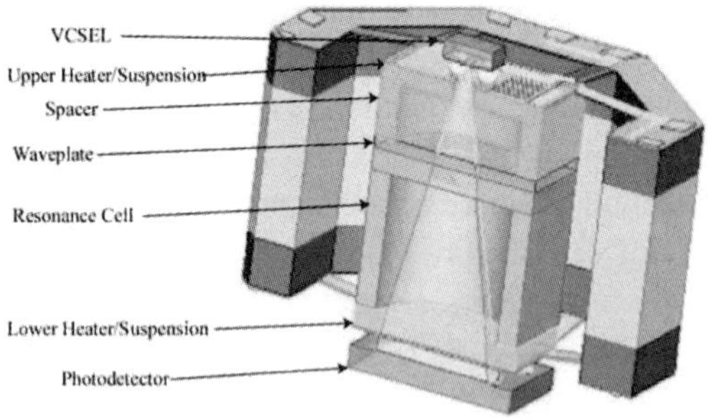

Figure 2: CSAC Physics package

For thermal isolation, the heated physics package assembly is suspended by polyimide tethers, shown in orange in **Figure 2**. Electrical connections to and from the physics package are patterned directly onto the polyimide, thus allowing their characteristic dimensions, and thereby thermal conductivity, to be determined by electrical, rather than mechanical, requirements. The physics package assembly is sealed in a getter-pumped vacuum package to eliminate convection and conduction to the ambient. Once sealed, the overall thermal resistance from the heated assembly to ambient is 7°C/mW, dominated by radiation from the heated surfaces. Thus the physics package consumes less than 10 mW when operating at 85°C in a 25°C ambient[12].

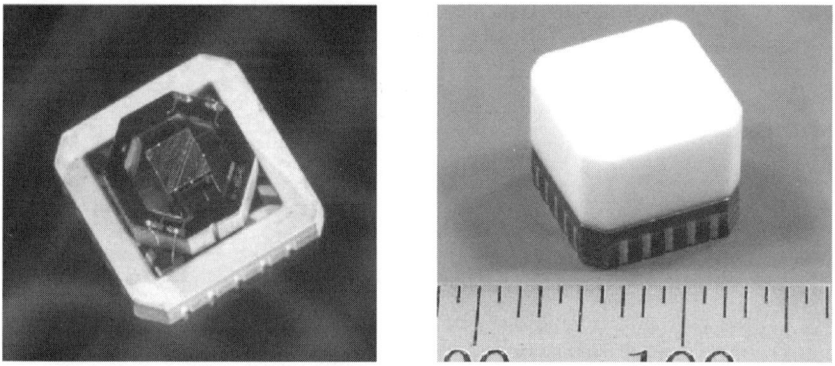

(a) (b)

Figure 3: Physics package assembly (a) mounted in carrier (b) vacuum sealed

Figure 3 shows photographs of the completed physics package assembly (a) mounted in a ceramic leadless chip carrier and (b) following vacuum sealing. The completed physics package is about 1 cm on a side.

We have shown that the power consumption could be reduced by a factor of 3X by coating the heated surfaces with a low emissivity material[12]. This is undesirable, though, for the following reason. There is residual heating, applied to the physics package, even when the heaters are turned off, due to the dissipation of the VCSEL. The upper operating ambient temperature limit is thus (stabilized temperature) – (VCSEL dissipation x thermal resistance). Sandia has developed VCSELs with relatively low threshold currents for CSAC in order to reduce the VCSEL power dissipation and thereby increase the upper operating ambient temperature without requiring unacceptably high stabilization temperatures[10]. **Figure 4** shows the efficiency of a CSAC VCSEL as a function of operating current. The threshold current for lasing is 350 μA with a slope efficiency of 50 μW/mA and forward voltage drop of ≈2V. Thus the VCSEL delivers up to 50 μW of optical power for CSAC

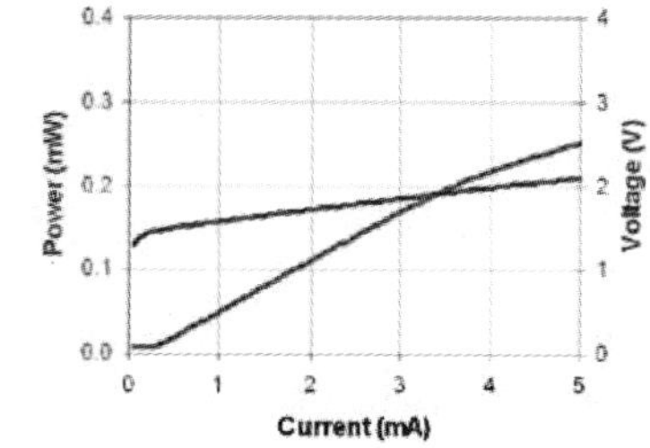

Figure 4: VCSEL power and voltage drop vs. bias current

operation while dissipating <2mW. Nonetheless, at a stabilization temperature of 85°C, this limits the CSAC operating ambient to ≈ 70°C.

Figure 5 shows typical physics package performance, measured with optimized laboratory electronics. Typical linewidths are $\gamma \approx 1.5$ kHz, with a signal contrast of ≈1%. **Figure 5(b)** shows the measured Allan deviation of 10 prototype physics packages, typically $\sigma_y(\tau) = 2\text{-}5 \times 10^{-11}\ \tau^{-1/2}$.

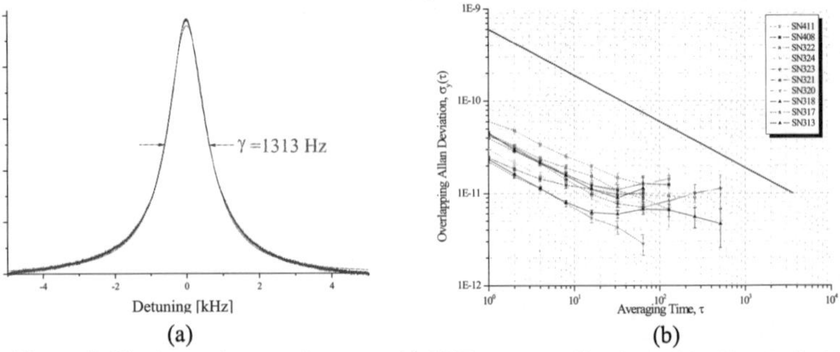

Figure 5: Physics package performance (a) CPT resonance lineshape (b) Allan deviation

3.2 Electronics

Reducing the size and power of the CSAC control electronics requires novel approaches to signal recovery, servo implementation, and microwave synthesis. Much of the complexity of conventional atomic clock electronics has been implemented in the CSAC firmware, thereby reducing parts count and power consumption. Multiplexing of the modulation sources allows multiple servos to simultaneously stabilize the laser wavelength, laser power, microwave power, and microwave frequency.

The microwave synthesizer consists of a 4.6 GHz voltage-controlled oscillator which is phase-locked to a 10 MHz TCXO, utilizing a commercially-available fractional-N PLL synthesizer. Modulation and tuning are applied to the microwave source by periodically reprogramming the synthesizer dividers. Ultimately, this provides tuning resolution of $\Delta f \approx 1\text{x}10^{-12}$.

As shown in **Table 1**, the CSAC power consumption is 125 mW. The microwave synthesizer, at 60 mW, remains the greatest area for improvement and we are hopeful that ongoing advances in low-power synthesis, driven by the mobile telecommunications industry, will permit further power reduction in the future.

Subsystem	Power
Physics	11 mW
Control Electronics	41 mW
μWave Synthesis	60 mW
Regulation & Passives	13 mW
Total	125 mW

Table 1: Power budget

4 CSAC Performance

In 2007, we performed a "pre-production build" of 10 nearly identical CSACs, with the principal objective of investigating device-to-device variation.

Figure 6 shows the short-term stability of the 10 CSACs. Generally, all devices exhibit $\sigma_y(\tau)=1\text{-}3\times10^{-10}\tau^{-1/2}$. This is ≈ 3X worse than the performance of the 10 physics packages when measured with optimal electronics (see **Figure 5(b)**). This degradation is caused primarily by the higher level of phase noise on the microwave interrogation signal with the low-power CSAC electronics.

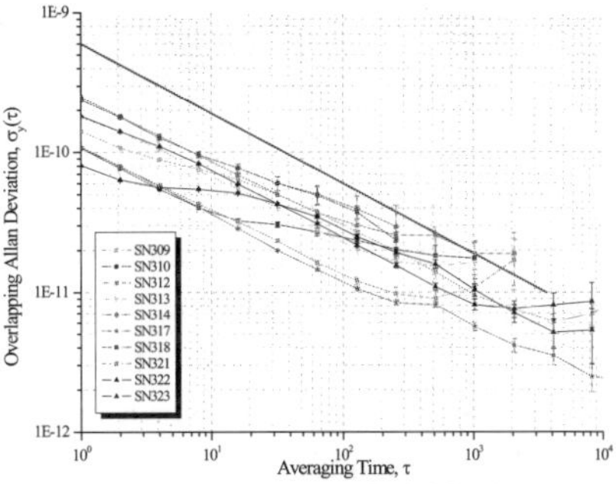

Figure 6: ADEV of 10 CSACs

One CSAC was power cycled 40 times over a five-day period in order to measure “retrace error,” i.e. frequency error when power to the unit is cycled. **Figure 7** shows a histogram of this measurement, with an RMS error of $+/-2\times10^{-11}$. This is competitive with RBO retrace, typically specified at $+/-\ 5\times10^{-11}$.

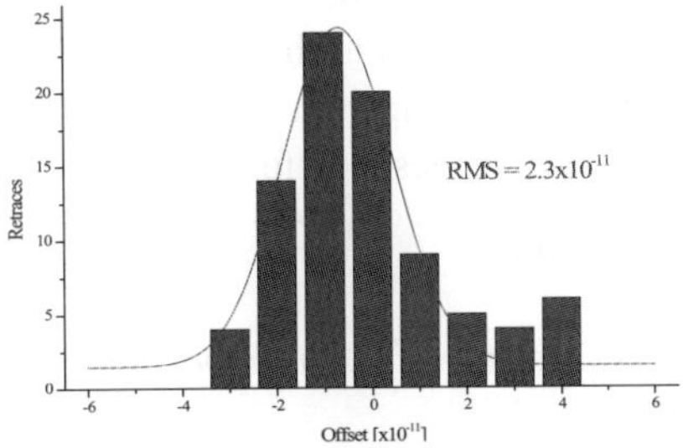

Figure 7: Retrace error

Figure 8 shows CSAC performance as a function of ambient temperature.

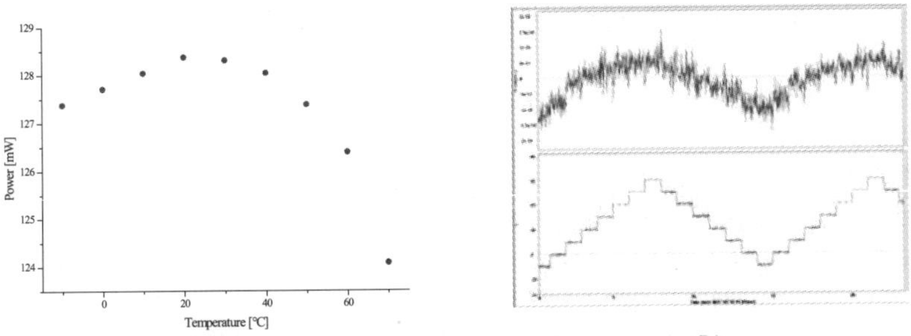

(a) **(b)**

Figure 8: (a) Power consumption and (b) frequency error over temperature

As shown in **Figure 8(a)**, at elevated temperature, the power consumption behaves as expected, reflecting the 7 mW/10°C decrease in physics package heater power with increasing ambient. Surprisingly, at lower temperatures, the increased efficiency of the microelectronics results in an overall *reduction* in CSAC power consumption. As such, the total power varies by only 4 mW (3%) over the entire operating range. **Figure 8(b)** shows the output frequency, measured over the temperature range of -10°C to +60°C. Typically, the CSACs exhibit temperature sensitivity of 1 -3×10^{-10} over 70°C, or about 3×10^{-12}/°C. This is comparable to RBO performance, typically specified as 2×10^{-10} over a somewhat wider operating range.

5 Conclusions

We have completed development of a chip-scale atomic clock, consuming 125 mW of power and exhibiting short-term stability of $\sigma_y(\tau)=1\text{-}3\times10^{-10}\tau^{-1/2}$. The CSAC is self-starting, fully autonomous, and capable of operating unattended for extended deployments over a wide range of ambient temperatures. We have built and tested ten identical CSACs of this design and found the performance to be consistent and highly reproducible from device to device.

Acknowledgements

The authors wish to thank DARPA for their continuous support and guidance as well as our numerous beta testers who have provided valuable feedback, test data, and suggestions for improvement.

References

1. R. Lutwak, *et al,* in "*Proc. of the 39th PTTI,*" Nov. 26-29, 2007, Long Beach, CA
2. P.J. Chantry, *et al,* in "*Proc. of the 46th FCS,*" May 27-29, 1992, Ft. Monmouth, NJ, p. 114.
3. P.J. Chantry, *et al,* in "*Proc. of the 50th FCS,*" June 5-7, 1996, Honolulu, HI, p. 1002.
4 . J. Kitching, *et al,* in "*Proc. of the 6th FSM,*" Sept. 9-14, 2001, St. Andrews, Scotland, p. 167.
5. R. Lutwak, *et al,* in "*Proc. of the 2005 Joint FCS and PTTI,*" August 29-31, 2005, Vancouver, BC, Canada, p. 752.
6. N. Cyr, *et al,* "*All-Optical Microwave Frequency Standard: A Proposal,*" IEEE Trans. Inst. Meas., **42**, 2, 1993.
7. R. Lutwak, *et al* in "*Proc. of the 35th PTTI,*" Dec. 2-4, 2003. San Diego, CA
8. M. Stahler, *et al, Optics Letters,* vol. **27**, 2002.
9. US Patent 6,359,916 (March 19, 2002).
10. D. K. Serkland, *et al,* in Proc. SPIE, vol. 6132, pp. 613208-1 to -11, 2006.
11. L. Liew, et al, *"Microfabricated alkali atom vapor cells,"* Appl. Phys. Lett. **48**, 2004
12. M. Mescher *et al,* in *"Proc. of the 13th Int. Conf. on Solid-State Sensors, Actuators and Microsystems,"* June 5-9, 2005, Seoul, Korea, pp. 311-316.

REACHING A FEW $10^{-13}\,\tau^{-1/2}$ STABILITY LEVEL WITH A COMPACT COLD ATOM CLOCK

F.X. ESNAULT, S. PERRIN, D. HOLLEVILLE, S. GUERANDEL, N. DIMARCQ
J. DELPORTE*

SYRTE, Observatoire de Paris, 61 av. de l'Observatoire, 75014 PARIS, FRANCE.
** CNES, 18 av. Edouard Belin, 31055 Toulouse Cedex, FRANCE*
E-mail: stephane.guerandel@obspm.fr

HORACE is a compact cold cesium atomic clock designed for space applications and onboard systems. The operation of this clock is different from fountains since the laser cooling, the microwave interrogation and the detection are sequentially performed inside the spherical microwave cavity. A short term relative frequency stability of $2.2 \cdot 10^{-13}\,\tau^{-1/2}$ is reported. Preliminary results on mid term show that a level of $4 \cdot 10^{-15}$ is reached after $5 \cdot 10^{3}$ s integration time. Limitations are investigated.

1. Introduction

The compact clock HORACE uses atomic laser cooling techniques and is designed for space operation and onboard systems, potentially having a physics package size of a few liters.[1] The target performance is a short term frequency stability of a few 10^{-13} at 1 s, a long term operation in the 10^{-15} range with an accuracy of a few 10^{-15}. The clock is operated at a high repetition rate (≈ 10 Hz) in order to reduce phase noise requirements of the local oscillator. Such a cold atom clock will take advantages of microgravity environment where interrogation time are longer.[2]

2. Experimental setup

The experimental setup shown in Fig. 1(a) is extensively described elsewhere.[3] Only the main features are recalled here. The specificity of HORACE is that all interactions (laser cooling, microwave spectroscopy and detection) are performed at the same place, inside a spherical microwave cavity. Atomic laser cooling is achieved with a nearly isotropic light field created inside the cavity from optical fibers in the (1,1,1) configuration.[4]

(a)

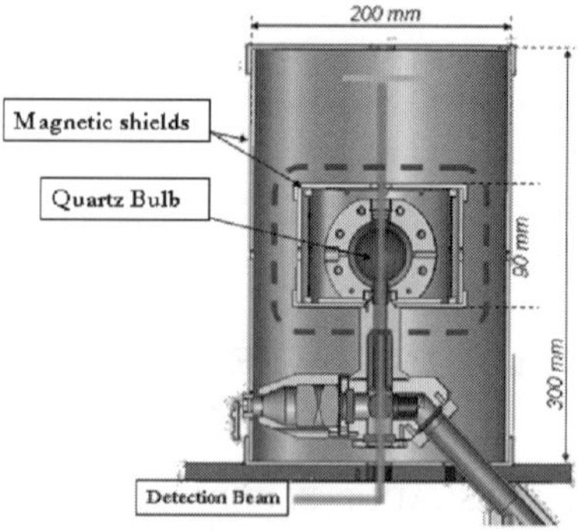

(b)

Fig. 1. (a): Physics package without magnetic shields. (b): Vertical cut of the whole vacuum chamber with magnetic shields. The useful part during clock operation is circled with the red dashed line.

The optically polished microwave cavity acts as a reflecting sphere for the laser field and averages power imbalances between each fiber. Two slits (1×4 mm) provide symmetrical microwave coupling. The cavity operates a TEM_{011} mode with quality factor $Q \approx 5000$. This cavity is surrounded by a solenoid creating a 5 mGauss vertical static magnetic field raising degeneracy of the different Zeeman sub-levels. Two holes of diameter $\phi = 8$ mm aligned in the vertical direction allow to perform *in situ* detection of atomic absorption thanks to a resonant probe beam shown on Fig. 1(b). It crosses the cavity vertically and is retroreflected by a mirror, creating a standing wave which prevents cold atoms from being heated up. The photodetector is located below.

3. The cooling sequence

For a high stability pulsed cold atom clock, the cooling sequence efficiency can be improved by recycling the cold atoms from a cycle to another. Indeed, a high stability results from a high atomic signal-to-noise ratio obtained in a short time, in order to reduce the sensitivity to the local oscillator phase noise. If the interrogation time T_{int} and the detection time T_{det} are not too long, a significant fraction of the cold atom cloud can be recaptured after about 10 cycles. Since no Zeeman selection process is done before interrogation, many cold atoms remain in all of the levels $|F = 3, m_F \neq 0\rangle$ during the entire clock cycle. Their contribution to the recapture process is thus very important. The recapture efficiency is characterized on Fig. 2(a). In Fig. 2(b) the number of useful atoms is plotted with or without recapture

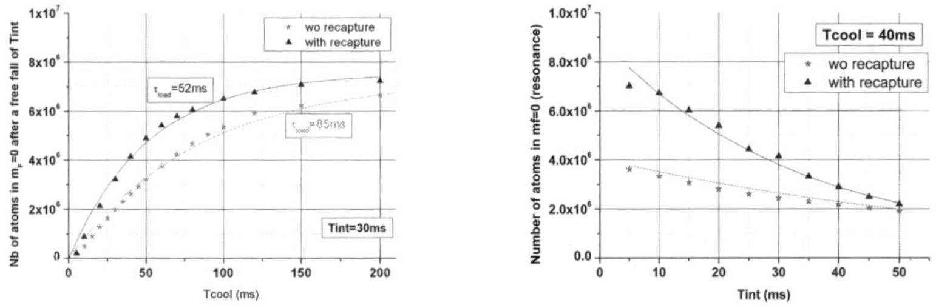

(a) (b)

Fig. 2. (a) Molasses loading time with and without recapture. With recapture, the cold atoms are recycled, reducing the loading time of next optical molasses. (b) Number of atoms in $m_F = 0$ vs T_{int} with and without recapture. Without recapture, the loss processes (70 ms time constant) are strongly dominated by collisions with Cs background pressure.

("without" means that a powerful blue-detuned pulse blows away all of the atoms after detection). With recapture, the gain of cold atoms is about a factor 2. The total number of atoms involved in the clock transition is $N_{at} \sim 3.5 \cdot 10^6$ typically. The recapture does not help for long interrogation times since the falling atomic cloud goes out of the capture zone.

4. The detection sequence

The N_4 atomic population is measured by resonant absorption of a 852 nm laser beam across the microwave cavity, which is convenient for a microgravity operation where the atoms stay in the center. Moreover, by increasing the cesium vapor pressure, the clock is able to run faster than in our previous work.[5] Although polarization, cooling light intensity and power imbalances between fibers are not actively controlled, the cold atom number N_{at} is surprisingly stable from shot to shot. We think that the combination of the high operation rate of our device (12 Hz) and the isotropic cooling scheme greatly reduce the influence of slow thermal-induced drifts on cooling parameters. The shot-to-shot relative fluctuations of the cold atom number are about $2 \cdot 10^{-4}$. Scaling as $1/\sqrt{N_{at}}$, these fluctuations are limited by the atomic shot noise. The different contributions to the detection noise are estimated in Fig. 3(a) where the shot-to-shot fluctuations of the detected signal are plotted against the detected number of cold atoms N_{det} after a $\pi/2$ interaction. This cancels the sensitivity to frequency noise but preserves other detection noises contributions, like technical noises or atomic shot noise. The curve on Fig. 3(a) exhibits two regimes respectively

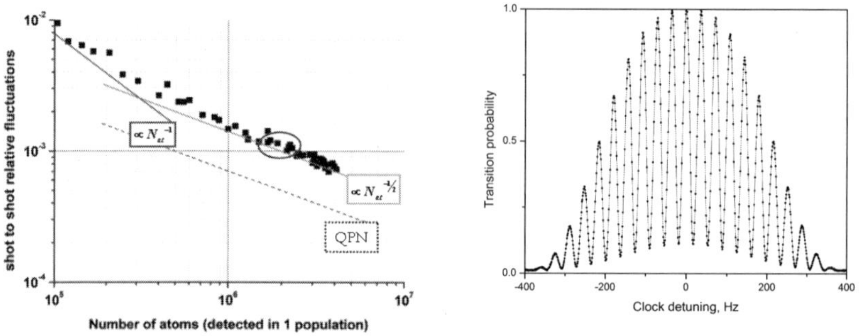

(a) (b)

Fig. 3. (a) Short term noise budget. The number of detected atoms is $2 \cdot 10^6$ leading to an absorption signal of 1.8%. (b) Ramsey fringes about 20 Hz wide measured with a signal-to-noise 1000 per cycle.

scaling as $1/N_{det}$ and $1/\sqrt{N_{det}}$. For $N_{det} < 3 \cdot 10^5$, the detection noise is dominated by technical noises. In our setup where the probe laser intensity is greater than $2\,\mu$W, the optical shot noise is the main technical limitation. For $N_{det} > 3 \cdot 10^5$, the fluctuations are dominated by atomic noises. Since N_{det} is lower than $5 \cdot 10^6$, the probe beam frequency noise does not contribute to the shot-to-shot fluctuations.

5. Frequency stability measurements

Frequency stability measurements against a reference oscillator[6] have been performed using either Rabi or Ramsey interactions. Typical central fringes have 23 Hz and 18 Hz FWHM with a contrast of 80%, respectively for a 35 ms Rabi interaction and 5+25+5 ms Ramsey interaction. The SNR is about 1000 per cycle. From Fig. 4(a), the short term frequency stability is $2.2 \cdot 10^{-13}\,\tau^{-1/2}$ and reaches $4 \cdot 10^{-15}$ at about $5 \cdot 10^3$ sec. Table 4(b) is a budget of the main noise sources and their contributions to the short term frequency stability. The main contribution is the atomic shot noise. An efficient normalization scheme for the cold atom number would nearly cancel this noise, but at the price of a longer time sequence and a bad recapture

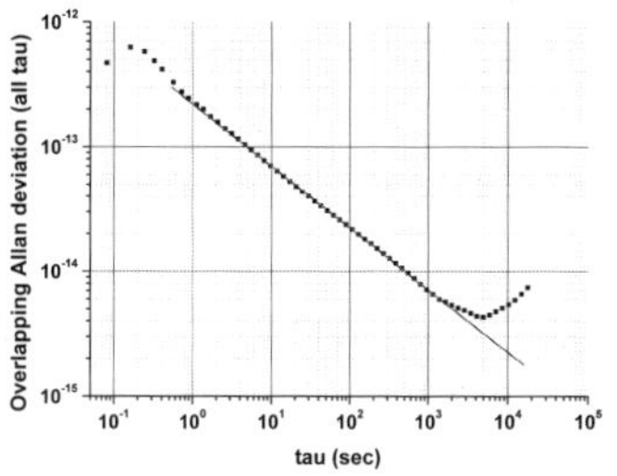

(a)

	Noise sources	$\frac{\sigma_{Signal}}{Signal}$ @ 1shot
TECHNICAL NOISES	Acq. Board + Photodiodes	0.1 10^{-3}
	Laser frequency noise	0.2 10^{-3}
	Optical shot noise	0.4 10^{-3}
ATOMIC NOISES	Quantum Projection Noise	0.5 10^{-3}
	Atomic shot noise	0.8 10^{-3} ⚠
Quadratic Sum (measured with cryogenic oscillator)		**1.0 10^{-3}**
Frequency Stability @ 1s ➡ 2.2 10^{-13}		
MICRO-WAVE	Dick Effect (Wentzel blue top)	0.45 10^{-3}
Quadratic Sum (Prediction)		**1.2 10^{-3}**
Frequency Stability @ 1s ➡ 2.4 10^{-13}		

(b)

Fig. 4. (a) The Allan deviation is fitted by $2.2 \cdot 10^{-13}\,\tau^{-1/2}$ with good agreement with experimental data. The bump appearing after $5 \cdot 10^3$ sec is attributed to daily temperature variations. (b) Short term noise budget. The number of detected atoms is $2 \cdot 10^6$ leading to an absorption signal of 1.8%.

efficiency. The laser intensity noise is at the level of the optical shot noise. This contribution will hardly be reduced. Indeed, the chosen intensity and duration of the detection pulse results from a trade-off between the optical shot noise and loss of absorption due to the heating of cold atoms. The microwave noise (Dick effect) of the local oscillator is an important issue for a pulsed device. Calculations[2] taking into account the present temporal sequence show that its contribution can be reduced to the 10^{-13} level using a commercial quartz oscillator optimized for our repetition rate (10 Hz).

6. Conclusions

With a short term frequency stability of $2.2 \cdot 10^{-13}\,\tau^{-1/2}$, HORACE reaches better performances than any compact Rb or thermal Cs beam clock, and is almost at the level of the fountain clocks. However, systematic effects affecting long term operation need to be further investigated. Moreover, the reduction of the optical bench is still a key issue towards a realistic space clock.[1]

References

1. S. Perrin *et al.*, Proc. of the 21th EFTF-IFCS, Geneva, Switzerland (2007).
2. S. Tremine *et al.*, Proc. of the IFCS, Vancouver, Canada (2005).
3. FX. Esnault *et al.*, Proc. of the 20th EFTF, Braunschweig, Germany (2006).
4. P.E. Pottie *et al.*, Optics Letters, Vol. 26 No. 21 1639-1641 (2001).
5. FX. Esnault *et al.*, Proc. of the 21th EFTF-IFCS, Geneva, Switzerland (2007).
6. D. Chambon *et al.*, Rev. Sci. Instrum. **76**, 094704 (2005).

EVALUATION OF LIN||LIN CPT FOR COMPACT AND HIGH PERFORMANCE FREQUENCY STANDARD

E. BRESCHI[†], G. MILETI

Laboratoire Temps-Fréquence, Physics Department, Université de Neuchâtel, Neuchâtel 2000 Switzerland

G. KAZAKOV, B. MATISOV

St.-Petersburg State Polytechnic University, St.-Petersburg, 195251, Russia

R. LAMMEGGER, L. WINDHOLZ

Institute of Experimental Physics TU-Graz Graz, 8010, Austria

We have investigated the lin||lin CPT signal in the ^{87}Rb D1 manifold when the atoms are contained in low pressure buffer gas cells. We predict the achievable clock frequency stability, basing our considerations on the signal-to-noise measurements. We show that a short-term stability of about $2\cdot10^{-11}\ \tau^{1/2}$ may be reached in a compact system using a modulated VCSEL, this value is mainly limited by the detection noise level and can be improved up to a factor 4 by using high frequency phase sensitive detection. Under the same experimental condition a challenging short-term stability of $1\text{-}3\cdot10^{-13}\ \tau^{1/2}$ can be achieved by using the PL ECDLs.

1. Introduction

Contrarily to the microwave-optical double resonance scheme, Coherent Population Trapping (CPT) does not require a microwave cavity, which may be an advantage in compact clocks. CPT usually suffers from smaller signal contrast, even though different techniques have been recently proposed to overcome this potential limitation [1].

In this communication we present our study on the use of CPT in vapour cell frequency standards. We have investigated the case of ^{87}Rb D_1 when a longitudinal magnetic field is applied and the CPT is excited by a linearly polarized, unidirectional and multi-frequency light field (the so called lin||lin

[†] This work has been supported by: Swiss National Science Foundation (SNRF project 200020-105624); Fund for Non-Profit 'Programs Dinastiya'; Austrian science Found (FWF project L300N02); INTAS CNES NSAU (project 06-1000024-9321).

CPT). The experiments were performed by using alternately two different light sources: a pair of phase-locked ECDLs or a current modulated VCSEL. The influence of the light source spectral quality on the short-term stability has been investigated.

Below the experimental setup is shortly described. In section 2 we discuss the clock signal in this configuration and the influence of the laser spectrum on it. In section 3 we discuss the detection noise and we predict the short-term stability achievable in optimized experimental conditions.

Figure 1 shows the block diagram of the experimental setup. The Rb atoms are confined in a cylindrical glass cell (volume 1 cm^3, 1.5 kPa of nitrogen as buffer gas) placed in a cylindrical magnetic shied. A longitudinal magnetic field (B_l) is provided by a solenoid placed inside the magnetic shield, B_l is parallel to the laser beam propagation vector.

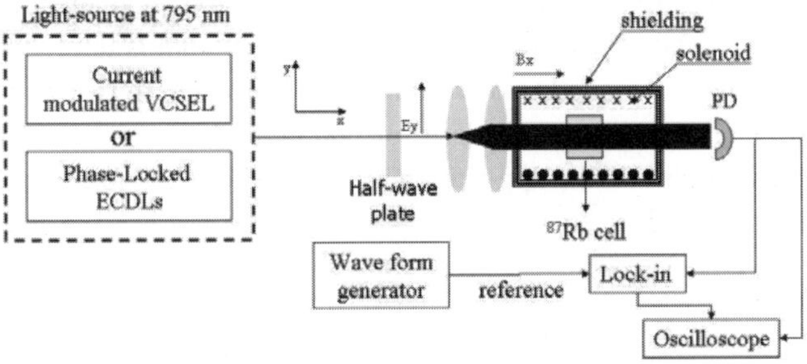

Figure 1: Block diagram of the experimental setup used to perform lin || lin CPT experiments: the modulated VCSEL and the Phase-Locked ECDLs are used alternately.

The multi-frequency optical spectrum is generated either, via modulation of the injection current of a single mode VCSEL chip emitting around 795 nm, or by phase-locking two lasers (in this case two ECDLs). In figure 2 the spectral characteristics of the light sources have been represented. The current modulated VCSEL has a broad multi-frequency spectrum (line-width 100 MHz) while the PL ECDLs have a narrow dichromatic spectrum (line-width 40 kHz).

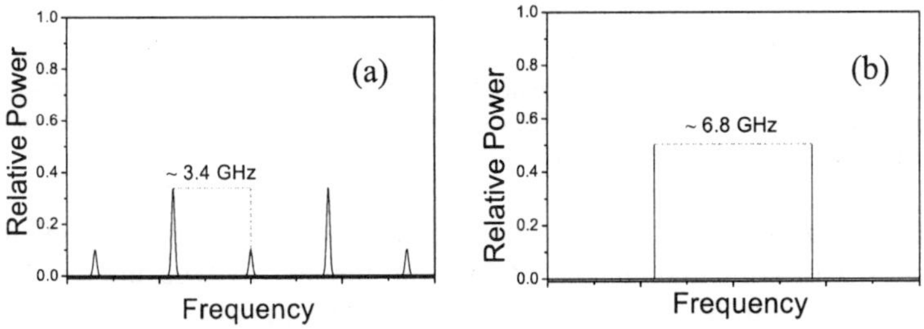

Figure 2: Schematic representation of the relative power distribution in the spectral components for (a) current modulated VCSEL and (b) PL ECDLs.

2. Clock signal in lin || lin configuration

Let us consider an atom, in a static magnetic field parallel to the laser propagation vector, illuminated with a laser light which has the two frequency components *linearly* polarized in the *same* plane. In this case the light-atom interaction scheme is the combination of σ^+ and σ^- transitions. The case of ^{87}Rb D_l line when the laser is tuned on F_e=1 is represented in figure 3.
Usually, the clock signal in the CPT-based atomic clocks is due to the coherence between the sublevels |1> and |2>, via (σ^+ - σ^+) or (σ^- - σ^-) excitation (represented by the bold lines in figure 3). In the lin||lin CPT, these two schemes occur simultaneously and cancel each other. However at the unperturbed hyperfine frequency, a CPT resonance due to the (σ^+ - σ^-) transitions occurs (dotted line in figure 3), and the clock signal is due to the coherence between the sublevels $|1a\rangle$-$|2a\rangle$ and $|1b\rangle$-$|2b\rangle$. Note that this signal is degenerate at low B_l and corresponds to a quadrupolar coupling ($|\Delta m| = 2$), thus it is suppressed at high buffer gas pressure because of excited state mixing. This intrinsic characteristic prohibits the application of the lin||lin CPT in mini-clocks unless wall-coating is used.

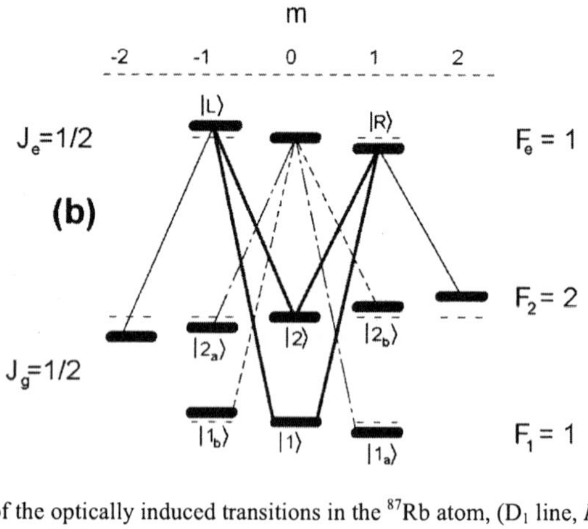

Figure 3: Scheme of the optically induced transitions in the ^{87}Rb atom, (D_l line, F_e=1). The signa that we propose to use as a clock reference is due to the coherence created between the hyperfine ground state Zeeman sub-levels $|1a\rangle$ -$|2a\rangle$ and $|1b\rangle$ -$|2b\rangle$.

Figure 4 represents the CPT signal obtained by sweeping the microwave frequency when the modulated VCSEL or the PL ECDLs are used in the experiments, figure 4 (a) and (b) respectively. The signal is plotted in terms of contrast, defined as the photo-detector current normalized with its value at zero Raman detuning. The contrast achieved with the PL ECDLs is about 7%, i.e. noticeably larger than the value obtained with modulated VCSEL (few %). We observe that this difference is mainly due to the photo-detector background levels caused by the different spectrum of the two laser systems. In the PL ECDLs about 100% of the laser power is contained within the frequency components responsible for the CPT resonance excitation. In the modulated VCSEL a fraction of about 68% of the total laser power is within the 1st order sidebands used for CPT resonance preparation. The residual 32% of the spectral power is distributed among the carrier and higher order sidebands which increase the background light on the photo-detector.

The CPT line-width is dominated by the power broadening and no influence due to the laser spectrum can be recorded in these experimental conditions.

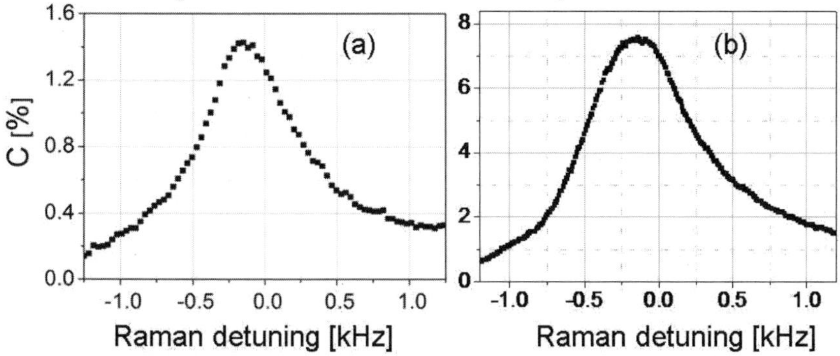

Figure 4: Clock signal in lin||lin configuration prepared with (a) modulated VCSEL and (b) PL ECDLs. The signal has the same amplitude (environ 300 nA) and line-width (about $2\pi\cdot900$ Hz), but the background level on the photo-detector changes dramatically. Thus the contrast obtained with PL ECDLs is noticeably larger than the contrast obtained with modulated VCSEL.

3. Detection noise and expected short-term stability

We measured the detection noise with the modulated VCSEL and PL ECDLs at the clock operating condition. The values are reported in Table 1 and compared with the shot-noise limit, calculated from the laser power measured after the cell. Remark that the detection noise level of PL ECDLs is only slightly worse than the shot-noise limit, while for modulated VCSEL is about two orders of magnitude worse. The detection noise for the modulated VCSEL can be improved by stabilizing the VCSEL on the absorption profile. However due to the VCSEL line-width (100 MHz) the FM-to-AM noise can not be reduced unless a very high speed servo loop for the laser locking is implemented, making the setup more complicated and less adequate in view of commercial applications. Another possibility to reduce the noise level in VCSEL-based clock is to increase the detection frequency. The pure optical detection of typical CPT experiments, allows using modulation frequency up to tens of kHz, instead of hundreds of Hz. In this way both the signal and the noise level decrease, however we record an enhancement of a factor 4 of the signal-to-noise ratio by increasing the detection frequency from 400 Hz to 40 kHz.

Table 1. Measured detection noise at the Fourier frequency of 1.5 kHz in the clock operating conditions.

	Shot-noise [A/√Hz]	Measured detection noise [A/√Hz]
PL ECDLs	$1.3 \cdot 10^{-12}$	$1.7 \cdot 10^{-12}$
Mod. VCSEL	$2 \cdot 10^{-12}$	$1 \cdot 10^{-10}$

The short-term stability is evaluated by the formula [2]: $\sigma_y(\tau)=0.2\cdot Q^{-1}\cdot(S/N)^{-1}$, where Q is the resonance quality factor and S/N is the signal-to-noise ratio.
In Table 2 the results of our analysis are summarized. The model in reference [3] predicts $\sigma_y(\tau)=2\cdot 10^{-13}\ \tau^{-1/2}$ for pure dichromatic laser spectrum, which is in very good agreement with the results we obtained with PL ECDLs. This challenging prediction inserts the lin||lin CPT signal as a candidate for high performances atomic clock. The $\sigma_y(\tau)$ for modulated VCSEL using high frequency detection is comparable with the state of the art of $\sigma_y(\tau)$ measured in continuous CPT interrogation with PL ECDLs ($\sigma_y(\tau)=1.4\cdot 10^{-12}\ \tau^{-1/2}$ [4]); and it is noticeably better than the value demonstrated in CPT-clocks based on modulated VCSEL [5].

Table 2. Predicted short-term stability, $\sigma_y(\tau)$ at 1 s.

Light source	Detection frequency	
	1.5 kHz	40 kHz
PL ECDLs	$3\text{-}4\cdot 10^{-12}$	$3\text{-}4\cdot 10^{-12}$
Mod. VCSEL	$1\text{-}2\cdot 10^{-11}$	$5\text{-}6\cdot 10^{-12}$

Acknowledgments

The authors thank for the useful discussion Dr. J. Delporte, Prof. Thomann and the colleagues of the Laboratoire Temps-Fréquence.

References

1 J. Vanier *Appl. Phys. B* **81**, 421 (2005).
2 J. Vanier and L. G. Bernier, *IEEE Tran. Instr.. Meas.* IM-30, **4** 277 (1981).
3 G. Kazakov, B. Matisov, I. Matzet, *et al.*, *Phys. Rev. A* **72**, 063408 (2005).
4 M. Zhu, *IEEE FCS. And PDA Exhibition Jointly 17th EFTF* 16 (2003)
5 J. Deng, P. Vlitas, D. Taylor, L. Perletz *et al.*, *proceedings EFTF* (2008).

Part XI

Time Scales

ATOMIC TIME SCALES TAI AND TT(BIPM): PRESENT STATUS AND PROSPECTS

GÉRARD PETIT

Bureau International des Poids et Mesures
92312 Sèvres, France

We review the stability and accuracy achieved by the reference atomic time scales TAI and TT(BIPM). We show that they presently are at the level of a few 10^{-16} in relative value, based on the performance of primary standards, of the ensemble time scale and of the time transfer techniques. We consider how the 1×10^{-16} value could be reached or superseded and which are the present limitations to attain this goal.

1. Introduction

International Atomic Time TAI gets its stability from some 350 atomic clocks worldwide that generate the free atomic scale EAL and its accuracy from a small number of primary frequency standards (PFS) which frequency measurements are used to steer the EAL frequency. Because TAI is computed in "real-time" (every month) and has operational constraints, it is not optimal and the BIPM computes in deferred time another time scale TT(BIPM), which is based on a weighted average of the evaluations of TAI frequency by the PFS.

The stability of EAL (and therefore that of TAI) depends on the performance of the clocks on which the ensemble average is based, and on the time transfer techniques used to link the clocks. On the other hand, the accuracy of TAI and TT(BIPM) depends on that of the PFS. Presently (2008) the stability of atomic time scales, the accuracy of primary frequency standards, and the capabilities of frequency transfer are at a similar level, about a few parts in 10^{16} in relative value. In Sec. 2 we present TT(BIPM), BIPM's realization of Terrestrial Time, and use it to review the present performances of time scales and PFS. We also briefly review the present performances of time transfer techniques.

To reach or supersede 1×10^{-16} in stability and accuracy, we need to obtain a similar uncertainty in each of the three fields, however they are in various stages of advancement towards this aim. While the present and future of frequency standards look promising, time scale formation and time transfer may have difficulties to keep up. In Sec. 3 we examine the possible progresses in these fields.

2. Presently achieving sub 10^{-15}

The stability of atomic time scales, the accuracy of primary frequency standards, and the capabilities of frequency transfer all achieve about a few parts in 10^{16} in relative frequency stability. This is evidenced in the following sections.

2.1. *TT(BIPM), BIPM's best realization of Terrestrial Time*

Terrestrial Time is a coordinate time in the geocentric reference system defined by Resolutions of the International Astronomical Union. TAI provides one realization of TT but, because it is computed in "real-time" and has operational constraints, it does not provide an optimal realization. The BIPM therefore computes in deferred time another realization TT(BIPM) [1], which is based on a weighted average of the evaluations of TAI frequency by the PFS. The present procedure for computing TT(BIPM) is described in [2] and a yearly computation is performed each January, the latest available being TT(BIPM07) available at ftp://ftp2.bipm.org/pub/tai/scale/TTBIPM/ttbipm.07. The basic features of the procedure for computing TT(BIPM) are the following:

- The computation starts in 1993 and uses all PFS measurements since 1992.
- The frequency of EAL with respect to the PFS is estimated for each month since 1993 following the algorithm in [3], using different estimations for the stability model of EAL according to the period considered. The resulting values of f(EAL-TT) are shown in Figure 1.
- The series of monthly values f(EAL-TT) is smoothed, interpolated and integrated with a 5-day step since MJD 48984 (28 Dec 1992), at which epoch continuity is ensured with previous realizations.

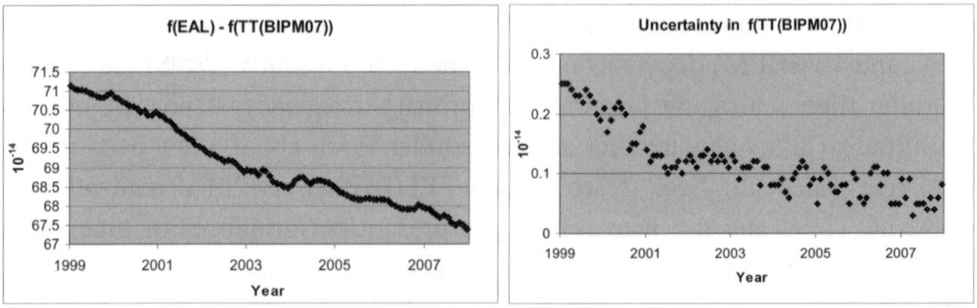

Figure 1: Monthly value (left) and standard uncertainty (right) of the frequency difference between EAL and TT(BIPM07).

The uncertainty of the monthly estimations f(EAL-TT), shown in Figure 1, also provide the estimated accuracy of TT(BIPM07). One can see that, over 2007, it is close to $5\text{x}10^{-16}$ on average. This is due to the ever increasing number of Cs fountain evaluations (about 120 since 1999, 29 of them in 2007), and to the improved accuracy of each fountain evaluation.

2.2. *Time scale and primary frequency standards*

TT(BIPM07) may be considered as the best time reference to be used to compute the stability of TAI and EAL. Here we estimate their stability over the 7-year period 2001-2007 (Figure 2), a period over which the performance of the ensemble time scale EAL is more or less constant (see Figure 3). For the short term (1 month) TT(BIPM07) is correlated to EAL and TAI so the corresponding values in Figure 2 are not significant. For the long term (one year averaging and above), one can see that the behavior of EAL is red noise and about corresponds to a drift which is obviously present (see Figure 1). The long-term instability of TAI is between $1\text{x}10^{-15}$ and $2\text{x}10^{-15}$, a factor two or three worse that the value for TT(BIPM07).

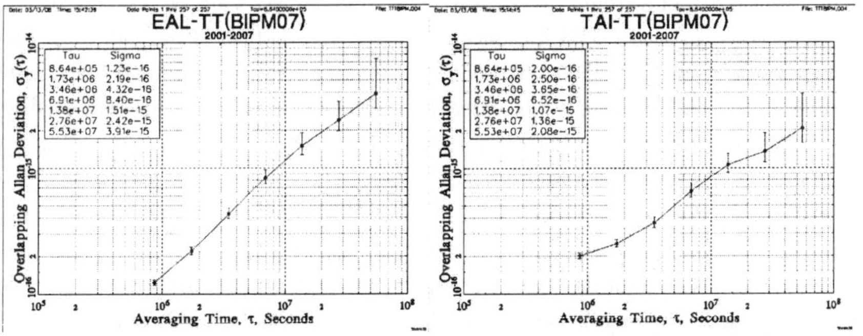

Figure 2: Fractional frequency instability of EAL (left) and TAI (right) vs. TT(BIPM07) over the period 2001-2007.

The 1-month instability of EAL, on the other hand, can be directly estimated from the instability of the clocks used in the ensemble average because the clocks are numerous and the maximum weight that one clock can reach is small, so that each clock can be considered independent of EAL [4]. Computing the 1-month instability over the period 1999-2007, one can see (Figure 3) that it is around $4\text{x}10^{-16}$ since 2002.

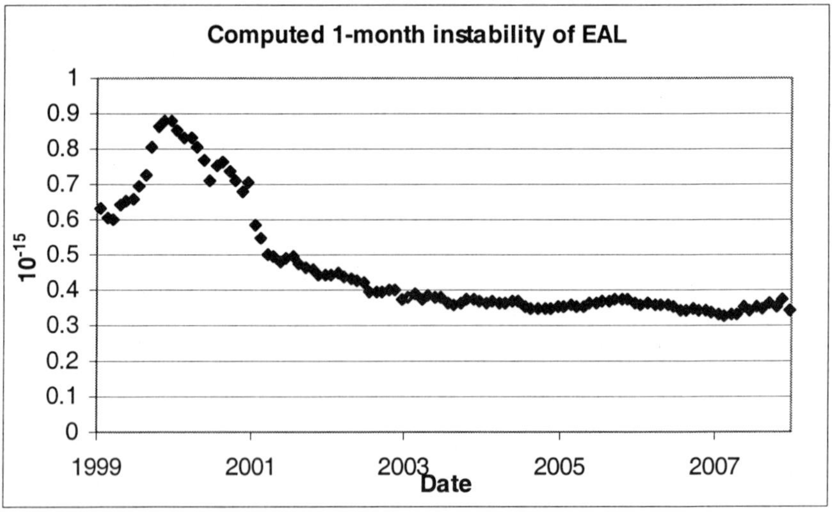

Figure 3: 1-month fractional frequency instability of EAL over the period 1999-2007.

Finally TT(BIPM07) may be used to estimate the quality of the PFS evaluations that contribute to it: First, as TT(BIPM) can be considered a weighted average of the PFS measurements, the distribution of the values of the frequency differences f(PFS)-f(TT(BIPM07)) should fulfill statistical tests, e.g. we check that the reduced χ^2 is close to 1, see [5] for a detailed study using TT(BIPM05). Second, for each individual PFS with a sufficient number of evaluations, it is possible to statistically compare its evaluations with a TT-like scale, here noted Tp, computed like TT(BIPM) but without considering data from the PFS under examination. It has been shown [5] that while some PFS show slightly significant deviations from Tp, the stated uncertainties of the PFS evaluations are, in general, statistically sound. This conclusion is confirmed by another approach [6] in which the evaluations of two or more PFS are compared when they are close enough in time to be directly linked. These studies bring confidence in the stated uncertainties of the PFS evaluations, which in turn directly yield the uncertainty estimates of TT(BIPM), therefore the stability and accuracy estimates presented in this section.

2.3. *Time and frequency transfer*

Two time and frequency transfer techniques are presently used in TAI computation: one uses the Global Positioning System (GPS) code measurements in a mode called “all in view” [7] which produces best results when using dual-frequency receivers, the other one is Two way satellite time transfer (TW) using telecommunication satellites. A further development of the GPS technique is to use phase and code measurements, its implementation in TAI computation is under way [8]. Since some years it has been shown, see e.g. [9], that these techniques are capable of a frequency transfer uncertainty of 1×10^{-15} in one day (GPS code and phase, TW at short distance) or in a few days (GPS code only, TW at long distance). For their use in TAI computation, where the optimal stability is searched at 1-month averaging time, the contribution of the frequency transfer uncertainty is expected to be at a few parts in 10^{16}.

3. Aiming at 1×10^{-16} and beyond

From numerous recent publications, see e.g. [10,11,12], it is clear that some frequency standards have reached a level where all systematic effects may be estimated with an uncertainty close to 1×10^{-16} or below, and the number of potential candidates in this group is expected to increase continuously. Without considering when and in which way this situation can result in a new definition of the second, it indicates that PFS will have the capacity to reach the level of 1×10^{-16} and to surpass it. In the following sections, we briefly review the situation for the other two items relevant to TAI computation, i.e. the ensemble time scale formation (Sec. 3.1) and time and frequency transfer (Sec. 3.2).

3.1. *Ensemble time scale*

As we have seen above (Figure 3 in Sec. 2.2), the instability of the ensemble time scale EAL has nearly reached a plateau over the past few years. This is mostly due to the fact that the industrial Cs clocks, which are averaged to form EAL, have not significantly changed over the past 15 years. Progresses in the ensemble scale EAL over this period are due to two factors: an increase in the number of clocks considered (see Figure 4), and changes in the algorithm used to compute the ensemble scale. Although some progress can still be expected in both directions, it is unlikely that this can lead to 1-month instability of 1×10^{-16} any time soon, as this would require a 10-fold increase in the number of clocks used.

A possible solution would be the emergence of a new generation of stable clocks, which would provide a similar kind of step-wise improvement as did the introduction of the present generation in the early 1990s. The required properties for this new generation are to show a stability performance in the low 10^{-16} at 1-month averaging time and to allow seamless continuous operation for at least one year. Possible candidates for this new generation of clocks may be emerging, see e.g. [13,14]. If this does not happen, new approaches to the ensemble time scale system, e.g. based on the use of non-continuous primary standards, may have to be developed.

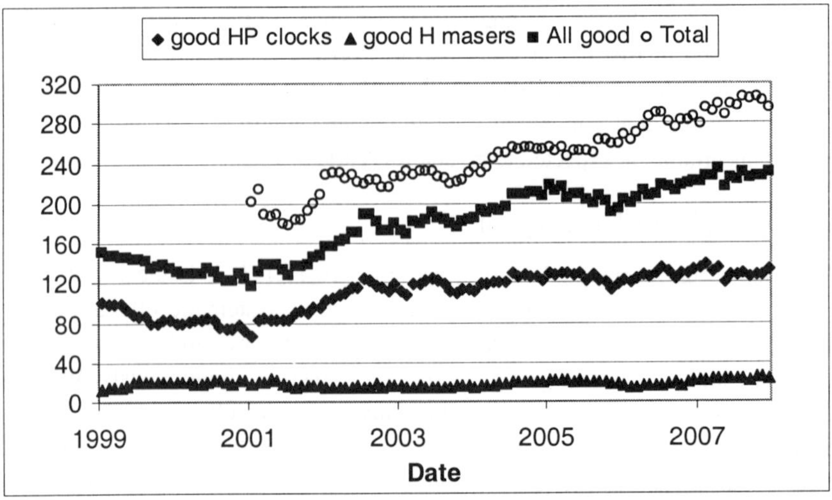

Figure 4: Number of clocks used in TAI computation over the period 1999-2007. Here 'good' applies to a clock with at least 11-month continuous operation and with an average weight at least 10% of the maximum possible weight.

3.2. *Time and frequency transfer*

As already mentioned, the optimal stability for TAI computation is searched at 1-month averaging time. Although it is now difficult to prove this, it is possible that the techniques presently used for time and frequency transfer may reach an uncertainty of 1×10^{-16} for such an averaging time. For techniques based on global navigation satellite systems, progresses are possible with new or modernized systems including more signals at several frequencies and enhanced code performances. The TW technique may also benefit for new hardware designs and from the mere increase in the number of data points. In addition, it is also possible to compute time links by mixing the two techniques to take advantage of both [15]. Ultimately, however, these techniques will find their limitations and in any case they will be insufficient to compare new PFS operating for short durations.

In recent years several new techniques have emerged that should be able to improve on the current limitations and to provide frequency transfer uncertainty in the low 10^{-17} region and possibly below. One is based on a microwave link to a low Earth orbit payload with one uplink and two downlink signals [16] and another one (T2L2) on a two-way optical link to a low Earth orbit payload [17]. Both promise a frequency uncertainty below 1×10^{-16} at 1-day averaging, if the clock of the space payload is stable enough. Another promisisng technique transfers a stable laser frequency over a standard fiber link [18], reaching a frequency uncertainty below 1×10^{-18} at 1-day averaging when the fiber-induced phase noise is measured and cancelled. Although the latter technique may be limited in spatial extension, one can expect these techniques to meet the future needs for averaging times of 1 to a few days and above.

4. Conclusions

We have shown that the present performance of the reference atomic time scales TAI and TT(BIPM) is in the region of a few parts in 10^{16} in stability and accuracy. Improving to 1×10^{-16} and below is not out of reach but may require the development of a new generation of commercial atomic clocks or a new approach to the problem of elaborating an ensemble time scale, or both.

References

1 B. Guinot, *Astron. Astrophys.*, 192, pp. 370-373, (1988)

2. G. Petit, A new realization of Terrestrial Time, *Proc. 35th PTTI,* pp. 307-316 (2003).

3. J. Azoubib, M. Granveaud, B. Guinot, *Metrologia* 13, 87 (1977).

4. G. Petit, The long-term stability and accuracy of EAL and TAI (revisited), *Proc. 21st EFTF,* pp.391-394 (2007).

5. P. Wolf, G. Petit, E. Peik et al, Comparing high accuracy frequency standards via TAI, *Proc. 20th EFTF,* pp.476-485 (2006).

6. T. Parker, Status of Cs fountain primary frequency standards, these proceedings (2008).

7. G. Petit, Z. Jiang, GPS all in view time transfer for TAI computation, *Metrologia* 45-1, 35-45 (2008).

8. G. Petit, Z. Jiang, Precise point positioning for TAI computation, *IJNO* Article ID 562878, doi:10.1155/2008/562878 (2008).

9. A. Bauch, J. Achkar, S. Bize, et al., Comparison between frequency standards in Europe and the USA at the 10^{-15} uncertainty level, *Metrologia* 43, 109-120 (2006).

10. A.D. Ludlow, T. Zelevinsky, G.K. Campbell et al., *Science* 319, 1805-1808 (2008).

11. W.H. Oskay, S.A. Diddams, E.A.Donley et al., *PRL* 97, 020801 (2006).

12 T. Rosenband, P.O. Schmidt, D.B. Hume et al., *PRL* 98, 220801, (2008).

13. E.A. Burt, J.D. Prestage, R.L. Tjoelker, these proceedings (2008).
14. C. Ekstrom, S. Peil, T. Swanson, S. Crane, these proceedings (2008).
15. Z. Jiang, G. Petit, Combination of TWSTFT and GNSS for Accurate UTC Time Transfer, *Metrologia* submitted.
16. A Seidel, M.P. Hess, J. Kehrer et al., *Proc 22nd EFTF* (2008).
17. E. Samain, J. Weick, P. Vrancken et al., International Journal of Modern Physics D 17-7, 1043 (2008)
18. X. Lopez, A. Amy-Klein, C. Daussy, *Eur. Phys. J.* D 48, 35-41 (2008).

WEIGHT FUNCTIONS FOR BIASES IN ATOMIC FREQUENCY STANDARDS*

JON H. SHIRLEY

National Institute of Standards and Technology, Boulder, CO 80305 USA

We present a unified treatment of frequency-standard biases that vary significantly during the period of measurement. We introduce three time-dependent weight functions built from the solution of the unperturbed equations of motion for a two-level system. By integrating a weight function together with the time dependence of a perturbation over the excitation period we find the change in the lineshape and can deduce any biases. The same weight function may be used for treating more than one cause of a bias.

1. Introduction of weight functions

We summarize a formulation based on the time-dependent Schrödinger equation as presented in [1]. A corresponding analysis may also be done by use of the three-component equations derived from the density matrix [2, 3].

The Hamiltonian for a two-level system excited by radiation at frequency ω can be written $\mathcal{H} = -\Delta\sigma_z + b\sigma_x$, where the σ_i are the Pauli spin matrices. We have introduced the abbreviation $\Delta = \frac{1}{2}(\omega - \omega_0)$ for one half the detuning from the atomic resonance frequency ω_0. Similarly, b is one half the Rabi frequency associated with the excitation. The coefficients b and Δ are both real and may be time-dependent. We have applied the rotating-wave approximation and redefined the phases of the wave-function components to eliminate the rapid time dependence of the coupling coefficient b [1].

The components α and β of the wave function obey the time-dependent Schrödinger equation,

$$\begin{aligned} i d\alpha / dt &= -\Delta\alpha + b\beta \\ i d\beta / dt &= b\alpha + \Delta\beta \end{aligned} , \tag{1}$$

with the initial conditions $\alpha(0) = 1$ and $\beta(0) = 0$. The probability that a transition has occurred after excitation for a period τ is then $P = |\beta(t)|^2$.

A two-level perturbation Hamiltonian can be represented by $\mathcal{H}_1 = \hbar\sum_i G_i(t)\sigma_i$, where the index i ranges over x, y, z and the $G_i(t)$ are real. The x component of the sum represents a perturbation in the amplitude of the exciting field. The y component represents a perturbation in the phase of the exciting field. The z component represents a perturbation in either the frequency of the exciting field or the energy-level separation.

The presence of a perturbation alters the wave function by ψ_1, which satisfies $i\hbar \frac{d}{dt}\psi_1 = \mathcal{H}\psi_1 + \mathcal{H}_1\psi$ to first order in the perturbation. A formal solution to this equation at time τ is

$$\psi_1(\tau) = -(i/\hbar)U(\tau,0)\int_0^\tau U^{-1}(t,0)\mathcal{H}_1(t)\psi(t)\,dt\,,$$

where $U(t,0)$ is the evolution matrix for the unperturbed system. The wave function correction ψ_1 is thus expressed as an integral over the time dependence of the perturbation with functions known from the solution of the unperturbed Schrödinger equation.

To first order the correction to the transition probability is also an integral over the perturbation with a function built up from solutions of (1):

$$P_1(\tau) = \int_0^\tau \sum_i W_i(\tau,t)G_i(t)dt\,. \qquad (2)$$

We call the real functions $W_i(\tau,t)$ weight functions, because they weight the averaging of the perturbation time dependence. Each weight function depends on intermediate and final times, as well as the unperturbed excitation amplitude b and detuning Δ. Each weight function can be expressed in terms of the unperturbed probability amplitudes α and β satisfying (1) as follows:

Amplitude weight function: $W_x(\tau,t) = \mathrm{Im}\left\{2\alpha(\tau)^*\beta(\tau)^*\left[\alpha(t)^2 - \beta(t)^2\right]\right\}$ (3)

Phase weight function: $W_y(\tau,t) = \mathrm{Re}\left\{2\alpha(\tau)^*\beta(\tau)^*\left[\alpha(t)^2 + \beta(t)^2\right]\right\}$ (4)

Detuning weight function: $W_z(\tau,t) = -\,\mathrm{Im}\left[4\alpha(\tau)^*\beta(\tau)^*\alpha(t)\beta(t)\right]$. (5)

The detuning weight function is equivalent to the sensitivity function used in analysis of the Dick effect [3,4].

In figure 1 we plot the three weight functions as a function of t for half-sine-wave Rabi excitation $b(t) = (\pi/2)b_0\sin(\pi t/\tau)$. These functions were found by inserting numerical solutions of (1) for α and β into (3)-(5).

1.1. *Properties of Weight Functions*

From the symmetries of the real and imaginary parts of α and β, we find that W_x is an even function of the detuning Δ, while W_y and W_z are odd functions. Thus an amplitude perturbation adds a symmetric contribution to the lineshape but does not shift the resonance frequency. Phase and detuning perturbations add asymmetric contributions to the lineshape and may shift the resonance frequency.

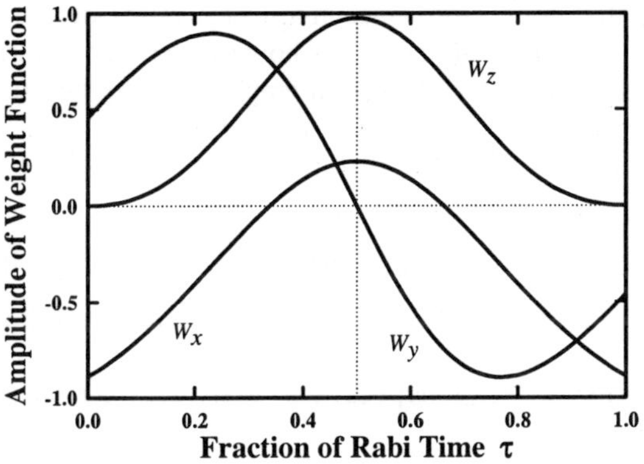

Fig. 1. Weight functions for Rabi excitation evaluated at $b_0\tau = 0.50\pi$ (optimum excitation) and $\Delta\tau = 0.544\pi$ (halfwidth of the Rabi lineshape).

From the time symmetry of the unperturbed Schrödinger equation we find that if the unperturbed excitation amplitude $b(t)$ is symmetric about the mid-excitation time $\tau/2$, then W_x and W_z are symmetric about $\tau/2$, while W_y is anti-symmetric. Conversely, if $b(t)$ is antisymmetric about time $\tau/2$, then W_y and W_z are symmetric about $\tau/2$, while W_x is antisymmetric. These symmetries combine with the time symmetries of perturbations to null certain effects.

1.2. *Relations of weight functions*

From (1) and the definitions (3)-(5) we deduce that the weight functions obey the following differential equations:

$$\begin{aligned} \partial W_x / \partial t &= 2\Delta W_y \\ \partial W_y / \partial t &= -2\Delta W_x - 2bW_z \; . \\ \partial W_z / \partial t &= 2bW_y \end{aligned}$$

These equations have the same form as the three-component equations [2,3], but have different initial conditions.

2. Biases in Rabi excitation

Consider slow square-wave modulation of the exciting frequency with amplitude ω_m. We define the resonance position by that detuning $\delta\omega$ that makes the signals at the detunings $\omega_m+\delta\omega$ and $-\omega_m+\delta\omega$ equal. To first order, this resonance position becomes $\delta\omega = -P_1(\omega_m)/\partial P(\omega_m)/\partial\omega$. To obtain a bias, we compute the numerator from (2) and the denominator from the unperturbed lineshape.

Suppose the atomic resonance frequency varies during the period of excitation. The z-component of (2) tells us how to average the variation. Figure 1 shows that the detuning weight function emphasizes the mid-time part of the variation. Any asymmetric part of the variation averages to zero.

As an example of a phase perturbation, consider phase modulation at frequency Ω with half Rabi amplitude b_1 and initial phase φ. It can be represented by the perturbation function $G_y = b_1 \sin(\Omega t + \varphi)$. If we add and subtract $\frac{1}{2}\Omega\tau$ to the argument of the sine we can rewrite it as follows:

$$\sin(\Omega t + \varphi) = \sin\Omega\left(t - \tfrac{1}{2}\tau\right)\cos\left(\tfrac{1}{2}\Omega\tau + \varphi\right) + \cos\Omega\left(t - \tfrac{1}{2}\tau\right)\sin\left(\tfrac{1}{2}\Omega\tau + \varphi\right).$$

For excitations symmetric about $t = \frac{1}{2}\tau$, the phase weight function W_y is anti-symmetric. The second term is symmetric, hence its integral with W_y vanishes. The factor $\cos\left(\frac{1}{2}\Omega\tau + \varphi\right)$ in the first term is independent of t, hence may be removed from the integral. The change in lineshape reduces to

$$P_1 = \int_0^\tau W_y(\tau, t)\, b_1 \sin\Omega\left(t - \tfrac{1}{2}\tau\right) dt \cos\left(\tfrac{1}{2}\Omega\tau + \varphi\right).$$

The final cosine factor completely describes the dependence of P_1 and the corresponding shift on the initial phase. As discussed in [5] any thorough average over φ eliminates this first order shift. For constant excitation we can perform the integration. The result agrees with that given in (7) of [5].

3. Biases in Ramsey excitation

Ramsey excitation consists of two short periods of excitation separated by a relatively long "drift" period T free of excitation. Weight functions for Ramsey excitation are still defined by (3)-(5), but the probability amplitudes now refer to the complete cycle spanning both excitations. The integral (2) must span both excitations. The symmetries of the weight functions are the same as given before. But note that time symmetry is now about the midpoint of the drift time.

Figure 2 shows the shape of the three weight functions as a function of time for $T = 4\tau$ and 2 dB below optimum excitation. All of W_x and the central part of W_y vanish when the excitation is exactly optimum.

Consider first that the atomic resonance ω_0 is changing, as it might in an inhomogeneous magnetic field. During the drift time, Fig. 2 shows that W_z is constant. Hence the unweighted time average of ω_0 is adequate . Changes in ω_0 during excitation are underweighted toward the beginning of the first excitation and the end of the second excitation.

The following four causes of frequency biases can all be interpreted as perturbations of phase and all treated using the <u>*same*</u> weight function G_y: end-to-end cavity phase shift, microwave leakage, distributed cavity phase shift and

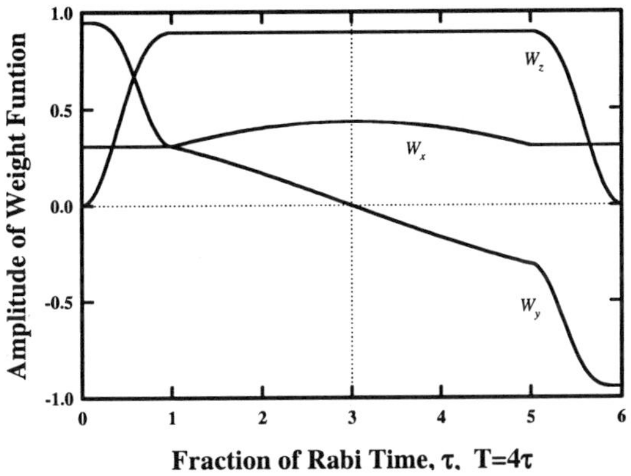

Fig. 2. Weight functions for Ramsey excitation evaluated at $2b_0\tau = 0.40\pi$ and $2\Delta T = .50\pi$ (halfwidth of a Ramsey fringe).

sidebands or spurs on the exciting radiation. They differ in the regions over which the integral in (2) is zero.

For example, the perturbation effect of phase modulation is found by use of the same perturbation function, as in the Rabi case. But the integration now extends over both excitation regions. For symmetric excitation about the mid-time $t = \frac{1}{2}T + \tau$, we can factor out the initial phase dependence, just as we did in the Rabi case. We then find the perturbation transition probability

$$P_1 = 2\int_0^{\tau} W_y\left(\tau, t\right)\sin\Omega\left(t - \tau - \tfrac{1}{2}T\right)dt\cos\left(\tfrac{1}{2}\Omega T + \Omega\tau + \varphi\right).$$

The symmetry of the integrand assures us that the integration result is the same for both excitation regions. For constant excitation the integral can be evaluated. The result is in agreement with (15) in [5], but the derivation here is much shorter.

A more detailed paper with additional examples is planned for publication elsewhere.

References

1. J. H. Shirley, W. D. Lee, R. E. Drullinger, *Metrologia* **38**, 427-458 (2001).
2. J. Vanier and C. Audoin, *The Quantum Physics of Atomic Frequency Standards* (Adam Hilger, Ed., IOP Publishing Ltd., Bristol, England, 1989).
3. P. Lemonde, G. Santarelli, Ph.Laurent, F. P. Dos Santos, A. Clairon, C. Salomon, *Proc. IEEE Int. Freq. Control Symp.*, 110-115 (1998).
4. G. J. Dick, *Proc. 19th PTTI*, 133-147 (1987).
5. J. H. Shirley, T. P. Heavner, S. R. Jefferts, to appear in *IEEE Trans. Inst. Meas.*

Part XII

Interferometers

DEFINITION AND CONSTRUCTION OF NOISE BUDGET IN ATOM INTERFEROMETRY

Erika D'Ambrosio

Istituto Nazionale di Fisica Nucleare, Sez. di Firenze
Via Bruno Rossi 1/3 - 50019 Sesto Fiorentino, Italia

Given the scientific potential of established and evolving quantum technologies and the new proposed detector for gravitational wave astronomy MIGO (Matter-wave Gravitational Wave Observatory) it is timely and beneficial to characterize some "actual" prototype and gather important in-depth knowledge in atom interferometry.

Of first and foremost importance is distinguishing the interferometric approach of *real space* detectors (based on an apparatus layout whose elements are massive optics or diffraction gratings that fix the ends of the interferometer arms) and the inertial sensors based on a superposition of atomic states, where no arm end is assigned by any optics, but rather the superposition starts and ends at **given times** meanwhile accruing a relative phase between the different momentum states.

The studies presented at the VII Symposium on Frequency Standards and Metrology aim at identifying the potential and sensitivity limits of atom interferometers, based on demonstrated concepts. Their sensitivity is determined by the competition between the signal induced by various disturbances versus that induced by the external fields of interest; the effects depend on the configuration, since different schemes respond differently to the same excitations.

The result is a feasibility exploration that demonstrates the actual capability of operating systems. It identifies both the properties and limitations that are characteristic of atom interferometers. The behaviour of such systems must be fully understood in order to have a basis for the development of the next generation of atom interferometers, as detectors applicable in tests of general relativity and as sensors of gravitational waves.

1. Introduction

Because of the high-accuracy they enable, interferometric techniques have always been used in the search for elusive entities and to measure small perturbations; one of the earliest applications being the renowned Michelson-Morley experiment. The basic principle is simple: the time-varying change of the phase accrued by a propagating wave is the observable quantity and the information is extracted by the superposition with an independent wave. The interferometric technique relies on the phase accrual being a distinctive property of each wave, so that a change in the trajectory of one of them is revealed by the beat with an unaffected one.[1]

In its basic form, this is the operating principle of long-baseline ground-based gravitational wave detectors. As illustrated in Fig.1(a) the ends of the phase accrual are fixed by the local ends of the interferometer arms, identified in real space by x_A and x_B and embodied by optical elements. Perpendicular arms are distinctively affected by gravitational waves, which makes the geometry of Michelson interferometers ideal for the purpose. All laser-light ground-based gravitational wave interferometers are variants of this configuration. Currently VIRGO and the detectors comprising LIGO (Laser Interferometer Gravitational-wave Observatory) are the most sensitive interferometers of this class and their data are being analysed in a joint search. Scientific informations are extracted from the error signal at the output port, that is also fed back to the interferometer arms. This is due to the fact that the observation of small perturbations refers to a null point identified by a resonant condition x_A and x_B must satisfy.[2]

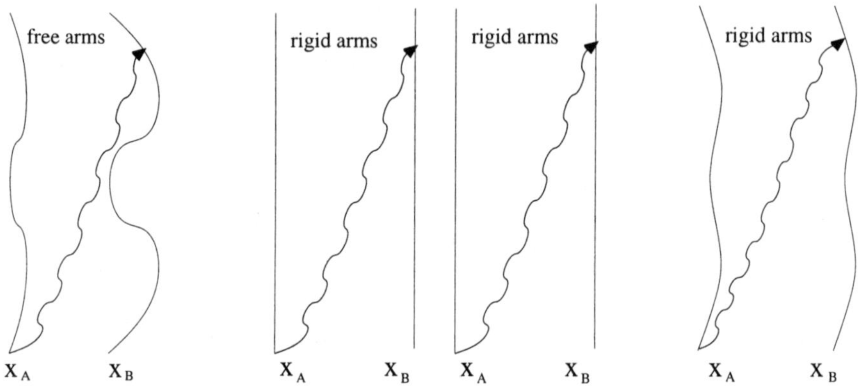

(a) The distinction between LIGO and MIGO is that the arm ends are suspended optics in the former and mirrors rigidly attached to the frame of the apparatus in the latter. This means that the distance between two mirrors in MIGO is constant and the world lines of x_A and x_B only have to satisfy the rigidity constraint

(b) Both the description of the propagating wave and the world lines of the ends of the arms of MIGO depend on the choice of the coordinate system, but this dependence is such that the total expression for the output signal of the rigid interferometer is invariant and does not change with the chosen reference frame

Fig. 1. The ends of the arms of LIGO and MIGO identify the start and conclusion of the wave propagation

An alternative to LIGO and its space-based complement LISA (Laser Interferometer Space Antenna) has also been proposed and its capabilities are under scrutiny by the gravitational wave community. The De Broglie wavelength and the particle-wave duality of the atoms is used in this concept. The beam in Fig.1(b) is a matter-wave that propagates from x_A to x_B which are the coordinates of the optical elements. These identify the real space structure of MIGO (Matter-wave Interferometric Gravitational-wave Observatory) fixing the ends of its arms and the boundaries of the path-integral that renders the phase.[3]

Distinguished from this proposal and affected by different problems, is the atom interferometer based on the internal interference of two atomic states, where no mirror fixes any arm.[4] Instead of watching the propagation between two local optical elements, the wave function of a microscopic object is manipulated in this system so as to extract, from the relative evolution of two states, their dynamics and any deviation from flatness that is encoded in the non uniform progression of the associated phase.[5]

This technique is of great interest for its implications, as it provides a novel approach to gravitational wave astronomy, making use of demonstrated concepts in atom manipulation and interferometry.[6]

The fundamental distinction between the operating principles of the two classes of interferometers also results in totally unrelated instrumental perturbations. In the next section, we explicitly show the construction of the noise budget and use the gravitational wave transfer function, to convert the induced phase shift in units of equivalent astrophysical excitation. This means that the sensitivity of the detector can be inferred as the gravitational wave strength, necessary to generate the same effect as the noise source dominating in the frequency bin of interest. The competition between signal and noise is determined by the ratio of their respective transfer function (in addition to the relative amplitude of the corresponding excitations) and this varies with the interferometer configuration and with the frequency.

2. Definition of noise budget

In real space interferometers

$$e^{-ik\hat{r}}|\bar{q}>= e^{-ik\bar{q}}|\bar{q}>$$

and the phase accumulates

$$\Phi = \int_{x_A}^{x_B} \mathrm{d}\Phi(q)$$

between the local ends of the detector arms. Momentum space interferometers rely on two atomic states

$$|\bar{p} - \hbar k/2, \ldots > \qquad \text{and} \qquad |\bar{p} + \hbar k/2, \ldots >$$

connected by the manipulation

$$e^{-ik\hat{r}}|\bar{p} - \hbar k/2, \ldots >= |\bar{p} + \hbar k/2, \ldots >$$

by means of the induced transition between the internal atomic energy levels. In this class of interferometers, the relative evolving phase of the quantum states that form the superposition is the quantity of interest: it encodes the relative dynamics of the two momentum states, whose progression is a manifestation of the spacetime curvature. Variations of such a phase modulate the relative population of the two mixed states, in a similar way as in real space interferometers there is a modulation in the relative intensity of the two output ports of the system.

In both cases, particle counting fluctuations limit the accuracy of the inferred phase, as described by

$$< \delta\Phi^*(t)\delta\Phi(t') > = C(t'-t) = \text{correlation function} = \int_{-\infty}^{+\infty} S(\omega)e^{-i\omega(t'-t)}\frac{\mathrm{d}\omega}{2\pi}$$

$$< \delta\Phi^*(\omega)\delta\Phi(\omega') > = \int_{-\infty}^{+\infty}\int_{-\infty}^{+\infty} C(t'-t)e^{i\omega' t'}e^{-i\omega t}\mathrm{d}t\mathrm{d}t' = 2\pi\delta(\omega-\omega')S(\omega')$$

where $S(\omega)$ is the power spectral density. In general $S(\omega)$ provides a quantitative assessment of the energy released per frequency bin. When it is a a constant $C(t'-t) \propto \delta(t'-t)$ as in the case of counting fluctuations.

The gravitational wave amplitude

$$\sqrt{< \tilde{h}^2(\omega) >_{\mathrm{PSD}}} = \frac{1}{2\vec{k}\cdot\vec{v}Tf(\omega)} \times \frac{1}{\sqrt{\dot{N}}} \xrightarrow[\omega\to 0]{} \frac{3}{ikL(\omega T/2)^3} \times \frac{1}{\sqrt{\dot{N}}}$$

identifies the corresponding shot noise limit of the sensitivity and represents the strength of the astrophysical excitation, necessary to induce the same phase shift as the Poissonian fluctuations around the mean atomic flux, that makes for the interferometer observable quantity.

We recognize the scaling parameter $L = vT$ that is the average traveled distance. This is a parameter that can be specified by sensitivity requirements and whose value does not identify the fundamental properties of a detector, other than its size. The suitability of a certain concept is instead better studied in terms of its transfer functions. In the above equation $f(\omega)$ describes the response of the detector to an incident gravitational wave and is specific of the scheme.

For example, using the same configuration but different parameters MIGO emulates the sensitivity of both LIGO and LISA with average flux, interferometer size and De Broglie wavelength satisfying

$$N_{atom}\left(\frac{L_{atom}}{\lambda_{atom}}\right)^2 = N_{photon}\left(\frac{L_{photon}}{\lambda_{photon}}\right)^2 \Rightarrow \text{design requirements}$$

based on the assumption that all such real space systems are shot-noise limited. Note that shot-noise is white noise, that means the perturbation it generates is a frequency independent term. Were it not, its transfer function would enter the above form for the equivalent smallest detectable gravitational wave. More specifically it would multiply the numerator of the equation for the noise limited sensitivity, that would thus become proportional to the ratio of the two transfer functions. This occurs for the instrumental disturbances illustrated in Sec.4 and Sec.5 following Sec.3 on $f(\omega)$.

3. Mathematical formalism

We refer to Fig.2(a) for notation and illustration purposes. The quantum superposition is manipulated through

$$\begin{pmatrix} c_2(t+\tau) \\ c_1(t+\tau) \end{pmatrix} = \begin{pmatrix} \cos\frac{\Omega_r\tau}{2} & -i\mathcal{U}^{-1}_{\bar{p}+\hbar k/2}(t)\mathcal{U}_{\bar{p}-\hbar k/2}(t)\sin\frac{\Omega_r\tau}{2} \\ -i\mathcal{U}^{-1}_{\bar{p}-\hbar k/2}(t)\mathcal{U}_{\bar{p}+\hbar k/2}(t)\sin\frac{\Omega_r\tau}{2} & \cos\frac{\Omega_r\tau}{2} \end{pmatrix} \begin{pmatrix} c_2(t) \\ c_1(t) \end{pmatrix}$$

where $\Omega_r \equiv 2\sqrt{\Omega_2^{AC}\Omega_1^{AC}}$. We assume the basis

$$\{|\psi_2, \bar{p}+\hbar k/2>, |\psi_1, \bar{p}-\hbar k/2>\}$$

with dressed atomic frequencies

$$\bar{\omega}_1 = \Omega_1^{AC} + \omega_1 \qquad \Omega_1^{AC} \equiv \frac{|<\psi_3|\vec{d}\cdot\vec{E}_+|\psi_1>|^2}{4\Delta\hbar^2}$$

$$\bar{\omega}_2 = \Omega_2^{AC} + \omega_2 \qquad \Omega_2^{AC} \equiv \frac{|<\psi_3|\vec{d}\cdot\vec{E}_-|\psi_2>|^2}{4\Delta\hbar^2}$$

where $\vec{d}$ is the atomic dipole and Ω_i^{AC} the AC Stark shifts. A resonant condition applies to

$$\omega_+ - \omega_- - \vec{p}\cdot\frac{\vec{k}}{m} = \bar{\omega}_2 - \bar{\omega}_1 \qquad \vec{k} = \vec{k}_+ - \vec{k}_- \tag{1}$$

relating the atomic transition and the electromagnetic field relative frequency. The atom moves in

$$\vec{E}_+\cos(\vec{k}_+\cdot\vec{r} - \omega_+ t - \varphi_+(t)) + \vec{E}_-\cos(\vec{k}_-\cdot\vec{r} - \omega_- t - \varphi_-(t)) \tag{2}$$

during impulsive interactions. Over a larger time scale $t \gg \tau$ there is free evolution $<p|\Psi> = e^{\frac{iF(p)}{\hbar}}\Psi$ as in Fig.3. The Bloch vector rotates in the horizontal plane, by an angle that is the accrued relative phase of the two components, as determined by their respective momentum dynamics

$$i\hbar\frac{\mathrm{d}|\Psi>}{\mathrm{d}t} = \hat{H}|\Psi> \Rightarrow \frac{p^2}{2m} + V(-\frac{\partial F}{\partial p}) + \frac{\partial F}{\partial t} = 0$$

that is by the Hamilton-Jacobi equation applied to $\mathcal{U}_p(t)$. Any momentum progression results in an accrued relative phase, which in turn is manifested by the non ideal overlap of the final and initial Bloch vector in Fig.3.

Changes of the parameters, operating point or provisions that identify the ideal realization, modify the fringe pattern of reference of these interferometers and generate an equivalent shift. When this perturbation of the phase is not distinguishable from the effect of a gravitational wave, it limits the smallest signal the instrument can distinguish and the competition between the impact of instrumental noises and gravitational waves on the observable shift determines the noise budget of the interferometer.

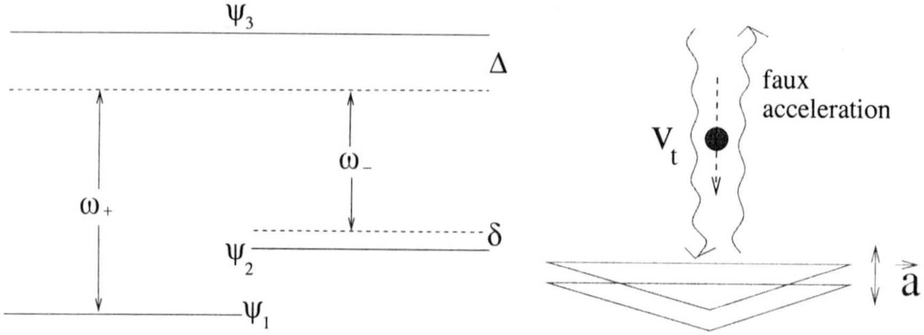

(a) The intermediate state is never populated in the two-photon transitions connecting the two hyperfine ground states ψ_1 and ψ_2. The population of these states is so manipulated. The Raman detuning of the laser fields can be adjusted to enable a complete population inversion. This is possible only if the resonant condition between the differential frequency of the two fields and the total relative energy of ψ_2 and ψ_1 is satisfied. The latter includes the momentum dependent contribution as well as the AC Stark shift of each state

(b) The vibrational motion of the retroreflecting mirror adds a fluctuating phase to the observable phase $\delta\Phi$. Its contribution depends on the type of configuration. The impact on the phase of a single-loop interferometer is the same as the effect of a momentum change and is perceived as a faux acceleration

Fig. 2. As in the text v_t is the velocity of the atomic beam within the apparatus. We use t and v_t to fi a reference in (2). All perturbations have an impact on the phase accrual, if they modify the unperturbe operating point that the resonant condition identifies.

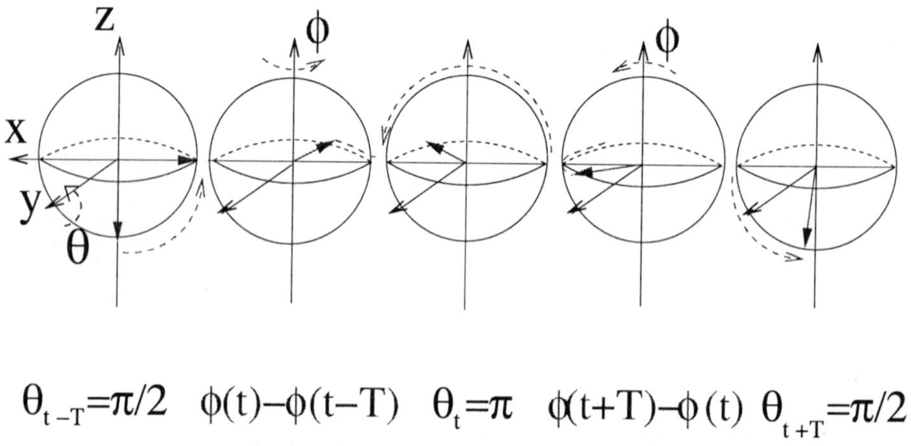

Fig. 3. We use the Bloch sphere to represent the action of three pulses and the interspersed relative free evolution of the mixed states: the former by rotating the Bloch vector around $\hat{y}$ and the latter around $\hat{z}$. The coordinates $\{x, y, z\}$ identify $\{c_1, c_2\}$. For example $c_2 = 0$ is the ground state $z = -1$ and $c_1 = 0$ the excited state $z = 1$. The perfectly split combination is represented by $z = 0$. In the plane $z = 0$ the free evolution phase has its maximum effect

Somehow these imperfections introduce noise in this class of interferometers, in a similar way as unwanted variations of the configuration would do in the other (based on local elements).

Matter-wave interferometers with rigid arms (relying on the wave-particle nature of atomic beams and their De Broglie wave vector propagation) and the two states represented in Fig.3 are different systems with different properties. We can describe their operating principles as follows: the unperturbed reference of MIGO is provided by the extension of its arms, related to the De Broglie wave vector. The unperturbed reference of the class of interferometers presented here is instead provided by the atomic transition frequency, related to the frequency difference of the two counterpropagating Raman fields. Changes in either term of the resonant condition (1) generate a perturbation of the phase shift.

Last thing we notice is that the mutual nature of a change in the atomic energy gap or laser frequency in (1) is what makes the phase gauge invariant. This is thoroughly analysed in[5] where a general discussion is provided. The basic concept can be summarized as follows: the phase is due to the relative evolution of the two states, with respect to the field differential frequency in (2) which represents the reference for the observer in the laboratory. Conversely for an observer comoving with the atom, the transition frequency is a reference for the evolution of the laser frequency, that is perceived as Doppler shifting out of resonance.

4. Vibrational noise

As already mentioned, the unperturbed reference for the class of interferometers in momentum space is (1) which locks $\vec{p}$ to other quantities. Any change of $\vec{p}$ therefore spoils the resonant condition and alters the phase.

If the relative phase of the Raman laser fields and the relative evolving phase of the two atomic states match each other, the action of the three pulses is the rotation of the Bloch vector by $\theta = \theta_{t+T} + \theta_t + \theta_{t-T}$. This would correspond to Fig.3 were the accruing phase ϕ is unvaried. The system is therefore back in its initial state, at the end of the single-loop $\{\pi/2, \pi, \pi/2\}$. Any deviation from the ideal unperturbed match of the two relative phases results in an accruing residual, that makes the final state different. Perturbations are due to either changes of $\vec{k}$ or $\vec{p}$ that is either variations of the Raman detuning or in the relative evolution of the two mixed states. Once they are launched, atoms are free falling and interact with nothing but the electromagnetic field (2). Hence changes in $\vec{k} = \vec{k}_+ - \vec{k}_-$ modify the relation between the relative evolving phase of the two mixed states and $\omega_+ - \omega_-$ in the resonant condition (1). For the reasons explained in Sec.3 the system is sensitive to non uniform variations. Whether they regard $\vec{k}$ or $\vec{p}$ the interferometer generates an output. In Fig.2(b) the mechanism is presented as a time-varying Doppler effect. This type of alteration has the same effect as a piston-like vibrational motion. The result is an equivalent faux acceleration $\vec{a}$ not necessarily constant. Expressed in its Fourier constituents in fact, this generates the following term

$$\delta\Phi_a(t) = -T^2 \int_{-\infty}^{+\infty} \frac{\sin^2 \frac{\omega T}{2}}{(\frac{\omega T}{2})^2} \vec{k} \cdot \tilde{a}(\omega) e^{-i\omega t} \frac{d\omega}{2\pi} \underset{\omega\to 0}{\longrightarrow} -T^2 \vec{k} \cdot \vec{a} \qquad \delta\Phi_a^{DD} \underset{\omega\to 0}{\longrightarrow} 0$$

and were the whole system mounted on a free falling elevator $\vec{a} = \vec{g}$. In this case $\delta\Phi_a$ would cancel $T^2\vec{k} \cdot \vec{g}$ due to $\vec{p}(t)$. This is a direct consequence of the principle of equivalence. Note that there is no formal correspondance between this and the vibrational noise of MIGO or LIGO.

A different mechanism for phase perturbations is discussed next: the wave vector $\vec{k}$ is randomly rotated. This generates a deviation from (1) that results in additional noise.

5. Rotational noise

There are two forms of rotational noise we analyse in this section; both are due to direction changes of $\vec{k}$ around $\vec{\Omega}$ at the rate Ω fixes. One is formally similar to the Coriolis acceleration, as can be seen in the term

$$\delta\Phi_\Omega(t) = 4T^2\vec{k} \cdot \int_{-\infty}^{+\infty} \frac{\sin \frac{\omega T}{2}}{\omega T} \cos \frac{\omega T}{2} \vec{v}_t \times \tilde{\Omega}(\omega) e^{-i\omega t} \frac{d\omega}{2\pi} \underset{\omega\to 0}{\longrightarrow} T^2 \vec{k} \cdot (2\vec{v}_t \times \vec{\Omega})$$

while the other depends on $\dot{\Omega}$ that is the time derivative of Ω. This induces the shift

$$\delta\Phi_{\dot{\Omega}}(t) = -4iT^2\vec{k} \cdot \int_{-\infty}^{+\infty} \frac{\sin^2 \frac{\omega T}{2}}{\omega T} \frac{\vec{r}_t}{T} \times \tilde{\Omega}(\omega) e^{-i\omega t} \frac{d\omega}{2\pi} \underset{\omega\to 0}{\longrightarrow} 0$$

in the superposition of the momentum states that generates the output signal, at the end of the rotations in Fig.3 identifying the single-loop scheme.

The observed interference phase is thus a consequence of the resonant condition being spoiled. The output signal of LIGO and similar laser-light interferometers is generated by deviations from the resonant condition that relates the optical length of the Fabry-Perot cavities to the laser frequency. Because of the nature of the Michelson configuration, the system is sensitive to any perturbation that distinctively alters the operating point of the two arms, as gravitational waves are supposed to do.[2,7] They compete with other excitations and the result depends on both the relative strength and on the respective transfer functions, that describe the response of the interferometer to them.

Therefore in the expression

$$\sqrt{<\tilde{h}^2(\omega)>_{\text{PSD}}} = \frac{1}{2\vec{k} \cdot \vec{p}/mTf(\omega)} \sqrt{\sum S_{noises}} = \frac{1}{2\vec{k} \cdot \vec{p}/mTf(\omega)} \sqrt{\sum g^2(\omega)\tilde{n}^2(\omega)}$$

the impact of the various noise sources depends on their own transfer functions. As shown in Tab.1 these are different for the two schemes we analyse. The computed transfer functions form the basis of our detector characterization models and enable an extensive comparison of the two considered interferometers, under the same circumstances. The resulting noise budget of the two realizations is shown in Sec.6. We emphasize our interest in the sensitivity enhancement due to the response of the interferometer, rather than in the common gain that can come from mitigation and suppression techniques, acting on the environmental and instrumental sources of noise. These modify the amplitude of the excitations and the same improvement would apply to all curves of Sec.6 as a result. For any configuration the noise budget scales with the strength of the dominating disturbances and so does the sensitivity curve.

We use the notations $\vec{v}_t =< \vec{v}(t) >$ and $\vec{r}_t =< \vec{r}(t) >$. We again emphasize the sensitivity of the single-loop interferometer to any acceleration that affects either the momentum states or the wave vectors of the laser fields. As it measures deviations from uniform acceleration, the double-diamond detector is sensitive to any variability in the force field, either due to metric changes or generated by environmental and instrumental disturbances. In Fig.3 this detector would be $\{\pi/2, \pi, \pi, \pi/2\}$. The response of the double-diamond configuration to a number of noise sources is compared with that of the single-loop one in Tab.1 for $\omega \to 0$. We note the extra factor $\sim \omega$ originated by the differentiator nature of the former.

The gravitational wave transfer function has an extra factor $1/\omega$ instead. It can be interpreted as the non zero derivative of a function that is zero. The low frequency sensitivity of the double-diamond interferometer is thus enhanced.

Table 1. The low frequency phase, from a number of alternating regression and progression stages that combine in the observable output of two different configurations of momentum space interferometer (specified as sequences of atom manipulation steps and stretches of interspersed free evolution intervals, marked by the inversions of population) is reported.

$\omega \to 0$	**single-loop configuration**	**double-diamond configuration**
$\vec{g}$ coupling	$\vec{k} \cdot \vec{g}T^2(1-(\omega T)^2/24)\tilde{h}(\omega))$	$\vec{k} \cdot \vec{g}T^2(3i(\omega T)^3/4)\tilde{h}(\omega)$
vibrational noise	$-\vec{k} \cdot \tilde{a}(\omega)T^2$	$\vec{k} \cdot \tilde{a}(\omega)T^2(2i\omega T)$
rotational noise	$\vec{k} \cdot (\vec{r}_t \times \tilde{\Omega}(\omega))(-i\omega T^2)$	$\vec{k} \cdot (\vec{r}_t \times \tilde{\Omega}(\omega))(2\omega^3 T^3)$
Coriolis noise	$\vec{k} \cdot (\vec{v}_t \times \tilde{\Omega}(\omega))(2T^2)$	$\vec{k} \cdot (\vec{v}_t \times \tilde{\Omega}(\omega)(6i\omega T^3))$
gravitational wave	$\vec{k} \cdot \vec{v}_t(i\omega^3 T^4/12)\tilde{h}(\omega)$	$\vec{k} \cdot \vec{v}_t(-\omega^2 T^3)\tilde{h}(\omega)$

6. Conclusions

The results reported in this section are based on our detector characterization models for atom interferometers in momentum space, consisting of a superposition of two quantum states and accumulating the relative evolving phase between them. We have considered two different realizations of atom interferometer: both their responses to noise excitations and to gravitational waves have been derived and their sensitivity curves constructed. The instrumental limitations of such systems are due to the phase shift induced by alterations of the provisions that define the unperturbed resonant condition (1) and identify the ideal set of atom manipulation steps. Each set is a different interferometer realization.

Any modification of the operating point results in a perturbation of the output signal, in a similar way as displacements or distortions of the optical surfaces do in real space interferometers, where phase accrual occurs between the ends of each arm of the detector. The actual sources are completely different and so are the transfer functions, which correspond to totally distinguished mechanisms and behaviours. The most relevant ones are studied and discussed in Sec.4 and Sec.5 with reference to the provision (1). This defines the unperturbed interferometer in momentum space. From the computed noise budget Fig.4(a) and Fig.4(b) are constructed for two different schemes. In both sets of sensitivity curves $1/T$ denotes the natural unit for frequency.

The plots of Sec.6 are the result of our instrumental noise analysis.

Although the present generation is not suitable for the search of gravitational waves, basically because direct detection relies on the availability of geodesic trajectories that are prevented by a variety of non-gravitational forces (at the needed level of sensitivity) we are interested in the characterization of actual systems, available as prototypes.

Many optimizations can be done and we expect that the technological progress of this growing field, which has resulted in the determination of the gravitational

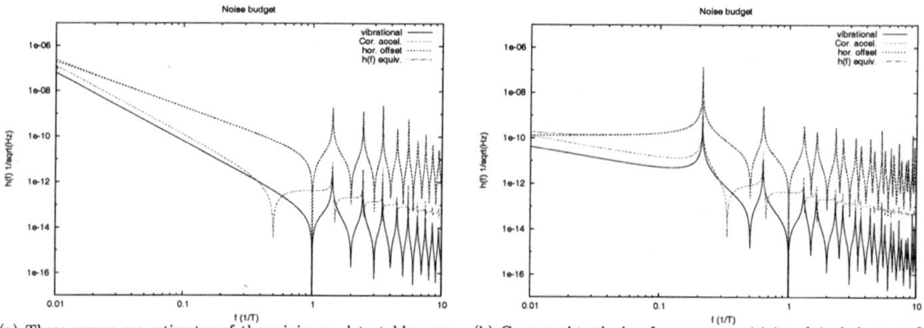

(a) These curves are estimates of the minimum detectable gravitational wave amplitude, identified in each frequency bin as the astrophysical excitation that would induce the same phase shift as the noise source having the largest impact

(b) Compared to the low frequency sensitivity of single-loop configurations and under identical circumstances, the response of double-diamond configurations provides a significant improvement

Fig. 4. For $T = 200ms$ the average traveled distance would be $2v_tT = 6m$ and $4v_tT = 12m$. Except for this quantity, all other parameters are identical and the sensitivity curves of the two interferometers reflect a fundamental difference in behaviour.

acceleration with unprecedented sensitivity, will keep on advancing and enabling further general improvements by means of noise reduction and control techniques. Feasibility investigations like these have already proved to be instrumental in the development of high-sensitivity gravitational wave detectors. The design and deployment of LIGO has been preceded by the operation of a smaller scale system, used to test new concepts and to gain a wealth of knowledge, the construction of the long baseline version has logically built on. The Caltech 40 Meter Prototype is currently serving as an experimental testbed for the subsystems that will be employed in Advanced LIGO.[8]

Moreover, since atom interferometry is a developing field and published data are showing a constantly improving sensitivity in all its applications, we think the margins for further progress are potentially broad beyond imagination. Meanwhile a complete understanding of the couplings and interactions that distinguish the response of atom interferometers, relying on a quantum superposition of momentum states, is in order. The behaviour of these systems needs to be timely well modelled, especially with a view to the deployment of such sensitive detectors in microgravity enviroment.

Acknowledgments

The work of E.D. is supported by the European Union as part of the FrameProgram FP6 through the Marie Curie Action EIF for the Panel PHY N° 518675. Words are not enough to acknowledge the inspiring and visionary mentorship of Lute Maleki * and Flavio Vetrano †.

References

1. M.Rakhmanov *Phys. Rev. D* 71, 084003, 2005
2. P.Saulson *Fundamentals of Interferometric Gravitational Wave Detectors* World Scientific Publishing Company, Incorporated, 1994
3. A.Roura,D.R.Brill,B.L.Hu,C.W.Misner,W.D.Phillips *Phys. Rev. D* 73, 084018, 2006
4. J.R.Kellogg,N.Yu,J.M.Kohel,R.Thompson, D.C.Aveline,E.D'Ambrosio,D.Strekalov and L.Maleki *Development of a quantum gravity gradiomenter for gravity measurement from space* Division of Atomic, Molecular and Optical Physics conference (DAMOP) June 5-9 2007, Calgary, Alberta, Canada
5. G.M.Tino, F.Vetrano *Classical and Quantum Gravity* 24, 2167-2178, 2007
6. E.D'Ambrosio, N.Yu, L.Maleki *Phys. Rev. D* 76, 122001, 2007
7. A.Giazotto *Physics Reports* 182, 365-424, 1989
8. Updated reports submitted on http://dziban.ligo.caltech.edu:40/40m/

*For the pioneering work of Lute Maleki please visit the website of the Quantum Science and Technology Group http://horology.jpl.nasa.gov/research.html

†Scientist-in-charge of the E.U. financed project on gravitational wave atom interferometry

CHARACTERIZATION OF A COLD ATOM GYROSCOPE

A. LANDRAGIN*, A. GAUGUET, T. LÉVÈQUE AND W. CHAIBI
LNE-SYRTE, UMR 8630 CNRS, Observatoire de Paris, UPMC
61 avenue de l'Observatoire, 75014 Paris, FRANCE
**E-mail: arnaud.landragin@obspm.fr*
http://syrte.obspm.fr/

We present the evaluation of the performances of our cold atom gyroscope. The gyroscope is based on two cold Cesium atom sources, which are manipulated thanks to Raman transitions. We show that the short term sensitivity is limited by the quantum projection noise and the long term sensitivity by wave front distortion of the Raman lasers. A study of the bias and scaling factor completes the characterization of the sensor.

Keywords: Atom interferometry; Raman transition; gyroscope.

1. Introduction

Inertial sensors are useful devices in both science and industry. Higher precision sensors could find scientific applications in the areas of general relativity[1] , geodesy and geology. There are also important applications of such devices in the field of navigation. Matter-wave interferometry is envisaged to be an extremely sensitive probe for inertial forces[2] . In 1991, atom interferometry techniques have been used in proof-of-principle work to measure rotations[3] and accelerations[4] . In the following years, many theoretical and experimental works have been performed to investigate this new kind of inertial sensors[5] . In 1997, a new period showed the first demonstrations of very highly sensitive gyroscope[6] and accelerometer[7] and then the achievement of sensors at the level of the state of the art[8,9] . From this time, most of the new atom interferometry experiments are based on the same principle: cold atoms manipulated by two photon transitions, and especially two photon Raman transitions using internal state labeling[10] . This new generation of experiments is dedicated to test for possible applications, in inertial navigation, geophysics, or fundamental physics (gradiometer for the measurement of G[11] , gravimeter for the watt balance experiment[12] , interferometer for the measurement of the fine structure constant thanks to h/m[13]) . Our gyroscope is the first of this new generation of sensors[14] .

We will give the results of the first full characterization of a cold atom gyroscope. As most of the expected applications need improved long term stability compared to standard technologies, the design has been chosen to be a compromise between short term sensitivity and long term performances. It enables to measure the full basis of inertia (three components of acceleration and rotation) by changing the direction of propagation of the Raman laser beams and the time sequence[14] (three or four pulses configuration).

2. Experimental setup

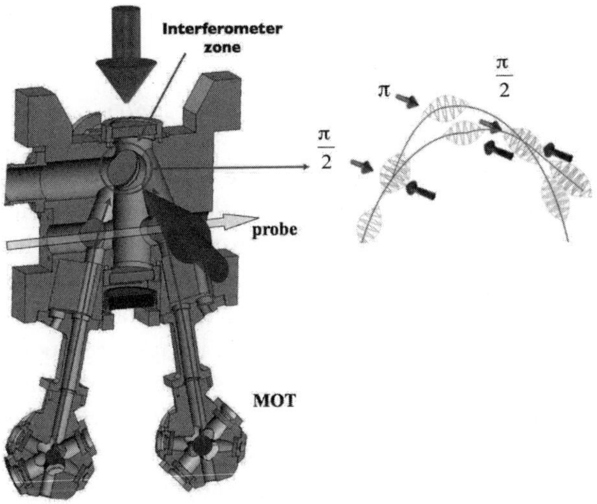

Fig. 1. Scheme of the gyroscope setup. The two atomic clouds of cold Cesium atoms are produced in two magneto-optical traps (MOT) and launched in two parabolic trajectories. The interferometers are realized at the apogees with a single pair of Raman laser beams oriented in the horizontal or vertical direction. The scheme on the right represents the trajectories of the atomic wave packets in the interferometer when horizontal Raman laser beams are used. The total interaction time between the first and third pulse is at maximum 80ms. The total cycling time is 580 ms.

In our experiment, cesium atoms are trapped from a vapour in magneto-optical traps during 125 ms, and cooled down to 1.2 μK. The Cesium clouds are launched along parabolic trajectories using moving molasses at 2.4 m.s^{-1}, with an angle of 8 degrees with respect to the vertical direction (see Fig. 1). Then the atoms are prepared in the state $|6S_{1/2}, F = 3, m_F = 0\rangle$ before entering in the interferometer zone at the top of their trajectory, where they interact with the Raman lasers. The interrogation sequence is achieved with a single pair of Raman beams covering the whole interrogation zone, which provides an easy way to change the pulse sequence. The beams are switched on during 12 μs to realize each Raman pulse. In the following, measurements have been performed with vertical Raman laser beams in a classical sequence of three pulses $(\pi/2 - \pi - \pi/2)$[4] . At the output of the interferometer, the transition probability P from one hyperfine state to the other is given by the well-known relation for a two wave interferometer: $P = \frac{1}{2}(1 + C\cos\Delta\Phi)$, where C is the interferometer contrast, and $\Delta\Phi$ the difference of the atomic phases accumulated along the two paths. This phase shift depends on the acceleration $\vec{a}$ and the rotation rate $\vec{\Omega}$ experienced by the atoms. It can be written as

$$\Delta\Phi = \phi(0) - 2\phi(T) + \phi(2T) = \vec{k}_{eff} \cdot (\vec{a} - 2(\Omega \vec{\times} \vec{v}))T^2 \qquad (1)$$

where $\phi(0, T, 2T)$ is the laser phase difference at the location of the center of the atomic wavepackets, for each of the three pulses[15] . Here $\vec{k}_{eff}$ is the effective wave vector (difference between the two Raman lasers), T is the time interval between two consecutive pulses and $\vec{v}$ the mean atom velocity in the laboratory frame. With vertical Raman laser beams, the interferometers are sensitive to one horizontal rotation component and to the vertical acceleration. To discriminate between acceleration and rotation signals, we use the dependence of the rotation signal with the direction of propagation of the atoms. By using two atomic clouds following the same trajectory but in opposite directions, the acceleration (respectively the rotation) phase shift is deduced from the sum (respectively the difference)of the two interferometer phase shifts.

The Raman lasers are two extended cavity diode lasers based on the design of Ref. 16, which are amplified by a tapered amplifier. Their frequency difference is phase locked onto a low phase noise microwave reference source. The two overlapped beams are injected in a polarizing fiber, and guided towards the vacuum chamber. We achieve counter-propagating beams by placing a mirror and a quarterwave plate at the bottom of the experiment. Four beams are actually sent onto the atoms. However, only two will drive one counter-propagating Raman transition, while the others are Doppler shifted thanks to the atoms free fall[19] .

3. Sensitivity

Short term sensitivity of atomic interferometers may be limited by various sources of noise: inertial noise (aliasing effect on rotations or vibrations), detection noise, Raman laser phase noise, light shift, etc... As the two atomic clouds experiment the same Raman laser beams, most of these sources of noise are cancelled on the rotation signal and appears only on the acceleration.

3.1. *Rotation: quantum projection noise limit*

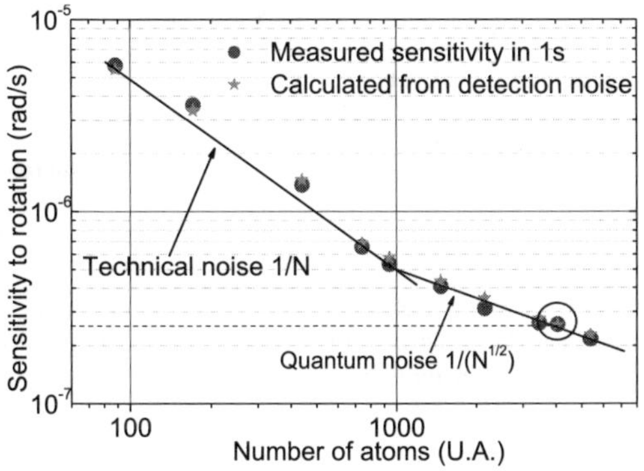

Fig. 2. Sensitivity to rotation rate as a function of the number of atoms. The dots represent the experimental measurements and the stars the calculated sensitivity taking into account the technical noise of the detection and the quantum projection noise, due to the finite number of atoms and the fringe contrasts of 30%. The circle shows the usual parameters of the experiment corresponding to 5×10^5 atoms per shot.

Since the experiment is placed on a passive isolation plate-form, the limiting source of noise on the rotation signal comes from the detection noise. Fig. 2 shows this limit as a function of the detected atom number. This limit comes from the photodiode for low number of atoms and from the quantum projection noise for high number of atoms. Quantum projection noise limit constitutes the intrinsic limit of sensitivity and scales as $C/\sqrt{N}$, where N is the number of detected atoms per shot. In the usual conditions, the sensitivity is 2.4×10^{-7} rad.s^{-1}.Hz$^{-1/2}$, limited by the quantum projection noise.

3.2. *Acceleration: vibration limits*

The main limit to acceleration sensitivity is due to residual vibrations of the plate-form as for atomic gravimeter[17,18] . Our sensitivity is 5.5×10^{-7} m.s^{-2}.Hz$^{-1/2}$, slightly smaller because of the reduced interrogation time and/or smaller repetition rate.

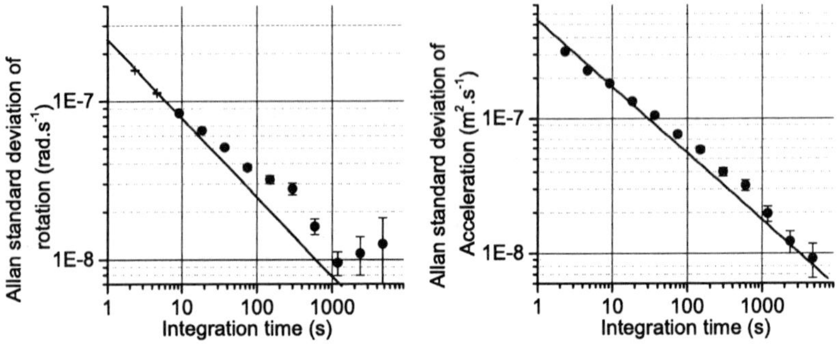

Fig. 3. Allan standard deviation of rotation and acceleration (resp. on left and right). The stabilities improve with the integration time up to 1000s. They have been calculated from a 36 hours continuous measurements

3.3. *Long term sensitivity: laser wave front distortions*

Acquisition of few days have been performed in order to evaluated the stability of the sensor. Fig. 3 shows the Allan standard deviations for the rotation and the acceleration. They show an improvement of the sensitivity almost as $1/\sqrt{\tau}$ as expected, where τ is the integration time. The sensitivity to rotation reaches a minimum for 1000 s integration time at about 10^{-8} rad.s^{-1}. This result is reproducible and linked to day to night temperature fluctuations.

We tested many possible sources of drifts to explained it. First, we implemented the k-reversal technic, which allows to remove all effects linked to the difference of internal states between the two arms of the interferometer, as magnetic field gradient and one photon light shift. Thanks to a two axes tiltmeter and two position sensors, we also controlled precisely the orientation of the experiment in the three dimensions. Then we have characterized the two photon light shift due to the off-resonance Raman transition[19] . The only remaining effect, which can explain the bias, is related to the wave front distortions of the Raman laser beam[20] . Since the atoms moved in the profile of the Raman lasers, they experiment a change in the difference of phase of the Raman lasers due to the distortions. If the trajectories of the two counter-propagating clouds are overlapped, they experiment a similar shift and the bias appears on the acceleration signal. If they do not match perfectly, a residual bias appears on the rotation signal. Therefore, if the atomic sources are fluctuating in initial position or velocity, this bias changes over time.

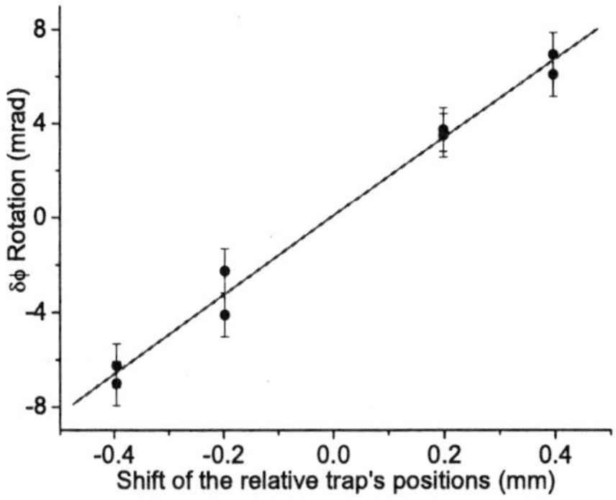

Fig. 4. Rotation like signal induced by the relative displacement between the two atomic clouds in the interferometer. The results have been obtained by differential measurements compared to the usual case at zero displacement. The line shows the linear fit of the data.

To quantify this effect, we performed two kind of measurements. First, we study the position stability of the two clouds in the interferometer. For this purpose, we have both measured: the stability of position in the detection probe to be below 10 μm at 1000 s using the time of flight method, and the stability of launching velocity using our Raman lasers to be typically 40 μm.s^{-1} at 1000 s. In a second time, we have estimated the dependence of the rotation shift with the relative displacement between the two clouds, in the direction of the mean velocity, by changing the timing sequence. Fig. 4 shows a clear linear dependence with the relative displacement, which has been fitted by a straight line of slope 17 mrad.mm^{-1}. From these measurements, we estimate the stability of the rotation signal to be limited in the range of 10^{-8} rad.s^{-1}, has observed in Fig. 3. As the fluctuations are not correlated between the two atomic sources, similar limitations appear on the acceleration stability, but is hidden on this result by the higher noise on the acceleration signal.

4. Bias and scaling factor

4.1. *Variation with the rotation rate*

One of the important features of gyroscope is the knowledge of the bias and the scaling factor, which need to be accurately measured in order to infer the rotation rate from the phase shift. Since we measure the projection of the Earth rotation rate on one horizontal component, we can turn the experiment around the vertical axis and benefit from its modulation to determine accurately these two parameters. Fig. 5 shows the sinusoidal modulation of the rotation phase shift compared to the West orientation. The fit of the data by a sinusoidal function gives a bias of 28.7 ± 0.7 mrad, corresponding to a rotation rate of $1.9{\times}10^{-6}$ rad.s^{-1}, and a scaling factor of 15124 ± 10 rad/(rad.s^{-1}). Another important criterion is the linearity of the scaling factor. The residue between data and the sinusoidal fit is displayed on Fig. 5 and does not exhibit any specific features. A fit including a quadratic term demonstrates than the contribution of a quadratic term is below 10^{-5} in the range of measurement.

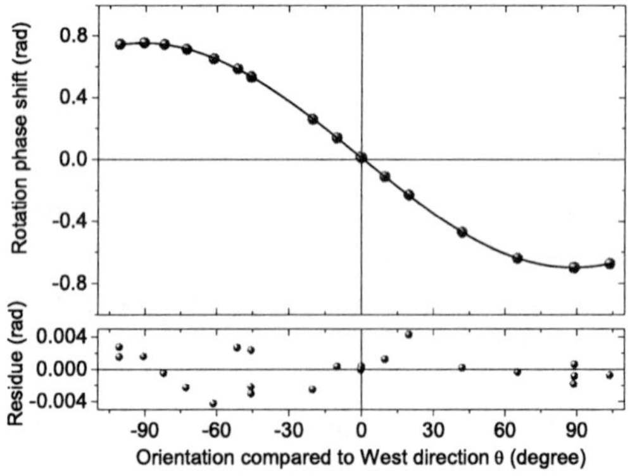

Fig. 5. On the top, variation of the rotation as a function of the orientation of the sensitive axis of the gyroscope compared to the West direction and in the horizontal plane. The points correspond to the data and the curve to a sinusoidal fit. The difference between the two is plotted on the bottom graph.

4.2. *Dependence on the interrogation time*

Thanks to the very good linearity, we can use the measurements at north and south directions only in order to determine the bias and the scaling factor. We then repeat measurement with different total interrogation time 2T from 20 to 80 ms (see Fig. 6). The behavior of the scaling factor is in very good agreement with the expected one as T^2. The bias clearly rises with the interaction time, which is in accordance with the increase of optical wavefront distortion at the edge of the optics (window, wave-plate and mirror) when using maximum interaction time.[20]

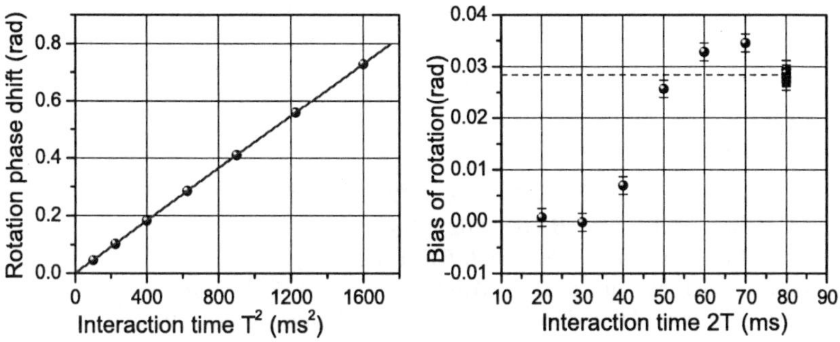

Fig. 6. Scaling factor (on right) and bias (on left) as a function of the interaction time. On left the line shows the evolution as T^2.

5. Conculsion

We have performed the first characterization of a cold atom gyroscope performances. The sensitivity to the rotation is limited at short term by the quantum projection noise and shows performances similar to the best fiber gyroscope. Long term performances are limited by the fluctuations of the atomic trajectories and the wave front distortions of the Raman lasers. The measurements of the bias and of the scaling factor were shown to be well controlled and stable, which is a key feature for applications. The limits at long term is not fundamental and can be overcome by at least one order of magnitude by different means. We can first reduce the spurious shifts by improving the quality of the optics and the stability of the launching velocities. Moreover, we can reduce its impact in term of rotation rate by optimizing the geometry. Indeed, atoms can be launched in a more straight trajectory with a much higher longitudinal velocity as in Ref. 21 or by using the four pulses configuration, previously demonstrated[14] . In both cases, the area of the interferometer increases faster than the shifts from the wave front distortions. Finally, it should be noted that similar limits should appears on the other dual cold atoms interferometers (based on molasses technics), as in gravity gradiometer or in test of the universality of free fall by comparing acceleration of two clouds of different species.

Acknowledgments

We would like to thank the Institut Francilien pour la Recherche sur les Atomes Froids (IFRAF), the European Union (FINAQS contract), the Délégation Générale pour l'Armement DGA and the Centre National d'Études Spatiales CNES for financial supports.

References

1. W. W. Chow *et al*, *Rev. Mod. Phys.***72**, 61 (1985).
2. J.F. Clauser, *Physica B* **151**, 262 (1988).
3. F. Riehle *et al*, *Phys. Rev. Lett.* **67**, 177 (1991).
4. M. Kasevich M. and S. Chu, *Appl. Phys. B* **54**, 321 (1992).
5. P. B. Berman, *Atom Interferometry*, (Academic Press, 1997).
6. T. L. Gustavson, P. Bouyer and M. A. Kasevich, *Phys. Rev. Lett.* **78**, 2046 (1997).
7. A. Peters, K. Y. Chung and S. Chu, *Nature* **400**, 849 (1999).
8. T. L. Gustavson, A. Landragin and M. A. Kasevich, *Class. Quantum Grav.* **17**, 1 (2000);
9. A. Peters, K. Y. Chung and S. Chu, *Metrologia* **38**, 25 (2001);
10. Ch. J. Bordé, *Phys. Lett. A* **140**, 140 (1989).
11. A. Bertoldi *et al*, *Euro. Phys. J. D* **40**, 271 (2006); J. B. Fixler *et al*, *Science* **315**, 74 (2007).
12. G. Geneves *et al*, *IEEE Trans. Instrum. Meas.***54**, 850 (2005).
13. A. Wicht *et al*, *Physica Scripta* **T102**, 82 (2002); P. Cladé, E. de Mirandes *et al*, *Phys. Rev. A* **74**, 052109 (2006).
14. B. Canuel *et al*, *Phys. Rev. Lett.* **97**, 010402 (2006).
15. Ch. Antoine and Ch. J. Bordé, *Phys. Lett. A* **306**, 277 (2003).
16. X. Baillard *et al*, *Optics Communications* **266**, 609 (2006).
17. J. Le Gouët *et al*, *Appl. Phys. B* **92**, 133-144 (2008).
18. H. Müller *et al*, *Phys. Rev. Lett.* **100**, 031101 (2008).
19. A. Gauguet *et al*, *Phys. Rev. A* **78**, 043615 (2008).
20. J. Fils *et al*, *Eur. Phys. J. D* **36**, 257 (2005).
21. T. Müller *et al*, arXiv:0806.0956.

A MOBILE ATOM INTERFEROMETER FOR HIGH PRECISION MEASUREMENTS OF LOCAL GRAVITY

M. SCHMIDT, A. SENGER, T. GORKHOVER, S. GREDE, E. V. KOVALCHUK, and A. PETERS

AG Optische Metrologie, Humboldt-Universität zu Berlin
Hausvogteiplatz 5-7, 10117 Berlin, Germany
E-mail: malte.schmidt@physik.hu-berlin.de, achim.peters@physik.hu-berlin.de
http://www.physik.hu-berlin.de/qom

We present a new design for the mobile and robust gravimeter GAIN (Gravimetric Atom Interferometer), which is based on interfering ensembles of laser cooled ^{87}Rb atoms in an atomic fountain configuration. With a targeted accuracy of a few parts in 10^{10} for the measurement of local gravity, g, this instrument would offer about an order of magnitude improvement in performance over the best currently available absolute gravimeters. Together with the capability to perform measurements directly at sites of geophysical interest, this will open up the possibility for a number of interesting applications. We report on important subsystems of this atom interferometer, including a rack-mounted laser system and a compact vacuum chamber. Furthermore, a high flux 2-dimensional Magneto-optical trap capable of providing up to 10^{12} atoms/second and a high-power laser system providing 6.4 W at 780 nm are presented.

Keywords: atom interferometry; compact laser system; inertial sensor; 2D MOT.

1. Introduction

Since first experimental demonstrations in 1991,[1–4] atom interferometry has developed into a powerful tool for the ultra precise measurement of accelerations and rotations. It is now used in various laboratories for experiments in the fields of fundamental physics[5] and metrology.[6–9] Most of these setups, however, are laboratory-based and cannot be used for on-site research such as gravity mapping, geophysics, seismology or navigation, even though the accuracy of atom interferometers could substantially exceed that of classical gravimeters.[10] Our transportable high-precision gravimeter will therefore open up new possibilities in all of these fields and might in the future also be used on satellite missions.[11,12]

The working principle of a gravimetric atom interferometer has been described in length elsewhere.[6] In summary, a cloud of laser-cooled Rubidium-87 atoms is prepared in a 3D Magneto-Optical Trap (MOT) and launched upwards. During their parabolic flight, the atoms are subjected to three Raman pulses from counterpropagating laser beams that transfer them between the two hyperfine ground states. The sequence consists of one $\frac{\pi}{2}$-, one π-, and finally another $\frac{\pi}{2}$-pulse, which constitute an atom optic beam splitter, mirror and recombiner, respectively, so that the atomic wave packet is split into two parts that travel on different trajectories due to momentum transfer from the photons. At the output of the atom interferometer, the transition probability P from one hyperfine state to the other is given by $P = \frac{1}{2}(1 + C\cos\Delta\Phi)$, where $\Delta\Phi$ is the accumulated phase difference between both wave packets. This includes the acceleration contribution of $\Delta\Phi_g = kgT^2$ with T being the time between Raman pulses, k the Raman laser wavenumber and g local acceleration. In the following, we present the overall GAIN design and first characterizations of subcomponents.

2. Laser System

Naturally, the laser system for a mobile atom interferometer has to be mobile and compact as well. Since standard laser systems for cold atom experiments are not only quite complex, but also notoriously sensitive to mechanical and electrical disturbances, almost all optical mounts had to be designed from scratch, since commercially available products rarely offer sufficient mechanical stability and are simply too large for our purposes.

Our laser system consists of four separate modules with a base area of 42 by 42 cm and a height of a few centimeters, so that they can be mounted in a standard electronics rack as shown in Figure 1. For easy access, these modules are mounted on telescopic rails. Light is transferred between these modules by polarization maintaining fibers. Some optical mount designs are inherited from the QUANTUS drop tower project[13] and have proven their stability even under extreme accelerations. For use in these modules we have built compact external cavity diode lasers (ECDL) that include an interference filter as a wavelength selector and have been adapted from a design developed at SYRTE.[14] The output power is up to 50 mW with a linewidth of less than 100 kHz.

The functions of the four laser modules are organized as follows: One module houses a *Reference Laser* that is stabilized close to the rubidium $F = 2 \rightarrow F' = 3$ transition using MTS spectroscopy.[15] In the *Cooling*

Laser Module, laser light at both frequency classes required for the cooling of Rb atoms ($F = 1 \to F'$ and $F = 2 \to F'$) is produced by phase locking two ECDLs to the reference laser. The $F = 2 \to F'$ light is amplified to 2 Watts using tapered amplifiers. Since we need to launch the atoms by detuning the cooling light and we also need state selection and detection light, Acousto-Optic Modulators (AOMs) in the *Distribution Module* are used to frequency shift and switch both lasers between different functions within the interferometer sequence.

A fourth module constitutes the *Raman Laser System* that houses two ECDLs that generate the Raman pulses for splitting and recombination of the atoms. Since in our experiment we cannot discern between $\Delta\Phi_g$ and a phase difference that is due to Raman laser phase noise, the latter will directly translate onto our gravity measurement and could potentially limit our gravimeter sensitivity. Using a digital phase-frequency-detector and direct laser diode modulation, we have, however, been able to achieve a phase lock with a phase noise spectral density of less than 1 μrad/Hz$^{1/2}$ between 100 Hz and 60 kHz, which would, if it were the main contributing factor, limit our single shot sensitivity $\frac{\Delta g}{g}$ to approximately $2 \cdot 10^{-10}$. However, the frequency reference for the phase lock and mechanical vibrations are expected to give higher contributions to our sensitivity limit.

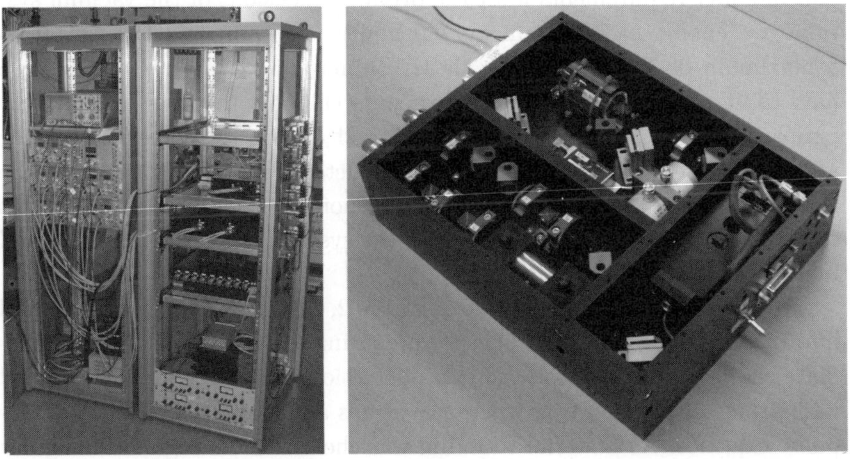

Fig. 1. Rack-mounted laser system (left), Reference laser module (right)

Fig. 2. Vacuum chamber (left), complete mechanical setup in proportion to an adult person (right)

3. Mechanical Setup

Our mobility requirements limit the overall dimensions of the vacuum system to a cuboid of 1900 mm by 700 mm by 700 mm and the weight to a maximum of 300 kg. Within these volume and weight restrictions, we have to fit a robust, versatile UHV vacuum chamber together with a high-performance vibration-isolation system and the optics necessary for the MOT, state selection, Raman-pulses and detection. As in the case of the laser system, these special demands could not be met with standard parts commercially available, so a new compact system had to be designed and built.

As shown on the left-hand side of Figure 2, the vacuum system is divided into three parts: a lower chamber for preparing the cold atoms, an intermediate chamber for state-selection and detection, and the carefully shielded interferometer zone on top, where the atoms are submitted to the Raman-pulse sequence. As material titanium was chosen for its high stiffness, low density and non-magnetic properties.

Our custom design also includes very compact fibre collimators for the MOT-beams, an advanced differential detection scheme based on noise cancellation in a balanced beam pair[16] and an experimental vibration-isolation platform. The vacuum chamber is suspended in an ultra-stiff frame (Fig. 2 right) and can be rotated 360 degrees for Coriolis force compensation.

4. Loading MOT and High Power Laser

For future fast loading (and therefore higher possible repetition rates) of the main chamber's 3D-MOT, a 2-dimensional loading MOT has been built. The atoms are caught in a moving molasses by two large (d = 45 mm) cooling laser beam pairs. The beams are at an angle of 45° relative to the atomic motion (Figure 3) and slightly detuned in respect to each other.[17]

In its most basic configuration at relatively low laser power of 90 mW per beam, an output flux of some 10^9 atoms/second has been achieved. Currently, efforts are underway to triple the loading MOT's active area by four additional laser beam pairs and increase cooling laser power by at least one order of magnitude. Extensive numerical simulations predict an output rate of $> 10^{12}$ atoms/second in that configuration.

To provide the high optical cooling power required by the full three-stage loading MOT, a high power laser system has been developed. Laser light at 1560 nm from a tunable ECDL is amplified to 15 Watts using a polarization maintaining Erbium Fiber Amplifier (IPG Laser EAR-15-1560-LP-SF) and then frequency-doubled to 780 nm by a 38 mm long PPLN crystal in a double-pass configuration. Currently, we have reached 6.4 Watts at 780 nm in this configuration (Figure 3). Using improved crystals, an output power of 10 Watts are theoretically possible.

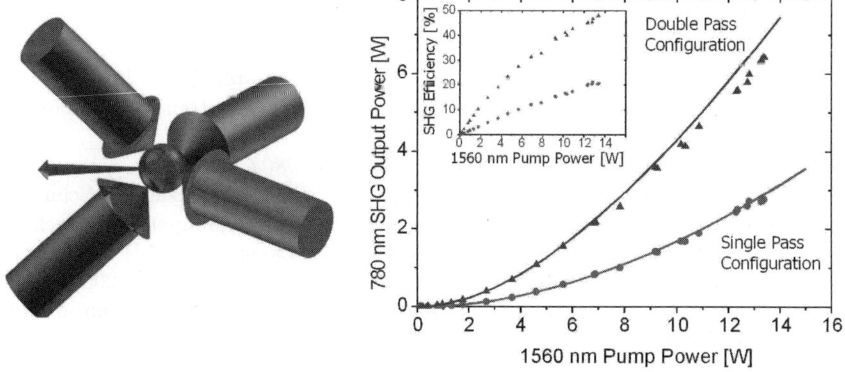

Fig. 3. Loading MOT concept with two beam pairs (left), High power laser system PPLN output (right)

5. Outlook

Our transportable atom interferometer opens up the possibility for a variety of on-site high precision gravity measurements. Collaborations with other fields of research such as vulcanology, seismology and geology have already been initiated.

6. Acknowledgement

The work is supported by the FINAQS (Future Inertial Atomic Quantum Sensors) cooperation of the European Commission, Contr. No. 012986-2 NEST.

References

1. O. Carnal and J. Mlynek, *Phys. Rev. Lett.* **66**, 2689 (1991).
2. F. Riehle, T. Kisters, A. Witte, J. Helmcke and C. Borde, *Phys. Rev. Lett.* **67**, 177 (1991).
3. D. Keith, C. Ekstrom, Q. Turchette and D. Pritchard, *Phys. Rev. Lett.* **66**, 2693 (1991).
4. M. Kasevich and S. Chu, *Appl. Phys. B* **54**, 321 (1992).
5. D. Weiss, B. Young and S. Chu, *Phys. Rev. Lett.* **70**, 2706 (1993).
6. A. Peters, K. Chung and S. Chu, *Metrologia* **38**, 25 (2001).
7. T. Gustavson, A. Landragin and M. Kasevich, *Classical and Quantum Gravity* **17**, 2385 (2000).
8. M. Snadden, J. McGuirk, P. Bouyer, K. Haritos and M. Kasevich, *Phys. Rev. Lett.* **81**, 971 (1998).
9. J. McGuirk, G. Foster, J. Fixler, M. Snadden and M. Kasevich, *Phys. Rev. A* **65** (2002).
10. J. Le Gouet, T. E. Mehlstaeubler, J. Kim, S. Merlet, A. Clairon, A. Landragin and F. P. Dos Santos, *Appl. Phys. B* **92**, 133 (2008).
11. S. G. Turyshev, U. E. Israelsson, M. Shao, N. Yu, A. Kusenko, E. L. Wright, C. W. F. Everitt, M. Kasevich, J. A. Lipa, J. C. Mester, R. D. Reasenberg, R. L. Walsworth, N. Ashby, H. Gould and H. J. Paik, *International Journal of Modern Physics D* **16**, 1879 (2007).
12. A. Landragin and F. Pereira Dos Santos, *arXiv:0808.3837v1* .
13. A. Vogel, M. Schmidt, K. Sengstock, K. Bongs, W. Lewoczko, T. Schuldt, A. Peters, T. Van Zoest, W. Ertmer, E. Rasel, T. Steinmetz, J. Reichel, T. Koenemann, W. Brinkmann, E. Goeklue, C. Laemmerzahl, H. J. Dittus, G. Nandi, W. P. Schleich and R. Walser, *Appl. Phys. B* **84**, 663 (2006).
14. X. Baillard, A. Gauguet, S. Bize, P. Lemonde, P. Laurent, A. Clairon and P. Rosenbusch, *Opt. Comm.* **266**, 609 (2006).
15. J. Supplee, E. Whittaker and W. Lenth, *Appl. Opt.* **33**, 6294 (1994).
16. P. C. D. Hobbs, *Appl. Optics* **36**, 903 (1997).
17. K. Dieckmann, R. Spreeuw, M. Weidemuller and J. Walraven, *Phys. Rev. A* **58**, 3891 (1998).

DEMONSTRATION OF ATOM INTERFEROMETER COMPRISED OF GEOMETRIC BEAM SPLITTERS

HIROMITSU IMAI AND ATSUO MORINAGA

Department of Physics, Faculty of Science and Technology, Tokyo University of Science, 2641 Yamazaki, Noda-shi, Chiba 278-8510, Japan

A time domain Ramsey atom interferometer was constructed using two geometric beam splitters and a geometric phase shifter, which were formed by laser-controlled pulses with a relative phase.

1. Introduction

Light-pulse atom interferometers are attractive tools for precise measurements of fundamental physical constants or present frequency standards such as a Cs fountain. Generally, a time domain Ramsey atom interferometer comprises of the two $\pi/2$ pulses, which are separated by a time interval T, as shown in Fig. 1(a), where the first light-pulse interacts with atoms in the ground state and separates them into the excited state and the ground state. The second pulse with a phase shift interacts with and splits them again. Then two excited states or two ground states interfere and interference fringes are obtained by changing a phase between the two $\pi/2$ pulses [1]. However, we can construct the time domain atom interferometer with geometric operations.

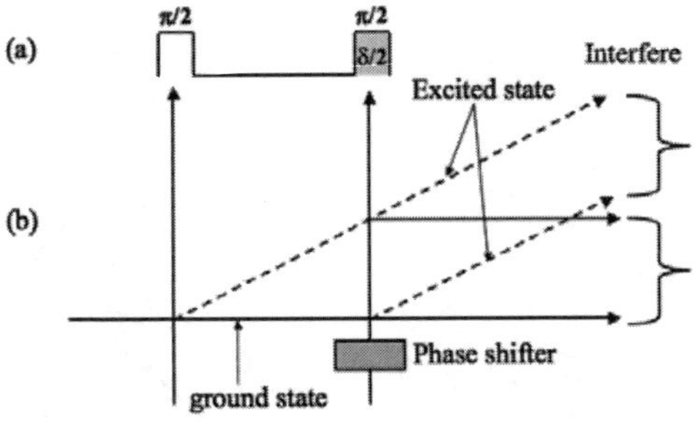

Figure 1. (a) Time sequence and (b) trace of atom for a conventional the typical time domain Ramsey atom interferometer.

In 1984, Berry proposed that quantum system evolving through a cyclic path gains a geometric phase in addition to the dynamic phase. It only depends on the amount of the solid angle enclosed by the evolution path, but does not on the details of the path, the time spent, the driving Hamiltonian, or the initial and final states of the evolution [2]. Therefore, it is said that geometric operation is more robust than dynamic one such as Rabi excitation. In 2004, Tian et al. [3] proposed that the Rabi excitation and phase shift between wave functions, which are corresponding to rotations around axes 2 and 3 on the Bloch sphere, are accomplished using resonant laser-controlled pulses with a relative phase.

Recently, we have developed geometric manipulations of rotations around axes 3[4] and 2[5] with an ensemble of cold atoms in order to perform a geometric universal single qubit operation in quantum computation [6]. We attempted to construct a geometric atom interferometer using these rotations. In this paper, we report the principles and the results of geometric atom interferometer.

2. Principle

According to the proposal presented by Tian et al. [3], the rotations of a two-level atom around axes 3 and 2 (U_3 and U_2) in the Bloch sphere can be accomplished geometrically by plural resonant laser-controlled pulses with a relative phase between them. Figure 2 shows the time sequence of resonant laser-controlled pulses and the trace of the Bloch vector on the Bloch sphere for (a) U_3 and (b) U_2.

For U_3, two resonant π–pulses whose second pulse has a relative phase of $\delta/2$ are applied to the wave function. Then the Bloch vector of a ground state $\mathbf{r}_{-3}$ rotates on the Bloch sphere, tracing an enclosed circuit with a solid angle of $2(\pi-\delta/2)$ and gains a phase shift of $-(\pi-\delta/2)$ because Berry's phase is defined as $-\Omega/2$ for SU(2) [1], where Ω is a solid angle. Similarly, by the same pulses the excited state gains a phase shift of $(\pi-\delta/2)$. These produce a matrix,

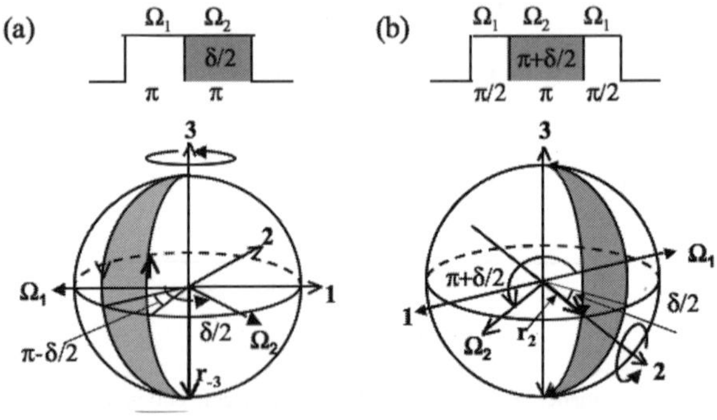

Figure 2. (a) and (b) are resonant controlled pulses and traces of the Bloch vector for manipulations of rotation around axes 3 and 2, where Ω_1 and Ω_2 represent Rabi vectors.

$$U_3(\delta) = e^{-i\pi}\begin{pmatrix} e^{i\delta/2} & 0 \\ 0 & e^{-i\delta/2} \end{pmatrix}, \tag{1}$$

where $e^{-i\pi}$ is called a global phase and can be ignored as it has no effect on the observations. Therefore, this manipulation is equivalent to a rotation around axis 3.

For U_2, three resonant pulses with pulse areas of $\pi/2$, π and $\pi/2$, whose second pulse has a relative phase of $\pi+\delta/2$, are applied to the Bloch vector $\mathbf{r}_2$. Then the vector rotates tracing a closed circuit with a solid angle of δ. Therefore, the wave function gains a phase shift of $-\delta/2$. Similarly, an initial vector of $\mathbf{r}_{-2}$ gains a phase of $\delta/2$. If we convert those basis vectors to $|0\rangle$ and $|1\rangle$, these produce a matrix

$$U_2(\delta) = \begin{pmatrix} \cos\left(\delta/2\right) & \sin\left(\delta/2\right) \\ -\sin\left(\delta/2\right) & \cos\left(\delta/2\right) \end{pmatrix}, \tag{2}$$

which is equivalent to a rotation around axis 2.

With these resonant laser-controlled pulses, we can rotate geometrically the wave function of two-level atoms around axes 3 and 2 on the Bloch sphere.

3. Experiments

3.1. *Setup*

We employed an ensemble of cold sodium atoms. The energy diagram of sodium atom and the experimental setup are written in reference [4-6]. The ground hyperfine levels of $3S_{1/2}$ F=1, m_F=0 and F=2, m_F=0 were used as the two states of atom. The transition between the two states was generated by the two photon stimulated Raman transition via an intermediate state of $3P_{3/2}$, F'=1 or 2 with detuning below about 500 MHz [6,7]. The two photons were produced through an electro-optic modulator (EOM) driven at 1.772 GHz by a synthesizer. The frequency is equivalent to a frequency difference between the F=1 and F=2 states. The synthesizer was used to sweep the relative phase of $\delta/2$ or the frequency difference between two photons. The RF power of each photon was adjusted to cancel out the ac stark shift [8].

The present experimental procedure is briefly summarized as follows; (i) Sodium atoms were trapped by Magneto-optical Trap (MOT) at 1mK. After release from the trap, those atoms were cooled down to 200 μK by polarization gradient cooling (PGC). (ii) After being initialized to the F=1 state by an initializing beam (F=2→F=1), those atoms were manipulated by laser-controlled pulses. (iii) The population probability of the excited state was measured from the transmittances of a probe beam (F=2→F'=3) before and after the geometric manipulations. The resonant laser-controlled pulses were prepared carefully so as to be free from the ac Stark shift, to have a precise pulse area and to have little distortion and decoherence.

3.2. *Results*

Figure 3 (a) and (b) show typical results of geometric rotations around axes 3 and 2, observed with resonant laser-controlled pulses in Fig. 2(a) and (b). The phase shift between the two states caused by the rotation around axis 3 was measured by an atom interferometer adding two $\pi/2$ pulses to both sides of the laser-controlled pulses of U_3. The phase shift was found to be $(2.01\pm0.02)\times\delta/2$ or just double the phase difference of control. On the other hand, the geometric rotation around axis 2 was observed as the variation of the population probability of the excited state. It varies according to a sinusoidal function of δ with an uncertainty of 1%, that is, $(1.98\pm0.02)\times\delta/2$. The rotation around axis 2 is equivalent to the Rabi oscillation with an excitation pulse, but it should be noted that the oscillation due to geometric rotation is more robust than dynamic one over 2π.

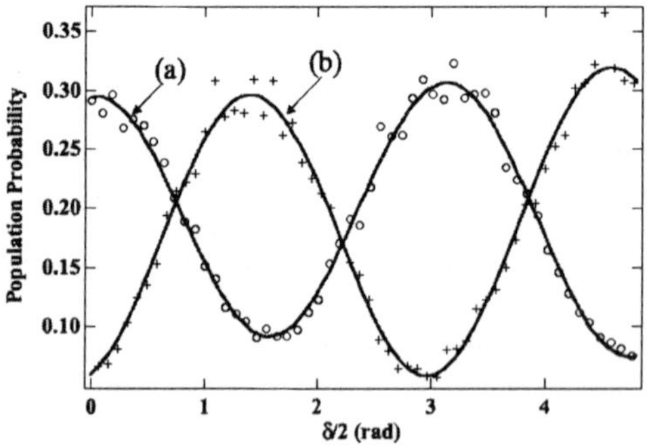

Figure 3. (a) Phase shift by a geometric rotation around axis 3. (b) Variation of a population probability by a geometric rotation around axis 2.

Last, we constructed a complete geometric atom interferometer using U_2 as a beam splitters and U_3 as a phase shifter. The Bloch vector was first rotated around axis 2 by δ_2 and next rotated around axis 3 by δ_3. In the present experiment, δ_2 was fixed to $\pi/2$ and δ_3 was changed continuously. Then we can observe the phase difference of δ_3 with an atom interferometer formed by adding the geometric rotation around axis 2 by $\pi/2$ after the rotation of δ_3.

Figure 4(a) shows a time sequence of the resonant laser-controlled pulses. Figure 4(b) shows the observed phase shift of $(2.01\pm0.02)\times\delta/2$. This signal corresponds to Fig. 3(a), observed by the atom interferometer composed of two $\pi/2$ Rabi excitation pulses. The phase shift is appropriate, except that the amplitude of fringes was reduced so that the total amount of pulse area was 6π.

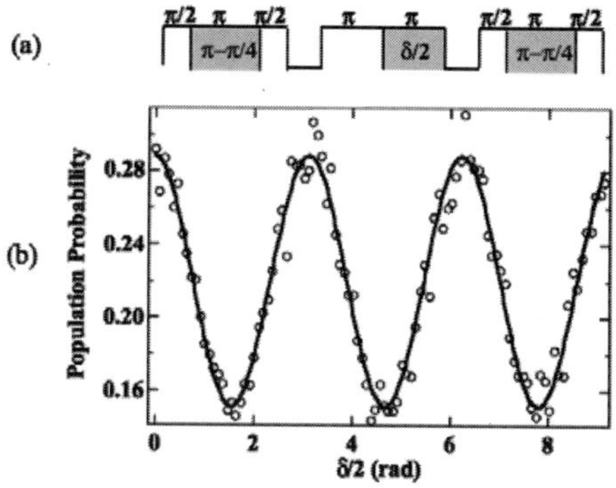

Figure 4. (a) Laser-controlled-pulses of geometric atom interferometer comprised of two beam splitters and a phase shifter. (b) Interference fringes observed using geometric atom interferometer.

4. Summary

We first examined two rotations around axes 3 and 2 on an ensemble of cold sodium atoms. The results show twice phase difference of control with an uncertainty of 1%. Next, a complete geometric atom interferometer was accomplished by two geometric beam splitters and a geometric phase shifter with carefully prepared resonant laser-controlled pulses. The interference fringes with an appropriate phase shift were demonstrated.

We are now trying to measure the second-order Zeeman shift due to the Scalar Aharonov-Bohm effect using this geometric atom interferometer.

References

1. T. Aoki, K. Shinohara and A. Morinaga, *Phys. Rev.* **A63**, 063611 (2001).
2. M. V. Berry, Proc. R. *Soc. London, Ser.* **A392**, 45 (1984).
3. M. Tian, Z. W. Barber, J. A. Fischer and Wm. Randall Babbitt, *Phys. Rev.* **A69**, 050301(R), (2004).
4. H. Imai, Y. Otsubo and A. Morinaga, *Phys. Rev.* **A76**, 012116 (2007).
5. H. Imai, and A. Morinaga, *Phys. Rev.* **A76**, 062111 (2007).

6. H. Imai and A. Morinaga, *Phys. Rev.* **A78**, 010302(R) (2008).
7. M. Kasevich and S. Chu, *Appl. Phys.* **B54**, 321 (1992).
8. D. S. Weiss, B. C. Young and S. Chu, *Appl. Phys.* **B59**, 217 (1994).

Part XIII

New Directions

ACTIVE OPTICAL CLOCKS*

JINGBIAO CHEN

Institute of Quantum electronics, School of Electronics Engineering and Computer Science, Peking University
Beijing, 100871, P. R. China

An active optical clock is a special bad cavity laser with perturbation-free medium lasing transition. Spectrum narrowing due to laser mechanism expressed by modified Schawlow-Townes formula improves its high stability of the center frequency of active optical clock by several orders of magnitude is expected. The ultimate limit factor caused by Johnson thermal noise in conventional ultra-stable cavity can be reduced by cavity-pulling effect. In this paper, we will discuss the main features of different configurations of active optical clock, including thermal atomic beam, laser slowed atomic beam, optical lattice and magneto-optical trap trapped atoms. The applications of active optical clock in sub-natural linewidth laser spectroscopy and ultra-long coherence-time laser are discussed here also. Besides the active optical clock, several new concepts including Ramsey laser---a laser based on Ramsey separated fields method, active atom interferometry, and kilosecond laser are introduced.

1. Introduction

I first introduced the concept of active optical clock [1-8] three years ago. After a three-year long-time consideration and many stimulating discussions with my colleague and students, I would like to give a brief review on this topic.

Up to now, all realized high-performance optical clocks are based on the absorption laser spectroscopy. This means we have to use quantum transition line with ultra-narrow linewidth, and ultra-stable cavity stabilized laser as a local oscillator. However, recently, it is demonstrated experimentally and theoretically that, ultra-stable cavity has a Johnson thermal noise limit [9,10]. Moreover, the long-term operation of cryogenic cavity is impractical now. Thus, it is obvious that, in the past ten years and in the near future, to reach a sub-Hz linewidth clock transition was and will be limited by the available super-narrow linewidth laser [11,12]. I believe the active optical clock will play a role in this field.

Except in the Hydrogen maser, the active microwave atomic clock, the stimulated emission procedure does not play an important role in the field of

* This work is supported by the National Fundamental Research Programme of China under Grants No. 2005CB724500 and by NSFC under Grand No.10874009.

atomic clock and precision laser spectroscopy. Another point interested me is, the Ramsey separated fields method has never been used in stimulated emission procedure. Thus there are shoulders of two giants for us to stand. A Ramsey laser has the advantages of spectrum narrowing due to laser mechanism expressed by modified Schawlow-Townes formula [13] and spectrum narrowing due to Ramsey separated fields method [14].

The mechanism of active optical clocks provides the possibility of many potential applications: high-stability optical clock, sub-Hz linewidth laser, active atom interferometry, sub-natural laser spectroscopy, kilo-second laser and related metrology.

2. Different Configurations of Active Optical Clock

The two main features are common to different active optical clocks [1-8]: the first one is its linewidth of the output laser is determined by the modified Schawlow-Townes formula[7], can be much narrower than 1 Hz, the second feature is the center frequency of the output laser of an active optical clock affected by the cavity via cavity-pulling effect, which can be reduced by several orders of magnitude by using bad cavity with cavity mode linewidth is much wider than the laser gain profile. Active optical clocks based on thermal atomic beam, laser slowed atomic beam, optical lattice and magneto-optical trap trapped atoms have been discussed in detail before and will be reviewed briefly as follows. Certainly, other possible schemes are attractive also, like miniature active optical clock based on solid lasing gain medium.

2.1. *Active optical clocks with thermal atomic beam*

Besides the common features of active optical clocks mentioned above, the advantages and disadvantages of thermal beam based active optical clock are:

(a). By using atomic thermal beam, it is possible to built a small compact optical clock with linewidth at Hz level [1]. Like in the commercial Cs clock, the size of atomic beam vacuum tube can be around 2 liters.
(b). Any two-level quantum system with suitable lifetime of excited state can be used in this configuration.
(c). Light shift due to pumping laser can be reduced by spatial separation of pumping region and lasing region.
(d). The high stability can be used to evaluate the accuracy with smaller uncertainty.
(e). The Doppler effect will be the ultimate limit of accuracy.

The second order Doppler shift can be as large as 1 kHz, and it is quadratically depends on the particle velocities. Thus using laser slowed atomic beam can reduce the Doppler effect.

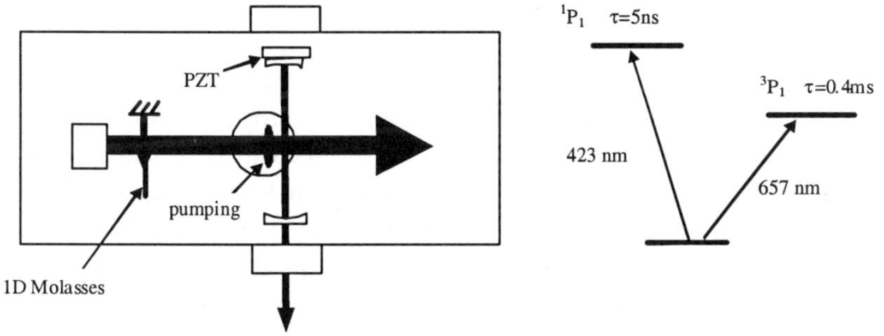

Figure 1. Experimental scheme of active optical clock based on thermal Ca beam with related energy levels.

2.2. *Active Optical Clocks with Slowed Atomic Beam*

Once slowed atomic beam is used, the Doppler effect related shifts will decrease, thus accuracy can be improved accordingly. An accuracy of 1×10^{-16} is expected for active optical clock with slowed beam [3]. However, a longer lifetime of excited state is required as Mg atom to ensure that the spontaneous decay does not affect the population inversion too seriously.

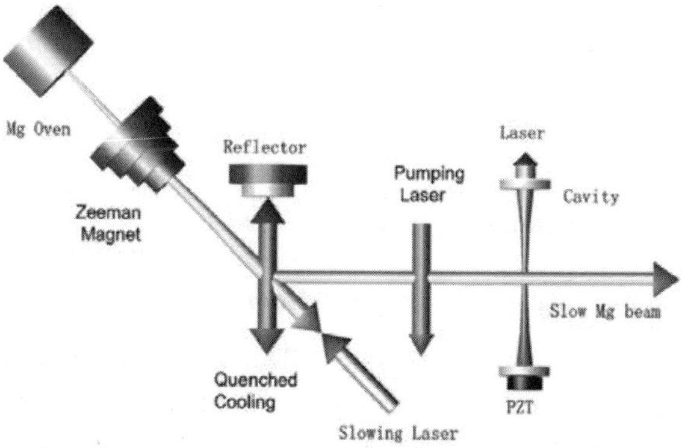

Figure 2. Experimental scheme of active optical clock based on slowed Mg atomic beam [3].

Since a laser slowed atomic flux of 1×10^{-13} has been reached [15], the power of the output laser of active optical clock can be nW.

2.3. *Ramsey Laser and Active Atom Interferometry*

The well-known Ramsey separated fields method [14] has been playing a great role in the field of atomic clock [16] and atom interferometry [17]. However, to the best of my knowledge, Ramsey separated fields method is only used in absorption spectroscopy[14], it has never been applied in the procedure of stimulated emission. Comparing the active optical clock with one stimulated interaction region [1], atoms in Ramsey laser [4] will interact with the cavity mode twice. On the other hand, comparing the absorption process in a conventional Ramsey separated fields method, atoms in this scheme experience an stimulated emission process. Thus, an active optical clock with Ramsey separated fields method combines the advantages of active optical clock and Ramsey separated fields method, and provides very small linewidth of the output laser light. When considering the external states, a Ramsey laser will be an active atom interferometry, a matter wave atom interferometry with extreme phase sensitivity. More detailed configurations include conventional linear Ramsey method with two-standing wave, and nonlinear multi-wave configuration.

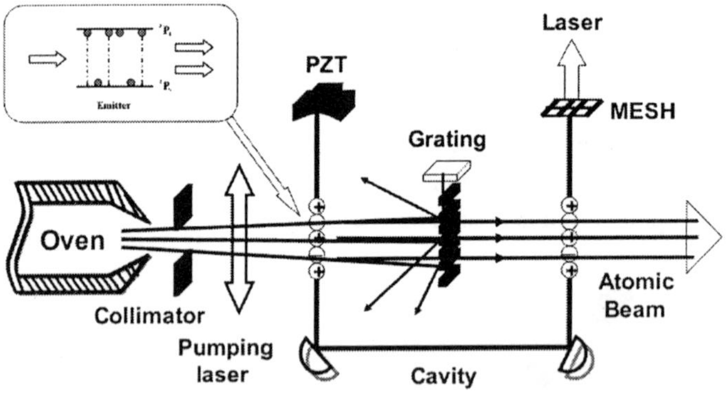

Figure 3. An experimental scheme of Ramsey Laser [4], which combines the Ramsey separated fields method and the stimulated emission procedure of laser mechanism.

2.4. *Active Optical Clocks with Trapped Atoms*

In the lattice case [2,4,5], the Doppler effect is almost removed away. This means the linewidth of gain profile can be very narrow, but the challenge will be the light shift caused by pumping laser. Thus a 4-level quantum system will be the best candidate. An example of 4-level quantum system of Yb atom for active optical clock based on optical lattice trapped atoms has been proposed [4,5].

The wavelength of the transition between $^3P_2(m=0)$ to $^3P_1(m=0)$ in Yb atom is 5813.7 nm, this transition can be the lasing transition in an active optical clock. Clearly, the pumping lasers at 398.9 nm, 1311.2 nm, 649.1 nm are far-detuned to the lasing transition at 5813.7 nm, thus the light shifts due to these pumping lasers is suppressed.

Table 1. The potential $^3P_2$ state to $^3P_1$ state lasing transition parameters for trapped atoms [4,5].

Atom	Wavelength(μm)	Frequency(GHz)	Einstein Coefficient $A(s^{-1})$
Mg	245.616	1221.42	9.106×10^{-7}
Ca	94.447	3176.40	1.612×10^{-5}
Sr	25.367	11826.4	8.266×10^{-4}
Yb	5.819	51551.4	6.848×10^{-2}

For active optical clock based on magneto-optical trap trapped atoms, 4-level atomic system will be easy for operation. The potential lasing transitions of candidates of atom are listed in Table 1. Taking the Yb atom as an example, the proposed lasing transition is the 6s6p $^3P_2$ to 6s6p $^3P_1$, which has a wavelength of 5819 nm. In a 398.9nm blue MOT, some atoms are pumped to 6s7s $^3S_1$ state, and via fast decay, are accumulated on the 6s6p $^3P_2$ state since this state has 15s lifetime. Because the atom in 6s6p $^3P_1$ state has a lifetime much shorter than that of 6s6p $^3P_2$ state, the population inversion between 6s6p $^3P_2$ and 6s6p $^3P_1$ states is built up. Therefore the lasing transition can be coupled out via a laser cavity, and the output laser light can be used as an optical clock directly.

2.5. *Potential Application of Active Optical Clock in Sub-natural linewidth Spectroscopy*

The conventional sub-natural linewidth spectroscopy is to use Ramsey separated fields method to select the long-lived atom in the excited state for detection. However, the excited state atoms number decay exponentially via time, the signal to noise ratio of detection degraded dramatically once atoms with a lifetime several times longer than the averaged lifetime. Thus the recorded sub-natural linewidth is only about 6 times narrower than the natural linewidth of atomic transition.

In the thermal Ca beam case, the calculated result showed the 657 nm output laser from active optical clock can be much narrower than 1 Hz comparing to 400 Hz natural linewidth. Thus in the view of sub-natural linewidth laser spectroscopy, the active optical clock mechanism is superior to the conventional sub-natural spectroscopy by several orders of magnitude. But, in order to pump atoms to the excited state and to be stimulated emit photon with very short lifetime will limit this new method only applicable to some available transitions.

2.6. *Kilosecond Laser*

Recently, in the ultra-short laser field, attosecond laser has been developing from femtosecond laser, attosecond science is rapidly becoming a new frontier of ultra-short lasers. However, in its long time counterpart, the long time coherent laser is only around one second for one decade [12]. And the mechanism used solely for decades to reach this narrow linewidth (thus long coherent time) laser is to lock the laser frequency to one mode of an ultra-stable Fabry-Perot cavity with temperature and vibration isolations kept in quiet room [11]. Our calculation results showed that, mHz linewidth, thus thousand-second coherent time, of laser cab be reached via active optical clock. Therefore, it is possible to push one-second coherent time laser to ten seconds coherent time laser [3], even to thousand-second coherent time laser, which can be named as kilosecond laser. A kilosecond laser is a must-have tool for next-generation optical atomic clock with linewidth as narrow as mHz, and will be extremely important for precision metrology.

3. Summary

It is obvious that, there are some basic analogies between Hydrogen maser and active optical clock in technology and mechanism. However, just as an optical clock is superior to microwave atomic clock, active optical clock will be superior to Hydrogen maser is expected. An active optical clock use spectrum narrowing by several orders of magnitude due to laser mechanism expressed by modified Schawlow-Townes formula to improve its high stability of the center frequency. Since it is a special bad cavity laser with perturbation-free medium lasing transition, the ultimate limit factor caused by Johnson thermal noise in conventional ultra-stable cavity can be reduced by cavity-pulling effect as in Hydrogen maser. In this paper, we have discussed the main features of different configurations of active optical clock, including thermal atomic beam, laser slowed atomic beam, optical lattice and magneto-optical trap trapped atoms. The applications of active optical clock in sub-natural linewidth laser spectroscopy and ultra-long coherence-time laser are discussed also. Besides the active optical clock, several new concepts including Ramsey laser-- laser based on Ramsey separated fields method, active atom interferometry, and kilosecond laser have been introduced.

Two active optical clock systems based on Ca beam are under developing in Peking University, the detailed experimental set-up and preliminary experimental results will appear elsewhere.

Acknowledgments

I would like to thank Yiqiu Wang, Donghai Yang, Xuzong Chen, Hong Guo, Xiaoji Zhou for many discussions. My students, particularly Wei Zhuang, Deshui Yu, Zhenhui Chen have contributed to this research. This work is supported by MOST of China under Grant No. 2005CB3724500, and NSFC under Grand No. 10874009.

References

1. J. Chen, *arXiv:physics*/0512096
2. J. Chen, X. Chen, *Proc. of 2005 IEEE IFCS*, 608(2005).
3. W. Zhuang, J. Chen, *Proc. of EFTF 2006*, 373(2006).
4. W. Zhuang, D. Yu, and J. Chen, *Proc. of 2006 IEEE-IFCS*, 277(2006).
5. W. Zhuang, D. Yu, Z. Chen, K. Huang, and J. Chen, *Proc. of 2007 EFTF& IEEE IFCS,* 96(2007).
6. Z. Chen, W. Zhuang, J. Chen, unpublished.
7. D. Yu and J. Chen, *Phys. Rev.* A78, 013846 (2008).
8. W. Zhuang, D. Yu, Z. Chen, K. Huang, and J. Chen, *Proc. of 2007 Time and Frequency Symposium of China*, 47 (2007), in Chinese.
9. K. Numata, A. Kemery, and J. Camp, *Phys. Rev. Lett.* 93, 250602 (2004).
10. M. Notcutt, L. -S. Ma, J. Ye, and J. L. Hall,*Opt. Lett.* 30, 1815 (2005).
11. D. Hils, and J. L. Hall, In *Frequency Standards and Metrology* (Springer-Verlag Berlin, Heidelberg 1989, ed. A. DeMarchi,) 162-173.
12. B. C. Young et al., *Phys. Rev. Lett.* 82, 3799 (1999).
13. A. L. Schawlow, and C. H. Townes, *Phys. Rev.* 112, 1940 (1958).
14. N. F. Ramsey, *Appl. Phys.* B60, 85(1995).
15. http://www.atomoptics.ru/
16. F. Riehle, *Frequency Standards: Basics and Applications* (WILEY-VCH Verlag GmbH & Co. KGaA, Weinheim, 2004).
17. P. Berman (ed.), *Atom Interferometry* (San Diego, CA: Academic, 1997).

PROSPECTS FOR A NUCLEAR OPTICAL FREQUENCY STANDARD BASED ON THORIUM-229

E. PEIK,* K. ZIMMERMANN, M. OKHAPKIN and CHR. TAMM

Physikalisch-Technische Bundesanstalt,
Bundesallee 100, 38116 Braunschweig, Germany
** E-mail: ekkehard.peik@ptb.de*

The 7.6-eV-isomer of Thorium-229 offers the opportunity to perform high resolution laser spectroscopy of a nuclear transition. We give a brief review of the investigations of this isomer. The nuclear resonance connecting ground state and isomer may be used as the reference of an optical clock of very high accuracy using trapped and laser-cooled thorium ions, or in a compact solid-state optical frequency standard of high stability.

1. The low-lying isomer of ^{229}Th

Thorium-229 seems to be a unique system in nuclear physics in that it possesses the only known isomer with an excitation energy in the range of optical photon energies and in the range of outer-shell electronic transitions. ^{229}Th is part of the decay chain of ^{233}U and undergoes α-decay with a halflife of 7880 years. Its energy level structure was studied by the group of C. W. Reich at the Idaho National Engineering Laboratory since the 1970s, mainly relying on spectroscopy of the γ-radiation emitted after the α-decay of ^{233}U [1,2]. It was noted that the lower energy levels belong to two rotational bands whose band heads must be very close, one being the ground state, the other the isomer. Evaluating several γ-decay cascades, the value (3.5 ± 1.0) eV was obtained for the isomer energy [2]. Further studies confirmed and extended the knowledge on the overall nuclear level scheme of ^{229}Th [3–5] and also supported the value for the isomer excitation energy [4]. The isomer may decay to the ground state under the emission of magnetic dipole radiation with an estimated lifetime of a few 1000 s [5,6] in an isolated nucleus.

These results on the extremely low excitation energy of the isomer inspired a number of theoretical studies, investigating the decay modes of the isomer in different chemical surroundings and possible ways of exciting it with radiation (see Refs. 7,8 for reviews). On the experimental side, two false optical detections of the decay of the isomer were reported [9,10], but it was quickly clarified that the observed light was luminescence induced by the background of α-radiation [11,12]. Fluorescence experiments with radioactive samples and ultraviolet light sources may be severely affected by thermoluminescence, Cherenkov radiation etc. [13,14]. All further attempts at a direct observation of the optical transition connecting ground state and isomer failed.

A reanalysis of the data presented in Ref. 2 indicated a possible shift to higher energies by about +2 eV [15]. New experimental data on the transition energy became available only after a group at LLNL used a high resolution γ-spectrometer and measured two decay cascades very precisely. The result on the transition energy is (7.6 ± 0.5) eV [17], placing the transition in the vacuum ultraviolet at about 160 nm. Radiation at this wavelength is not transmitted through air, water or optical glasses, which explained the failure of most of the previous attempts to detect the radiation emitted in the decay of the isomer. The energy of 7.6 eV is higher than the ionization potential of the Th atom but lower than that of Th^+. This opens internal conversion as an alternative decay channel in neutral thorium.

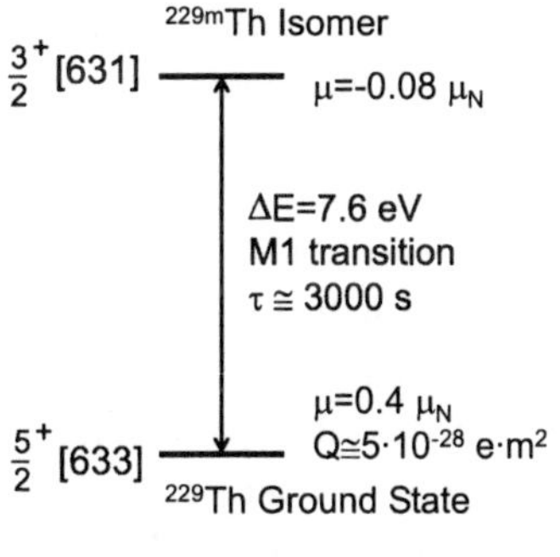

Fig. 1. The ground state and lowest excited state of the ^{229}Th nucleus with the level classification in the Nilsson model, radiative lifetime[5,6] for the magnetic dipole transition, magnetic moments in nuclear magnetons and quadrupole moment of the ground state.[6,18]

We have extended the search for light emission from ^{229}Th isomers produced in the α-decay of U-233 using photodetectors with sensitivity in the VUV range. Light emission from a solid ^{233}U-nitrate sample as well as from freshly produced ^{229}Th recoil nuclei in a catcher foil was investigated

– without seeing the expected signals [16]. In the experiment with ^{233}U a spectrometer was used but the noise level on the detector was too high to reach the required sensitivity. The experiment with the recoil nuclei showed negligible background and the negative outcome may be interpreted as a sign for a rapid nonradiative decay of the isomer in the solid.

The expected transition wavelength is in a part of the VUV range that is accessible by frequency upconversion of narrow-bandwidth continuous and phase-coherent femtosecond laser sources. Thus it appears possible to measure the nuclear energy level splitting of ^{229}Th with the precision afforded by high-resolution laser spectroscopy and optical frequency metrology. Figure 1 summarizes the relevant spectroscopic information that is available today.

2. Nuclear optical clock with trapped ions

Nuclear transition frequencies are much more stable against external perturbations than transition frequencies of the electron shell because the characteristic nuclear dimensions are small compared to the atomic dimension. Therefore nuclear transitions are attractive as highly accurate frequency references with small field-induced shifts [19]. Apart from motional frequency shifts that can be well controlled e.g. in laser-cooled trapped ions, the interaction with ambient electric or magnetic fields usually is the dominant source of systematic uncertainty in optical frequency standards. Estimates on the magnitude of systematic frequency shifts must, however, also consider the coupling of the nuclear and electronic energy level systems through the Coulomb and hyperfine interactions. For external electric field gradients the electron shell may acutally lead to an enhancement at the nucleus (Sternheimer anti-shielding). It will therefore be important to select a suitable electronic state for the nuclear excitation.

In order to illustrate the role of hyperfine interactions in nuclear spectroscopy of an isolated atom or ion, let us consider the Zeeman and Stark shifts of the nuclear transition frequency. In an LS coupling scheme the eigenstates of the coupled electronic and nuclear system are characterised by sets of quantum numbers $|\alpha, I; \beta, L, S, J; F, m_F\rangle$, where I denotes the nuclear spin, L, S, J the orbital, spin and total electronic angular momenta, and F and m_F the total atomic angular momentum and its orientation. α and β label the nuclear and electronic configurations. In the nuclear transition, the nuclear and total angular momentum quantum numbers (α, I, F, m_F) can change, while the purely electronic quantum numbers (β, L, S, J) remain constant. The nuclear transition frequency is independent of all mechanisms that produce level shifts depending only on the electronic quantum numbers (β, L, S, J), because these do not change and consequently the upper and the lower state of the transition are affected in the same way. This applies to the scalar part of the quadratic Stark effect, which typically is the dominant mechanism for the shift of electronic transition frequencies due to static electric fields, electromagnetic radiation, and collisions. The observed nuclear transition frequency is however shifted by the hyperfine Stark shift, which depends on F and m_F, and has been studied in microwave atomic clocks. In the optical frequency range, a relative magnitude of typically 10^{-19} may be expected for the hyperfine Stark shift caused by the ≈ 10 V/cm room temperature blackbody radiation field. In order to avoid the influence of the linear Zeeman effect, an electronic state can be chosen such that F is an integer. In this case a Zeeman component $m_F = 0 \rightarrow 0$ is available, that shows only a small quadratic Zeeman effect around zero magnetic field. Since this shift depends similarly on the electronic and the nuclear g-factor its magnitude will be comparable to those in other atomic clocks. Further field dependent shifts may arise from the tensor part of the quadratic Stark effect and from the quadrupole interaction between the atomic quadrupole moment and electric field gradients. Both these shifts can be expressed as a product of J-dependent and F-dependent terms and vanish if either $J < 1$ or $F < 1$.

From these general considerations it can be seen that for every radiative nuclear transition, an electronic state can be selected which makes the hyperfine coupled nuclear transition frequency immune against the linear Zeeman effect and the quadratic Stark effect as well as the quadrupole shift. For electronic transitions, this combination of advantageous features can not be obtained. Since the selected electronic state has to be stable or at least long-lived, the choice could be made among the ground states of the differently charged positive ions of the element in question. In the case of a half integer nuclear spin (like in ^{229}Th), the optimal electronic states are $^2S_{1/2}$ or $^2P_{1/2}$, and for an integer nuclear spin the states $^1S_0$ or $^3P_0$ fulfill all criteria.

For a high precision nuclear clock, the case of trapped $^{229}Th^{3+}$-ions seems to be especially promising [19] because its electronic level structure is suitable for laser cooling. The sensitive detection of excitation to the isomeric state will be possible using a double resonance scheme that probes the hyperfine structure of a transition in the electron shell. No electric dipole transitions originate from the electronic ground state of Th^{3+} in the range of 1.8 – 15 eV so that resonant coupling between electronic and nuclear excitations is not expected to play an important role for the decay of the isomeric state. The $5f\ ^2F_{5/2}$ ground state of Th^{3+} does not fulfill the condition $J < 1$ for elimination of the tensor Stark effect and the quadrupole shift, but a metastable $7s\ ^2S_{1/2}$ state of lifetime ≈ 1 s is also available. Alternatively, the method of quantum logic spectroscopy [20] with an auxiliary ion may be applied to other charge states of ^{229}Th that can not be laser-cooled directly.

3. A solid-state nuclear frequency standard

A nuclear transition may also provide a resonance with very high resolution if the nuclei are embedded in a solid, as it is observed in Mößbauer spectroscopy. Thorium-229 opens the possibility to perform optical Mößbauer spectroscopy using a laser as a tuneable, coherent source of radiation. This may provide a compact and simple reference for an optical frequency standard with performance much superior to what is available in simple atomic systems like vapor cells.

The host crystal should be transparent at the nuclear resonance wavelength λ_0, a criterion that is fulfilled by a number of candidates like the fluorides of the alkaline earths. It would be only lightly doped with ^{229}Th. If the broadening is dominantly homogeneous one nucleus per λ_0^3 may be used in order to avoid strong radiation trapping. Still, this would allow to handle 10^{11} nuclei in a cube of 1 mm dimension. With this number of nuclei direct fluorescence detection of the resonance radiation would be possible even if the resonant scattering rate is only of the order 10^{-4}/s per nucleus.

The uncertainty budget of such a solid-state nuclear clock will be quite different from that of a realization with trapped ions considered above. The crystal field shifts of the nuclear resonance frequency will be dominantly due to electric fields if a diamagnetic host is used. In insulators with high bandgap like fluorides rather high internal electric fields and field gradients will be found. The electron charge density at the position of the nucleus will lead to the isomer shift $\Delta f_{iso} = Ze^2\rho_0\langle r^2\rangle/(h\epsilon_0)$, where Ze is the nuclear charge, ρ_0 the electron density at the nucleus and $\langle r^2\rangle$ the mean squared nuclear charge radius. The contribution of a 7s electron in thorium would shift the nuclear ground state by $\Delta f_{iso} \approx 1$ GHz with respect to its energy in a bare nucleus. An electric field gradient will produce a quadrupole shift that may be of comparable magnitude: In the tetragonal crystal ThB_4, for example, the field gradient along the principal axis is about 5×10^{21} V/m^2. Coupling to the ground state quadrupole moment of ^{229}Th of 5×10^{-28} m^2 would produce a quadrupole shift of 0.6 GHz. The field gradient can be avoided in a crystal lattice of higher symmetry, like a cubic one.

Both these shifts would be of less concern if they would be constant, which would only be the case if the positions of all charges in the lattice would be rigorously fixed. Thermal motion, however, will lead to a temperature-dependent broadening and shift of the line, where the line shape will depend on phonon frequencies and correlation times. Much information on these effects has been obtained in conventional (γ-ray) Mößbauer spectroscopy already. While the relativistic Doppler shift will lead to a temperature dependent relative frequency shift of about 10^{-15}/K, the temperature dependence of the crystal field will critically depend on the choice of crystal host and may be significantly bigger. For a solid state nuclear clock of high accuracy (beyond 10^{-15}) the temperature dependence may be eliminated if the crystal is cryogenically cooled to well below the Debye temperature, so that the influence of phonons is effectively frozen out.

4. Conclusion

Nuclear laser spectroscopy of ^{229}Th seems to offer great potential for frequency metrology and promises to open a new field of research at the borderline between nuclear and atomic physics, shedding new light on familiar phenomena like nuclear radiative decay or hyperfine interactions. It may allow improved tests of fundamental physics, as it was recently shown that the resonance frequency would be the most sensitive probe in the search for temporal variations of the fundamental coupling constants [21,22].

Acknowledgments

This work is supported by DFG through SFB 407 and the cluster of excellence QUEST.

References

1. L. A. Kroger and C. W. Reich, Nucl. Phys. A **259**, 29 (1976).
2. R. G. Helmer and C. W. Reich, Phys. Rev. C **49**, 1845 (1994).
3. K. Gulda et al., Nucl. Phys. A **703**, 45 (2002).
4. V. Barci et al, Phys. Rev. C **68**, 034329 (2003).
5. E. Ruchowska et al., Phys. Rev. C **73**, 044326 (2006).
6. A. M. Dykhne and E. V. Tkalya, Pis'ma Zh. Eksp. Teor. Fiz. **67**, 233 (1998) [JETP Lett. **67**, 251 (1998)].
7. S. Matinyan, Phys. Rep. **298**, 199 (1998).
8. E. V. Tkalya, Physics Uspekhi **46**, 315 (2003).
9. G. M. Irwin and K. H. Kim, Phys. Rev. Lett. **79**, 990 (1997).

10. D. S. Richardson et al., Phys. Rev. Lett. **80**, 3206 (1998).
11. S. B. Utter et al., Phys. Rev. Lett. **82**, 505 (1999).
12. R. W. Shaw, J. P. Young, S. P. Cooper and O. F. Webb, Phys. Rev. Lett. **82**, 1109 (1999).
13. J. P. Young, R. W. Shaw and O. F. Webb, Inorg. Chem. **38**, 5192 (1999).
14. E. Peik, J. Phys. D: Appl. Phys. **40**, 3330 (2007).
15. Z. O. Guimaraes-Filho, O. Helene, Phys. Rev. C **71**, 044303 (2005).
16. K. Zimmermann et al., to be published.
17. B. R. Beck et al., Phys. Rev. Lett. **98**, 142501 (2007).
18. S. Gerstenkorn et al. J. Physique (Paris) **35**, 483 (1974).
19. E. Peik and Chr. Tamm, Europhys. Lett. **61**, 181 (2003).
20. P. O. Schmidt, T. Rosenband, C. Langer, W. M. Itano, J. C. Bergquist and D. J. Wineland, Science **309**, 749 (2005).
21. V. V. Flambaum, Phys. Rev. Lett. **97**, 092502 (2006).
22. V. V. Flambaum, N. Auerbach and V. F. Dmitriev, arXiv:0807.3218; V. V. Flambaum and R. B. Wiringa, arXiv:0807.4943.

WHISPERING GALLERY MODE OSCILLATORS AND OPTICAL COMB GENERATORS

A. B. MATSKO*, A. A. SAVCHENKOV, W. LIANG,
V. S. ILCHENKO, D. SEIDEL, and L. MALEKI

OEwaves Inc., 1010 E. Union str.,
Padadena, CA 91106, USA
**E-mail: andrey.matsko@oewaves.com*
www.oewaves.com

We discuss an approach for generating optical combs using four-wave mixing process in a nonlinear whispering gallery mode resonator. We show that pumping the resonator with strong enough continuous wave coherent light results in appearance of optical sidebands around the optical carrier. The pumping threshold of the oscillation can be in a few microWatt range for resonators with an ultra-high quality factor. Natural cascading of the the nonlinear process leads to mode locking of the sidebands and emergence of an optical frequency comb. Demodulation of the comb by means of a fast photodiode produces a high frequency spectrally pure RF signal, the frequency of which is given by the resonator morphology.

Keywords: Four-Wave Mixing, Optical Frequency Comb, Optical Soliton, Mode Locking.

1. Introduction

It is difficult to underestimate the impact of optical frequency combs on the development of frequency standards and metrology. The comb bridges several orders of magnitude of frequency gap between radio frequency and optical frequency range, resulting in metrological studies with unprecedented accuracy. The comb also represents a great potential for fabrication of the most accurate clocks and the most stable frequency standards. Being an excellent scientific tool the comb still is not widely used. Utilization of the comb beyond well equipped laboratories is impeded by the large size, power consumption, and cost of the mode locked lasers that are the basis of the optical frequency comb. A standard comb generator based on a titanium-doped sapphire laser system consumes tens of watts of power and has feet-scale dimensions. In this paper we discuss a possibility of fabricating compact and power efficient optical frequency comb generators using four wave mixing in whispering gallery mode (WGM) resonators made of transparent optical materials. We discuss an application of this type of optical frequency comb generators for producing highly spectrally pure RF signals.

It is well known that propagation of cw light in a medium possessing refractive $\chi^{(3)}$ nonlinearity and anomalous group velocity dispersion can be unstable, if the power of the light exceeds a specific threshold.[1,2] The observed instability, dubbed as modulational instability, results in transformation of the cw light into a train of short optical pulses.[3] Earlier it was proposed to use the modulation instability for fabrication of a tunable infrared source[1] and a laser.[4,5] The modulational instability occurs for both positive and negative dispersion if the light is confined in a resonator. Using that property, an additive modulational instability ring laser was predicted and demonstrated.[6–8] The same effect of modulational instability is also iat the core of fiber optical parametric oscillators.[9–12] In fact, it was noted in[9] that the modulational instability laser is really a fiber optical parametric oscillator, in which pump, signal, and idler waves are resonated inside a fiber ring cavity.

The nonlinearity of commonly available fibers is low, so the modulation instability fiber lasers require optical pumping with watt-range powers. A high fiber resonator with enough finesse allows reducing the power to hundred milliwatt range;[8] however the frequency spectral width of the generated pulses does not exceed a THz. Recently it was shown that the efficiency of four wave mixing process can be significantly enhanced in dielectric WGM resonators.[13,14] The power of the cw light circulating in resonators pumped with 100 mW light can exceed megawatts because of the large finesse of the resonators.[15] Such an enhancement of the power results in broadening of the optical spectrum, ultimately leading to generation of optical frequency combs.[16–18]

Unlike the modulation instability fiber lasers, WGM-resonator based devices do not suffer from Brillouin scattering. The finesse of the resonators is high in a very broad range of optical wavelengths. The morphology of the WGM resonator allows engineering the spectrum of the modes and, hence, the resonator dispersion. The resonators have no moving parts and coatings, and are environmentally stable. They are polarization selective. Those features make WGM resonators attractive for optical frequency comb generation. Generation of spectrally pure RF signals at any desired frequency is one of the promising applications of four-wave mixing based oscillation in an optical WGM resonator.[14,19–23] To obtain the RF signal, one needs to demodulate the optical signal on a fast photodiode or rectify it by other means. Optical frequency comb is also promising here;[17,18] the spectral purity of the RF signal increases with increasing Q factor of the WGMs and the optical power of the generated optical sidebands. The resultant RF generators can be extremely compact and power efficient, as well as being high performance.

In this paper we study properties of optical oscillators and optical frequency comb generators based on the effect of four-wave mixing in high-Q WGM resonators. We review the basic formalism used for the theoretical description of the oscillator and discuss its application for description of the mode locking process in WGM resonators. We show that mode locking is the basic condition for the generation of the spectrally pure RF signals with the comb. We also describe the basic steps for frequency stabilization of combs and argue that it is possible to lock only one frequency harmonic of the optical comb generated in a WGM resonator to an atomic line to be able to accurately predict the frequency of the other harmonics.

2. Modulational instability in a WGM resonator

There are essentially two nonlinear processes that result in generation of a coherent frequency comb in an externally driven WGM resonator possessing $\chi^{(3)}$ nonlinearity. Those are hyper-parametric oscillation (modulational instability) and four wave mixing (Fig. 1). The oscillation process involves two pump photons and two sideband photons. The process starts when the cw pump power exceeds some threshold value.[21] Frequency symmetrical sidebands are then generated independently in this process. The threshold value depends on the dispersion and nonlinearity of the resonator, and generation of widely separated sidebands sometimes has a lower threshold compared with that of closely separated sidebands.[18] The initial frequency separations between the sideband pairs are statistically independent, and the wave mixing bounds them. This becomes important in the regime of well developed oscillation when the total power of generated sidebands approaches the power of the carrier. It is this process that locks the generated sidebands and the carrier together, making the frequency interval between the sidebands equal.[16] As the result, optical pulses can be generated in the resonator.

Let us consider the interaction of a pair of any two equidistant sidebands (Fig. 1a) with the carrier, assuming that the carrier does not experience power depletion. Following the discussion presented in,[23] we find that the

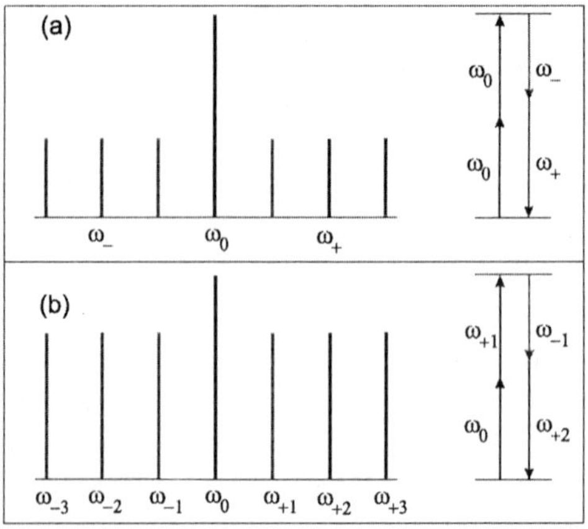

Fig. 1. (a) Wave mixing process resulting in the instability of a cw field inside an externally driven nonlinear WGM resonator. (b) Wave mixing process resulting in the mode locking inside an oscillating externally driven nonlinear WGM resonator.

pump wave acquires the phase shift

$$A_{Gout} = A_{Gin} e^{iG} \tag{1}$$

after passing through the nonlinear medium, where $A_{Gin} = |A_{Gin}| \exp(i\phi_A)$ and A_{Gout} is the pump amplitude at the exit and entrance of the of the coupling prism (Fig. 2). The phase shift can be represented in the form $G = 2\pi R\gamma |A_{Gin}|^2 = \gamma_\Sigma |A_{Gin}|^2$, if $|A_{Gin}|^2$ has the dimensionality of power, γ is the nonlinearity parameter ($\gamma_\Sigma = 2\pi R\gamma$), and R is the radius of the resonator.

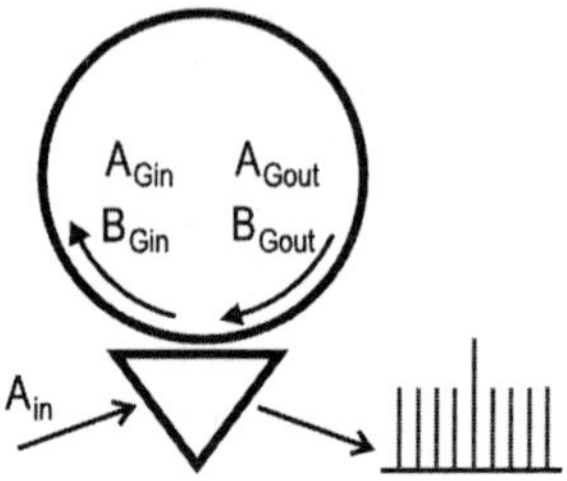

Fig. 2. Schematic of a WGM resonator with a coupling prism.

To find the oscillation condition for the sideband modes we write equations connecting the field amplitudes of the modes at the exit and entrance of the coupling prism (Fig. 2)

$$B_{Gout+} = \left[B_{Gin+} + iG(B_{Gin+} + B^*_{Gin-}e^{2i\phi_A})\right] e^{iG}, \tag{2}$$

$$B^*_{Gout-} = \left[B^*_{Gin-} - iG(B_{Gin+}e^{-2i\phi_A} + B^*_{Gin-})\right] e^{-iG}, \tag{3}$$

where B_+ and B_- are the slow amplitudes of the sidebands. Frequencies of the sidebands (ω_+ and ω_-) obey to the energy conservation law (Fig. 1a)

$$2\omega_0 = \omega_+ + \omega_-, \tag{4}$$

We also introduce additional conditions for the field components resulting from the intrinsic loss in the resonator as well as external pumping of the resonator

$$A_{Gin} = A_{Gout}e^{i\beta_{0\Sigma}-\alpha_\Sigma}\sqrt{1-T_c} + \sqrt{T_c}A_{in}, \tag{5}$$

$$B_{Gin\pm} = B_{Gout\pm}e^{i\beta_{\pm\Sigma}-\alpha_\Sigma}\sqrt{1-T_c}, \tag{6}$$

where $\beta_{0\Sigma} = 2\pi R\omega_0 n_0/c$ and $\beta_{\pm\Sigma} = 2\pi R\omega_\pm n_\pm/c$ are the phase shifts for the carrier and sidebands, $\alpha_\Sigma \ll 1$ is the round trip frequency independent light absorption in the resonator, $T_c \ll 1$ is the coupling factor with the external world through the prism coupler.

Combining (1) and (5) we get equation for the optical pump

$$|A_{Gin}|e^{i\phi_A}\left\{1 - e^{-\alpha_\Sigma + i(G+\beta_{0\Sigma})}\sqrt{1-T_c}\right\} = \sqrt{T_c}A_{in}; \tag{7}$$

combining (2-3) and (6) we derive equations for the sidebands

$$B_{Gin+}\left[\frac{e^{\alpha_\Sigma - i\beta_{+\Sigma}}}{\sqrt{1-T_c}} - (1+iG)e^{iG}\right] = iGe^{i(2\phi_A+G)}B^*_{Gin-}, \tag{8}$$

$$B^*_{Gin-}\left[\frac{e^{\alpha_\Sigma + i\beta_{-\Sigma}}}{\sqrt{1-T_c}} - (1-iG)e^{-iG}\right] = -iGe^{-i(2\phi_A+G)}B_{Gin+}. \tag{9}$$

Set (8) and (9) has a nontrivial solution if

$$1 - T_c + e^{2\alpha_\Sigma} = \sqrt{1-T_c}e^{\alpha_\Sigma}\cos\left[\frac{\beta_{+\Sigma}-\beta_{-\Sigma}}{2}\right] \times \tag{10}$$

$$\left[e^{-i\left(\frac{\beta_{+\Sigma}+\beta_{-\Sigma}}{2}+G\right)}(1-iG) + c.c.\right], \quad \text{and}$$

$$\sin\left[\frac{\beta_{+\Sigma}-\beta_{-\Sigma}}{2}\right]\left\{\sqrt{1-T_c}\left[e^{-i\left(\frac{\beta_{+\Sigma}+\beta_{-\Sigma}}{2}+G\right)}(1-iG) + c.c.\right] - \right. \tag{11}$$

$$\left. 2e^{\alpha_\Sigma}\cos\left[\frac{\beta_{+\Sigma}-\beta_{-\Sigma}}{2}\right]\right\} = 0.$$

This set can be transformed into two independent sets of equations

$$1 - T_c + e^{2\alpha_\Sigma} = \sqrt{1-T_c}e^{\alpha_\Sigma}\left[e^{-i\left(\frac{\beta_{+\Sigma}+\beta_{-\Sigma}}{2}+G\right)}(1-iG)+c.c.\right], \quad (12)$$

$$\sin\left[\frac{\beta_{+\Sigma}-\beta_{-\Sigma}}{2}\right] = 0, \quad (13)$$

and

$$\cos^2\left[\frac{\beta_{+\Sigma}-\beta_{-\Sigma}}{2}\right] = \frac{1}{2}\left[1+(1-T_c)e^{-2\alpha_\Sigma}\right], \quad (14)$$

$$\sqrt{1-T_c}\left[e^{-i\left(\frac{\beta_{+\Sigma}+\beta_{-\Sigma}}{2}+G\right)}(1-iG)+c.c.\right] = 2e^{\alpha_\Sigma}\cos\left[\frac{\beta_{+\Sigma}-\beta_{-\Sigma}}{2}\right]. \quad (15)$$

The second set of equations (14,15) is dropped. Eq. (14) requires the frequency difference between the generated sidebands to depend on the loss in the resonator, which is unphysical.

We find from Eq. (13)

$$\beta_{+\Sigma} - \beta_{-\Sigma} = 4\pi m_1, \quad (16)$$

where m_1 is an integer number (here we take into account that the sidebands are separated by even number of the free spectral ranges of the resonator). Assuming that

$$\beta_{+\Sigma} + \beta_{-\Sigma} = 4\pi m_2 + 2\delta, \quad (17)$$

where $\delta \ll 1$, we transform Eq. (12), in the limit of the small loss and gain, to

$$(2G+\delta)^2 + \left(\alpha_\Sigma + \frac{T_c}{2}\right)^2 = G^2. \quad (18)$$

Solving (18) allows us to find the threshold conditions of the oscillations.[21] The oscillation (the modulational instability of the cw background in the resonator) can occur only if $-\sqrt{3}\,(\alpha_\Sigma + T_c/2) > \delta$ and $G > (2/\sqrt{3})\,(\alpha_\Sigma + T_c/2)$. To find the oscillation condition with respect to the external pump power we have to solve Eq. (7), which takes form

$$G\left[\left(\alpha_\Sigma + \frac{T_c}{2}\right)^2 + (G+\delta_0)^2\right] = T_c G_{in}, \quad (19)$$

in the case of small gain/loss, where $G_{in} = \gamma_\Sigma |A_{in}|^2$, and

$$\beta_{0\Sigma} = 2\pi m_2 + \delta_0. \quad (20)$$

Let us describe phase shifts δ and δ_0 in terms of the dispersion of the modes in the resonator as well as detuning of the carrier frequency of the cw pump light from the corresponding optical mode. We assume that

$$\beta_{0\Sigma} = \frac{2\pi R n_0}{c}(\omega_0 - \omega_{m2}) + \frac{2\pi R n_0}{c}\omega_{m2}, \tag{21}$$

where ω_{m2} is the resonant frequency of the optically pumped mode

$$\omega_{m2} = \frac{c m_2}{R n_0}, \tag{22}$$

and

$$\delta_0 = \frac{2\pi R n_0}{c}(\omega_0 - \omega_{m2}). \tag{23}$$

Using similar reasoning we derive for the sidebands

$$\beta_{\pm\Sigma} = \frac{2\pi R n_\pm}{c}(\omega_\pm - \omega_{m2\pm m1}) + \frac{2\pi R n_\pm}{c}\omega_{m2\pm m1}, \tag{24}$$

where $\omega_{m2\pm m1}$ is the resonant frequency of the sideband

$$\omega_{m2\pm m1} = \frac{c(m_2 \pm m_1)}{R n_\pm}, \tag{25}$$

and

$$\delta = \frac{\pi R}{c}\left[n_+(\omega_+ - \omega_{m2+m1}) + n_-(\omega_- - \omega_{m2-m1})\right]. \tag{26}$$

On the other hand, we can decompose the sideband phase shifts as

$$\beta_{\pm\Sigma} \approx \beta_{0\Sigma} + \beta_{1\Sigma}(\omega_\pm - \omega_0) + \frac{1}{2}\beta_{2\Sigma}(\omega_\pm - \omega_0)^2 + \frac{1}{6}\beta_{3\Sigma}(\omega_\pm - \omega_0)^3 + \ldots, \tag{27}$$

where $\beta_{i\Sigma} = 2\pi R\beta_i$, and β_i (i is a natural number) are usual dispersive characteristics of the modes of the resonator that take into account both geometrical and material dispersion. It is easy to find

$$\delta = \delta_0 + \frac{1}{8}\beta_{2\Sigma}(\omega_+ - \omega_-)^2 + \frac{1}{384}\beta_{4\Sigma}(\omega_+ - \omega_-)^4. \tag{28}$$

Hence, each pair of the sidebands generated in the resonator is characterized with its own δ. Generally, the larger is the difference between δ and some optimal phase shift $\delta_{optimal}$,[21] the lower is the sideband power. In the case of large or zero absolute value of the group velocity dispersion $|\beta_{2\Sigma}|$, the material/geometrical contribution to δ is small only for the sidebands in the nearest vicinity of the carrier. The oscillations usually involve those sidebands.[14] However, if signs of $\beta_{2\Sigma}$ and $\beta_{4\Sigma}$ are the opposite, the material/geometrical part of δ becomes minimal for some particular nonzero frequency detuning given by expression

$$(\omega_+ - \omega_-)^2 = 48 \left| \frac{\beta_{2\Sigma}(\omega_0)}{\beta_{4\Sigma}(\omega_0)} \right| . \qquad (29)$$

The oscillation starts at higher frequencies in this case (the sidebands are generated many FSRs from the carrier). Such a behavior was observed experimentally.[18] To find a more rigorous expression for the expected oscillation frequency one have to find $\delta_{optimal}$ analyzing the value of the oscillation threshold and then solve equation (28).

Using Eq. (16) and condition $2\omega_0 = \omega_+ + \omega_-$ we also get equation

$$\beta_{1\Sigma}(\omega_+ - \omega_-) + \frac{1}{6}\beta_{3\Sigma}(\omega_+ - \omega_-)^3 = 4\pi m_1, \qquad (30)$$

which shows that the sidebands are spaced nearly equidistantly with respect to each other. The equidistant position of the sidebands is distorted due to third order dispersion, and the difference can be significant. For instance, for a CaF_2 WGM resonator with 25 GHz free spectral range at 1.55 μm the separation between the carrier and the first sideband and the separation between the forty ninth and fiftieth sidebands is as large as 28 kHz (we took into account that in this case $\beta_{1\Sigma} = 4 \times 10^{-11}$ s^{-1}, $\beta_{3\Sigma} = 3.7 \times 10^{-43}$ s^{-3}, $2\pi R \simeq 0.84$ cm). It is worth noting here that each sideband pair is symmetric with respect to the carrier because of the energy conservation law ($\omega_+ - \omega_0 = \omega_0 - \omega_-$). The inequidistance takes place between the sidebands belonging to different sideband pairs. In the case of mode locking the frequency difference between the pairs can be compensated, as is shown in.[16] The separation of the sidebands (30) does not depend on the optical power. The power dependence can occur indirectly due to light absorption and subsequent thermal heating of the resonator that changes both $\beta_{1\Sigma}$ and $\beta_{3\Sigma}$.

One is able to describe the threshold and averaged frequency of the oscillation in the resonator, made of a material with known dispersion and nonlinearity, using equations (18), (19), (28), and (30). This technique allows studying generation of just one dominating pair of sidebands, which can be produced either in the case of large enough dispersion when the mode spectrum is significantly non-equidistant,[14] or in the case of a significantly nonlinear dispersion curve and operation in the vicinity of zero dispersion point.[18,24] The technique also works for description of low power multiple sidebands up to a point when the sideband pairs start to influence each other. The generation of the low power sidebands is a stable process and there is an apparent attractor for the oscillator operating close to threshold.[6] Beating any sideband pair and the pump light on a fast photodiode produces a spectrally pure RF signal.[23] The situation is different for the multi-sideband oscillation far enough from the threshold. The generated sidebands are not exactly equidistant and are not phase locked. It means that an RF signal produced by light exiting the resonator on a fast photodiode has a comparably wide linewidth. For the example of CaF_2 resonator with 25 GHz FSR considered above the linewidth of RF signal should be on the order of tens of kHz. Mode locking of multiple sidebands is possible. It fixes the situation and significantly improves the quality of the RF signal.[18]

3. Forced nonlinear Schrodinger equation (NLSE)

To describe advanced modulational instability oscillation in a WGM resonator one can use the master equation for the optical field inside the resonator[6]

$$T_R \frac{\partial A}{\partial T} + \frac{i}{2}\beta_{2\Sigma}\frac{\partial^2 A}{\partial t^2} = i\gamma_\Sigma |A|^2 A - \left(\alpha_\Sigma + \frac{T_c}{2} - i\delta_0\right) A + \sqrt{T_c} A_{in}, \quad (31)$$

where $A(T,t)$ is the slowly varying envelope of the electric field, $T = z/V_g$, z is the coordinate along the resonator circumference, $V_g = 1/\beta_1$ is the group velocity, and t is the retarded time. By definition, time scale T is much longer than the pulse round trip time $T_R = 2\pi R/V_g$, $2\pi R$ is the length of the resonator circumference, α_Σ is the amplitude attenuation per round trip, $T_c/2$ is the coupling loss per one round trip.

Equation (31) is utilized to study a large number of physical processes. For instance, it is used to describe spatial pattern formation in cavities containing $\chi^{(3)}$ nonlinear medium.[25] The dynamic behavior of the solutions of Eq. (31) was widely studied. Various kinds of attractors, including global, have been found in the solution of the master equation.[26–28] It was shown, on the other hand, that if A_{in} represents a train of pulses, a chaotic solution of the equation can occur.[29] The detailed study of the solutions of Eq. (31) is out of the scope of the current paper. We show that Eq. (31) correctly describes the modulation instability threshold for a pair of sidebands (the case considered in the previous section). Noting that there is a solution of a lossless nonlinear Schrodinger equation with negative dispersion showing coexistence of the optical pulse and cw background,[30,31] we use the method similar to the standard moment technique[32] used for description of properties of the zero-background pulses generated in mode locked lasers and find an approximate solution of the forced equation. Because the solution shows the possibility of generation of ultrashort pulses in the system, we take into account higher order dispersive and nonlinear properties of the resonator.

3.1. *Two-mode instability*

Let us present the steady state solution of Eq. (31) in the form[6]

$$A = A_{Gin} + B_{Gin+}e^{-i\Omega t} + B_{Gin-}e^{i\Omega t}, \tag{32}$$

where $2\Omega = \omega_+ - \omega_-$. Substituting (32) into Eq. (31) we get the set of equations for the cw background

$$A_{Gin}\left[\alpha_\Sigma + \frac{T_c}{2} - i\left(\delta_0 + \gamma_\Sigma |A_{Gin}|^2\right)\right] = \sqrt{T_c}A_{in}, \tag{33}$$

and sidebands

$$\left[\alpha + \frac{T_c}{2} - i\left(\delta_0 + \frac{1}{2}\beta_{2\Sigma}\Omega^2 + 2\gamma_\Sigma |A_{Gin}|^2\right)\right] B_{Gin-} - \tag{34}$$
$$i\gamma_\Sigma |A_{Gin}|^2 e^{2i\phi_a} B^*_{Gin+} = 0,$$
$$\left[\alpha + \frac{T_c}{2} - i\left(\delta_0 + \frac{1}{2}\beta_{2\Sigma}\Omega^2 + 2\gamma_\Sigma |A_{Gin}|^2\right)\right] B_{Gin+} - \tag{35}$$
$$i\gamma_\Sigma |A_{Gin}|^2 e^{2i\phi_a} B^*_{Gin-} = 0,$$

in the limit of small sidebands. It is easy to verify that those equations generate equations (18) and (19). Hence, the two approaches are identical up to the point of independent sideband pairs. The forced NLSE is much more general, though. It allows finding the mode-locked attractors of the oscillator.

3.2. *An approximate solution of the forced NLSE*

The nonzero cw background always present in the resonator is the basic difference of the cw-pumped resonator-based four wave mixing oscillator and a passively mode locked laser. This background interacts with the optical pulses generated in the resonator if the oscillator operates in the mode locked regime. There exists an analytical solution of the lossless master equation with negative dispersion ($\beta_{2\Sigma} < 0$) showing coexistence of the optical pulse and a cw background.[30,31] This solution can be roughly approximated by an autosoliton pulse and a cw background. We here are interested of more general case of oscillations, when the second order dispersion is allowed to have any sign and value. We have also to consider losses in the system, so that the previously developed approach[30,31] is not applicable in our case. In what follows we try to adopt the formalism generally used for description of mode locking to get some understanding of properties of the hyper-parametric oscillator. We show that the formalism give quantitatively reasonable result, though it cannot be used for strict description of the process of mode locking in the oscillator.

Let us assume that the WGM resonator possesses positive dispersion, which is true for small enough fused CaF_2 WGM resonators pumped with 1550 nm cw light (as is shown in the following section). We look for the Gaussian pulse solution of Eq. (31) and use the variational method.[33]

$$\begin{aligned} A(T,t) &= A_c + A_p(T,t), \\ A_c &= \sqrt{P_c} e^{i\phi_c}, \\ A_p(T,t) &= \left[\frac{P_p}{\sqrt{\pi}}\right]^{1/2} \left[\exp\left(\frac{t-\xi}{\sqrt{2}\tau}\right)^2\right]^{-1-iq} e^{i\Omega(t-\xi)+i\phi_p}, \end{aligned}$$

where P_c is the power of the cw background, ϕ_c is the phase of the background wave, P_p is the pulse peak power ($E = P_p\tau$ is the pulse energy), ξ is the temporal shift (timing), Ω is the frequency shift, q is the chirp, and τ is the pulse duration, and ϕ_p is the phase of the pulse. As can be seen from[18] the optical comb center frequency is sometimes shifted with respect to the pumping frequency, so the frequency shift should be taken into account.

We assume at the next step that the constant background is a solution of an algebraic equation

$$A_c\left[\alpha_{\Sigma 0} + \frac{T_{c0}}{2} - i\left(\delta_0 + \gamma_\Sigma\left(|A_c|^2 + 2P_p\frac{\tau}{T_R}\right)\right)\right] = \sqrt{T_c}A_{in}, \tag{36}$$

and the pulse part is a solution of a perturbed NLSE

$$\begin{aligned} &T_R\frac{\partial A_p}{\partial T} + \frac{i}{2}\beta_{2\Sigma}\frac{\partial^2 A_p}{\partial t^2} - i\gamma_\Sigma|A_p|^2A_p = \frac{1}{6}\beta_{3\Sigma}\frac{\partial^3 A}{\partial t^3} + \frac{i}{24}\beta_{4\Sigma}\frac{\partial^4 A}{\partial t^4} - \\ &\frac{\gamma_\Sigma}{\omega_0}\frac{\partial}{\partial t}(|A_p|^2A_p) - \left(\alpha_{\Sigma 0}\left[1 + \frac{4i}{\omega_0}\frac{\partial}{\partial t} - \frac{6}{\omega_0^2}\frac{\partial^2}{\partial t^2}\right] + \frac{T_{c0}}{2}\left[1 - \frac{3i}{2\omega_0}\frac{\partial}{\partial t} - \frac{15}{8\omega_0^2}\frac{\partial^2}{\partial t^2}\right] - \right. \\ &\left. i(\delta_0 + 2\gamma_\Sigma|A_c|^2)\right)A_p + i\gamma_\Sigma\left[A_c^2A_p^* + A_p^2A_c^* + 2A_c(|A_p|^2 - P_p\frac{\tau}{T_R})\right], \end{aligned} \tag{37}$$

where the perturbation is in the right hand side of the equation. The equation for the constant background (36) takes into account the average cross-phase modulation of the background due to the pulse. Equation for the pulse (37) contains more terms than Eq. (31). Those terms describe the finite response time of the nonlinearity of the material, dispersion of the loss, coupling, as well as high order dispersion of the WGM resonator. We assume that the loss is given by Rayleigh scattering, so that in the vicinity of the carrier frequency of the pump

$$\alpha_\Sigma = \alpha_{\Sigma 0}\frac{\omega^4}{\omega_0^4} \approx \alpha_{\Sigma 0}\left[1 + \frac{4i}{\omega_0}\frac{\partial}{\partial t} - \frac{6}{\omega_0^2}\frac{\partial^2}{\partial t^2}\right]. \tag{38}$$

The coupling loss also has a dispersion. In the case of zero gap between the coupling prism and the resonator the frequency behavior of the resonator loading can be approximated as

$$\frac{T_c}{2} = \frac{T_{c0}}{2}\frac{\omega_0^{3/2}}{\omega^{3/2}} \approx \frac{T_{c0}}{2}\left[1 - \frac{3i}{2\omega_0}\frac{\partial}{\partial t} - \frac{15}{8\omega_0^2}\frac{\partial^2}{\partial t^2}\right]. \tag{39}$$

Analyzing the solution of Eq. (37) with variational method we obtain equations for the energy and phase of the pulse

$$T_R\frac{\partial(P_p\tau)}{\partial T} = -P_p\tau\left\{2\alpha_\Sigma + T_{c0} - \left[\frac{i\gamma_\Sigma A_c^2}{\sqrt{1-iq}}\exp\left(-2i\phi_p - i\frac{\Omega^2\tau^2}{i+q}\right) + \right.\right.$$
$$\left.\left.\frac{i\sqrt{2}\gamma_\Sigma A_c\sqrt{P_p}}{{}^{1/4}\!\sqrt[4]{\pi}\sqrt{3-iq}}\exp\left(-i\phi_p - i\frac{\Omega^2\tau^2}{2(3i+q)}\right) + c.c.\right]\right\}, \tag{40}$$

$$T_R\frac{\partial\phi_p}{\partial T} = \delta_0 + \frac{\beta_{2\Sigma}}{2\tau^2}(1-\Omega^2\tau^2) +$$
$$\frac{\beta_{3\Sigma}\Omega}{12\tau^2}(3q^2 + 4\Omega^2\tau^2 - 3) + \frac{\beta_{4\Sigma}}{32\tau^4}[(1+q^2)(3-q^2) - 4\Omega^2\tau^2(2q^2+\Omega^2\tau^2)] +$$
$$2\gamma_\Sigma|A_c|^2 + \frac{5\gamma_\Sigma P_p}{4\sqrt{2\pi}} + \left[\frac{(2i+3q)\gamma_\Sigma A_c^2}{4i(1-iq)^{3/2}}\exp\left(-2i\phi_p - i\frac{\Omega^2\tau^2}{i+q}\right) + \right. \tag{41}$$
$$\left.\frac{3(7i+3q)\gamma_\Sigma A_c\sqrt{P_p}}{2\sqrt{2}\,{}^{1/4}\!\sqrt[4]{\pi}i(3-iq)^{3/2}}\exp\left(-i\phi_p - i\frac{\Omega^2\tau^2}{2(3i+q)}\right) + c.c.\right]\Bigg\}.$$

Deriving Eqs. (40) and (41) we have assumed that $\tau\omega_0 > 1$ and have neglected the influence of the dispersion of the loss on phase and amplitude. Equations for the energy and phase taken in steady state, along with Eq. (36) resemble Eqs. (18) and (19). Eqs. (40) and (41) allow to calculate the pulse power and phase if other parameters of the pulse are known. The parameters can be found as described below.

Let us assume that the energy of the pulse reaches its steady state, i.e. $\partial(P_p\tau)/\partial T = 0$. For the frequency of the pulse we obtain

$$T_R\frac{\partial\Omega}{\partial T} = -(2\alpha_{\Sigma 0} + T_{c0})\Omega + \frac{16\alpha_{\Sigma 0} - 3T_{c0}}{4\tau^2\omega_0}(1 + q^2 + 2\Omega^2\tau^2) - \frac{\gamma_\Sigma P_p q}{\sqrt{2\pi}\tau^2\omega_0}. \tag{42}$$

The value of the frequency converges to its steady state

$$\Omega = -\frac{1}{4\tau^2\omega_0}\left[\frac{3T_{c0} - 16\alpha_{\Sigma 0}}{2\alpha_{\Sigma 0} + T_{c0}} + \frac{4}{\sqrt{2\pi}}\frac{\gamma_\Sigma P_p q}{2\alpha_{\Sigma 0} + T_{c0}}\right]. \tag{43}$$

We next derive equations for chirp and duration of the pulse

$$T_R\frac{\partial q}{\partial T} = \left[\beta_{2\Sigma} - \Omega\beta_{3\Sigma} + \frac{\beta_{4\Sigma}}{4\tau^2}(1 + q^2 + 2\Omega^2\tau^2)\right]\frac{1+q^2}{\tau^2} + \tag{44}$$

$$\frac{\gamma_\Sigma P_p}{\sqrt{2\pi}} - \frac{3q(1+q^2)}{8\tau^2\omega_0^2}(32\alpha_{\Sigma 0} + 5T_{c0}) +$$

$$\left[-\frac{i\gamma_\Sigma A_c^2 q}{\sqrt{1-iq}}\exp(-2i\phi_p) + \frac{\sqrt{2}\gamma_\Sigma A_c\sqrt{P_p}(1-iq)}{{}^{1/4}\!\sqrt{\pi}\sqrt{3-iq}}\exp(-i\phi_p) + c.c.\right]$$

$$T_R\tau\frac{\partial \tau}{\partial T} = q\left[\beta_{2\Sigma} - \Omega\beta_{3\Sigma} + \frac{\beta_{4\Sigma}}{4\tau^2}(1 + q^2 + 2\Omega^2\tau^2)\right] - \frac{3}{16\omega_0^2}(q^2-1)(32\alpha_{\Sigma 0} + 5T_{c0}) -$$

$$\frac{\tau^2}{2}\left[\frac{\gamma_\Sigma A_c^2 q}{(1-iq)^{3/2}}\exp(-2i\phi_p) + \frac{i\sqrt{2}\gamma_\Sigma A_c\sqrt{P_p}(1-iq)}{{}^{1/4}\!\sqrt{\pi}(3-iq)^{3/2}}\exp(-i\phi_p) + c.c.\right] \tag{45}$$

Assuming that the fourth order dispersion is negative and that it results in the leading term in the decomposition of the refractive index (the second order dispersion is close to zero, the impact of the third order dispersion is small due to the small detuning Ω, $\tau^2 \ll |\beta_{4\Sigma}|/(\beta_{2\Sigma} - \Omega\beta_{3\Sigma})$, contribution of the sixth and higher orders dispersion is negligible) we find from Eq. (44)

$$\tau \approx \left[\frac{\sqrt{2\pi}}{4}\frac{|\beta_{4\Sigma}|}{\gamma_\Sigma P_p}\right]^{1/4}. \tag{46}$$

The value of chirp can be estimated from the steady state solution of Eq. (45), where the dispersion term in square brackets is substituted from Eq. (44). We assume that $|q| \ll 1$. Taking into account that the parametric term is on the order of the loss of the resonator (as follows from Eq. (40))

$$\frac{i\sqrt{2}\gamma_\Sigma A_c\sqrt{P_p}(1-iq)}{{}^{1/4}\!\sqrt{\pi}(3-iq)^{3/2}}\exp(-i\phi_p) + c.c. \approx \frac{1}{3}(2\alpha_{\Sigma 0} + T_{c0}), \tag{47}$$

we find

$$q \approx -\frac{\sqrt{2\pi}}{6}\frac{2\alpha_{\Sigma 0} + T_{c0}}{\gamma_\Sigma P_p}. \tag{48}$$

Finally, for the timing of the pulse we get

$$T_R\frac{\partial \xi}{\partial T} \approx \frac{\beta_{3\Sigma}}{4\tau^2} \approx \frac{\beta_{3\Sigma}}{2}\sqrt{\frac{\gamma_\Sigma P_p}{\sqrt{2\pi}|\beta_{4\Sigma}|}}. \tag{49}$$

Therefore, the timing of the pulse propagating in the resonator is not zero. The group velocity of the pulse depends on the pulse power. Eq. (49) basically means that the period of the pulse traveling through the resonator is

$$T_{period} \approx T_R + \frac{\beta_{3\Sigma}}{2}\sqrt{\frac{\gamma_\Sigma P_p}{\sqrt{2\pi}|\beta_{4\Sigma}|}}. \tag{50}$$

According to the analytical calculations presented above the oscillator based on a WGM resonator can operate in the mode locked regime. However, the analytical model is rather deficient. The model asserts that the duration of the optical pulse can be less than a few femtoseconds with strong enough optical pumping. In this case the decomposition of the material dispersion in the vicinity of the pumping frequency is invalid for some materials. A more rigorous study is necessary here.

Let us make some numerical estimations for a particular example of a CaF_2 resonator with 25 GHz FSR at 1550 nm. To find the dispersive properties of the resonator we use Sellmeiers dispersion equation[34] for the bulk material as well as the expression for the geometrical dispersion.[35] We assume that the resonator is overcoupled $T_{c0} \approx 10^{-5} \gg \alpha_{\Sigma 0}$. The nonlinearity is $\gamma_\Sigma = 2 \times 10^{-8}$ mW^{-1} and pulse power is $P_p \approx 10^6$ mW. The power of the pulse can be found from Eqs. 40 and 41, however we do not perform the strict evaluation here. Physically, the power of the pulse can be much higher because of the high finesse of the resonator and the short duration of the pulse. We do not consider this case because it cannot be directly analyzed with the formalism provided. Even for the given parameters, the simplified expression for the pulse duration (Eq. (46)) is not valid. By solving equations numerically we find $\tau \approx 2.6$ fs, $q \approx -2 \times 10^{-4}$, and $\Omega \approx -1.6$ THz. Those numbers can be used for qualitative understanding only. The solutions show that the spectrum of the comb generated in the CaF_2 resonator shifts to shorter wavelengths compared with the frequency of the pump; that the chirp of the pulse is small; and that the duration of the pulse is a few femtosecond. The possibility of generation of femtosecond pulses in the resonator contradicts our earlier prediction about restrictions of the frequency comb spectral width that was made without taking into account the mode locking process.[23] Apparently, mode locking allows broadening the spectrum of the generated light beyond the dispersion limits. A more rigorous description of the oscillation is needed to find quantitatively correct values.

It is also interesting to find the frequency of the sidebands that nullify

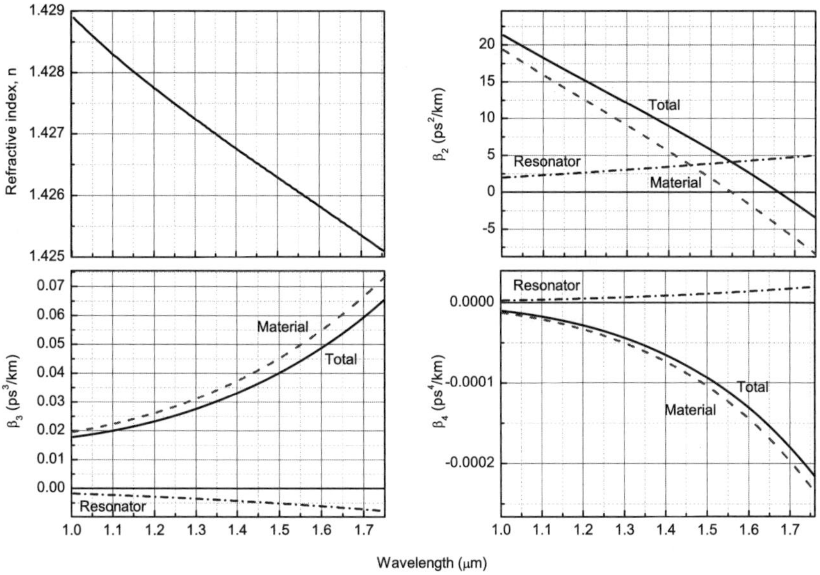

Fig. 3. Refractive index of calcium fluoride and dispersion of calcium fluoride whispering gallery resonator with 25 GHz free spectral range.

the resonator and the material dispersion. Using Eq. (29) and the results of the dispersion calculated for the particular resonator, we find the dependence shown in Fig. (4).

4. Discussion

The stability of the optical signal generated in the resonator is determined by two processes. One is the fundamental oscillator fluctuations given by the intrinsic properties of the oscillator under the condition of the absolute stability of the resonator. For instance, it was shown[21] that the Townes-like formula is valid for the oscillation frequency linewidth, and the diffusion of the oscillator phase is determined by the power of generated optical sidebands and the Q-factor of the WGMs. Those fluctuations are expected to be small because of the high quality factor of the resonator, as well as high power of the generated signal. The oscillator by itself represents a stable secondary frequency standard.

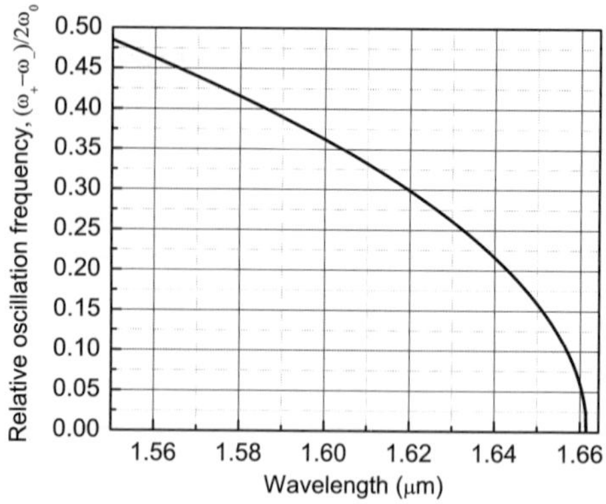

Fig. 4. The frequency detuning at which the sidebands and the carrier belong to zero-dispersion line ($\beta_{+\Sigma} + \beta_{-\Sigma} - 2\beta_{0\Sigma} = 0$) in a calcium fluoride resonator with 25 GHz free spectral range (see Eq. (29)).

The other process that determines the stability of the oscillator is connected with the fundamental, as well as technical, fluctuations of the resonator parameters like temperature, strain, and geometrical shape and size. It is possible to show that the modes that are involved in generation of the optical frequency comb have nearly perfect spatial overlap and nearly the same polarization (less than a few percent difference for two WGMs separated by an octave). The frequency dispersion of the thermorefractive coefficient of the material the resonator is made from is small, as well. Thus all the WGMs involved in the oscillation process drift together with changing resonator parameters. It is enough to lock the frequency of only one WGM to an absolute frequency reference (e.g. atomic transition) to stabilize the entire WGM spectrum.

One of the basic issues in realization of the frequency reference out of the WGM-based oscillator is the locking of frequency of a cw pumping laser to the corresponding resonator mode. Due to the high finesse of the WGM resonator, a small amount of pump power can build up a huge optical intensity inside the resonator. This high optical intensity combined with the absorption of the material leads to a significant thermal drift of the modes and even oscillatory instability. The optical mode as well as the entire mode family drifts due to thermal heating. Such a drift results in a change of the relative frequency detuning of the laser and the mode leading to modification of the oscillation conditions.

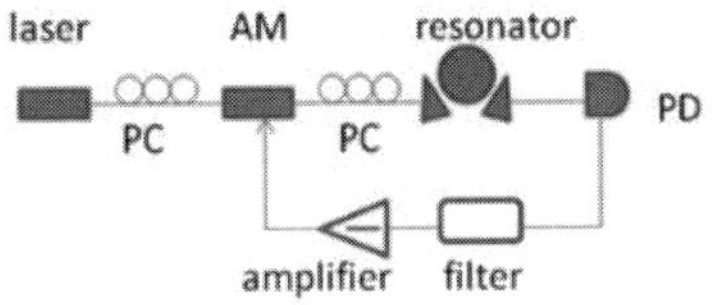

Fig. 5. Schematic diagram of the negative feedback technique employing amplitude modulation.

Here we propose to use a negative feedback technique employing amplitude modulation to suppress this detrimental thermal-induced oscillation. With the thermal oscillation being suppressed, the resonator can be further stabilized using, e.g., the well studied PDH technique. A schematic diagram of the negative feedback technique is shown in Fig. (5). Emission of the laser is sent through an amplitude modulator followed by an optical resonator of high Q factor. A photodiode (PD) is used to detector the transmitted signal. The output of the PD is then filtered, amplified and fed back to the amplitude modulator. When this negative feedback loop is turned on, it suppresses any intensity fluctuation seen by the photodiode within the bandwidth of the feedback loop. Hence if the frequency of the thermal oscillation falls into the bandwidth of the feedback loop, it will be suppressed by the loop.

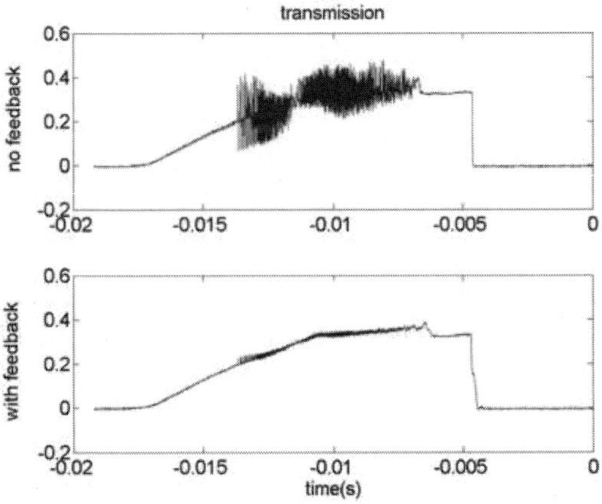

Fig. 6. Transmission of the resonator as the frequency of the laser is swept: (a) thermal oscillation is present; (b) thermal oscillation is suppressed by the negative feedback loop employing amplitude modulation.

We have demonstrated the suppression of thermal oscillation with the negative feedback loop experimentally. In the experiment we used a Koheras fiber laser of a few kHz linewidth and CaF_2 WGM resonator. We swept the frequency of the laser across the resonance frequency of the resonator and measured the transmission of the resonator. Fig. (6a) shows a typical transmission as a function of the sweep time. Significant thermal oscillation appeared when the optical intensity inside the resonator was above the threshold. The thermal oscillation was significantly suppressed when we turned on the negative feedback loop, as seen in Fig. (6b).

5. Conclusion

We have theoretically studied whispering gallery mode oscillators and comb generators, and described the application of the oscillator for generation of spectrally pure RF signals based on the four wave mixing process in non-linear WGM resonators. We have shown that there are two regimes for oscillation. In one regime, with low enough optical pump power, the four wave mixing results in generation of multiple optical harmonics distributed symmetrically with respect to the pumping carrier frequency. These harmonics are not equidistant with respect to each other because of the dispersion of the optical system. The RF signal generated by demodulation of the light exiting the resonator on a fast photodiode is inhomogeneously broadened. In the other regime, for a high enough optical pump power, the optical harmonics phase lock and the output optical signal transforms to a train of ultrashort pulses overlapped with cw background. Demodulation of these pulses on a fast photodiode results in generation of narrowband spectrally pure RF signal. The RF signal can be stabilized by stabilizing the whispering gallery mode resonator. The resonator can be stabilized by locking one of its modes to a reference; only one locking point is necessary to stabilize all the frequencies. Our analysis is rather qualitative, and a more rigorous study of the two oscillation regimes, especially their stability and exchange, is necessary. The frequency stability of an RF signal generated with the oscillator operating in the mode locked regime should be evaluated.

References

1. A. Hasegawa and W. F. Brinkman, "Tunable coherent IR and FIR sources utilizing modulational instability," IEEE J. Quantum Electron. 16, 694-697 (1980).
2. G. P. Agrawal, *"Nonlinear fiber optics,"* (Academic Press, San Diego, 2001).

3. K. Tai, A. Hasegawa, and A. Tomita, "Observation of modulational instability in optical fibers," Phys. Rev. Lett. 56, 135 - 138 (1986).
4. M. Nakazawa, K. Suzuki, and H. A. Haus, "The modulational instability laser-Part II: Experiment," IEEE J. Quantum Electron. 25, 2036-2044 (1989).
5. M. Nakazawa, K. Suzuki, H. Kubota, and H. A. Haus, "The modulational instability laser-Part II: Theory," IEEE J. Quantum Electron. 25, 2045-2052 (1989).
6. M. Haelterman, S. Trillo, and S. Wabnitz, "Additive-modulation-instability ring laser in the normal dispersion regime of a fiber," Opt. Lett. **17**, 745-747 (1992)
7. S. Coen and M. Haelterman, "Modulational instability induced by cavity boundary conditions in a normally dispersive optical fiber," Phys. Rev. Lett. **79**, 4139-4142 (1997).
8. S. Coen and M. Haelterman, "Continuous-wave ultrahigh-repetition-rate pulse-train generation through modulational instability in a passive fiber cavity," Opt. Lett. **26**, 39-41 (2001).
9. D. K. Serkland and P. Kumar, "Tunable fiber-optic parametric oscillator," Opt. Lett. 24, 92-94 (1999).
10. J. E. Sharping, M. Fiorentino, P. Kumar, and R. S. Windeler, "Optical parametric oscillator based on four wave mixing in microstructure fiber," Opt. Lett. 27, 1675-1677 (2002).
11. C. J. S. De Matos, J. R. Taylor, K. P. Hansen, "Continuous-wave, totally fiber integrated optical parametric oscillator using holey fiber," Opt. Lett. 29, 983-985 (2004).
12. Y. Deng, Q. Lin, F. Liu, G. P. Agrawal, and W. H. Know, "Broadly tunable femtosecond parametric oscillator using a photonic crystal fiber," Opt. Lett. 30, 1234-1236 (2005).
13. T. J. Kippenberg, S. M. Spillane, and K. J. Vahala, "Kerr-Nonlinearity Optical Parametric Oscillation in an Ultrahigh-Q Toroid Microcavity," Phys. Rev. Lett. **93**, 083904 (2004).
14. A. A. Savchenkov, A. B. Matsko, D. Strekalov, M. Mohageg, V. S. Ilchenko, and L. Maleki, "Low Threshold Optical Oscillations in a Whispering Gallery Mode CaF_2 Resonator," Phys. Rev. Lett. **93**, 243905 (2004).
15. A. A. Savchenkov, A. B. Matsko, V. S. Ilchenko, and L. Maleki, "Optical resonators with ten million finesse," Opt. Express 15, 6768-6773 (2007).
16. P. Del'Haye, A. Schliesser, O. Arcizet, T. Wilken, R. Holzwarth, and T. J. Kippenberg, "Optical frequency comb generation from a monolithic microresonator," Nature (London) 450, 1214-1217 (2007).
17. P. DelHaye, O. Arcizet, A. Schliesser, R. Holzwarth, and T. J. Kippenberg, "Full Stabilization of a Microresonator-Based Optical Frequency Comb," Phys. Rev. Lett. 101, 053903 (2008).
18. A. A. Savchenkov, A. B. Matsko, V. S. Ilchenko, I. Solomatine, D. Seidel, and L. Maleki, "Tunable Optical Frequency Comb with a Crystalline Whispering Gallery Mode Resonator," Phys. Rev. Lett. 101, 093902 (2008).
19. A. B. Matsko, A. A. Savchenkov, V. S. Ilchenko, and L. Maleki, "Nonlinear optics with WGM crystalline resonators: advances and puzzles," 2004 Digest

of the LEOS Summer Topical Meeting "WGM Microcavities" (2004).
20. A. A. Savchenkov, A. B. Matsko, D. Strekalov, M. Mohageg, V. S. Ilchenko, and L. Maleki, "All-optical photonic oscillator with high-Q whispering gallery mode resonators," Proc. IEEE International Topical Meeting on Microwave Photonics, 205-208 (2004).
21. A. B. Matsko, A. A. Savchenkov, D. Strekalov, V. S. Ilchenko, and L. Maleki, "Optical hyper-parametric oscillations in a whispering gallery mode resonator: threshold and phase diffusion," Phys. Rev. A **71**, 033804 (2005).
22. A. B. Matsko, A. A. Savchenkov, D. Strekalov, and L Maleki, "High frequency photonic microwave oscillators based on WGM resonators," 2005 Digest of the LEOS Summer Topical Meetings, 113-114 (2005).
23. A. A. Savchenkov, E. Rubiola, A. B. Matsko, V. S. Ilchenko, and L. Maleki, "Phase noise of whispering gallery photonic hyper-parametric microwave oscillators," Opt. Express 16, 4130-4144 (2008).
24. C. Lin, W. A. Reed, A. D. Pearson, and H. -. Shang, "Phase matching in the minimum-chromatic-dispersion region of single-mode fibers for stimulated four-photon mixing," Opt. Lett. 6, 493-495 (1981).
25. L. A. Lugiato and R. Lefever, "Spatial dissipative structures in passive optical systems," Phys. Rev. Lett. 58, 2209-2211 (1987).
26. J.-M. Ghidaglia, "Finite dimensional behavior for weakly damped driven Schrodinger equation," Ann. Inst. Henri Poincare 5, 365-405 (1988).
27. N. I. Karachalios and N. M. Stavrakakis, "Global attractor for the weakly damped driven Schrodinger equation in H2(R)," Nonlinear Differ. Equ. Appl. 9 347360 (2002).
28. C. Zhu, "Attractor of the nonlinear Schrodinger equation," Commun. Math. Analysis 4, 6775 (2008).
29. K. J. Blow and N. J. Doran, "Global and Local Chaos in the Pumped Nonlinear Schrodinger Equation," Phys. Rev. Lett. 52, 526-529 (1984)
30. Q.-H. Park and H. J. Shin, "Parametric control of soliton light traffic by cw traffic light," Phys. Rev. Lett. 82, 4432-4435 (1999).
31. S. Li, L. Li, Z. Li, and G. Zhou, "Properties of soliton solutions on a cw background in optical fibers with higher-order effects," J. Opt. Soc. Am. B 21, 2089-2094 (2004).
32. H. A. Haus and A. Mecozzi, "Noise of mode-locked lasers", IEEE J. Quantum. Electron. **29**, 983-996 (1993).
33. A. Hasegawa, "Soliton-based optical communications: an overwiew," IEEE J. Sel. Top. Quantum Electron. **6**, 1161-1172 (2000).
34. M. Daimon and A. Masumura, "High-accuracy measurements of the refractive index and its temperature coefficient of calcium fluoride in a wide wavelength range from 138 to 2326 nm," Appl. Opt. **41**, 5275-5281 (2002).
35. L. Maleki, A. A. Savchenkov, V. S. Ilchenko, and A. B. Matsko, "Whispering gallery mode lithium niobate micro-resonators for photonics applications," Proc. SPIE **5104**, 1-13 (2003).

FREQUENCY COMPARISON USING ENERGY-TIME ENTANGLED PHOTONS

A. Stefanov

Federal Office of Metrology METAS,
Laboratory Time and Frequency,
Lindenweg 50, CH-3003 Bern-Wabern
E-mail: andre.stefanov@metas.ch

Several theoretical schemes have been proposed to improve clock synchronization making advantage of distributed entanglement and quantum states. We propose here to use the quantum correlations in pairs of energy-time entangled photons for frequency comparison.

1. Introduction

Entanglement in quantum systems[1] manifests itself through correlations stronger than purely classical ones. Several schemes have been proposed to make use of such correlations for clock synchronization, expecting an advantage over classical schemes. So far all proposed quantum protocols assume that the frequencies of the clocks to synchronize are equal and are hence designed to determine the phase difference between the clocks. Here we propose a different protocol based on the frequency entanglement that appears in photon pairs generated by parametric down-conversion in non-linear crystal. Instead of measuring phase differences, this protocol measures frequency differences.

2. Quantum clock synchronization

Similarly to classical time comparison methods, quantum protocols can be classified into two main categories: comparison by transportable clocks and exchange of electromagnetic signals. A quantum version of the first is realized using ticking qubits as transportable clocks.[2] Because it is assumed that both parties share a common frequency standard, the problem of synchronization is then reduced to the problem of aligning a phase reference. Quantum phase estimation techniques can be implemented for that specific application.[3] The efficiency of a protocol is measured by the scaling relation of the uncertainty with the number of exchanged qubits N. It has been shown[4] that exchanging independent qubits doesn't show any gain in comparison to the shot noise limit (or quantum standard limit). In both cases the uncertainty scales with $N^{-1/2}$. On the other hand exchanging a N-qubits entangled state will give an improved uncertainty scaling of N^{-1} . However the standard quantum limits can be beaten by non-entangled state, using N coherent exchanges of one qubit. Such an improved protocol gives a $\frac{lnN}{N}$ scaling of the uncertainty. From that point of view, quantum states can provide some improvements with respect to classical protocols. However if the channel is too noisy to preclude synchronization by classical protocols, synchronization with quantum protocols is not possible neither.[5] Some other protocols make use of entanglement by sharing the entanglement between both parties,[6] instead of sending an N-entangled state from one party to the other. However, it is not clear if shared entanglement between two parties does effectively provide an advantage for clock synchronization.[3,7,8]

Another type of protocols makes use of the precise timing of light signals exchanged between parties. By using quantum states of light instead of classical light, more accurate timing measurements can be achieved thanks to the bunching properties of photons[9] or simply by using the correlations in emission time of photon pairs from parametric downconversion.[10] The photons do not need to be in an entangled state, the important feature being the single photon state. The entanglement can however be advantageous to cancel the effects of the dispersion in the medium.[11] In the following we describe a protocol based on the energy correlation properties of photon pairs instead of timing correlations.

3. Non-local frequency comparison

Two partners, Alice and Bob, want to compare their local oscillators by optical means. A classical way[12] is when Alice modulates her light at the frequency Ω_a and sends the modulated light to Bob. Bob can then modulate the light with a frequency $-\Omega_b$ and the resulting signal shows a beat at the sum frequency $\Omega_a - \Omega_b$. A similar scheme can be implemented for non-local two-photon interferometers[13] where photon pairs created in parametric downconversion show energy-time entanglement. They are correlated in a non-classical way because their frequencies sum up to the frequency of the pump laser, although each photon exhibits a broad energy spectrum. The entanglement of such states can be proved by launching each photon into a strongly unbalanced interferometer [fig. 1] and observing interferences in the coincidence rate as a function of the local phases of each interferometer. A contrast higher than 71% is a witness of genuine entanglement within the system.

Similarly to the phenomenon of beats for classical interferometry, the concept of two-photon interferometry can be extended to two-photon beats. When Alice and Bob modulate one arm of their interferometers with frequencies Ω_a and $-\Omega_b$ respectively the coincidence rate is modulated by $\Omega_a - \Omega_b$. Fig.2 shows the results of a proof of principle experiment[14] using the experimental scheme of fig. 1. By measuring the distribution of the time difference between the coincidence events, a signature of the two-photon beat at the expected frequency appears. This would allow Alice and Bob to compare their frequencies non-locally. One should note that Alice and Bob need already a common clock in order to be able to measure coincidences, by time tagging the single detection events. However the uncertainty needed to perform this task can be very large. They only have to agree on the time ordering of the detection events. Alice and Bob can receive the photons from a third party and because of the non-local nature of entanglement, they could perform the frequency comparison independently of the distribution of the entangled resources. Hence, frequency comparison between space-like separated clocks could be performed. Because this protocol aims to compare frequencies and not phases, it may not be affected by the same limitations as quantum clock synchronization protocols with entangled states. In particular purification of entanglement could be applied in case of noisy channels. Further investigations are still needed to rigorously demonstrate its advantage over classical protocols.

4. Conclusion

Using quantum states can theoretically provide higher efficiency with respect to classical protocols. This has been proposed for clock synchronization but also in other field of metrology[15] where the standard quantum limit can be improved by using entangled states. However practical realizations are still to be developed. The main concern being the generation and coherent manipulation of states with a degree of entanglement high enough to achieve a practical gain. Classically the uncertainty scales with N^{-1} and N can be very large, on the other hand for quantum states it scales with $N^{-1/2}$ but N is limited to a few qubits for the time being. Manipulations of such few qubits entangled states have allowed important fundamental tests of quantum mechanics and demonstrations of the principles of quantum in-

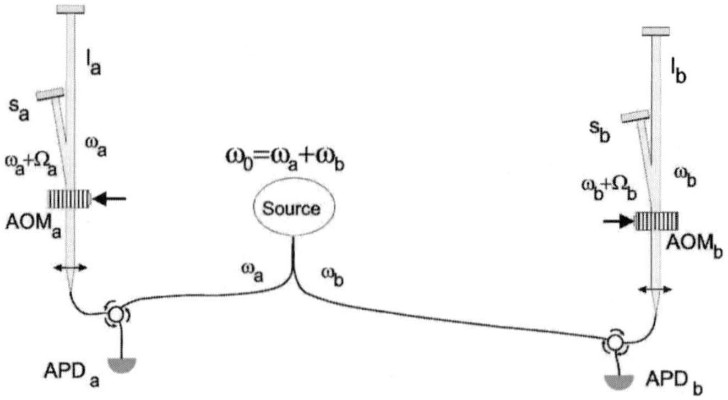

Fig. 1. Scheme of the experiment. Photon pairs are created by parametric downconversion from a continuous wave laser. The sum energy of both photons is fixed by the pump laser energy ω_0, although the energy dispersion of each individual photon can be large. Each photon is launched into a Michelson interferometer with a short arm s_i and a long one l_i. The length difference $l_i - s_i$ is larger than the coherence length of each photon, such that no one-photon interferences can occur. An acousto-optic modulator (AOM) is used as a beamsplitter such that the waves traveling through the short arm are frequency shifted by Ω_j. The output of the interferometers are collected after a circulator into single photon detector and the coincidence rate is observed.

formation methods, but, as for quantum computers, a practical realization would still require a technological breakthrough.

Another domain where quantum methods can be applied with success is applications where there is no classical equivalent. The best example is quantum cryptography, only single qubit manipulations are needed in order to achieve secret key exchange, which is not possible by classical means. It is thus already possible to outperform classical cryptography with today's technology. Similarly they may be tasks for clock synchronization that could only be performed by quantum techniques.

References

1. R. Horodecki, et al., arXiv:quant-ph/0702225v2 (2007).
2. I. L. Chuang,*Phys. Rev. Lett.* **85**, 2006 (2000).
3. S. D. Bartlett, et al., *Rev. Mod. Phys.* **79**, 555 (2007).
4. M. de Burgh and S. D. Bartlett, *Phys. Rev. A* **72**, 042301 (2005).
5. V. Giovannetti, et al., *Phys. Rev. A* **65**, 062319-1 (2002).
6. R. Jozsa, et al.,*Phys. Rev. Lett.* **85**, 2010 (2000).
7. E. A. Burt, et al., *Phys. Rev. Lett.* **87**, 129801-1 (2001).
8. J. Preskill, arXiv:quant-ph/0010098v1 (2000)
9. T. B. Bahder and W M. Golding, arXiv:quant-ph/0311162v1 (2003)
10. A. Valencia, et al.,*Appl. Phys. Lett.* **13**, 2655 (2004).

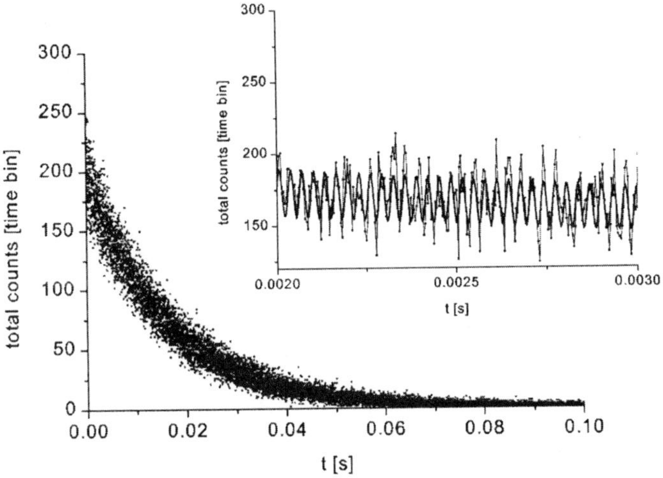

Fig. 2. Histogram of the time difference between successive coincidences. The time intervals between two successive coincidence events are recorded. The corresponding histograms are plotted. Left: the exponential decay corresponds to the poissonian process of photon-pair emission, right: by zooming a beat frequency of 31.25 KHz can be observed, corresponding to the applied frequency difference $\Omega_a - \Omega_b$.

11. J. D. Franson, *Phys. Rev. A* **45**, 3126 (1992).
12. S. M. Foerman, et al., *Rev. Sci. Instr.* **78**, 021101 (2007).
13. J. D. Franson, *Phys. Rev. Lett.* **62**, 2205 (1989).
14. A. Stefanov, et al., *Phys. Rev. A* **67**, 042115 (2003).
15. V. Giovannetti, et al.,*Phys. Rev. Lett.* **96**, 010401 (2006).

List of Participants

Christoph Affolderbach
James Bergquist
John Bernard
Sébastien Bize
Rainer Blatt
Rodolphe Boudot
Dmitry Budker
Eric Burt
Luigi Cacciapuoti
James Camparo
Jingbiao Chen
Sheng-wey Chiow
Arman Cingoz
Giovanni Antonio Costanzo
Scott Crane
Anne Curtis
Erika D'Ambrosio
Andrea De Marchi
Edoardo Detoma
Scott Diddams
Sheperd Doeleman
Elizabeth Donley
Pierre Dubé
Gabriel Durkin
Christopher Ekstrom
Daphna Enzer
Victor Flambaum
Vladislav Gerginov
Kurt Gibble
Patrick Gill
Gesine Grosche
Stephane Guerandel
William Happer
Huijuan He
Thomas Heavner
Dr. Michael Holmes
Ronald Holzwarth
David Howe
Hiromitsu Imai
Bernardo Jaduszliwer
Steven Jefferts
Hidetoshi Katori
Yann Kersale
John Kitching
Daniel Kleppner
Svenja Knappe
Brahim Lamine
Arnaud Landragin
Steve Lecomte
Pierre Lemonde
Filippo Levi
Tianchu Li
Liang Liu
Jose Mauricio Lopez-Romero
Luca Lorini
Andrew Ludlow
Robert Lutwak
Alan Madej
Lute Maleki
Cipriana Mandache
Helen Margolis
Andrey Matsko
Alain Maurissen
Barbara McGivney
Katharina Mohle
Atsuo Morinaga
Holger Mueller
Eamonn Murphy

Nils Nemitz
Chris Oates
Takeo Oita
Thomas Parker
Ekkehard Peik
Steven Peil
Franck Pereira Dos Santos
Achim Peters
Gérard Petit
John Prestage
Ernst Rasel
Stefania Romisch
Till Rosenband
Enrico Rubiola
Christophe Salomon
Giorgio Santarelli
Wolfgang Schaefer
Vishal Shah
Jon Shirley
Marco Siccardi
Sean Spillane
André Stefanov
Uwe Sterr
Thomas Swanson
Shervin Taghavi
Christian Tamm
Pierre Thomann
Massimo Tinto
Robert Tjoelker
Michael Tobar
Veronika Tsatourian
Philip Tuckey
Thomas Udem
Kerry Vahala
Tom Van Baak
Jacques Vanier
Ronald Walsworth
Rabi Wang
Yuzhu Wang
Rong Wei
David Wineland
Peter Wolf
Jun Ye
Nan Yu